Separation Chemistry: Concepts and Applications

Separation Chemistry: Concepts and Applications

Editor: Ethan Evans

NY RESEARCH
P R E S S

New York

Published by NY Research Press
118-35 Queens Blvd., Suite 400,
Forest Hills, NY 11375, USA
www.nyresearchpress.com

Separation Chemistry: Concepts and Applications
Edited by Ethan Evans

International Standard Book Number: 978-1-63238-835-3 (Hardback)

Cataloging-in-Publication Data

Separation chemistry : concepts and applications / edited by Ethan Evans.
    p. cm.
Includes bibliographical references and index.
ISBN 978-1-63238-835-3
1. Separation (Technology). 2. Chemistry, Analytic. 3. Chemistry, Technical. I. Evans, Ethan.
QD63.S4 S46 2022
543.089--dc23

Contents

Preface

The main aim of this book is to educate learners and enhance their research focus by presenting diverse topics covering this vast field. This is an advanced book which compiles significant studies by distinguished experts in the area of analysis. This book addresses successive solutions to the challenges arising in the area of application, along with it; the book provides scope for future developments.

The method that converts a solution or mixture of chemical substances into two or more distinct product mixtures is known as the separation process. It plays a significant role in the field of chemistry. Some chemical elements or compounds exist in an impure state of nature. Separation processes exploit differences in physical properties or chemical properties between the constituents of a mixture. They are classified according to the particular differences they use to achieve separation. Multiple operations are used if no single difference can accomplish the desired separation. For analytical or preparative purposes, separation can be done on a small scale such as in a laboratory. Large scale or intermediate scale is only considered for preparative purposes. Some of the processes used in separation chemistry include adsorption, chelation, chromatography, decantation, centrifugation, electrophoresis, fractional distillation, etc. This book explores all the important aspects of separation chemistry in the present day scenario. It strives to provide a fair idea about this discipline and to help develop a better understanding of the latest advances within this field. Coherent flow of topics, student-friendly language and extensive use of examples make this book an invaluable source of knowledge.

It was a great honour to edit this book, though there were challenges, as it involved a lot of communication and networking between me and the editorial team. However, the end result was this all-inclusive book covering diverse themes in the field.

Finally, it is important to acknowledge the efforts of the contributors for their excellent chapters, through which a wide variety of issues have been addressed. I would also like to thank my colleagues for their valuable feedback during the making of this book.

Editor

Analysis of Serum Sphingomyelin Species by Uflc-Ms/Ms in Patients Affected with Monoclonal Gammopathy

Lazzarini A[1], Floridi A[1], Pugliese L[1], Villani M[1], Cataldi S[1], Michela Codini[2], Lazzarini R[1], Beccari T[2], Ambesi-Impiombato FS[3], Curcio F[3] and Albi E[1]*

[1]Laboratory of Nuclear Lipid BioPathology, CRABiON, Perugia, Italy

[2]Department of Pharmaceutical Science, University of Perugia, Italy

[3]Department of Clinical and Biological Sciences, University of Udine, Italy

Abstract

Cancer cells are hungry of cholesterol incorporated from serum with avidity and used to favour the expressions of proteins involved in cell proliferation such as RNA polymerase II, STAT3, PKCz and cyclin D1. Numerous studies have shown that exists a strong interaction between unesterified cholesterol and saturated fatty acid sphingomyelin which arises from the Van der Waals interaction. Since sphingomyelin and cholesterol association is responsible for the formation of membrane lipid raft involved in cell signalling, we studied the possible hyposphingomyelinemia associated to hypocholesterolemia in the patients with cancer. The blood of 23 patients with monoclonal gammopathy were analyzed for cholesterol, 12:0 sphingomyelin, 16:0 sphingomyelin and 18:1sphingomyelin content. The results demonstrated that only the patients with very low level of cholesterol (65-99 mg/dl) had low amount of sphingomyelin and, in particular, of saturated sphingomyelin specie (16:0 sphingomyelin). The possibility that the hypocholesterolemia in cancer was secondary to hyposphingomyelinemia was discussed.

Keywords: Cholesterol; Sphingomyelin; MS/MS-UFLC; Blood lipids; monoclonal gammopathy

Introduction

Lipids are structural and functional molecules localized in cell membranes and in the nucleus where interact with active chromatin and play specific roles in cell proliferation, differentiation, apoptosis and cancer [1]. Hypocholesterolemia is common in patients with various malignant diseases since cancer cells are hungry of cholesterol (CHO) [2]. In fact, CHO is incorporated in cancer cells from serum with avidity and used to favour the expressions of proteins involved in cell proliferation such as RNA polymerase II, STAT3, PKCz and cyclin D1 [2]. In cells, CHO is not randomly distributed over the membranes but forms lipid microdomains thank to its link with saturated fatty acid of sphingomyelin (SM) by Van der Waals interaction [3]. The lipid microdomains, also known as lipid rafts, are platforms for specific proteins involved in cell signaling if they are localized in cell membrane and are platforms for DNA duplication and transcription if they are associated to the inner nuclear membrane [4,5]. It has been demonstrated that lipid rafts play important mediating roles in cell migration, metastasis, cell survival and tumor progression [6]. In particular CHO has an important role in regulating the synthesis of *invadopodia*: protrusion in the cell membrane of some cells, displaying properties of SM and CHO-enriched membranes or lipid rafts frequently seen in invasive and metastatic cancer cells that invade surrounding tissues [7]. It remains unclear how the rafts are formed in cancer cells, i.e., if using lipids that incorporate from the blood, reducing their physiological values. There is evidence that low levels of serum CHO-phospholipids are associated with the antiphospholipid antibodies in monoclonal gammopathy [8]. However, while the hypocholesterolemia is widely studied in cancer nothing is known about the possible change in the content of SM in the blood of cancer patients. Thus, we aimed to study the possible association of hypocholesterolemia with hyposphingomyelinemia in the patients with monoclonal gammopathy (MG). We analysed the main species of SM by SM/SM-UFLC in order to highlight the changes of saturated fatty acid SMs that interact specifically with CHO.

Materials and Methods

Patients

Blood samples from patients affected by monoclonal gammopathy (MG), with diagnosis from the "Laboratorio Centralizzato di analisi chimico-cliniche Ospedale Silvestrini, Perugia", were collected over a 24 month period. Young and without poor nutrition state or extremely dietary control patients were chosen as experimental model. From all patients (62), only those with a low level of total CHO (23 patients) were analyzed. The population was composed of 13 males and 10 females, average age was 36 yrs. The control group was composed of 20 healthy blood donors: 12 males and 8 females, average age 39 yrs, with normal serum CHO levels.

Lipid extraction

Lipid extraction was performed according to Matyash et al. [9]. 1 mL of serum was diluted with 1mL methanol. 3 mL ultra pure water and 3 mL methyl tert -butyl ether (MTBE) were added. Each Sample was vortexed for 1 min and centrifuged at 3000 g for 5 min. The supernatant was recovered. The extraction with MTBE was repeated on the pellet and the supernatant was added to the first. The organic phase was dried under nitrogen flow and resuspended in 500 μL of methanol.

*Corresponding author: Albi E, Laboratory of Nuclear Lipid BioPathology, CRABiON, Perugia, Italy; e-mail: elisabetta.albi@yahoo.com

TOTAL CHOLESTEROL		
160-220	100-159	65-99
C		P 1
	P 2	
	P 3	
	P 4	
	P 5	
	P 6	
		P 7
	P 8	
		P 9
	P10	
		P11
	P12	
	P13	
	P14	
	P15	
	P16	
	P17	
	P18	
		P19
	P20	
	P21	
	P22	
	P23	

Table 1: Content of cholesterol in control and patients affected with monoclonal gammopathy.

Ultra-Fast Liquid chromatography tandem mass spectrometry (UFLC- MS/MS)

Lipid standards SM 18:1 12:0, SM 18:1 16:0 and SM 18:1 SM 18:1 were prepared according to Matyash et al. [9]. Standards were dissolved in chloroform/methanol (9:1 v/v) at 10 μg/mL final concentration. The stock solutions were stored at -20°C. Working calibrators were prepared by diluting stock solutions with methanol to 500:1, 250:1, 100:1, 50:1 ng/ml final concentrations. 20 μL of standards or lipids extracted from serum were injected after purification with specific nylon filters (0.2 μm)

Analyses were carried out according to [10] by using Ultra Performance Liquid Chromatography system tandem Mass Spectometer Applaied biosistem (Shimadzu Italy s.r.l., Italy). The lipid species were separated, identified and analysed by following the methods of Rabagny et al. [10]

Statistical analysis

Data are expressed as mean ± S.D. of three analysis and t test was used for statistical analysis. $^*p< 0.001$.

Results and Discussion

MG patients with low level of CHO in serum were chosen as experimental model. The comparison of the data was performed with the serum of healthy blood donors with normal level of CHO. Of 23 GM patients, 18 (78%) were found to have low level of CHO (100-159 mg/dl) and 5 (22%) very low level of CHO (65-99 mg/dl), exactly the patients number 1,7,9,11,19 (Table 1).

The analysis of SM was performed by UFLC-MS/MS. Chromatograms displaying mass transitions for SM showed more than one peak. The correct peak of the main species under study were identified by using available standards (SM 12:0, SM 16:0 and SM:18:1). Since CHO is linked by saturated SM species [3], the species under

study were chosen to the aim to highlight the behavior of saturated and unsaturated fatty acids. Among saturated fatty acid SM species, were chosen one present in the cells in low concentration (SM:12:0) and another present in high concentration (SM:16:0). The analysis of total concentration of SM species showed that of the 18 patients with low level of CHO, none had low level of SM and three patients (number 4, 5, and 21) had values higher than control samples (Figure 1). Differently of the 5 patients (number 1,7,9,11 and 19) with very low level of CHO, 100% had values of SM lower than control patients (Figure 1). The value of 12:0 SM, 16:0 SM and 18:1 SM in serum of donor controls were 14.63 ± 6.54, 2975.00 ± 238.60 and 402.04 ± 124.25, respectively. Since the content of 12:0 SM was 203 times lower than that of 16:0 SM, its value has no bearing on the behavior of the total SM. The analysis

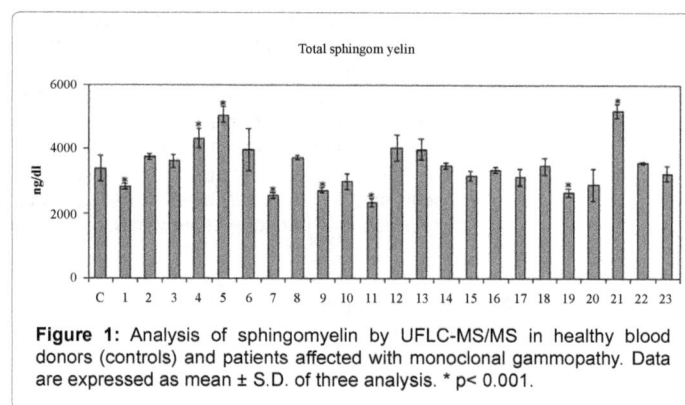

Figure 1: Analysis of sphingomyelin by UFLC-MS/MS in healthy blood donors (controls) and patients affected with monoclonal gammopathy. Data are expressed as mean ± S.D. of three analysis. $^*p< 0.001$.

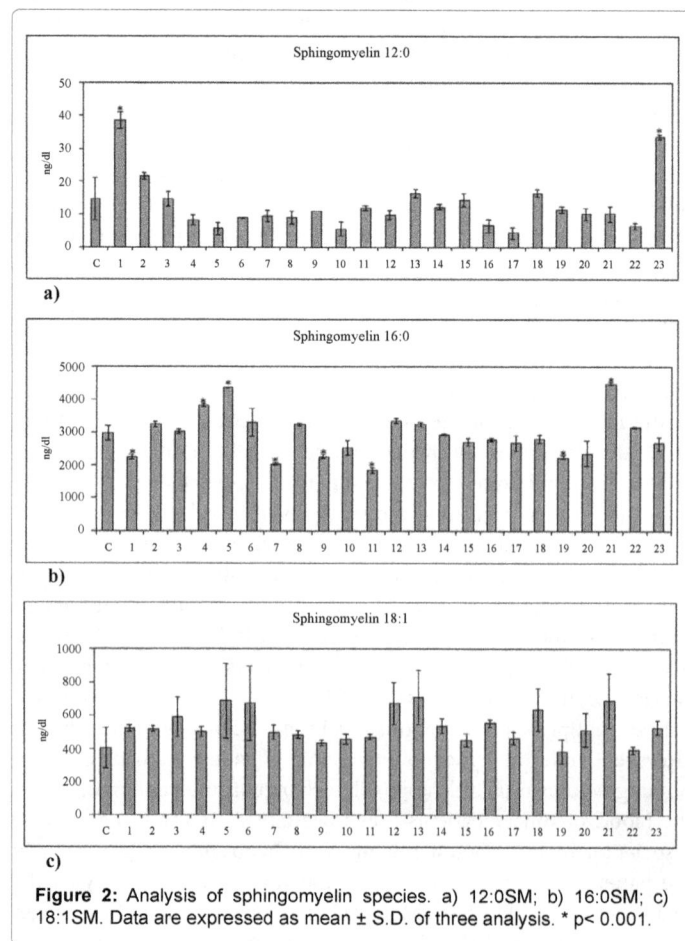

Figure 2: Analysis of sphingomyelin species. a) 12:0SM; b) 16:0SM; c) 18:1SM. Data are expressed as mean ± S.D. of three analysis. $^*p< 0.001$.

of 12:0 SM in GM patients showed that none sample had highly significant difference in comparison with controls with the exception of number 1 and 23 whose values increased 2.63 and 2.30 times, respectively (Figure 2a). The results of the 16:0 SM analysis showed a behavior similar to that obtained with the analysis of total SM, lower values in 1,7,9,11, and 19 and higher values in 4, 5, and 21 patients in comparison with donor controls (Figure 2b). No change was reported for 18:1 SM (Figure 2c). The increase in patients 1 and 23 of 12:0 SM was difficult to interpret at the time. It's really important to note that there was no change in the unsaturated SM (18:1), while the saturated and strongly represented SM (16:0) decreased significantly only in the samples with very low CHO content. Since CHO is bound by saturated SM [3], it is possible to hypothesize that in patients with cancer CHO follows the fate of saturated SM. It is well known that in patients with cancer hypocholesterolemia is due to a strong incorporation of CHO from the tumor cell [2]; our data suggest the possibility that the hypocholesterolemia is secondary to hyposphingomyelinemia (5/23 total patients have hypocholesterolemia and hyposphingomyelinemia). The idea of screening and following patients with malignancy by CHO and SM blood based tests is appealing from several points of view as its ease, economic advantage, non invasiveness, and possibility of repeated sampling.

Acknowledgement

We wish to acknowledge financial support from University of Udine, Italy.

References

1. Albi E (2011) The role of intranuclear lipids in health and disease. Clinical Lipidology 6: 59-69

2. Pugliese L, Bernardini I, Pacifico N, Peverini M, Damaskopoulou E, et al. (2010) Severe hypocholesterolaemia is often neglected in haematological malignancies. Eur J Cancer 46: 1735-1743.

3. Albi E, Magni MV (2002) The presence and the role of chromatin cholesterol in rat liver regeneration. J Hepatol 36: 395-400.

4. Albi E, Lazzarini A, Lazzarini R, Floridi A, Damaskopoulou E, et al. (2013) Nuclear lipid microdomain as place of interaction between sphingomyelin and DNA during liver regeneration. Int J Mol Sci 14: 6529-6541.

5. Cascianelli G, Villani M, Tosti M, Marini F, Bartoccini E, et al. (2008) Lipid microdomains in cell nucleus. Mol Biol Cell 19: 5289-5295.

6. Ma DW (2007) Review Lipid mediators in membrane rafts are important determinants of human health and disease. Appl Physiol Nutr Metab 32: 341-350.

7. Fairbanks KP, Witte LD, Goodman DS. (1984) Relationship between mevalonate and mitogenesis in human fibroblasts stimulated with platelet-derived growth factor. J Biol Chem 259: 1546-1551.

8. Pugliese L, Bernardini I, Viola-Magni MP, Albi E. (2006) Low levels of serum cholesterol/phospholipids are associated with the antiphospholipid antibodies in monoclonal gammopathy. Int J Immunopathol Pharmacol 19: 331-337.

9. Matyash V, Liebisch G, Kurzchalia TV, Shevchenko A, Schwudke D (2008) Lipid extraction by methyl-tert-butyl ether for high-throughput lipidomics. J Lipid Res 49: 1137-1146.

10. Rabagny Y, Herrmann W, Geisel J, Kirsch SH, Obeid R (2011) Quantification of plasma phospholipids by ultra performance liquid chromatography tandem mass spectrometry. Anal Bioanal Chem 401: 891-899.

HPLC and Densitometric TLC Methods for Simultaneous Determination of Gemifloxacin with Some Co-administered Drugs in Human Plasma

Nehad A Abdallah*

Experiments and Advanced Pharmaceutical Research Unit (EAPRU), Faculty of Pharmacy, Ain Shams University, Egypt

Abstract

Two chromatographic methods have been developed for determination of gemifloxacin in human plasma with three co-administered drugs, theophylline, warfarin and omeprazole. First method depends on reverse phase high performance liquid chromatography. The plasma sample was extracted using acetonitrile. The method was linear over the concentration range 0.05 to 6 µg/mL, 0.25 to 8 µg/mL, 0.1 to 10 µg/mL and 0.1 to 6 µg/mL of gemifloxacin mesylate, theophylline, warfarin and omeprazole, respectively. The mobile phase used was prepared by mixing acetonitrile and 0.02 mol L^{-1} potassium dihydrogen phosphate buffer (pH adjusted to 2.5 using ortho phosphoric acid) in a ratio 10:90 with the addition of 1% TEA and flow rate 1 mL/min in isocratic mode and UV-detection at wavelength 254 nm. Second method depends on densitometric thin layer chromatography. The method was linear over concentration range 0.1 to 3 µg/ml, 0.5 to 6 µg/mL, 0.2 to 2.5 µg/mL and 0.1 to 1.5 µg/mL of gemifloxacin mesylate, theophylline, warfarin and omeprazole, respectively. The mobile phase was selected as mixture of dichloromethane, methanol and ammonia in the ratio of (7: 5.5: 3 v/v/v) for the development of plates. Densitometric analysis was carried out at wavelength 254 nm. The stability of gemifloxacin mesylate and the co-administered drugs in plasma was confirmed during three freeze–thaw cycles (−20°C).

Keywords: Gemifloxacin; Theophylline; Warfarin; Omeprazole; Plasma; HPLC; TLC

Introduction

Gemifloxacin mesylate (R, S)-7(3-aminomethyl-4-syn-methoxyimino-1-pyrrolidinyl)-1-cyclopropyl-6-fluro-1, 4dihydro-4-oxo-1, 8naphthyridine-3-carboxylic acid is a new fluoroquinolone antibacterial compound with enhanced affinity for bacterial topoisomerase IV and it is used for the treatment of respiratory and urinary tract infection. The compound has broad spectrum of activity against gram-positive and gram-negative bacteria [1]. Gemifloxacin mesylate is not official in any pharmacopoeia. The literature survey revealed that analytical methods reported for the estimation of gemifloxacin mesylate in human plasma were Spectrofluorimetric method [2], Densitometric TLC [3], HPLC-Tandam mass [4], HPLC [3,5,6]. No method has been reported for determination of gemifloxacin mesylate in human plasma with co-administered drugs by HPLC and densitometric TLC. Some fluoroquinolones have the potential to modify the kinetics of co-administered drugs via CYP inhibition and may be via inhibition of drug transport [7]. The co-administration of gemifloxacin and theophylline was well tolerated, with no clinically significant changes seen in vital signs. Adverse events were generally transient, mild to moderate in nature [8]. Co-administration of warfarin with fluoroquinolones generally needs caution because of a possible increased anticoagulant response [9]. A study was designed to demonstrate the lack of effect of steady-state concentrations of gemifloxacin on the pharmacodynamic effects of warfarin. There were no changes of clinical significance in vital signs [10]. A study was designed to demonstrate the effect of omeprazole on the pharmacokinetics of oral gemifloxacin in healthy volunteers. Following co-administration of gemifloxacin and omeprazole the $AUC_{0-\infty}$ and C_{max} for gemifloxacin increased on average by 10% and 11%, respectively, although neither t_{max} nor $t_{1/2}$ appeared to be affected. However, none of these changes are considered to be clinically important and there were no recommendations to alter the dose of fluoroquinolones in the presence of omeprazole [11].

The proposed research work describes the estimation of gemifloxacin mesylate with three different co-administered drugs using only one separation system that cover the expected concentration ranges of all of the studied drugs in human plasma, Figure 1. Other published work describes the determination of gemifloxacin with only one co-administered drug.

Material and Methods

Instrumentation

HPLC Knauer instrument (Germany) equipped with K-501 pump, Knauer injector and UV-detector K-2501. Data acquisition was performed on Eurochrom 2000 software. The analytical column employed was X-terra LC-18-DB, (25 cm×4.6 mm×5 µm). The working temperature was 25°C.

DESAGA CD 60 HPTLC densitometer connected to IBM compatible computer fitted with Proquant evaluation software for Windows. (Sarstedt-Gruppe, Germany) with precoated silica gel Plate 60F254 (20 cm×20 cm) 250 µm thicknesses (E. Merck, Darmstadt, Germany) was used as stationary phase. Sample application was done by using DESAGA AS30 HPTLC Applicator. (Sarstedt-Gruppe, Germany).

Linear ascending development was carried out in 25 cm×25 cm glass chamber. Evaluation of chromatogram was done by using peak areas.

Rotatory vacuum evaporator, Buchi Rotavaper R-3000 (Germany).

*****Corresponding author:** Abdallah NA, Experiments and Advanced Pharmaceutical Research Unit (EAPRU), Faculty of Pharmacy, Ain Shams University, Egypt
E-mail: nehad_nany@hotmail.com

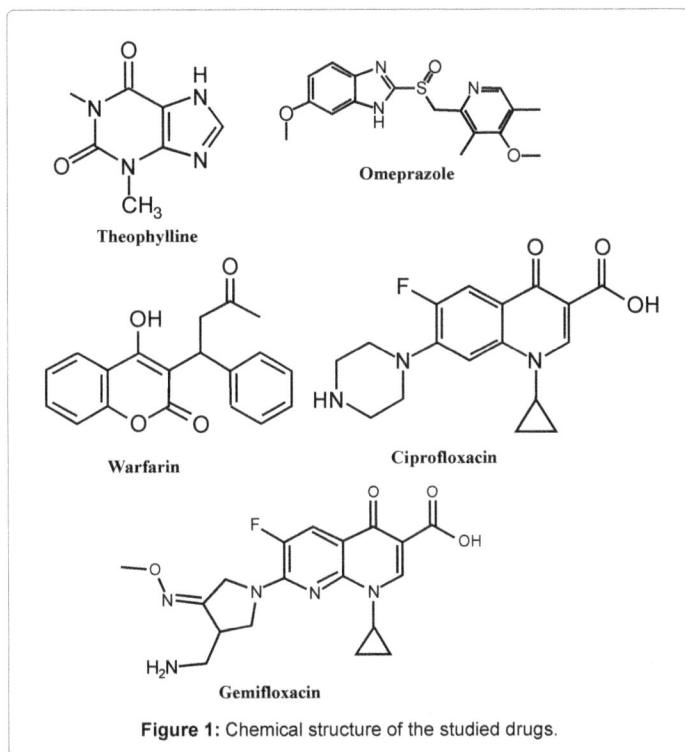

Figure 1: Chemical structure of the studied drugs.

Sonicator (Crest Ultrasonics, New York). Syringe filters (Gelman, Sigma-aldrich).

Chemicals

Gemifloxacin mesylate (Mediphar pharmaceutical company, Cairo, Egypt), theophylline (GlaxoSmithKline, Cairo, Egypt), Warfarin (Nile Company for pharmaceuticals and chemical industries), omeprazole and ciprofloxacin (Al-hekma for pharmaceuticals) were received having 99.20%, 99.20%, 99.60%, 99.53 and 99.65% purity, respectively. The HPLC grade methanol, acetonitrile, ortho phosphoric acid, dichloromethane, ammonia and water were purchased from (Sigma Gmbh, Germany). Analytical Reagent grade di-potassium hydrogen phosphate was used. Freshly isolated human plasma from collected blood used for research work was supplied by Vacsera, Cairo, Egypt.

Preparation of stock solutions and working standard solutions

Standard solutions preparation was conducted at room temperature. The solutions were protected from light with aluminum foil wrapping and stored at -20°C.

Stock solutions 1.00 mg/mL each of gemifloxacin mesylate, theophylline, warfarin omeprazole and ciprofloxacin were prepared in methanol.

The first working standard solutions of 0.1 mg/mL of gemifloxacin mesylate, theophylline, warfarin, omeprazole and ciprofloxacin hydrochloride were prepared by further dilution of stock solutions with mobile phase for HPLC and with methanol for TLC. Second working standard solutions (0.01 mg/mL) of gemifloxacin mesylate, theophylline, warfarin, omeprazole and ciprofloxacin hydrochloride were prepared by further dilution of stock solutions with mobile phase for HPLC and with methanol for TLC.

Preparation of plasma samples

For HPLC, in a stoppered centrifuge tube, an aliquot quantity of 500 μL plasma was added and spiked with 50 μL ciprofloxacin (internal standard 100 μg/mL) and 450 μL mixtures of gemifloxacin, theophylline, warfarin and omeprazole working solutions to provide concentrations of (0.05, 0.1, 0.5, 1, 2, 4, 6 μg/mL), (0.25, 0.5, 1, 2, 4, 6, 8 μg/mL), (0.1, 0.25, 0.5, 1, 2.5, 5, 10 μg/mL) and (0.1, 0.25, 0.5, 1, 2, 4, 6 μg/mL); respectively.

The quality control samples (QCs) were prepared in plasma concentration range (0.15, 1, 5 μg/mL), (0.75, 4, 7 μg/mL), (0.3, 2, 9 μg/mL) and (0.3, 1, 4 μg/mL) for gemifloxacin, theophylline, warfarin and omeprazole, respectively. Protein precipitation and extraction were carried out by using 1ml acetonitrile. The samples were sonicated for 10 minutes followed by vortex mixing for 5 minutes and centrifugation at 5000 rpm for 15 minutes. The organic layer was transferred to another centrifuge tube and evaporated to dryness at 40°C under vacuum. The residue was reconstituted in 0.1 mls mobile phase and 20 μl injected into HPLC system.

For densitometric TLC, in a stoppered centrifuge tube, an aliquot quantity of 500 μL plasma was spiked with 50 μl ciprofloxacin (internal standard 100 μg/mL) and 450 μl mixtures of gemifloxacin, theophylline, warfarin and omeprazole. Different aliquots of gemifloxacin were added to provide concentrations of (0.1, 0.2, 0.4, 0.8, 1, 1.5, 2, 2.5, 3 μg/mL). Different aliquots of theophylline were added to provide concentrations of (0.5, 1, 1.5, 2, 2.5, 3, 3.5, 4, 5, 6 μg/mL). Different aliquots of warfarin were added to provide concentrations of (0.2, 0.6, 1, 1.2, 1.6, 2, 2.5 μg/mL). Different aliquots of omeprazole were added to provide concentrations of (0.1, 0.2, 0.4, 0.6, 0.8, 1, 1.2, 1.5 μg/mL). The quality control samples (QCs) were prepared in plasma concentration range (0.3, 1.5, 2.5 μg/mL), (1.5, 3, 5 μg/mL), (0.6, 1, 2 μg/mL) and (0.3, 0.8, 1.2 μg/mL) for gemifloxacin, theophylline, warfarin and omeprazole, respectively. Protein precipitation and extraction were carried out as previously mentioned in HPLC method. The residue was reconstituted in 0.05 mLs methanol and 20 μL were applied to TLC plates.

Chromatographic Conditions

For HPLC, the mobile phase used was prepared by mixing acetonitrile and 0.02 mol L^{-1} potassium dihydrogen phosphate buffer (pH adjusted to 2.5 using ortho phosphoric acid) in a ratio 10:90 with the addition of 1% TEA. The mobile phase was freshly prepared and filtered by vacuum filtration through 0.45 μm filter and degassed by ultrasound sonication for 50 minutes just prior to use. The samples were also filtered using 0.45 μm syringe filters. The analysis was done under isocratic conditions at a flow rate 1 ml.min^{-1} and at room temperature using UV detector at 254 nm.

For densitometric TLC, the mobile phase was selected as mixture of dichloromethane, methanol and ammonia in the ratio of (7: 5.5: 3 v/v/v) for the development of plates. The densitometric scanning was performed at 254 nm. The samples were also filtered using 0.45 μm filters. Analysis was performed on precoated 20×20 cm silica gel 60 F$_{254}$ aluminum sheets (E.Merck). The plates were pre-washed with methanol and activated at 60°C for 5 min prior to chromatography. Samples were applied to the plates using a DESAGA AS30 Applicator (Germany). Spots were applied 1.5 cm apart from each other and 2 cm from the bottom edge. The chromatographic chamber was pre-saturated with the mobile phase for 45 min. the developing distance on TLC-plate was 180 mm.

Method Validation

The described methods were validated in terms of linearity, limit of detection (LOD), limit of quantification (LOQ), recovery, selectivity, stability, precision and accuracy according to FDA guidelines regarding standard bioanalytical method validation recommendation [12]

Linearity: The analytical range to be validated was chosen on the basis of the expected plasma concentrations of the studied drugs [13-18]. The calibration curve was done for each analyte in the biological sample. The calibration curve should consist of a blank sample (matrix sample processed without internal standard), a zero sample (matrix sample processed with internal standard), and six to eight non-zero samples covering the expected range, including LLOQ that were prepared by adding the required volume of working solution of analyte to blank plasma. The plasma samples were subjected to the sample preparation procedure and injected into the LC or applied on TLC plates.

Plasma calibration curve was prepared by taking area ratio of analyte to internal standard as Y-axis and concentration of analyte (µg/mL) as X-axis.

Accuracy and precision: Accuracy and precision were determined for LQC, MQC and HQC (Low, Middle and High Quality Control) samples with LLOQ. Five replicates of each concentration were analyzed on the same day to determine the within-run accuracy and precision of the method. To confirm the between-run accuracy and precision five replicates of each concentration were analyzed at three separate days.

For HPLC, the used concentrations were (0.05, 0.15, 1 and 5 µg/mL), (0.25, 0.75, 4 and 7 µg/mL), (0.1, 0.3, 2 and 9 µg/mL), (0.1, 0.3, 1 and 4 µg/mL) for gemifloxacin, theophylline, warfarin and omeprazole, respectively.

For TLC, the used concentrations were (0.1, 0.3, 1.5 and 2.5 µg/mL) equivalent to (0.04, 0.12, 0.6, 1 µg/spot), (0.5, 1.5, 3 and 5 µg/mL) equivalent to (0.2, 0.6, 1.2, 2 µg/spot), (0.2, 0.6, 1 and 2 µg/mL) equivalent to (0.08, 0.24, 0.4, 0.8 µg/spot) and (0.1, 0.3, 0.8 and 1.2 µg/mL) equivalent to (0.04, 0.12, 0.32, 0.48 µg/spot) for gemifloxacin, theophylline, warfarin and omeprazole, respectively.

Selectivity: The selectivity of the methods was investigated by analyzing six blank plasma samples. Each blank sample was tested for interference using proposed extraction procedure and the response of the endogenous compounds at the retention times of the studied drugs in plasma samples were compared with the response of LLOQ of the studied drugs.

Recovery: The extraction recovery of analytes was determined by measuring the peak areas of the drugs from the prepared plasma quality control samples. The peak areas of extracted LQC, MQC and HQC were compared to the absolute peak area of the unextracted samples in mobile phase for HPLC and in methanol for TLC containing the same concentration of the drug. To obtain good extraction efficiency the extraction recovery of gemifloxacin and its co-administered theophylline, warfarin and omeprazole was determined using five replicates of each QC samples.

Stability study: 4.6.5.1. Freeze and thaw stability: The stability of gemifloxacin together with co-administered theophylline, warfarin and omeprazole was determined after three freeze and thaw cycles. Five aliquots at each of the LLOQ, low, mid and high quality control concentrations were stored at -20°C for 24 hours and thawed unassisted at room temperature. When completely thawed, the samples were refrozen for 24 hours under the same conditions. The freeze–thaw cycle were repeated two more times, and then analyzed.

Short term temperature stability: Five aliquots of each of the LLOQ, low, mid and high quality control concentrations were thawed at room temperature and kept at this temperature for 6 hours and then analyzed.

Long term stability: Long-term stability was determined by storing five aliquots of each of the LLOQ, low, mid and high concentrations of the studied drugs at -20°C for 6 weeks. The concentrations of all the stability samples were compared to the mean values for the standards at the appropriate concentrations from the first day of long-term stability testing.

Stock solutions stability: The stability of stock solutions of each of the studied drugs and the internal standards used were evaluated at room temperature for 10 hours. After completion of the desired storage time, the stability was tested by comparing the instrument response with that of freshly prepared solutions.

Post-preparative stability: The stability of the processed samples was examined by keeping five replicates of the LLOQ, low, mid and high plasma quality control samples at room temperature for approximately 24 hours. The stability was tested by comparing the instrument response with that of freshly prepared samples.

Results

HPLC-UV detection and densitometric TLC methods were suggested for the simultaneous determination of each of gemifloxacin with three co-administered drugs; theophylline, warfarin and omeprazole. This work aimed to develop highly selective and sensitive methods with a quantitation limits that cover the expected concentration ranges of all of the studied drugs in human plasma to be able to be used in pharmacological, bioavailability, bioequivalence or other clinical studies to obtain certain pharmacokinetic information.

Extraction procedure optimization

One of the most difficult parts during the method development was to achieve a high and reproducible recovery from the solvent which is used for extraction of the drugs. Different solvents were tried for the extraction of gemifloxacin, theophylline, warfarin and omeprazole from human plasma. Methanol was tried for plasma precipitation and extraction of the studied drugs. It was found that the recoveries of the gemifloxacin and internal standard were acceptable but it was below 70% for theophylline and warfarin. Other solvents were tried as chloroform, ethyl acetate and n-hexane. The use of acetonitrile raised the recoveries of all the studied drugs above 80%.

Optimization of chromatographic condition

For HPLC, Initially the use of acetonitrile and potassium dihydrogen phosphate buffer (pH=3.5) in a ratio 30:70, the peaks of gemifloxacin, ciprofloxacin (IS) and omeprazole were well resolved. The peaks of theophylline and warfarin were overlapped. This was treated by gradual increasing the ratio of the buffer solution in the mobile phase to obtain reasonable resolution. Gemifloxacin peak tailing was treated by the addition of 1% TEA. The final mobile phase used for the simultaneous determination of gemifloxacin and its co-administered theophylline, warfarin and omeprazole was mixture of acetonitrile and 0.02 mol L^{-1} potassium dihydrogen phosphate buffer (pH adjusted to 2.5 using ortho phosphoric acid) in a ratio 10:90 with the addition of 1% TEA. The average retention time (minutes) ± SD, for 6 replicate injections of gemifloxacin, theophylline, warfarin, omeprazole and ciprofloxacin, were found to be 10.457 ± 0.07, 4.411 ± 0.05, 5.478 ± 0.08, 7.356 ± 0.04 and 3.308 ± 0.06, respectively, as shown in figures 2 and 3.

For densitometric TLC, Different solvent systems were tried. Initially systems like mixtures of cyclohexane and ethanol were used

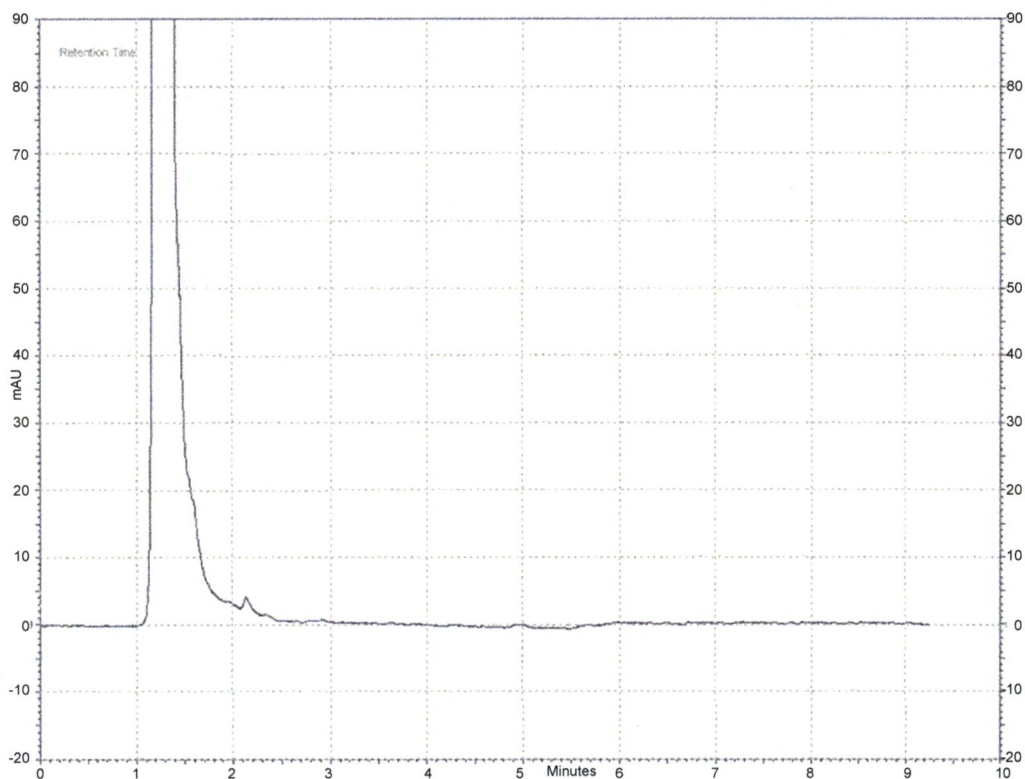

Figure 2: Chromatogram of blank human plasma sample.

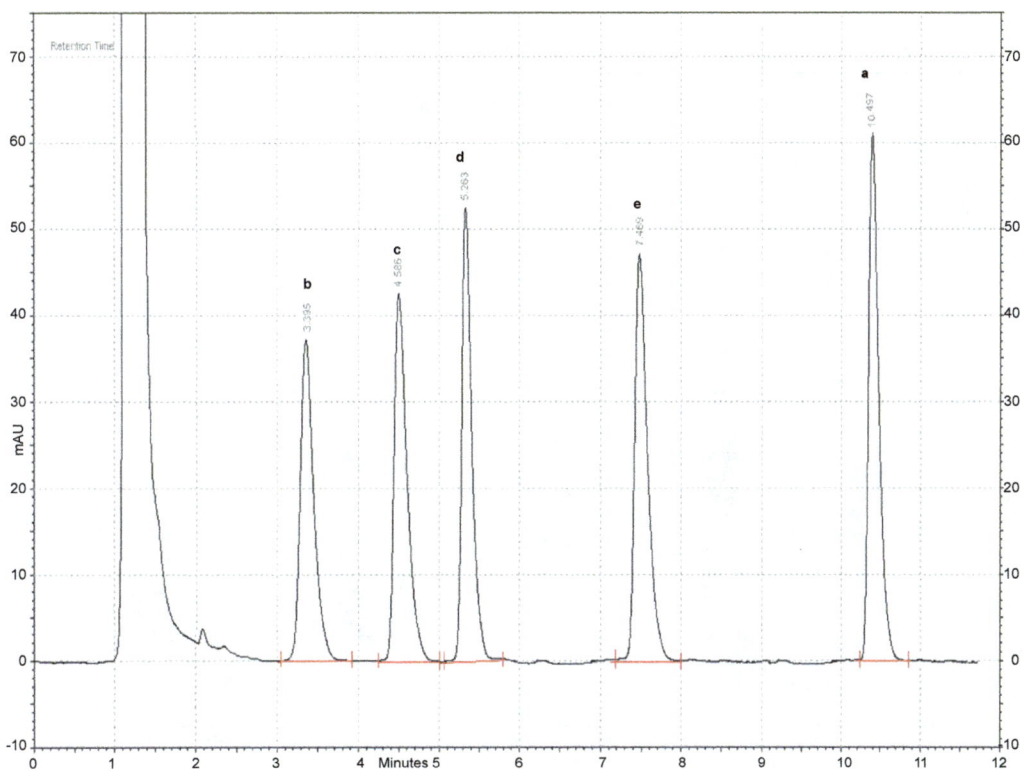

Figure 3: Chromatogram of human plasma sample spiked with (a) gemifloxacin, (b) ciprofloxacin, (c) theophylline, (d) warfarin and (e) omeprazole.

in varying ratios, but these systems showed excessive tailing and most peaks were too high near the solvent front. Other systems of mixtures of methanol and chloroform were tried. Better results were obtained but tailing was observed and peaks of theophylline and warfarin were still overlapped. Other mixture of dichloromethane, methanol and ammonia were tried and the best results were obtained by using dichloromethane, methanol and ammonia in the ratio of (7: 5.5: 3 v/v/v) for the development of plates.

The RF values were 0.16 ± 0.051, 0.28 ± 0.042, 0.53 ± 0.037, 0.58 ± 0.022 and 0.74 ± 0.031 for ciprofloxacin, gemifloxacin, theophylline, warfarin and omeprazole, respectively, Figures 4 and 5.

Method validation

Linearity: For HPLC, the seven point calibration curves were

constructed by plotting the peak area ratio of each of gemifloxacin, theophylline, warfarin and omeprazole to ciprofloxacin (IS) versus their concentrations in plasma. The mean equations of calibration curves consisting of seven points are y=0.3835C+0.0013 for gemifloxacin mesylate with a correlation coefficient 0.9998, y=15.412C+0.0498 for theophylline with a correlation coefficient 0.9997, y=0.0308C+0.0013 for warfarin with a correlation coefficient=0.9997 and y=0.1938C+8E-05 for omeprazole with correlation coefficient 0.9996. Where y represents the ratios of peak area of each drug to that of IS and C represents the plasma concentration of each drug. The values of correlation coefficient confirmed that the calibration curves were linear over the concentration ranges 0.05-6 µg/mL, 0.25-8 µg/mL, 0.1-10 µg/mL and 0.1-6 µg/mL for gemifloxacin, theophylline, warfarin and omeprazole, respectively.

For densitometric TLC, the calibration curves were constructed

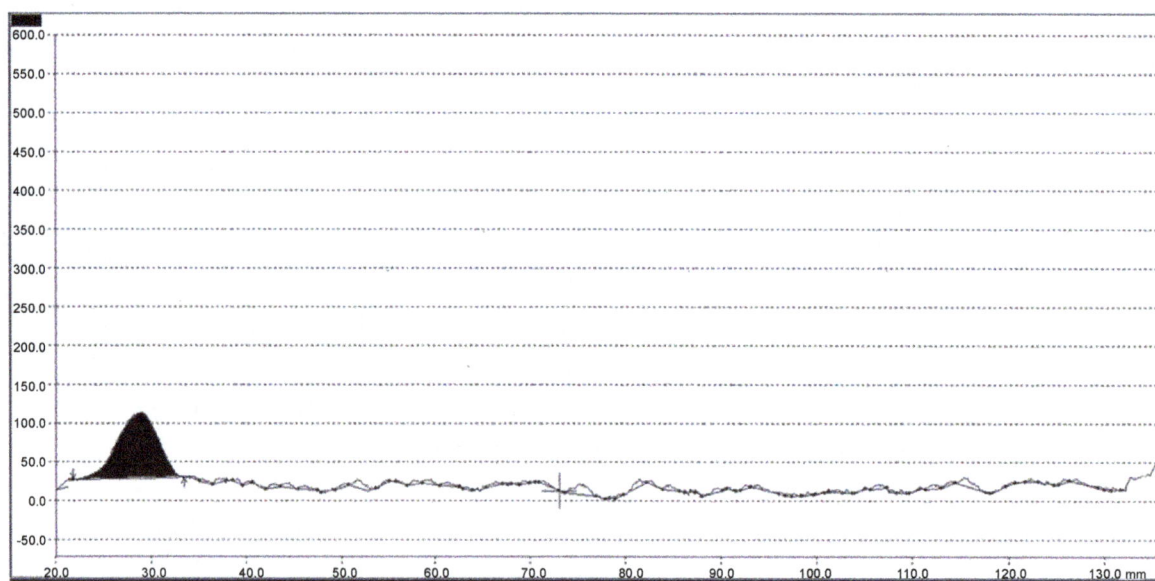

Figure 4: TLC densitogram of blank human plasma sample.

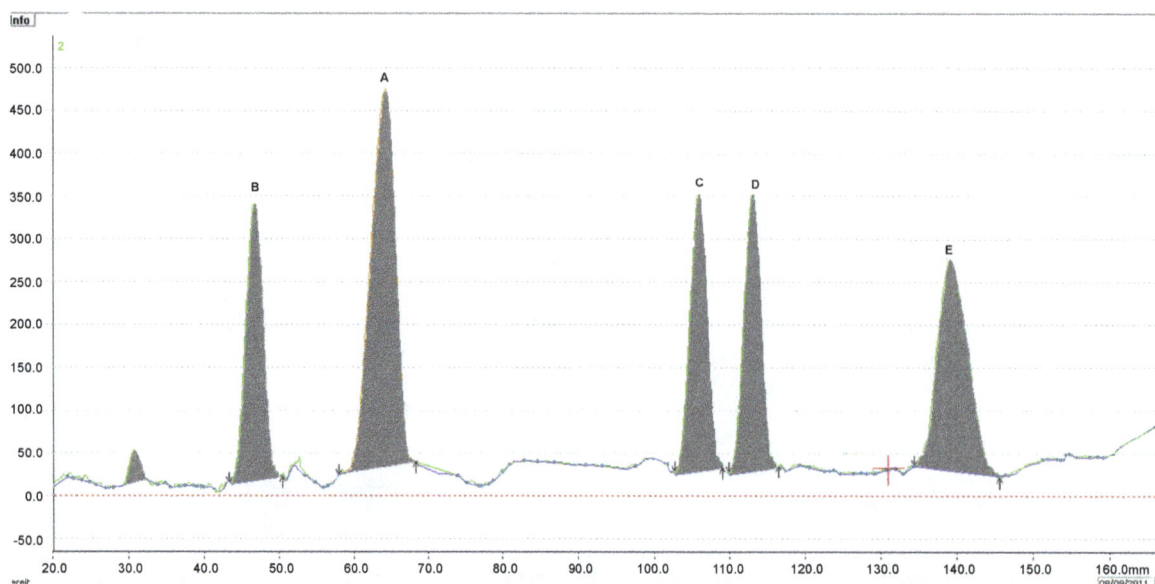

Figure 5: TLC densitogram of human plasma sample spiked with (A) gemifloxacin, (B) ciprofloxacin, (C) theophylline, (D) warfarin and (E) omeprazole.

by plotting the peak area ratio of each of gemifloxacin, theophylline, warfarin and omeprazole to ciprofloxacin (IS) versus their concentrations in plasma. The mean equations of calibration curves consisting of seven points y=1.1053C-0.0374 for gemifloxacin mesylate with a correlation coefficient of 0.9998, y=0.227C+0.012 for theophylline with correlation coefficient of 0.9995, y=0.5007C+0.00563 for warfarin with correlation coefficient=0.9991 and y=1.915C–0.0146 for omeprazole with correlation coefficient of 0.9993. Where y represents the ratios of peak area of each drug to that of IS and C represents the plasma concentration of each drug. The calibration curves were linear over the concentration ranges 0.1-3 µg/mL, 0.5-6 µg/mL, 0.2-2.5 µg/mL and 0.1-1.5 µg/mL for gemifloxacin, theophylline, warfarin and omeprazole; respectively.

Accuracy and precision: The accuracy of the method expressed in term of bias (percentage deviation from true value). The mean value should be within 15% of the actual value except at LLOQ, where it should not deviate by more than 20%. The precision determined at each concentration level should not exceed 15% of the coefficient of variation (CV) except for the LLOQ, where it should not exceed 20% of the CV [12]

Precision of the method was determined by repeatability (intraday) and intermediate precision (inter-day) and accuracy for set of quality control (QC) sample (low, mid, high) in replicate (n=5). The inter-day and intra-day precision and accuracy for pazufloxacin, cefoperazone and sulbactam sodium were evaluated by assaying the QC samples (low, mid, high) (n=5) in (%CV). In HPLC assay for gemifloxacin, the intra-run precision (CV %) was found to be in the range of 1.041-5.187% and the inter-run precision was 1.571-4.591%. For theophylline, the intra-run precision (CV %) was found to be in the range of 0.668-1.748% and the inter-run precision was 0.799-3.061%. For warfarin, the intra-run precision (CV %) was found to be in the range of 1.272-5.625% and the inter-run precision was 1.672-6.122%. For omeprazole, the intra-run precision (CV %) was found to be in the range of 1.208-4.947% and the inter-run precision was 1.972-5.414%. The accuracy values (%RE) were within the ranges of -0.70-1.50% for gemifloxacin, -1.6-1.729% for theophylline, -4-1.75% for warfarin and -5.8-1.4 for omeprazole.

The above values were within the acceptable range, they show that the HPLC method is accurate and precise.

For densitometric TLC for gemifloxacin, the intra-run precision (CV %) was found to be in the range of 1.751-6.186% and the inter-run precision was 1.849-5.119%. For theophylline, the intra-run precision (CV %) was found to be in the range of 2.717-6.695% and the inter-run precision was 3.273-5.556%. For warfarin, the intra-run precision (CV %) was found to be in the range of 2.176-4.725% and the inter-run precision was 3.290-5.380%. For omeprazole, the intra-run precision (CV %) was found to be in the range of 1.408-3.720% and the inter-run precision was 1.779-4.384%. The accuracy% values (%RE) were found to be in the range of -3.667-0.2% for gemifloxacin, -6.4-0.133% for theophylline, -7-0.333 for warfarin and -7--1.167 for omeprazole. The low percent coefficient of variation (%CV) and (%RE) were within the acceptable limit. The results of inter-day, intra-day precision and accuracy for gemifloxacin, theophylline, warfarin and omeprazole are shown in Table 1.

Selectivity: Selectivity was assessed to show that the intended analytes are measured and that their quantitation is not affected by the presence of the biological matrix. Figures 2-5 showed the typical chromatograms and densitograms of blank plasma and gemifloxacin with its co-administered drugs spiked with plasma

Recovery: Absolute recovery was calculated by comparing peak areas obtained from freshly prepared sample extracted with unextracted standard solutions of the same concentration. Recovery data was determined in triplicates at three concentrations (low, mid, high) as recommended by the FDA guidelines [12]. The average recovery of gemifloxacin, theophylline, warfarin and omeprazole for RP-HPLC, determined at the three concentrations (low, mid, high concentration) of each were found to be 89.555, 90.824, 89.574 and 88.471, respectively. For TLC the average recovery using the three concentrations for gemifloxacin, theophylline, warfarin and omeprazole were found to be 89.967, 89.050, 89.079 and 90.208, respectively as shown in table 2.

Sensitivity: Sensitivity of the method is defined as the lowest concentration that can be measured with an acceptable limit of accuracy and precision which is lower than 20% [12]. For HPLC, LLOQ were

Method	Gemifloxacin mesylate				Theophylline				Warfarin				Omeprazole			
RP-HPLC	Concentration (µg/ml)	SD (n=5)	CV%	%RE	Concentration (µg/ml)	SD (n=5)	CV%	%RE	Concentration (µg/ml)	SD (n=5)	CV%	%RE	Concentration (µg/ml)	SD (n=5)	CV%	%RE
intraday	0.05	0.0025	5.187	-3.60	0.25	0.0043	1.748	-1.6	0.1	0.0054	5.625	-4	0.1	0.0047	4.947	-5
	0.15	0.004	2.400	0.00	0.75	0.005	0.668	-0.267	0.3	0.007	2.034	-1.667	0.3	0.006	1.993	0.333
	1	0.012	1.206	-0.50	4	0.054	1.343	0.55	2	0.025	1.272	-1.7	1	0.012	1.268	-0.7
	5	0.052	1.041	0.30	7	0.112	1.603	-0.2	9	0.204	2.262	0.222	4	0.084	2.087	0.625
interday	0.05	0.0023	4.591	0.20	0.25	0.004	1.639	-2.4	0.1	0.006	6.122	-2	0.1	0.0051	5.414	-5.8
	0.15	0.003	2.000	0.00	0.75	0.006	0.799	0.133	0.3	0.005	1.672	-0.333	0.3	0.006	2.007	-0.333
	1	0.031	3.054	1.50	4	0.090	2.264	-0.6	2	0.036	1.769	1.75	1	0.020	1.972	1.4
	5	0.078	1.570	-0.70	7	0.218	3.061	1.729	9	0.196	2.165	0.578	4	0.132	3.313	-0.4
TLC																
intraday	0.1	0.006	6.186	-3	0.5	0.032	6.695	-4.4	0.2	0.006	3.125	-4	0.1	0.002	2.083	-4
	0.3	0.013	4.498	-3.667	1.5	0.058	3.862	0.133	0.6	0.016	2.658	0.333	0.3	0.004	1.408	-5.333
	1.5	0.026	1.751	-1	3	0.084	2.953	-5.167	1	0.021	2.176	-3.5	0.8	0.013	1.656	-1.875
	2.5	0.062	2.545	-2.56	5	0.133	2.717	-2.08	2	0.092	4.725	-2.65	1.2	0.043	3.720	-3.667
interday	0.1	0.005	4.990	0.2	0.5	0.026	5.556	-6.4	0.2	0.009	4.839	-7	0.1	0.004	4.301	-7
	0.3	0.015	5.119	-2.333	1.5	0.061	4.144	-1.867	0.6	0.023	3.879	-1.167	0.3	0.005	1.779	-6.333
	1.5	0.030	2.009	-0.467	3	0.092	3.273	-6.3	1	0.033	3.290	0.3	0.8	0.016	2.062	-3
	2.5	0.046	1.849	-0.48	5	0.162	3.401	-4.72	2	0.102	5.380	-5.2	1.2	0.052	4.384	-1.167

CV%: coefficient of variation
%RE: percent relative error

Table 1: Accuracy and precision of gemifloxacin mesylate, Theophylline, Warfarin and Omeprazole in human plasma.

Method	Gemifloxacin mesylate			Theophylline			Warfarin			Omeprazole		
	Concentration (µg/ml)	Recovery %	CV%	Concentration (µg/ml)	Recovery %	CV%	Concentration (µg/ml)	Recovery %	CV%	Concentration (µg/ml)	Recovery %	CV%
RP-HPLC	0.05	90.000	6.667	0.25	90.000	5.333	0.1	87.000	8.046	0.1	87.000	9.195
	0.15	92.000	5.797	0.75	92.667	5.899	0.3	91.333	7.664	0.3	88.333	9.811
	1	89.500	2.570	4	89.800	2.840	2	87.050	3.561	1	90.400	6.416
	5	86.720	4.935	7	90.829	6.055	9	92.911	4.329	4	88.150	3.318
TLC	0.1	88.000	6.818	0.5	86.400	7.176	0.2	87.000	4.598	0.1	91.000	7.692
	0.3	88.333	3.396	1.5	93.000	4.659	0.6	88.667	4.323	0.3	87.000	4.215
	1.5	90.133	2.663	3	91.500	3.716	1	89.600	5.134	0.8	92.625	4.318
	2.5	93.400	2.484	5	85.300	7.550	2	91.050	5.052	1.2	93.583	5.788

CV%: Coefficient of Variation %

Table 2: Results of recovery of gemifloxacin, theophylline, warfarin and omeprazole in human plasma.

	Gemifloxacin			Theophylline			Warfarin			Omeprazole		
	Added plasma concentration (µg/ml)	Found* (Mean ± SD)	CV%	Added plasma concentration (µg/ml)	Found* (Mean ± SD)	CV%	Added plasma concentration (µg/ml)	Found* (Mean ± SD)	CV%	Added plasma concentration (µg/ml)	Found* (Mean± SD)	CV%
Freeze-thaw	0.05	0.0502 ± 0.003	6.374	0.25	0.243 ± 0.005	1.893	0.1	0.098 ± 0.005	5.306	0.1	0.094 ± 0.006	6.064
	0.15	0.1489 ± 0.003	1.746	0.75	0.7493 ± 0.006	0.774	0.3	0.3011± 0.005	1.594	0.3	0.2966 ± 0.006	1.922
	1	1.003 ± 0.023	2.333	4	3.965 ± 0.069	1.735	2	1.969 ± 0.036	1.854	1	0.9885 ± 0.021	2.165
	5	5.035 ± 0.114	2.264	7	7.115 ± 0.174	2.446	9	8.885 ± 0.205	2.307	4	4.114 ± 0.132	3.209
Short term	0.05	0.0485 ± 0.003	5.979	0.25	0.2504 ± 0.004	1.518	0.1	0.101 ± 0.005	4.752	0.1	0.098 ± 0.005	5.408
	0.15	0.1503 ± 0.004	2.728	0.75	0.752 ± 0.007	0.904	0.3	0.312 ± 0.004	1.346	0.3	0.3025 ± 0.004	1.190
	1	1.008 ± 0.034	3.383	4	4.088 ± 0.091	2.228	2	2.052 ± 0.064	3.099	1	1.013 ± 0.019	1.935
	5	5.113 ± 0.201	3.931	7	7.066 ± 0.217	3.071	9	9.032 ± 0.265	2.934	4	4.052 ± 0.196	4.837
Long term	0.05	0.0492 ± 0.003	7.114	0.25	0.246 ± 0.004	1.707	0.1	0.095 ± 0.005	6.000	0.1	0.095 ± 0.005	5.895
	0.15	0.1496 ± 0.005	3.208	0.75	0.748 ± 0.009	1.229	0.3	0.298 ± 0.005	1.778	0.3	0.2963 ± 0.007	2.396
	1	0.989 ± 0.029	2.963	4	3.966 ± 0.078	1.979	2	1.974± 0.071	3.602	1	0.9932 ± 0.023	2.366
	5	4.989 ± 0.236	4.730	7	6.982 ± 0.306	4.383	9	8.963 ± 0.301	3.358	4	3.942 ± 0.223	5.657
Post preparative	0.05	0.0501 ± 0.003	6.786	0.25	0.247 ± 0.004	1.781	0.1	0.097± 0.005	5.567	0.1	0.101 ± 0.006	6.039
	0.15	0.1501 ± 0.004	2.798	0.75	0.752 ± 0.006	0.838	0.3	0.305 ± 0.006	1.869	0.3	0.301 ± 0.004	1.194
	1	0.985 ± 0.028	2.883	4	4.011 ± 0.0902	2.249	2	1.996 ± 0.046	2.305	1	1.011 ± 0.019	1.939
	5	4.988 ± 0.102	2.049	7	7.062 ± 0.141	1.997	9	9.05 ± 0.185	2.044	4	4.023 ± 0.096	2.399

• Mean of five determinations

Table 3: Stability study of gemifloxacin, theophylline, warfarin and omeprazole in human plasma by the proposed HPLC method.

	Gemifloxacin			Theophylline			Warfarin			Omeprazole		
	Added plasma concentration (µg/ml)	Found* (Mean ± SD)	CV%	Added plasma concentration (µg/ml)	Found* (Mean ± SD)	CV%	Added plasma concentration (µg/ml)	Found* (Mean ± SD)	CV%	Added plasma concentration (µg/ml)	Found* (Mean± SD)	CV%
Freeze-thaw	0.1	0.095 ± 0.006	6.316	0.5	0.477 ± 0.032	6.709	0.2	0.187 ± 0.006	3.209	0.1	0.093 ± 0.003	3.226
	0.3	0.285 ± 0.016	5.614	1.5	1.395 ± 0.056	4.014	0.6	0.574 ± 0.013	2.265	0.3	0.283 ± 0.005	1.767
	1.5	1.442 ± 0.032	2.219	3	2.805 ± 0.092	3.280	1	0.936 ± 0.046	4.915	0.8	0.781 ± 0.016	2.049
	2.5	2.415 ± 0.062	2.567	5	4.855 ± 0.202	4.161	2	1.885 ± 0.105	5.570	1.2	1.174 ± 0.036	3.066
Short term	0.1	0.097 ± 0.006	6.186	0.5	0.489 ± 0.026	5.317	0.2	0.192 ± 0.005	2.604	0.1	0.096 ± 0.003	3.125
	0.3	0.302 ± 0.014	4.636	1.5	1.450 ± 0.062	4.276	0.6	0.591 ± 0.018	3.046	0.3	0.291 ± 0.007	2.405
	1.5	1.486 ± 0.029	1.952	3	2.936 ± 0.102	3.474	1	0.942 ± 0.032	3.397	0.8	0.792 ± 0.017	2.146
	2.5	2.501 ± 0.058	2.319	5	4.962 ± 0.165	3.325	2	1.925 ± 0.101	5.247	1.2	1.184 ± 0.028	2.365
Long term	0.1	0.096 ± 0.005	5.208	0.5	0.478 ± 0.029	6.067	0.2	0.182 ± 0.008	4.396	0.1	0.094 ± 0.002	2.128
	0.3	0.281 ± 0.018	6.406	1.5	1.433 ± 0.081	5.652	0.6	0.584 ± 0.016	2.740	0.3	0.286 ± 0.009	3.147
	1.5	1.466 ± 0.036	2.456	3	2.856 ± 0.122	4.272	1	0.926 ± 0.062	6.695	0.8	0.783 ± 0.018	2.299
	2.5	2.425 ± 0.048	1.979	5	4.725 ± 0.205	4.339	2	1.900 ± 0.092	4.842	1.2	1.173 ± 0.034	2.899
Post preparative	0.1	0.101 ± 0.007	6.931	0.5	0.493 ± 0.021	4.260	0.2	0.196 ± 0.009	4.592	0.1	0.097 ± 0.004	4.124
	0.3	0.301 ± 0.011	3.654	1.5	1.485 ± 0.055	3.704	0.6	0.592 ± 0.011	1.858	0.3	0.289 ± 0.011	3.806
	1.5	1.495 ± 0.034	2.274	3	2.896 ± 0.065	2.244	1	0.965 ± 0.052	5.389	0.8	0.766 ± 0.025	3.264
	2.5	2.496 ± 0.006	2.644	5	4.905 ± 0.094	1.916	2	1.922 ± 0.102	5.307	1.2	1.154± 0.041	3.553

• Mean of five determinations

Table 4: Stability study of gemifloxacin, theophylline, warfarin and omeprazole in human plasma by the proposed TLC method.

found to be 0.05, 0.25, 0.1 and 0.1 µg/mL. LOD values were found to be 0.0165, 0.0825, 0.033 and 0.033 µg/mL for gemifloxacin, theophylline, warfarin and omeprazole, respectively. For TLC, LLOQ were found to be 0.1, 0.5, 0.2 and 0.1 µg/mL. LOD values were found to be 0.033, 0.15,

	Gemifloxacin		Theophylline		Warfarin		Omeprazole	
	Proposed HPLC method	Reported method[3]	Proposed HPLC method	Reported method[13]	Proposed HPLC method	Reported method[15]	Proposed HPLC method	Reported method[17]
Mean	98.088	98.784	98.228	100.022	98.035	99.845	97.968	98.816
SD	2.1625	2.3580	1.0535	1.5622	2.1709	2.6743	2.4952	2.8113
n	4	4	4	4	4	4	4	4
Variance	4.6764	5.5602	1.1099	2.405	4.7128	7.1519	6.2260	7.9034
t (2.447)*	0.44		1.9		1.05		0.46	
F (9.28) *	1.189		2.167		1.518		1.2694	

*The values between parentheses are the corresponding theoretical values of t and F at the 95% confidence level.

Table 5: Statistical comparison between the results obtained by the proposed HPLC method and the reported methods for the analysis of gemifloxacin, theophylline, warfarin and omeprazole in human plasma.

	Gemifloxacin		Theophylline		Warfarin		Omeprazole	
	Proposed HPLC method	Reported method[3]	Proposed HPLC method	Reported method[13]	Proposed HPLC method	Reported method[15]	Proposed HPLC method	Reported method[17]
Mean	97.790	98.784	98.922	100.022	97.346	99.845	96.896	98.816
SD	1.1709	2.358	1.4535	1.5622	2.2999	2.6743	1.4392	2.8113
n	4	4	4	4	4	4	4	4
Variance	1.3710	5.5602	2.1127	2.4405	5.285	7.1519	2.0713	7.9034
t (2.447)*	1.11		1.03		1.42		1.85	
F (9.28)*	4.056		1.155		1.352		3.816	

* The values between parentheses are the corresponding theoretical values of t and F at the 95% confidence level.

Table 6: Statistical comparison between the results obtained by the proposed TLC method and the reported methods for the analysis of gemifloxacin, theophylline, warfarin and omeprazole in human plasma.

Parameter	Gemifloxacin		Theophylline		Warfarin		Omeprazole	
	HPLC	TLC	HPLC	TLC	HPLC	TLC	HPLC	TLC
Capacity factor (K')	3.56	4.16	2.67	5.44	4.35	4.07	3.66	3.17
Selectivity factor (α)	1.44	1.08	1.36	1.24	1.76	1.56	1.08	1.56
Resolution (R)	3.83	2.65	2.08	1.78	2.55	1.98	4.55	3.78
Theoretical plates (N)	3589.56	2654.77	4554.76	2564.78	3786.88	3007.78	5641.67	3211.67
Tailing factor (t)	1.33	1.45	1.15	1.55	1.25	1.33	1.03	1.15

Table 7: System suitability results of the HPLC and TLC methods for determination gemifloxacin, theophylline, warfarin and omeprazole in spiked human plasma.

0.06 and 0.033 µg/mL for gemifloxacin, theophylline, warfarin and omeprazole, respectively.

Stability study: In stock solution stability the studied drugs with their internal standards samples were left at room temperature for 10 h. Comparison of the results with freshly prepared stock solution showed that there was no significant difference between response of freshly prepared solutions and samples of the studied drugs after 10 h.

Freeze–thaw stability was determined after three freezes–thaw cycles for five replicate of LLOQ, low, mid and high QC samples. The samples were stored at -20°C temperature for 24 h. Then thaw at room temperature. No significant difference between freeze–thaw samples and freshly prepared samples was observed.

Statistical analysis: Statistical comparison between the results of the proposed HPLC and TLC methods to those obtained by applying reported methods [3,13,15,17] showed that the calculated t and F values are less than the theoretical ones, indicating that there was no significant difference between the results obtained from the proposed methods and those obtained from the reported methods shown in tables 5 and 6.

System suitability testing: The results of the system suitability tests represented in table 7 assured the ability of the proposed methods for the routine analysis of the studied drugs.

Discussion

Using the proposed methods it was found that acetonitrile was the appropriate solvent for plasma precipitation and extraction that lead to high recovery % (above 70%). For HPLC, the chromatographic conditions, especially the composition of mobile phase, were optimized to achieve a good resolution and symmetric peak shapes for the analytes and the internal standard, as well as a short analytical time. For TLC, The proposed TLC method is based on the difference between Rf values of gemifloxacin mesylate and its co-administered drugs which differ in their polarities and consequently in their migration rates on TLC plates. The chromatographic conditions were optimized by spotting the drug with its co-administered drugs on TLC plates and developed in different solvent systems to achieve best separation. The proposed HPLC and TLC methods were accurate and precise as the accuracy % was not more than 15% even at LLOQ and the precision was not exceed 15% even also at LLOQ. Also the proposed methods were highly selective as there was no significant interference observed at the retention times of the analytes in the biological matrix. The result of stability experiments shows that no significant degradation occurred at ambient temperature for 6 h for short term stability, at -20°C for 6weeks for long term stability and for the post preparative stability for 24 h after comparing with freshly prepared sample. Results of stability for both RP-HPLC and TLC methods are shown in tables 3 and 4.

Conclusion

The proposed RP-HPLC and Densitometric-TLC methods for the estimation of gemifloxacin mesylate in human plasma in presence of three co-administered drugs; theophylline, warfarin and omeprazole, are selective and sensitive. Sensitivity of the methods are suitable for handling various therapeutic plasma levels of the mentioned drugs. The methods are economical and faster than earlier published methods. In future these methods can be used for bioequivalence and bioavailability studies or any other pharmacokinetic studies.

References

1. Krishna MV, Sanker DG (2008) Utility of σ and π acceptors for the spectrophotometric determination of gemifloxacin mesylate in pharmaceutical formulations. E-Journal of Chemistry 5: 493-498.

2. Tekkeli SE, Onal A (2011) Spectrofluorimetric Methods for the Determination of Gemifloxacin in Tablets and Spiked Plasma Samples. Journal of fluorescence 21: 1001-1007.

3. Rote AR, Pingle SP (2009) Reverse phase-HPLC and HPTLC methods for determination of gemifloxacin mesylate in human plasma. Journal of Chromatography B 877: 3719-3723.

4. Doyle E, FowlesSE, McDonnell DF, McCarthy R, White SA (2000) Rapid determination of gemifloxacin in human plasma by high-performance liquid chromatography–tandem mass spectrometry. Journal of Chromatography B746: 191-198.

5. Chakrabarty US, Das A, Bhaumik U, Chatterjee B, Ghosh A, et al. (2009) Rapid and Sensitive LC Method for the Analysis of Gemifloxacin in Human Plasma. Chromatographia 69: 853-858.

6. Al-Hadiyaa BM, Khadyb AA, Mostafa GA, (2010) Validated liquid chromatographic-fluorescence method for the quantitation of gemifloxacin in human plasma. Talanta 83: 110-116.

7. Leveque D, Jehl F (2009) Molecular Determinants of Fluoroquinolone Antibacterial Agents Pharmacokinetics. Curr Clin Pharmacol 4: 191-197.

8. Davy M, Allen A, Bird N, Rostt KL, Fuder, H (1999) Lack of effect of gemifloxacin on the steady-state pharmacokinetics of theophylline in healthy volunteers. Chemotherapy 45: 478-484.

9. Schelleman H, Bilker WB, Brensinger CM, Han X, Kimmel SE, et al. (2008) Warfarin-Fluoroquinolones, Sulfonamides, or Azole Antifungals Interactions and the Risk of Hospitalization for Gastrointestinal Bleeding. Clinical Pharmacology and Therapeutics 84: 581-588.

10. Davy M, Allen A, Bird N, Rostt KL, Fuder H (1999) Lack of effect of gemifloxacin on the steady-state pharmacodynamics of warfarin in healthy volunteers. Chemotherapy 45: 491-495.

11. Allena A, Vousdenb M, Lewis A (1999) Effect of omeprazole on the pharmacokinetics of oral gemifloxacin in healthy volunteers. Chemotherapy 45: 496-503.

12. http://www.fda.gov/downloads/Drugs/GuidanceComplianceRegulatoryInformation/Guidances/UCM368107.pdf

13. NirogiRV, Kandikere VN, Shukla M, Mudigonda K, Ajjala DR (2007) A simple and rapid HPLC/UV method for the simultaneous quantification of theophylline and etofylline in human plasma. J Chromatogr B Analyt Technol Biomed Life Sci 848: 271-276.

14. Mirfazaelian A, Goudarzi M, Tabatabaiefar M (2002) A quantitative thin layer chromatography method for determination of theophylline in plasma. J Pharm PharmSci 5: 131-134.

15. Sadrai S, Ghadam P, Sharifian R, Nematipour E (2008) Pharmacokinetic analysis of warfarin in Iranian warfarin sensitive patients. International Journal of Pharmacology 4: 149-152.

16. Huang Y, Liang M, Dong L, Yu Q, Qin Y (2001) Determination of warfarin in plasma by HPLC and an investigation of monitoring patients after cardiac valve replacement. Journal of Western China 32: 145-147.

17. Iuga C, Moldovan M, Popa A, Leucuta SE (2008) Validation of HPLC-UV method for analysis of omeprazole in presence of its metabolites in human plasma. Farmacia 6: 254-260.

18. RezkNL, Brown KC, Kashuba AD (2006) A simple and sensitive bioanalytical assay for simultaneous determination of omeprazole and its three major metabolites in human blood plasma using RP-HPLC after a simple liquid–liquid extraction procedure. Journal of Chromatography B 844: 314-321.

3

Development and Validation of HPLC Method for Simultaneous Estimation of Brimonidine Tartrate and Timolol Maleate in Bulk and Pharmaceutical Dosage Form

Abdullah A Elshanawane[1], Lobna M Abdelaziz[1], Mustafa S Mohram[2] and Hany M Hafez[2]*

[1]Medicinal Chemistry Department, Zagazig University, Zagazig, Egypt
[2]Quality control department, EIPICO, Egypt

Abstract

Brimonidine tartrate and Timolol Maleate are used in treatment of glaucoma by decreasing intra ocular pressure. A validated HPLC method was developed for the assay of them. The method was performed on BDS HYPERSIL Cyano column (250×4.6 mm, 5 µ) and the mobile phase consisted of Ammonium acetate (pH 5.0, 0.01 M) - Methanol (40:60, V/V) which pumped at a flow rate equals to 1.5 ml/min at ambient temperature. 20 µl of drugs sample solutions were monitored at two fixed wavelengths (lambda=254.0 nm for Brimonidine Tartrate and 300.0 nm for Timolol Maleate). The proposed method was validated in terms of linearity, accuracy, precision and limits of detection and quantitation according to ICH.

Keywords: Brimonidine Tartrate; Timolol Maleate

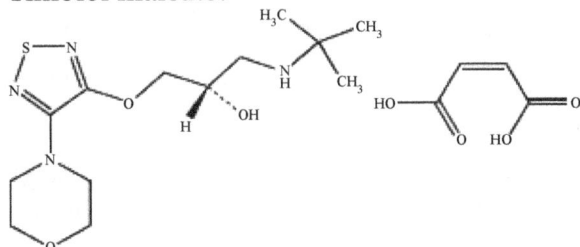

Abbreviations: HPLC: High Performance Liquid Chromatography

Introduction

Glaucoma describes a group of disorders characterized by a loss of visual field associated with cupping of the optic disc and optic nerve damage. Glaucoma is generally associated with raised intra-ocular pressure. Forms of glaucoma are primary open-angle glaucoma and primary angle closure glaucoma. Drugs that reduce intra-ocular pressure by different mechanisms are available for managing glaucoma. A topical beta-blocker or a prostaglandin analogue is usually the drug of first choice. It may be necessary to combine these drugs or add others, such as miotics, carbonic anhydrase inhibitors or *sympathomimetics* to control intra-ocular pressure [1].

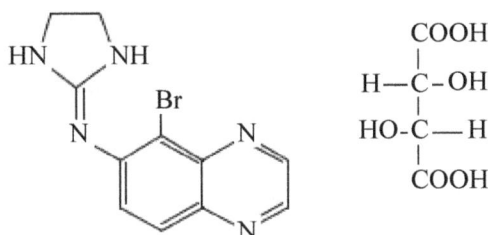

Timolol Maleate is (2S)-1-[(1, 1- dimethyl ethyl) amino]-3-[[4-(morpholin-4- yl)-1, 2, 5-thiadiazol-3-yl] oxy] propan-2-ol (Z)-butenedioate (Figure 1), it is a non-cardioselective beta blocker. It is reported to lack intrinsic sympathomimetic and membrane-stabilising activity. Timolol Maleate is used in the management of glaucoma, hypertension, angina pectoris and myocardial infarction [1]. Brimonidine is 5-Bromo-N-(4,5-dihydro-1H-imidazol-2-yl) quinoxalin-6-amine (Figure 1), it is a selective alpha -adrenoceptor agonist, is licensed for the reduction of intra-ocular pressure in open-angle glaucoma or ocular hypertension in patients for whom beta-blockers are inappropriate; it may also be used as adjunctive therapy when intra-ocular pressure is inadequately controlled by other anti-glaucoma therapy [1]. Eye drops (Combigan eye drop) containing Timolol Maleate equivalent to 0.5% of timolol and 0.2% of Brimonidine are instilled twice daily to reduce raised intra-ocular pressure in open-angle glaucoma and ocular hypertension.

Literature review reveals that different methods have been reported for estimation of both drugs individually like, RP-HPLC for the Analysis of Brimonidine in Ophthalmic Formulations [2,3], in blood serum and aqueous humor of the eye [4] and LC/MS/MS HPLC for the Analysis of Brimonidine in Ocular Fluids and Tissues [5]. Three papers discussed stability of Brimonidine tartrate [6-8], other analytical techniques were reported for determination of Brimonidine tartrate like HPTLC [9], GC/MS [10] and CE [11].

Timolol Maleate is listed in USP which described RP-HPLC for determination Timolol maleate in eye drops [12]. BP estimated Timolol Maleate potentiometrically [13].

Figure 1: Structures of Brimonidine Tartrate and Timolol Maleate respectively.

*Corresponding author: Hany M. Hafez, Bachelor degree of Pharmaceutical Science, Zagazig University, Zagazig, Egypt Quality control department, EIPICO, 10th Ramadan, Egypt, E-mail: hanyhaf_1982@yahoo.com

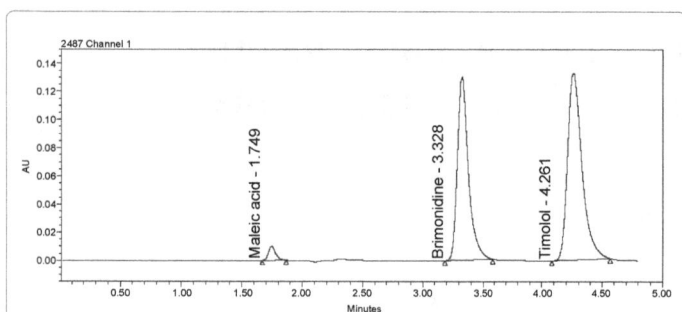

Figure 2: Typical HPLC chromatograms obtained from 20 μl injections of Brimonidine Tartrate (3.86 min.) and Timolol Maleate (3.86 min.) respectively under optimized chromatographic conditions.

Literature review reveals several methods have been reported for the estimation of Timolol Maleate in biological fluids [14-19] and there are some methods reported by voltammetry [20], spectroscopy [21-23], HPTLC [24] HPLC [25-28], UPLC [29,30] and capillary electrophoresis [31]. Two methods were only reported for estimation of this combination, the first is HPTLC [32] and the other is HPLC method [33]. The latter does not fulfill all requirements of validation which will be discussed later. Development of HPLC method for simultaneous estimation of this combination is a must and our scope that our new method will fulfill all requirements of validation according to ICH guidelines. It should be characterized by simplicity, fast analysis, sensitivity and suitability for routine pharmaceutical analysis.

Materials and Methods

Instrumentation

Analysis was performed on a chromatographic system of WATERS 2695 separation module connected to WATERS 2487 UV/VIS detector. The system equipped by Empower PC program. The chromatographic separation was achieved on BDS HYPERSIL Cyano column (250×4.6 mm, 5 μ).

Chemicals and reagents

All reagents used were of analytical grade or HPLC grade. Ammonium acetate and glacial acetic acid were supplied by (Merck, Darmstadt, Germany), Methanol HPLC grade was supplied by (Fischer scientific, U.K.) and Distilled water.

(Note: The water used in all the experiments was obtained from Milli-RO and Milli-Q systems (Millipore, Bedford, MA.)

Timolol Maleate and Brimonidine Tartrate working standard powders were kindly supplied by Egyptian international pharmaceutical industries company (EIPICO) (10th Ramadan, Egypt), and were used without further purification.

Pharmaceutical preparation

Combigan eye drops, Allergan (U.S.A) contains (Brimonidine Tartrate 0.2% and Timolol (as Maleate) 0.5%) B.NO: E64005.

Chromatographic condition

20 μl of drugs sample solutions were monitored at two fixed wavelengths (lambda=254.0 nm for Brimonidine Tartrate and 300.0 nm for Timolol Maleate). Liquid chromatography was performed on BDS HYPERSIL Cyano column (250×4.6 mm, 5 μ) and the mobile phase consisted of Ammonium acetate (pH 5.0, 0.01M) - Methanol (40:60, V/V) which pumped at a flow rate equals to 1.5 ml/min at ambient temperature.

Ammonium acetate (0.01 M) was prepared by dissolving 0.77 g Ammonium acetate in approximately 950 ml distilled water. The pH was adjusted to 5.0 with glacial acetic acid. Water was added to 1000 ml. Mobile phase was filtered through a 0.45 μl Nylon membrane filter (Millipore, Milford, MA, USA) under vacuum and degassed by ultrasonication (Cole Palmer, Vernon Hills, USA) before usage.

Preparation of stock standard solutions

Stock standard solutions containing 0.2, 0.68 mg/ml of Brimonidine Tartrate and Timolol Maleate (equivalent to 0.5 mg/ml of Timolol) respectively were prepared by dissolving 20, 68 mg of each in distilled water in 100 ml volumetric flask respectively. It was then sonicated for 5 minutes and the final volume of solutions was made up to 100 ml with distilled water to get stock standard solutions.

Preparation of calibration plot (working standard solutions)

To construct calibration plots, the stock standard solutions were diluted with distilled water to prepare working solutions in the concentration ranges (4-24 and 10-60 μg/ml) for of Brimonidine Tartrate and Timolol respectively. Each solution (n=5) was injected in triplicate and chromatographed under the mentioned conditions above. Linear relationships were obtained when average drug standard peak area were plotted against the corresponding concentrations for each drug. Regression equation was computed.

Sample preparation

Take 1 ml of Combigan E/D into 100 ml V.F. then complete with distilled water. Test solutions were analyzed under optimized chromatographic conditions and chromatogram is depicted in (Figure 2).

Method validation

Specificity: Specificity is the ability to assess unequivocally the analyte in the presence of components which may be expected to be present. Typically these might include impurities, degradants, matrix, etc [34]. A Bulk of Combigan E/D (solution contains excipients only) was prepared by mixing its excipients like benzalkonium chloride 0.005%; sodium phosphate, monobasic; sodium phosphate, dibasic and purified water then the bulk was injected under previous condition. Representative chromatogram showed that the bulk has negligible contribution after the void volume at the method detection wavelengths i.e. it did not interfere with developed method.

Linearity and range: The linearity of an analytical procedure is its ability (within a given range) to obtain test results which are directly proportional to the concentration (amount) of analyte in the sample. For the establishment of linearity, a minimum of 5 concentrations is recommended [34]. Five Concentrations were chosen in the ranges (4-24 and 10-60 μg/ml) for corresponding levels of 20-120% w/w of the nominal analytical concentration of Brimonidine Tartrate and Timolol respectively. The linearity of peak area responses versus concentrations was demonstrated by linear least square regression analysis. The linear regression equations were {Y=42167 X - 4937 (r=0.9999) and Y=22395X - 8914 (r=0.9998)} for Brimonidine Tartrate and Timolol Maleate respectively. Where **Y** is the peak area of standard solution and **X** is the drug concentration (Figure3).

Precision: The precision of the assay was investigated by measurement of both repeatability and Intermediate precision.

Repeatability: Repeatability was investigated by injecting a minimum of 6 determinations at 100% of the test concentration and percentage SD were calculated in Table 1.

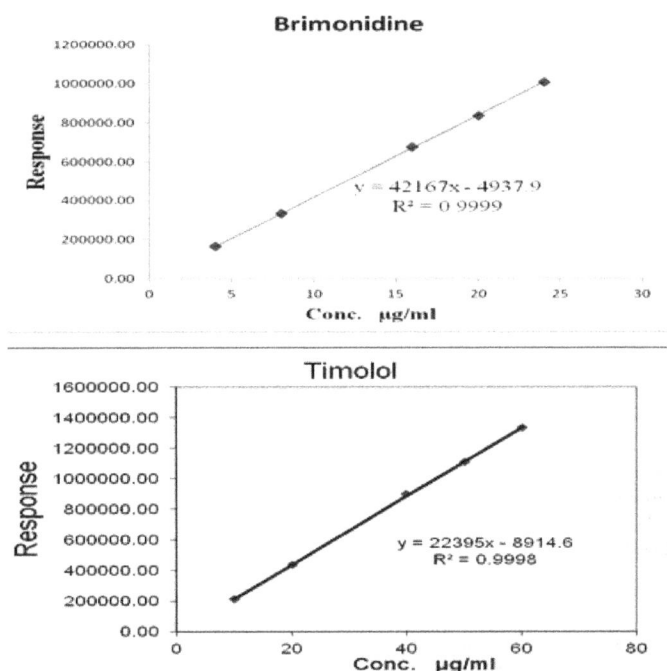

Figure 3: Calibration plot of Brimonidine Tartrate and Timolol Maleate.

Drug name		Brimonidine Tartrate		Timolol Maleate	
		AV ± SD mg/ml	AV ± SD %	AV ± SD mg/ml	AV ± SD %
Repeatability		19.98 ± 0.19	99.88 ± 0.94%	49.90 ± 0.50	99.81 ± 1.00%
Intermediate precision		19.99 ± 0.32	99.9 ± 1.59%	49.78 ± 0.84	99.56 ± 1.67%
Accuracy & Recovery %	80%	15.99 ± 0.13	99.94 ± 0.78%	40.21 ± 0.43	100.53 ± 1.06%
	100%	20.06 ± 0.15	100.32 ± 0.75%	50.01 ± 0.51	100.01 ± 1.02%
	120%	24.20 ± 0.19	100.83 ± 0.80%	60.34 ± 0.43	100.57 ± 0.72%

Table 1: Repeatability and Intermediate precision and Accuracy (Recovery %) of Brimonidine Tartrate and Timolol Maleate respectively.

Intermediate precision: In the inter-day studies, standard and sample solutions prepared as described above, were analyzed in triplicate on three consecutive days at 100% of the test concentration and percentage SD were calculated (Table 1).

Accuracy: Accuracy was assessed using 9 determinations over 3 concentration levels covering the specified range (80,100 and 120%). Accuracy was reported as percent recovery by the assay of known added amount of analytes in the sample (Table 1).

Limits of detection and Limits of quantitation: According to the ICH recommendations, determination of limits of detection and quantitation was based on the standard deviation of the y-intercepts of regression lines (n=3) and the slope of the calibration plots [34] (Table 2).

Robustness: Robustness of an analytical procedure is a measure of its capacity to remain unaffected by small variations in method parameters and provides an indication of its reliability during normal usage [34]. Robustness was tested by studying the effect of changing mobile phase pH by ± 0.5, the percentage of organic solvent (methanol) in the mobile phase by ± 5 %, temperature ± 5°C, wavelengths ± 5 nm and flow rate ± 0.1 ml/min had no significant effect on the chromatographic resolution of the method.

Stability of analytical solution: Also as part of evaluation of robustness, solution stability was evaluated by monitoring the peak area response. Standard stock solutions in methanol were analyzed right after its preparation 1, 2 and 3 days after at room temperature. The change in standard solution peak area response over 3 days was (1.01 and 0.89 %) for Brimonidine Tartrate and Timolol Maleate respectively. Their solutions were found to be stable for 3 days at room temperature at least.

Application on Pharmaceutical Preparation

The proposed methods were successfully used to determine Brimonidine Tartrate and Timolol Maleate respectively in Combigan E/D. Five replicate determinations were performed. Satisfactory results were obtained for each compound in good agreement with label claims. The results obtained were compared statistically with those from published method [33] by using Student's t-test and the variance ratio F-test. The results showed that the t and F values were smaller than the critical values. So, there were no significant differences between the results obtained from this method and published methods (Table 3).

Results and Discussion

Optimization of chromatographic condition

Several trials were carried out to obtain optimized chromatographic condition for simultaneous determination of Brimonidine Tartrate and Timolol Maleate in their pharmaceutical preparations. Firstly, maximum absorption wavelengths (254,300 nm) for Brimonidine Tartrate and

Item	Brimonidine Tartrate	Timolol Maleate
Linear range (µg/ml)	4-24	10-60
Detection limit (µg/ml)	0.05	0.09
Quantitation limit (µg/ml)	0.15	0.29
Regression data		
No.	5	5
slope (b)	42167	22395
Standard deviation of the slope	56.36	36.67
intercept (a)	4937	8914
Standard deviation of the intercept	637.86	641.36
correlation coefficient ®	0.9999	0.9998
Standard error of regression	0.09	0.3

(Y = a + bC, where C is the concentration of the compound (µg/ml) and Y is the drug peak area)

Table 2: Calibration data was resulted from method validation of Brimonidine Tartrate and Timolol Maleate respectively.

Drug name	Recovery ± SD		Calculated t- values	Calculated F- values
	Proposed methods	Reference method		
Brimonidine Tartrate	101.41 ± 0.95	100.85 ± 1.00	2.49	0.94
Timolol Maleate	101.40 ± 1.32	101.23 ± 1.16	1.13	1.28

(Where the Tabulated t-values and F -ratios at p= 0.05 are 2.57 and 5.05)

Tab.le 3: Statistical comparison of the proposed and published methods for determination of Brimonidine Tartrate and Timolol Maleate respectively in their dosage forms by reported method (T- student test) and (F –test for variance).

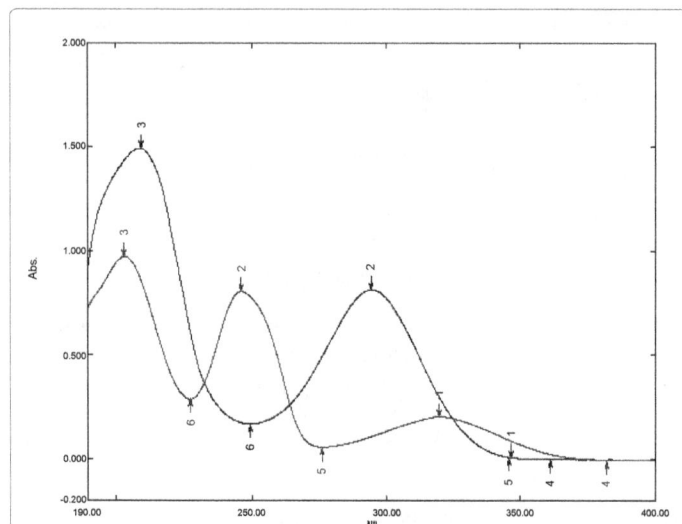

Figure 4: Typical UV spectrum of Brimonidine Tartrate and Timolol Maleate.

Timolol Maleate were selected by scanning from 350-200 nm under UV (Figure 4). Ammonium acetate buffer has no effect on absorption at wavelength more than 250 nm [35]. Low conc. of buffer (0.01M) is adequate for most reversed phase applications. This concentration is also low enough to avoid problems with precipitation when significant amounts of organic modifiers are used in the mobile phase [36] .PH of acetate buffer was examined, at pH=3 interference between maleic acid peak and solvent peak had occurred but interference disappeared at pH=5. Different percentages of organic solvent (methanol) were tried with acetate buffer to reach optimum separation with good resolution. 60% methanol in mobile phase gave adequate separation and resolution. Using Cyano column enabled us to produce sharp and symmetric peaks because it is better in separation of drug containing basic functionality like amino compounds [37].

References

1. BMJ Group (2011) British national formulary 62. (1stedn), Pharmaceutical Press, London.

2. Narendra A, Deepika D, Annapurna MM (2012) Liquid chromatographic method for the analysis of brimonidine in ophthalmic formulations. E-J Chem 9: 1327-1331.

3. Shirke RR, Pai N (2002) RP-HPLC determination of brimonidinetartrate in brimonidine tartrate eye drops. Ind Drugs 39: 484-486.

4. Karamanos NK, Lamari F, Katsimpris J, Gartaganis S (1999) Development of an HPLC method for determining the alpha2-adrenergic receptor agonist brimonidine in blood serum and aqueous humor of the eye. Biomed Chromatogr 13: 86-88.

5. Jiang S, Chappa SK, Proksch JW (2009) A Rapid and Sensitive LC/MS/MS Assay for the Quantitation of Brimonidine in Ocular Fluids and Tissues. J Chromatogr B Analyt Technol Biomed Life Sci 877: 107-114.

6. Ali MS, Khatri AR, Munir MI, Ghori M (2007) A Stability-Indicating Assay of Brimonidine Tartrate Ophthalmic Solution and Stress Testing Using HILIC. Chromatographia 70: 539–544.

7. Bhagav P, Deshpande P, Pandey S, Chandran S (2010) Development and Validation of Stability Indicating UV Spectrophotometric Method for the Estimation of Brimonidine Tartrate in Pure Form, Formulations and Preformulation Studies. Der Pharm Let 2: 106-122.

8. Sonanis MC, Rajput AP (2011) Development and validation of a new stability indicating analytical method for the determination of related components of brimonidine tartrate in drug substances and drug product using UPLC. Int. J. Pharm Pharm Sci 3: 156-160.

9. Anand M, Fonseca A, Santosh GV, Padmanabh DB (2010) Development and Validation of High Performance Thin Layer Chromatographic Method for Estimation of Brimonidine Tartrate as Bulk Drug and in Ophthalmic Solutions. Int. J. ChemTech Res 2: 1376-1379.

10. Acheampong A, Tang L, Diane DS (1995) Measurement of brimonidine concentrations in human plasma by a highly sensitive gas chromatography/mass spectrometric assay.J Pharm Biomed Anal 13: 995- 1002.

11. Tzovolou DN, Lamari F, Mela EK, Gartaganis SP, Karamanos NK (2000) Capillary electrophoretic analysis of brimonidine in aqueous humor of the eye and blood sera and relation of its levels with intraocular pressure. Biomed Chromatogr 14: 301-305.

12. http://www.healthwise.org

13. http://www.pharmacopoeia.co.uk

14. Nasir F, Iqbal Z, Khan A, Ahmad L, Shah Y, et al. (2011) Simultaneous determination of timolol maleate, rosuvastatin calcium and diclofenac sodium in pharmaceuticals and physiological fluids using HPLC-UV. J Chromatogr B Analyt Technol Biomed Life Sci 879: 3434-3443.

15. Maguregui MI, Jimenez RM, Alonso RM ,Akesolo U (2002) Quantitative determination of oxprenolol and timolol in urine by capillary zone electrophoresis. J Chromatogr A 949: 91–97.

16. Olah TV, Gilbert JD, Barrish A (1993) Determination of the β-adrenergic blocker Timolol in plasma by liquid chromatography-atmospheric pressure chemical ionization mass spectrometry. J Pharm Biomed Anal 11: 157–163.

17. Magureguia MI, Alonso RM ,Jiménez RM (1996) A Rapid Quantitative Analysis of the β-Blocker Timolol in Human Urine by HPLC-Electrochemical Detection. J Liq Chromatogr Rel Tech 19: 1643-1652.

18. He H, Edeki TI, Wood AJJ (1994) Determination of low plasma Timolol concentrations following topical application of timolol eye drops in humans by HPLC with electrochemical detection. J Chromatogr B Biomed Appl. 661: 351-356.

19. Gilbert JD, Olah TV, Morris MJ, Bortnick A, Brunner J (1998) The Use of Stable Isotope Labeling and LC/ MS To Simultaneously Determine the Oral and Ophthalmic Bioavailability of Timolol in Dogs. J Chromatogr Sci 36: 163-168.

20. Turkdemir MH, Erdogdu G, Aydemir T, Karagozler AA , Karagozler AE (2001) Voltammetric Determination of Timolol Maleate: A β-Adrenergic Blocking Agent. J Anal. Chem 56:1047-1050.

21. Erk N (2002) Simultaneous determination of dorzolamide HCL and Timolol maleate in eye drops by two different spectroscopic methods. J Pharm Biomed Anal 28: 391-397.

22. Ferraro MC (2004) Chemometric determination of amiloride hydrochloride, atenolol, hydrochlorothiazide and timolol maleate in synthetic mixtures and pharmaceutical formulations. J Pharm Biomed Anal 34:305-314.

23. Ayad MM, Shalaby A, Abdellatef HE , Hosny MM (2003) Spectrophotometric methods for determination of enalapril and timolol in bulk and in drug formulations. Anal Bioanal Chem 375: 556-560.

24. Kulkarni SP, Amin PD (2000) Stability indicating HPTLC determination of timolol maleate as bulk drug and in pharmaceutical preparations. J Pharm Biomed Anal 23: 983–987.

25. Erk N (2003) Rapid and sensitive HPLC method for the simultaneous determination of dorzolamide hydrochloride and timolol maleate in eye drops with diode-array and UV detection. Pharmazie 58: 491-493.

26. Nagori BP, Maru A, Muysuni P, Gupta S (2011) Method Development and Its Validation for Simultaneous Estimation of Timolol Maleate and Dorzolamide Hydrochloride as API and In Ophthalmic Solution Dosage Form by RP-HPLC. J Chem Pharm Res 3: 866-874.

27. Wanare RS, Kabra AO, Deshmukh A, Aher N (2012) Simultaneous Estimation of Dorzolamide Hydrochloride And Timolol Maleate In Eye Drops By a RP-HPLC Method. Int J Compreh Pharm 3: 1-4.

28. Rele RV, Mhatre VV, Parab JM, Warkar CB (2011) Simultaneous RP HPLC determination of Latanoprost and Timolol Maleate in combined pharmaceutical dosage form. J Chem Pharm Res 3:138-144.

29. Sharma N, Rao SS, Reddy AM (2012) A Novel and Rapid Validated Stability-Indicating UPLC Method of Related Substances for Dorzolamide Hydrochloride and Timolol Maleate in Ophthalmic Dosage Form. J Chromatogr Sci. 50:745-755.

30. Mehta J, Patel V, Kshatri N, Vyas N (2010) A versatile LC method for the simultaneous quantification of latanoprost, timolol and benzalkonium chloride and related substances in the presence of their degradation products in ophthalmic solution. Anal. Methods 2:1737-1744.

31. Marini RD et al. (2006) Interlaboratory study of a NACE method for the determination of R-timolol content in S-timolol maleate: assessment of uncertainty. Electrophoresis 27:2386-2399.

32. Jain PS, Khatal RN, Jivani HN , Surana SJ (2011) Development and Validation of TLC-densitometry Method for Simultaneous Estimation of Brimonidine tartrate and Timolol maleate in Bulk and Pharmaceutical Dosage Form. J Chromatograph Separat Techniq 2:113-117.

33. Phogat A, Kumar MS, Mahadevan N (2011) Simultaneous Estimation of Brimonidine Tartrate and Timolol Maleate in Nanoparticles Formulation by RP-HPLC. Int J Rec Adv Pharm Res 3: 31-36.

34. ICH Guidelines (1996) Validation of Analytical Procedures: Text and Methodology. ICH.

35. Sethi PD (2001) High Performance liquid Chromatography quantitative analysis of pharmaceutical formulations. (1stedn), India

36. http://www.mac-mod.com/pdf/TR-BufferedMP.pdf

37. Adamovics JA et al. (1996) chromatographic analysis of pharmaceuticals. (2ndedn),USA

4

Development of a Validated RP-HPLC Method for Simultaneous Estimation of Metformin Hydrochloride and Rosuvastatin Calcium in Bulk and In-House Formulation

Murthy TGK and Geethanjali J*

Bapatla College of Pharmacy, Bapatla, Andhra Pradesh, India

Abstract

A simple, precise, rapid, selective, and economic reversed phase high-performance liquid chromatography (RP-HPLC) method has been established for simultaneous analysis of Metformin hydrochloride and Rosuvastatin in bulk powder and In-House Formulation on a Phenomenex C18 (250×4.6 mm i.d) chromatographic column equilibrated with mobile phase containing Acetonitrile/0.02 M Sodium dihydrogen o-phosphate. Experimental conditions such as pH of mobile phase, ratio of organic phase, flow rate, wavelength, etc. were critically studied and the optimum conditions were selected. Efficient chromatographic separation was achieved with mobile phase containing combination of Phosphate buffer pH 2.8 and Acetonitrile in ratio of 65:35(v/v) adjusted to pH 3.8 at flow rate of 1.0 ml/min and eluents were monitored at 252 nm. 20 μl of sample was injected into chromatographic system and the total run time was 10 min. The retention time for Metformin and Rosuvastatin were 2.147 min and 3.80 min respectively. The method was linear in the range of 5μg mL^{-1}to 30μg mL^{-1}and 0.4 μg mL^{-1}to2.4μg mL^{-1}for Metformin and Rosuvastatin respectively. The proposed method was successfully applied to the analysis of Metformin and Rosuvastatin in bulk and in-house formulation without interference from other additives. The developed method was validated according to ICH guidelines. Linearity, regression value, recovery and % RSD of intra and interday precision values were found within the limits and the method was found to be satisfactory. This validatedHPLC procedure is economic, sensitive, user-friendly and less time consuming than other chromatographic procedures.

Keywords: RP-HPLC; Rosuvastatin; Metformin

Introduction

Diabetes mellitus, a metabolic disorder characterized by increased blood sugar levels also causes malfunctioning of major organs of the body and even enhances cholesterol biosynthesis and causes dyslipidemia. Hyperglycemia, increased levels of reactive oxygen species, production of advanced glycation end products and glycation of lipoproteins, and lipid abnormalities, such as increases in the levels of VLDL and their remnants, lead to diabetes induced dyslipidaemia. Diabetes is often associated with accelerated atherosclerosis and subsequent cardiovascular disease, people with diabetes are at two to fourfold greater risk of developing cardiovascular disease than people without diabetes, and the majority of type 2 diabetic patients die of atherosclerotic disease [1]. For many patients with diabetes induced dyslipidaemia, monotherapy with an oral antidiabetic agent is not sufficient to reach target glycemic goals and lipid levels therefore, a combination of 2 drugs i.e., an oral antidiabetic drug and a anti hyperlipidemic drug preferably a statin may be necessary to achieve adequate control. In such cases combination of metformin (MET) and Rosuvastatin (ROS) is used that lowers the blood glucose and lipid levels [2]. Metformin is an oral anti-diabetic drug in the biguanide class. It is the first-line drug of choice for the treatment of type-2

diabetes, particularly in overweight and obese people [3]. Rosuvastatin is a selective, competitive inhibitor of HMG-CoA reductase, the rate limiting enzyme that converts 3-hydroxy-3-methylglutaryl-coenzyme A to mevalonate, a precursor of sterols, including cholesterol. Hence, the combination of Rosuvastatin and Metformin complement each other and provides reduction in plasma cholesterol along with glycemic control thereby providing a comprehensive control of diabetes and associated dyslipidemia (Figures 1 and 2).

The literature survey reveals several analytical methods for quantitative estimation of Metformin alone in body fluids and in pharmaceutical formulations. Those methods include spectrophotometry, electrochemical methods, HPLC, liquid chromatography-electrospray ionization tandem mass spectrometry and electrophoresis [3-6]. Rosuvastatin was estimated in body fluids and in pharmaceutical formulations by spectrophotometry, HPLC methods and resonance light scattering technique [7-9]. Formulations containing combination of Metformin and statin derivative such as Atorvastatin are available in market. But the combinations of Metformin and Rosuvastatin are not yet available though this combination can be used to provide comprehensive control of Diabetes and associated

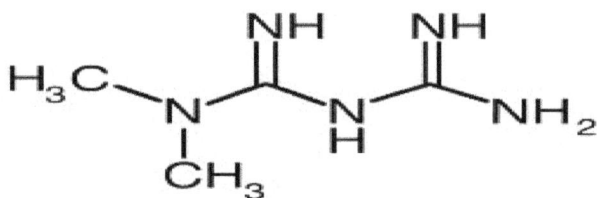

Figure 1: Chemical structure of Metformin.

***Corresponding author:** Geethanjali J, Bapatla College of Pharmacy, Pharmaceutical Analysis and Quality Assurance, D.No: 6-10-48/1, Srungarapuram, Bapatla, Guntur district, Bapatla, Andhra Pradesh, 522101 India
E-mail: geethanjalijakki@gmail.com

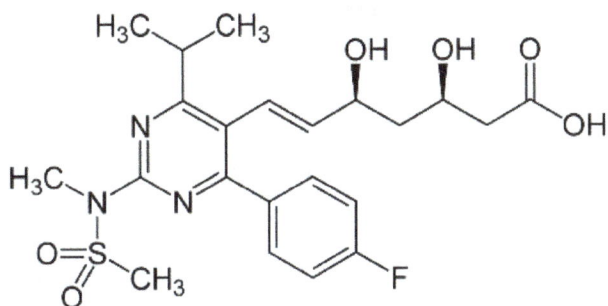

Figure 2: Chemical structure of Rosuvastatin.

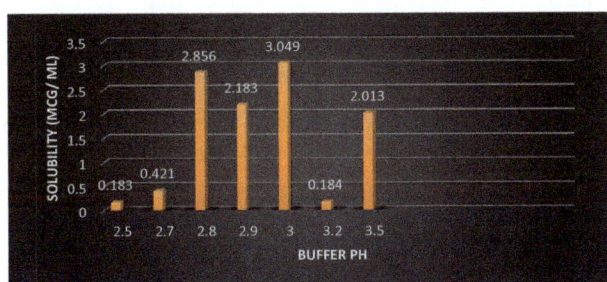

Figure 3: Solubility of Drug in Buffers of pH varying 2.5-3.5.

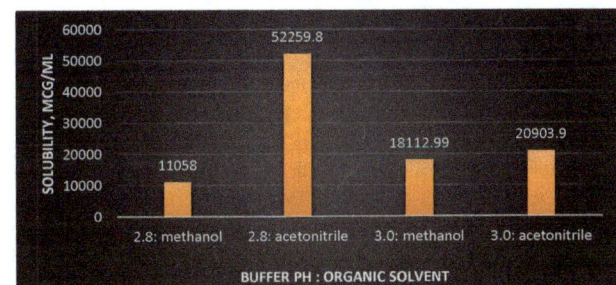

Figure 4: Solubility of Drug in a mixture of Buffer and Organic Solvent.

dyslipidemia. Up to our knowledge, no method was described for the simultaneous estimation of these drugs in bulk or laboratory prepared mixtures.Therefore an attempt was madeto develop a simple, accurate isocratic HPLC method for simultaneous determination of Metformin and Rosuvastatin in bulk and In-House formulation.

Materials and Methods

Instrumentation

HPLC system (Agilent technology, 1200 series) consisting of gradient pump, Auto sampler, column oven and photodiode array detector (PDA, Agilent technology) was employed for analysis. The analytical column employed was Phenomenex C_{18}(250×4.6 mm, 5.0 μm). Data acquisition was performed on EZ Chrom Elite software. An ELICO UV – spectrophotometer SL 164 was used for measuring absorbance of samples. A Mettler Toledo MP 225 pH meter 3510 pH/mV was used for pH adjustment.A vacuum filtration unit was employed to filter the mobile phase through 0.45 μm filter. A degasser, Crest Ultrasonics was used. Systronics electronic balance was used to weigh the samples.

Reagents and chemicals

Metformin Hydrochloride and Rosuvastatin active pharmaceutical ingredient were supplied by Cadila Healthcare Limited (Ahmadabad, India). The HPLC grade Acetonitrile, Methanol and Ortho Phosphoric acid were purchased from Sigma Gmbh, Germany. Analytical reagent grade disodium hydrogen phosphate and triethylamine were used.

Chromatographic Conditions

Column: Phenomenex C_{18}(250×4.6 mm, 5.0 μm)

Mobile phase: Acetonitrile and Phosphate buffer pH 2.8 in the ratio of 65:35 (total pH of mobile phase was adjusted to 3.8 using triethylamine)

Flow rate: 1.0 ml/min

Wavelength: 252 nm

Column temperature:ambient

Injection volume: 20 μl

Run time: 10 min

Preliminary Studies

Selection of mobile phase: A suitable mobile phase composition was selected based upon maximum solubility of drug in solvents employed viz., buffers and organic solvents like acetonitrile and methanol. Metformin is a hydrophilic drug which is freely soluble in most of organic solvents and buffers. Therefore, a suitable mobile phase was selected based on solubility of Rosuvastatin. Solubility of Rosuvastatin was determined by shake-flask method [10]. In this method, an excess amount of drug was added to the solvent medium so as to make a saturated solution in equilibrium with the solid phase. Then the two phases were separated by filtration and concentration of drug in saturated solution was determined by UV spectrophotometric method. Suitable solvents employed for determining solubility of Rosuvastatin were selected based upon pka of the drug (pka of Rosuvastatin: 4) [11]. Initially the solubility was tested in a series of buffers of pH ranging 2-5 which showed maximum solubility of the drug in buffers of pH 2 and 3. The solubility of drug was then determined in buffers ranging pH 2.5-3.5, which indicated maximum solubility of drug in buffers of pH 2.8 and 3. Solubility of drug was again determined in a mixture of buffer and organic solvent in a ratio of 1:1. The results of solubility studies were represented in the form of graphs (Figures 3 and 4) from the data obtained, it was concluded that Acetonitrile: 2.8 pH buffer was an appropriate mobile phase for simultaneous estimation of Metformin and Rosuvastatin.

Method development and optimization: Several trials were performed using Acetonitrile and 50 mm phosphoric acid (sodium) buffer solution (pH=2.8) in different proportions, under different pH conditions, by varying wavelength and flow rate conditions and a range was obtained for each factor and optimized chromatographic conditions were selected from fractional factorial design. The optimization parameters were listed in (Table 1).

S.NO	FACTOR	RANGE
1.	Mobile phase pH	3.8-3.9
2.	Organic phase proportion	60-65 %v/v
3.	Wavelength	251-253 nm
4.	Flow rate	1.0-1.2 ml/min

Table 1: Optimization parameters.

Formulation	Composition of tablet	Analyte	Label claim (mg)	% label claim estimated (*mean of 6 determinations)	Standard deviation
In-House Tablet formulation	Metformin Hydrochloride -79.6 % Rosuvastatin Calcium - 6.37 % Poly Vinyl Pyrrolidine - 5 % Sodium Starch Glycolate - 5 % Magnesium Sterate - 2 % Talc - 2 %	Metformin	500	98.5 %	0.950 %
		Rosuvastatin	40	99.2 %	1.547 %

Table 2: Analysis of In-House tablet formulation.

Procedure for analysis of tablets: The tablet containing 500 mg of Metformin and 40 mg of Rosuvastatin were prepared by wet granulation technique. The composition of tablet was given in (Table 2). Not fewer than 10 tablets were weighed and grounded to fine powder. Accurately weighed portion of this powder equivalent to 25mg of Metformin and 2mg of Rosuvastatin and transferred to a 100 mL volumetric flask containing 50 mL of mobile phase. The contents of the flask were allowed to stand for 10 minutes with intermittent sonication to ensure complete solubility of the drugs and made up to volume with mobile phase. The solution was then filtered through a 0.45 µm membrane filter. 1.0 mL of the filtrate solution was transferred into a 10 mL volumetric flask, and volume was made up to the mark with mobile phase.From this solution appropriate dilutions were made with mobile phase to obtain concentration in calibration range for both the drugs and this solution was used for estimation.

With the optimized chromatographic conditions, a steady baseline was recorded, the mixed working standard solution was injected and the chromatogram was recorded. The retention times of Metformin and Rosuvastatin were found to be 2.147 and 3.800 min respectively. The results of analysis shows that the amount of drug was in good agreement with the label claim of formulation (Table 2).

Validation of the Proposed Method

The proposed method was validated according to the International Conference on Harmonization (ICH) guidelines [10-15].

Specificity

The specificity of the method was evaluated by assessing whether excipients present in the formulations interfered with the analysis. A placebo for each tablet was prepared by mixing the respective excipients, and solutions of required concentrations were prepared.

Linearity

It is the ability of the method to obtain test results which are directly proportional to the concentration of analyte in the sample within a given range. Different levels of standard solutions were prepared and injected into the HPLC and the chromatograms were recorded. The peak area corresponding to different concentrations of selected drugs was observed and the data was analyzed for linearity by fitting the data in regression equation, $y = mx + b$.

Accuracy

The accuracy of the method was determined by recovery experiments. Known concentration of working standard was added to the fixed concentration of the pre-analyzed tablet solution. Percent recovery was calculated by comparing the area before and after the addition of working standard. For both the drugs, recovery was performed in the same way. The recovery studies were performed in triplicate. This standard addition method was performed at 80%, 100%, 120% level and the percentage recovery was calculated.

Precision

The precision of the method was demonstrated by inter day and intraday variation studies. In the intraday studies, solutions of standard and sample were repeated thrice in a day and percent relative standard deviation (%RSD) for response factor was calculated. In the interday variation studies, injections of standard and sample solutions were made on three consecutive days and % RSD was calculated.

Limit of detection and limit of quantification

The Limit of detection and quantification were calculated using standard deviation of the response (σ) and slope (S) of calibration curve.

Limit of Detection, LOD = 3.3 σ/S

Limit of quantification,LOQ = 10 σ/S

Robustness

Robustness of the method was checked by making slight deliberate changes in chromatographic conditions like mobile phase pH, organic phase ratio,and flow rate and detection wavelength.

Results and Discussion

A simple isocratic high-performance liquid chromatographic method was developed for the determination of Metformin and Rosuvastatin in pure form and in In-House tablet formulations using Phenomenex C$_{18}$ column. The mobile phase consisted of acetonitrile and buffer at pH 3.8 (65: 35 %v/v). The mobile phase was chosen after several trials to reach the optimum stationary/mobile-phase matching. The optimized chromatographic conditions were selected based on results obtained from fractional factorial design.

Fractional factorial design

The efficiency of chromatographic method may be influenced by pH of mobile phase, organic phase composition of mobile phase, flow rate and detection wavelength. So to study the effect of these parameters at 2 different levels on the efficiency of method, fractional factorial design with 3 replications was employed. The performance parameters such as asymmetry, efficiency and resolution of Metformin and Rosuvastatin peaks were observed for each experiment and the results were analyzed with MINITAB 16 version software. The effect of each factor was analyzed by observing the coefficients of polynomial equation and presented in the (Tables 3 and 4).

The basic Polynomial equation is given below:

$$Y = B_0 + B_1 X_1 + B_2 X_2 + B_3 X_3 + B_4 X_4 + B_{12} X_1 X_2 + B_{13} X_1 X_3 + B_{14} X_1 X_4$$

Where,

X_1 = pH of mobile phase B_0 = Constant B_4 = Coefficient of X_4

X_2 = organic phase ratio B_1 = Coefficient of X_1 B_{12} = Coefficient of X_1 and X_2

PARAMETER	EQUATION
Metformin peak asymmetry	$y=23.9-11.87X_1+3.61X_2+17.65X_3-1.04X_4-0.93X_1X_2-4.78X_1X_3+0.29X_1X_4$
Rosuvastatin peak asymmetry	$y=-45.69+14.28X_1+3.8X_2+38.43X_3-0.93X_4-0.99X_1X_2-10.22X_1X_3+0.23X_1X_4$
Metformin peak efficiency	$y=1474114-376578X_1+5651X_2-13967X_3-7164X_4-1456X_1X_2+3250X_1X_3+1837X_1X$
Rosuvastatin peak efficiency	$y=2223749-560236X_1+5697X_2-26259X_3-10115X_4-1484X_1X_2+5991X_1X_3+2565X_1X_4$
Resolution of Metformin & Rosuvastatin peaks	$y=23802-6136X_1-27X_2-272X_3-56.6X_4-7.12X_1X_2+67.32X_1X_3+22.36X_1X_4$

Table 3: Polynomial equations obtained from factorial design.

ANALYTICAL PARAMETER	COEFFICIENTS							
	B_0	B_1	B_2	B_3	B_4	B_{12}	B_{13}	B_{14}
Metformin Peak Asymmetry	23.9	-11.9	3.6	17.7	-1.04	-0.9	-4.8	0.92
Rosuvastatin Peak Asymmetry	-45.6	14.2	3.8	38.4	-0.92	-0.98	-10.2	0.23
Metformin Peak Efficiency	1474114	-376578	5651.2	-13967	-7164	-1456	3250	1536
Rosuvastatin Peak Efficiency	2223743	-560238	5697.7	-26259	-10115	-1484	5991	2565
Resolution	23802.7	-6136.8	-27.5	-272	-86.6	4.12	67	22.3

Table 4: Effect of various parameters on different analytical parameters.

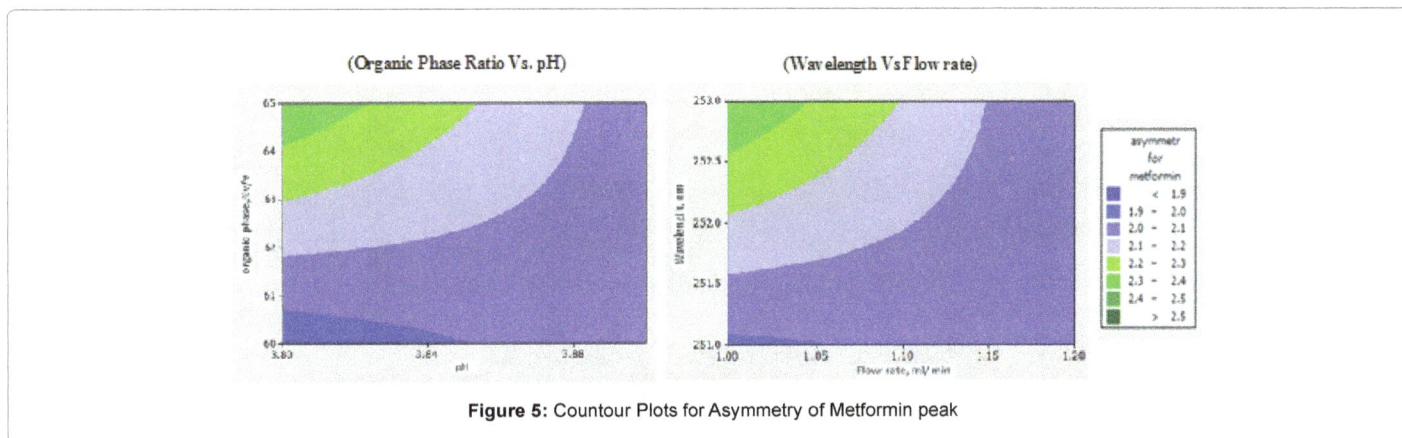

Figure 5: Countour Plots for Asymmetry of Metformin peak

X_3= flow rateB_2= Coefficient of $X_2 B_{13}$= Coefficient of X_1 andX_3

X_4= detection wavelengthB_3= Coefficient of $X_3 B_{14}$= Coefficient of X_1 andX_4

From the (Tables 3 and 4), the following conclusions were drawn:

The main factors, pH of the mobile phase and wavelength are having negative effect and organic phase ratio and flow rate are having positive effect on asymmetry of Metformin peak. The interaction effect existing among the factors pH of mobile phase-organic phase ratio, pH of mobile phase- flow rate are having negative effect and the interaction effect existing among the factors pH of mobile phase-detection wavelength was found to have positive effect.

The main factor detection wavelength is having negative effect and the factors, pH of the mobile phase, organic phase ratio and flow rate are having positive effect on asymmetry of Rosuvastatin peak. The interaction effect existing among the factors pH of mobile phase-organic phase ratio, pH of mobile phase- flow rate are having negative effect and the interaction effect existing among the factors pH of mobile phase-detection wavelength was found to have positive effect.

The main factors viz., pH of the mobile phase, flow rate and detection wavelength detection wavelength are having negative effect and organic phase ratio is having positive effect on efficiency of Metformin peak. The interaction effect existing among the factors pH of mobile phase-organic phase ratio is having negative effect and the interaction effect existing among the factors, pH of mobile phase- flow rate, pH of mobile phase-detection wavelength was found to have positive effect.

The main factors viz.,pH of mobile phase, flow rate, detection wavelength are having negative effect and factor, organic phase ratio is having positive effect on efficiency of Rosuvastatin peak. The interaction effect existing among the factors pH of mobile phase-organic phase ratio is having negative effect and the interaction effect existing among the factors pH of mobile phase- flow rate, pH of mobile phase-detection wavelength was found to have positive effect.

The main factors viz., pH of mobile phase, organic phase ratio, flow rate, detection wavelength is having negative effect on resolution of Metformin and Rosuvastatin peaks. The interaction effect existing among the factors viz., pH of mobile phase-organic phase ratio, pH of mobile phase- flow rate, pH of mobile phase-detection wavelength was found to have positive effect.

Optimizing of parameters based on countour plots

To optimize the selected factors, the observed response with respect to the selected factors was represented as countour plots (Figures 5-9).

Interpretation obtained from countour plots

The required asymmetry for the Metformin chromatogram peak (<2) is possible by maintaining the pH between 3.8-3.85, organic phase ratio between 60-60.5 %, flow rate at 1.0 ml/ min and wavelength at 251 nm.

The required asymmetry for the Rosuvastatin chromatogram peak (<2) is possible by maintaining the pH between 3.8-3.9, organic phase ratio between 60-64.5 %, flow rate between 1.0-1.2 ml/ min and wavelength between 251-253 nm.

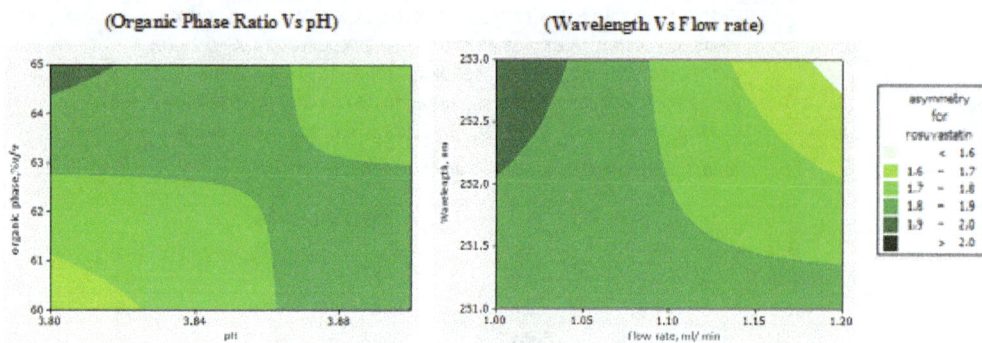

Figure 6: Countour Plots for Asymmetry of Rosuvastatin peak.

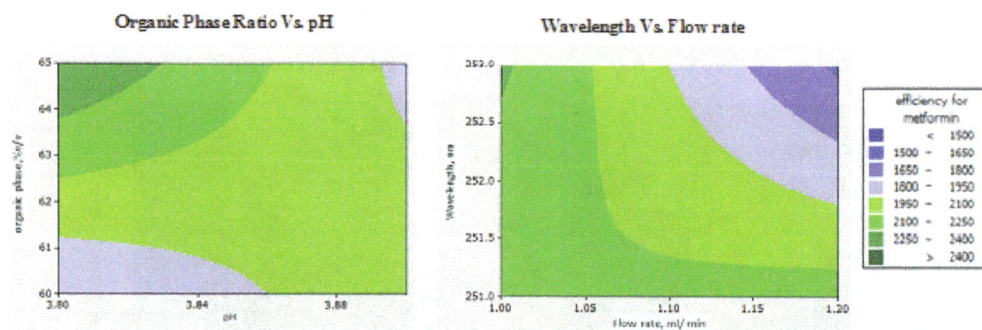

Figure 7: Countour Plots for Efficiency of Metformin peak .

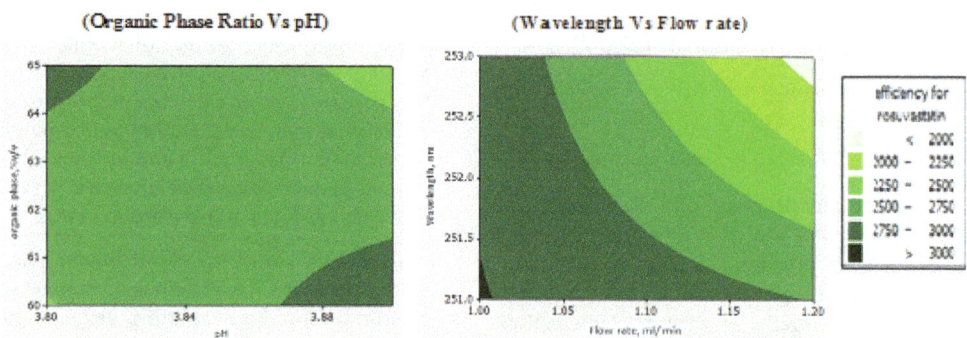

Figure 8: Countour Plots for Efficiency of Rosuvastatin peak.

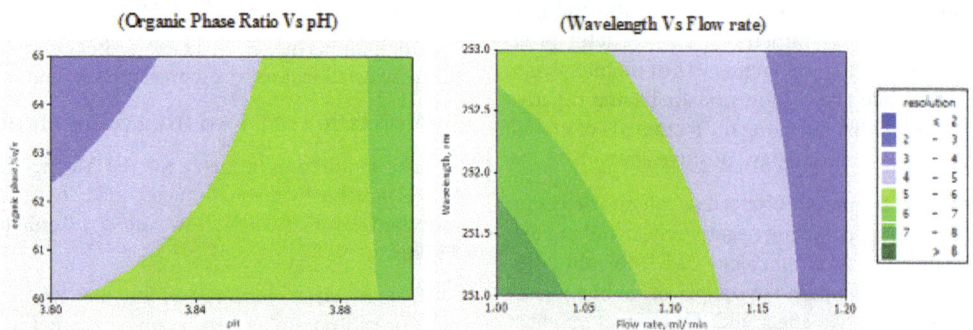

Figure 9: Countour Plots for Resolution of Metformin and Rosuvastatin peaks.

The required efficiency for the Metformin chromatogram peak (>2000) is possible by maintaining the pH between 3.85-3.9, organic phase ratio between 61-65 % v/v, flow rate at 1.0 – 1.1 ml/ min and wavelength between 251-253 nm.

The required efficiency for the Rosuvastatin chromatogram peak (>2000) is possible by maintaining the pH between 3.8-3.9, organic phase ratio between 60-65 % v/v, flow rate at 1.0 – 1.18 ml/ min and wavelength between 251-253 nm.

The required resolution for the Metformin and Rosuvastatin chromatogram peaks (>2) is possible by maintaining the pH between 3.8-3.9, organic phase ratio between 60-65 % v/v, flow rate at 1.0 – 1.2 ml/ min and wavelength between 251-253 nm.

Finally, based on the results obtained from fractional factorial design, Acetonitrile: phosphate buffer (65:35) pH adjusted to 3.8 with triethylamine at a flow rate of 1.0 ml/min was selected as an appropriate mobile phase and eluents were monitored at 252nm which gave good resolution, and acceptable system suitability parameters. The retention times of Metformin and Rosuvastatin were found to be 2.147 min and 3.800 min respectively.

S No	Area (mAU)	Area (mAU)	Other system suitability parameters		
1	388135	80501	USP Tailing	MET	ROS
2	390577	80580			
3	380323	79560		1.282	1.796
4	380211	80508	USP Plate count	MET	ROS
5	388563	78032			
6	398378	80520		6636	9051
Avg	387698.7	79950.17	USP Resolution	MET	ROS
StdDev	6837.32	1016.62		-	3.833
% RSD	1.76	1.27			

S.No.	System Suitability Parameters	RESULTS	
		Metformin	Rosuvastatin
1	Tailing factor(T$_f$)	1.282	1.796
2	Resolution (Rs)	-	3.833
3	Retention time(Rt)	2.140	3.800
4	Theoretical plates(N)	6636.38	9051.47

Method Validation

The method was specific since excipients in the formulation did not interfere in the estimation of MET and ROS. The proposed method was found to be linear in the concentration range of 5 μg /ml to 30 μg /ml for Metformin and 0.4 μg /ml to 2.4 μg /ml for Rosuvastatin. Accuracy of the method was indicated by recovery values of 98.3 % for Metformin and 99.3 % for Rosuvastatin. Precision is reflected by %RSD values less than 2. The intraday %RSD of Metformin and Rosuvastatin were found to be 0.836 and 0.146 respectively. The interday% RSD for Metformin and Rosuvastatin were found to be 0.577 and 0.767 respectively. From the data obtained the developed RP-HPLC method was found to be precise. The LOD for Metformin and Rosuvastatin were found to be 0.32 μg/ml and 0.05 μg/ml respectively. The LOQ were 0.μg/ml and 0.16 μg/ml for Metformin and Rosuvastatin respectively. It was observed that there were no marked changes in chromatograms obtained by altering mobile phase composition, pH, flow rate and detection wavelength, which demonstrated that the developed RP-HPLC method was robust.

The results of analysis of In-House formulations were shown in (Table 2) and validation parameters were summarized in (Table 5). A typical chromatogram of Metformin and Rosuvastatin observed from formulation is shown in (Figure 10).

PARAMETER		METFORMIN	ROSUVASTATIN
Linearity (µg/ml)		5-30 µg/ml	0.4-2.4 µg/ml
Correlation Coefficient		0.986	0.993
Intra Day Precision (% RSD)		0.836	0.146
Inter Day Precision (% RSD)		0.597	0.767
Accuracy	80 % level	98.01 %	99.80 %
	100 % level	99.02 %	99.40 %
	120 % level	98.00%	98.80 %
Limit Of Detection (µg/ml)		0.32	0.05
Limit Of Quantitation (µg/ml)		0.99	0.16
Robustness		Robust	Robust

Table 5: Summary of Validation Parameters.

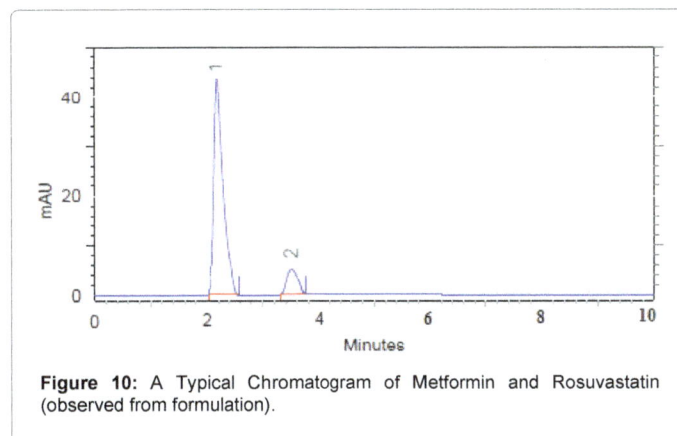

Figure 10: A Typical Chromatogram of Metformin and Rosuvastatin (observed from formulation).

Conclusion

This developed and validated method for simultaneous analysis of Metformin and Rosuvastatin is very rapid, accurate, and precise. The method was successfully applied for determination of MET and ROS in its In-House tablet formulations. The proposed method was simple and did not involve laborious time-consuming sample preparation. Moreover it has advantages of low costs of reagents used, short run time and the possibility of analysis of a large number of samples. Hence this method can be conveniently used for routine analysis and quality control of pharmaceutical preparations containing these drugs either as such or in combination.

Acknowledgement

The authors are thankful to Cadila Healthcare, India for providing gift samples of Metformin and Rosuvastatin for research. The authors are highly thankful to Bapatla College of Pharmacy, Bapatla, Andhra Pradesh, India for providing all the laboratory facilities to carry out the work.

References

1. Lamharzi N, Renard CB, Kramer F, Pennathur S, Heinecke JW, et al. (2004) Hyperlipidemia in concert with hyperglycemia stimulates the proliferation of macrophages in atherosclerotic lesions: potential role of glucose-oxidized LDL. Diabetes 53: 3217-3225.

2. Arayne MS, Sultana N, Tabassum A (2013) RP-LC simultaneous quantitation of co-administered drugs for (non-insulin dependent) diabetic mellitus induced dyslipidemia in active pharmaceutical ingredient, pharmaceutical formulations and human serum with UV-detector. Clinica Chimica Acta 425: 54-61.

3. Kumar TN, Rao KCN, Sreenivasulu R, Raju NSV, Pyreddy VR (2011) Novel Rp-Hplc Method For The Estimation Of Metformin Hydrochloride In Pharmaceutical Dosage Forms. International Journal of Science Innovations and Discoveries 1: 395-421.

4. Mubeen G, Noor K (2009) Spectrophotometric Method for Analysis of Metformin Hydrochloride. Indian journal of pharmaceutical sciences 7: 100–102

5. Hamdana I.I, A.K. Bani Jaberb, A.M. Abushoffa (2010) Development and validation of a stability indicating capillary electrophoresis method for the determination of metformin hydrochloride in tablets. Journal of Pharmaceutical and Biomedical Analysis 53: 1254–1257

6. Chen X, Qi Gu, FengQiu, Zhong D (2004) Rapid determination of metformin in human plasma by liquid chromatography–tandem mass spectrometry method. Journal of Chromatography B 802: 377–381

7. Kaila H.O, Ambasana M.A, Thakkar R.S, Saravaia H.T, Shah A.K (2010) A New Improved RP-HPLC Method for Assay of Rosuvastatin Calcium in Tablets. Indian Journal of Pharmaceutical Sciences 72: 592–598

8. Uma Devi S, PushpaLatha E, Nagendra Kumar Guptha CV , Ramalingam P (2011)Development and Validation of HPTLC method for Estimation of Rosuvastatin calcium In Bulk and Pharmaceutical Dosage Forms. International Journal of Pharma and Bio Sciences 2: 134-140

9. Gupta A, Mishra P, Shah K (2009) Simple UV Spectrophotometric Determination of Rosuvastatin Calcium in Pure Form and in Pharmaceutical Formulations. E-Journal of Chemistry 6: 89-92

10. Mohammad A, Fakhree, Experimental and Computational Methods Pertaining to Drug Solubility, Toxicity and Drug Testing, intech.com: 1-33.

11. www.drugbank.ca/drugs/DB01098

12. International conference on Harmonization (ICH) (1995) Guidelines on validation of analytical procedure definition & terminology. Federal Register 60: 11260.

13. ICH: Q2A, (1994) Text on validation of analytical procedure.

14. ICH: Q2B,(1996) Analytical validation – Methodology.

15. ICH Q2 (R1), (2005) Validation of analytical procedures text and methodology.

5

Gas Chromatographic Determination of Amino Acids and Polyamines in Human Skin Samples using Trifluoroacetylacetone and Isobutyl Chloroformate as Derivatizing Reagents

Sohail Ahmed Soomro*, Muhammad Yar Khuhawar[1] and Shahid Hussain Soomro[2]

[1]Institute of Advanced Research Studies in Chemical Sciences University of Sindh, Jamshoro, Pakistan
[2]Shaheed Mohatarma Benazeer Bhutto Medical University Larkana Sindh, Pakistan

Abstract

An analytical method has been developed with improved sensitivity for the determination of amino acids and polyamines after precolumn derivatization with trifluoroacetylacetone and isobutyl chloroformate. 28 compounds comprising of 20 amino acids, five polyamines (1,3 diaminopropane, putrescine, cadverine, spermindine and spermine) and three biogenic compounds (4-aminobutyric acid, histamine and dopamine) separated completely with linear calibration range 1-10 µg/ml and limits of detection 60 -200 ng/ml. The separation was obtained within 11 min from HP-5 (30 m × 0.32 mm id) with column temperature 100°C for 2 min, followed by ramping at the rate of 20°C / min upto 250°C. The nitrogen flow rate was 3.0 ml/min. The method was applied for the analysis of the compounds in human skin samples after acid hydrolysis. The variation in their contents were examined in the affected skin samples from pemphigus vulgaris, psoriasis, leishmaniasis and eczema patients and result were compared with unaffected skin samples from healthy volunteers. The extraction of amino acids and polyamines from samples calculated by standard addition was within 95-102 % with RSDs 1.23-6.75 %.

Keywords: Gas chromatography; Amino acids; Polyamines; Pemphigus vulgaris; Psoriasis; Leishmania; Eczema; Healthy volunteers; Derivatization

Introduction

The human skin is self-renewing protective layer, providing the body with an impenetrable cover. It is composed of keratinocytes, which form an ordered array, including hair follicles and their appendages, such as sebaceous glands [1]. The keratinocytes are composed of protein molecules, which are synthesized from free amino acids precursors [2]. The information concerning their amino acids contents and their alterations as a result of pathological effects is limited [3]. The amino acids alteration from the normal value may be considered as a result of some disorder [4]. Thus the identification and determination of a number of amino acids may be useful for diagnosis of diseases including skin disorders [5].

The aliphatic amines of low molecular weight (putrescine, cadverine, spermidine and spermine) play a vital role in the synthesis of proteins and nucleic acids and also in the regulation of cell growth and proliferation [6]. Polyamines are also known to be the most important compounds for early diagnosis and treatment of cancer [7].

Pemphigus vulgaris (Pv) is a rare but sever blistering disease that affects the skin and mucous membrane [8]. Histopathologically Pv results in the formation of acantholysis of keratinocytes and blister in the suprabasal layer of epidermis. The binding of auto anti body to keratinocytes causes a subsequent loss of cell adhesion [9].

Atopic eczema is a chronic inflammatory skin disease that shows a wide variety of clinical pictures and is often complicated by relapse caused by different kinds of food [10]. The pathogenesis of eczema is reported due to a complex interaction of genetic predisposition and environmental factors [11].

Leishmaniaisis is a disease caused by flagellated protozoan parasites of the genus leishmania, which belongs to the order kinetoplastida and family trypanosomatidae [12]. It is transmitted to their mammalian host by the bite of a hematophagous sandfly vector [13]. These parasites are the causative agents of a broad spectrum of human diseases ranging from simple self healing cutaneous lesions to a severe visceral dineminations [14].

Psoriasis is an inflammatory skin disease characterized by marked increases in keratinocytes proliferation, prominent alteration in dermal capillary vasculature and presence of dermal and epidermal T cells monocytes, macrophages and neutrophils [15,16].

The determination of amino acids and polyamines in biological materials continues to attract the attention of researches for physiopathalogical and clinical reasons. A large number of procedures have been developed for the determination of amino acids and polyamines but the procedures based on chromatographic methods and capillary electrophoresis [17-22] are more frequently reported. The analysis of amino acids by high performance liquid chromatography (HPLC) usually involve precolumn derivatization using various reagents including o-phthalaldehyde (OPA) [23], phenyl isothiocyanate [24], danyl chloride [25], or 9-fluorenylmethyl chloroformate (FMoC) [26]. The methods described include HPLC with photodiode array and fluoresence detection [27], HPLC-tandam mass spectrometry [28] and ultra high performance liquid chromatography tandam mass spectrometry [29]. However a number of difficulties have been reported to the procedures associated with HPLC and have been discussed [30]. The analysis of amino acids and polyamines by

*Corresponding author: Sohail Ahmed Soomro, Institute of Advanced Research Studies in Chemical Sciences University of Sindh, Jamshoro, Pakistan
E-mail: Sohail229@yahoo.com

GC requires derivatization due to their polar nature. Different silyl derivatives or chloroformates are used as derivatizing reagents to bind amino and carboxylic acid groups simultaneously. Chloroformates as derivatizing reagents in GC have advantages over silyating reagents because of short reaction time in aqueous phase at room temperature [31-33]. The amino acids containing primary amino group react with the trifluoroacetylacetone (FAA) to form the derivatives amiable to GC column. The sensitivity has been enhanced by the derivatization of carboxylic group of amino acids with ethyl chloroformate [34]. The work also examines the GC of two stage derivatization using FAA and isobutyl chloroformate (IBCF) as reagents for further enhancement in sensitivity using FID detection. The method is applied for the determination of amino acids and polyamines contents in different human skin disorders after hydrolysis of proteins and compares the results with normal skin samples from healthy volunteers.

Experimental

Chemicals and solutions

The compounds glycine (Gly), L-alanine (Ala), L-valine (Val), L-phenylalanine (Phe), tryptophan (Trp), (Sigma, Louis, USA), tyrosine (Tyr), serine (Ser), proline (Pro), leucine (Leu), isoleucine (Ile), methionine (Met), threonine (Thr), (Sigma, Reisenhofen, Germany), glutamic acid (Glu), glutamine (Gln), aspartic acid (Asp), asparagine (Asn), cystine (Cys), lysine (Lys), histidine (His), arginine (Arg) (Sigma GmbH, Germany), 1,3- diaminopropane (1,3 Dap), putrescine (Put), cadaverine (Cad), spermindine (Spd), spermine (Spm), 4-aminobutyric acid (GABA), histamine (Hst) dopamine (Dop), trifluoroacetylacetone (FAA) and isobutyl chloroformate (IBCF) (Fluka Buchs, Switzerland), methanol (Rdh, Chemicals Co.Spring valley, CA, USA) were used.

Guaranteed reagent grade hydrochloric acid (37%), potassium chloride, acetic acid, sodium acetate, ammonium acetate, sodium tetraborate, boric acid, sodium bicarbonate, sodium carbonate, ammonium chloride and ammonia solution were from E.Merk, Darmstadt, Germany.

Stock solutions of amino acids, polyamines and other biological active compounds containing 1 mg/ml were prepared in methanol and water. Further solutions were prepared by appropriate dilution. Buffer solutions (0.1 M) between pH 3-10 at unit and 0.5 unit were prepared from the following: acetic acid–sodium acetate (pH 3-6), ammonium acetate (pH 7), boric acid-sodium tetraborate (pH 7.5-8.5), sodium bicarbonate-sodium carbonate (pH 9), ammonium chloride-ammonia solution (pH 10).

Equipments

The pH measurements were made with an Orion 420A pH meter (Orion Research Inc. Boston, USA) with combined glass electrode and reference internal electrode, GC studies were carried out on an Agilant model 6890 network GC system, connected with flame ionization detector (FID) and split injector (Agilant Technologies Santa, Clara, CA, USA), hydrogen generator (Parker Balson) Analytical Gas system (H-90, Parker Hannif Haverhill, MA, USA) and pure nitrogen (British oxygen company (BOC) Karachi, Pakistan), computer with Chemstation software controlled the gas chromatograph. Capillary column HP-5 (30 m x 0.32 mm id) with film thickness 0.25 μm (J. W Scientific GC column Wilmington, Nc, USA) was used throughout the study.

GC Analytical procedure

The solution (0.2-3.0 ml) containing a mixture of amino acids,

polyamines, GABA, Hst and Dop (1–10 μg each) was added to 0.3 ml ammonium acetate buffer pH 7 and 0.3 ml FAA (2% v/v in methanol). The contents were heated on water bath (95°C) for 20 min. The mixture was allowed to cool at room temperature and 0.3 ml of solvent system (acetonitrile-water-methanol-pyridine 42:42:8:8 v/v) was added to it. The mixture was then added to 0.3 ml IBCF (2% in methanol) and carbonate buffer pH 9 (0.3 ml). The mixture was sonicated at room temperature (30°C) for 15 min. Chloroform (1.0 ml) was added and contents were mixed well. The layers were allowed to separate and an aliquot of organic layer was transferred to screw capped sample vial. The solution (1 μl) was injected in GC with split ratio 10:1 on the column HP-5 (30 m×0.32 mm id) with film thickness 0.25 μm at column temperature 100°C for 2 min., followed by ramping at the rate of 20°C/min up to 250°C and maximum temperature was maintained for 2 min. The nitrogen flow rate was 3 ml/min. The injector and detector temperatures were 270°C and 280°C respectively. The flow rates for FID were fixed for nitrogen as make up gas 45 ml/min, hydrogen 40 ml/min and air 450 ml/min.

Skin samples analysis

Dried skin sample (0.5 g) was added to 6 N hydrochloric acid (5 ml) in screw capped sample vial and contents were heated at 110°C for 24 h. The mixture was cooled and centrifuged for 20 min at 3000 g. The clear supernatant layer was separated and the residue was washed with de-ionized water (1 ml). The solvent from the combined extract was evaporated gently under nitrogen atmosphere. The residue was dissolved in water and volume was adjusted to 10 ml. The solution (3.0 ml) was taken and GC analytical procedure was followed. The quantitation was made from external calibration curve prepared from linear regression equation $Y=ax + b$.

Analysis of amino acids from skin samples using linear calibration curve with spiked samples

The dried skin sample (0.5g) from healthy volunteer, pemphigus vulgaris, leishmaniatic, psoriatic and eczema patients was treated as "skin samples analysis". Two portions of 3 ml from each of the sample were taken and a portion was processed as GC analytical procedure and the other was added to the solution of amino acids and polyamines (0.5 ml) containing 10 μg/ml each and GC analytical procedure was again followed. The quantitations were carried out from external calibration curves and from increases in the responses with added standards.

Procedure for sample collection

The skin's cut pieces during circumcision of normal children ages between 5 and 10 years with verbal/written permission of the parents of the children, were taken from Liaquat University of Medical and Health Sciences (LUMHS) Hospital, Hyderabad and were placed in separate screw capped tube containing formaldehyde solution (20%). New hand gloves were used for the collection of each sample. The skin pieces were washed thoroughly with de-ionized distilled water in the laboratory and dried at 60°C for 45 min. Alternatively, the feet of healthy volunteers who had not taken any medicine at least one preceding week with verbal/written consent were washed with de-ionized water and methanol and allowed to dry. The upper layer of the feet was collected by rubbing it gently with iron gauze pre washed with de-ionized water and methanol. Each of the samples was transported to the laboratory separately and was oven dried at 60°C for 45 min.

The skin samples of patients where possible were obtained by applied biopsy. The skin was washed thoroughly with de-ionized water and methanol. The punches were applied 4–8 mm deep at the lesions of the patients and were rotated clockwise and anti clockwise to cut the pieces of the skin. These pieces of the skin were placed separately in screw capped test tubes containing formaldehyde solution (20%) and were transferred to the laboratory for analysis. Each of the samples was thoroughly washed with de-ionized water in the laboratory and was dried at 60°C for 45 min.

In case of Pv, P and E the scales on the surface of the skin were also scratched with new surgical blade for each patient. To obtain the skin sample of L the xylocaine solution was injected at the infected part of the patient and then applied punch biopsy. The collected samples were placed in formaldehyde 20% solution and then transported to the laboratory for analysis.

The patients suffering from pemphigus vaulgaris, psoriasis, leishmaniasis and eczema were identified by the consultants of Dermatology wards of LUMHS, Hyderabad Hospital and Shaheed Mohatarma Benazeer Bhutto Medical University Hospital Larkana, Sindh, Pakistan.

Results and Discussion

The reagent FAA is known to react with primary amino groups of amino acids, polyamines and biological active compounds to form highly stable Schiff bases [34,35]. The reagent IBCF is also reported to react with carboxylic group of amino acid and to secondary amino groups to form derivative with extension of carbon chain length [36] and increase in the sensitivity of FID detection. Both the reactions are carried out in aqueous medium. The optimum derivatization conditions of amino acids with FAA and polyamines and GABA, Hst and Dop were checked as reported [34], followed by with IBCF. The reactions were monitored by GC and the condition which gave maximum response (peak height/peak area) was considered optimum. The reaction conditions were examined in terms of the effect of pH, concentration of derivatizing reagent, reaction time and temperature and reaction medium. The maximum reaction of FAA was observed at pH 7 with ammonium acetate buffer with 0.3 ml of (2% FAA in methanol) per determination as derivatizing reagent and warming time of 20 min at 95°C.

The amino acids after derivatization with FAA were examined for second derivatization with IBCF and an enhancement in GC-

FID response for each of the compound was observed and thus reaction conditions were optimized. The reactions of chloroformates are reported in alkaline medium and therefore the effect of pH was examined within 6-12 at 0.5 and unit interval. The maximum response was observed with carbonate buffer pH 9 and was selected. The volume of carbonate buffer was varied from 0.2 ml to 1.0 ml at an interval of 0.1 ml. The addition of the different volumes of the buffer was not significant and addition of 0.3 ml was selected. The addition of IBCF (2% v/v in methanol) was varied from 0.1 to 0.6 ml at an interval of 0.1 ml and it was observed that the addition of IBCF was not critical as long excess of IBCF was available in the reaction medium. The addition of 0.3 ml was selected. The addition of the solvents for reaction medium was examined. Water, water-methanol, water-acetonitrile, water-pyridine base and solvent system proposed by Husek [37] (acetonitrile-water-methanol-pyridine 42:42:8:8 v/v) were investigated. The solvent system proposed by Husek proved better and was used. The volume of the solvent was varied between 0.2-1.0 ml at an interval of 0.1 ml and the effect of different volumes of solvent was not significant and 0.3 ml was used. The solvent chloroform, amyl alcohol and ethyl acetate were examined for the extraction of the derivatives and chloroform proved better and was used [37]. The amino acids, polyamines and biogenic amines derivatives after their formation and after different intervals of time were examined on GC and no change in the response was observed upto 24 h.

The amino acids, polyamines, GABA, Hst and Dop derivatives formed, after necessary extraction in chloroform was examined for GC separation using different temperature elution programmes. Finally complete and base line separation of all 28 compounds examined was found from HP-5 (30 m x 0.32 mm id), when eluted at column temperature 100°C for 2 min, followed by heating rate 20°C/min upto 250°C. The maximum temperature was maintained for 2 min. The total run time was 11.5 min. The nitrogen flow rate was maintained at 3 ml/min. The order of the elution of amino acids was Gly, Ala, 1,3 Dap, Put, Ser, Cad, Pro, GABA, Val, Thr, Leu, Ile, Cys, Asn, Asp, Hst, Lys, Gln, Glu, Met, Spd, His, Phe, Dop, Arg, Tyr, Trp, Spm (Figure 1). The capacity factor (K') calculated for the amino acids varied within the range 0.87-5.87. The separation was repeatable in terms of retention time and peak height/peak area with RSD's (n=5) within 0.96-3.33 % and 2.30-4.84 % respectively.

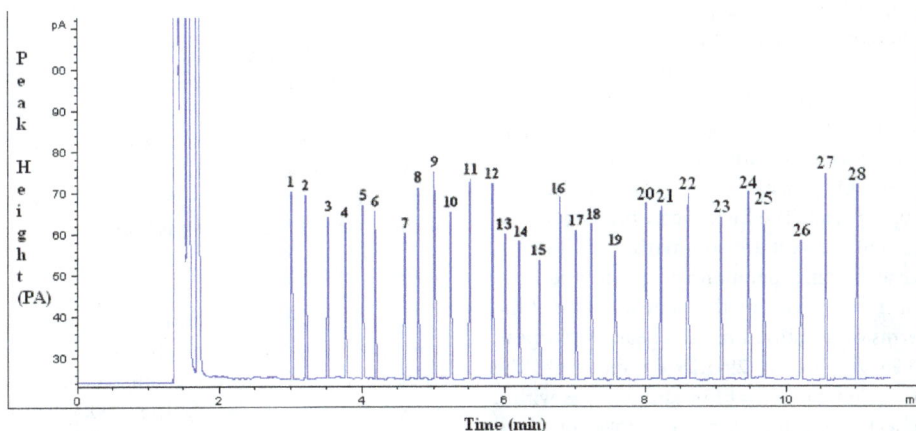

Figure 1: Seperation of amino acids and polyamines using FAA and IBCF as derivatizing reagents. 1 Gly, 2 Ala, 3 1,3 Dap, 4 Put, 5 Ser, 6 Cad, 7 Pro, 8 GABA, 9 Val, 10 Thr, 11 Leu, 12 Ile, 13 Cys, 14 Asn, 15 Asp, 16 Hst, 17 Lys, 18 Gln, 19 Glu, 20 Met, 21 Spd, 22 His, 23 Phe, 24 Dop, 25 Arg, 26 Tyr, 27 Trp, 28 Spm. GC conditions: column HP-5 (30 m x 0.32 mm id) with film thickness of 0.25 μm at column temperature 100 °C for 2 min with ramping of 20 °C/min up to 250 °C and stay at maximum temperature for 2 min. Nitrogen flow rate of 3 ml/min with split ratio 10:1. The injector and detector temperatures were 270 °C and 280 °C, respectively.

Amino Acid	LOQ (µg mL-1)	LOD (µg mL-1)	Retention time (min)	Linear calibration range	Coefficient of determination (r²)	Linear regression equation
Gly	0.18	0.06	3	1-10	0.9983	y=4.53x + 0.02
Ala	0.18	0.06	3.2	1-10	0.9999	y=4.405x + 0.17
1,3 Dap	0.18	0.06	3.5	1-10	0.9962	y=3.8357x + 0.0558
Put	0.18	0.06	3.8	1-10	0.9995	y=3.6979x + 0.2659
Ser	0.31	0.1	4	1-10	0.9992	y=4.145x - 0.05
Cad	0.18	0.06	4.2	1-10	0.9984	y=3.9778x + 0.0793
Pro	0.31	0.1	4.6	1-10	0.9933	y=3.41x + 0.24
GABA	0.18	0.06	4.8	1-10	0.9991	y=4.5704x + 0.2958
Val	0.31	0.1	5	1-10	0.9912	y=4.975x + 0.45
Thr	0.31	0.1	5.3	1-10	0.9946	y=3.92x + 0.34
Leu	0.61	0.2	5.5	1-10	0.9959	y=4.765x + 0.11
Ile	0.61	0.2	5.8	1-10	0.9993	y=4.64x + 0.02
Cys	0.18	0.06	6	1-10	0.9978	y=3.525x - 0.03
Asn	0.18	0.06	6.2	1-10	0.9998	y=3.235x + 0.17
Asp	0.18	0.06	6.5	1-10	0.9965	y=2.835x + 1.21
Hst	0.31	0.1	6.8	1-10	0.9945	y=4.3707x - 0.6397
Lys	0.31	0.1	7	1-10	0.9919	y=3.65x - 0.2
Gln	0.31	0.1	7.3	1-10	0.9992	y=3.72x + 0.02
Glu	0.18	0.06	7.6	1-10	0.9999	y=3.03x + 0.1
Met	0.61	0.2	8	1-10	0.9968	y=4.23x + 0.04
Spd	0.31	0.1	8.2	1-10	0.9953	y=4.1267x + 0.49
His	0.61	0.2	8.6	1-10	0.9966	y=4.305x + 0.07
Phe	0.31	0.1	9.1	1-10	0.9987	y=3.825x - 0.01
Dop	0.61	0.2	9.4	1-10	0.9933	y=4.3824x + 0.069
Arg	0.31	0.1	9.7	1-10	0.998	y=4.01x + 0.54
Tyr	0.31	0.1	10.2	1-10	0.9995	y=3.325x + 0.07
Trp	0.61	0.2	10.6	1-10	0.9977	y=4.85x + 0.62
Spm	0.61	0.2	11	1-10	0.9949	y=4.5971x + 0.9761

Table 1: Analytical data for the amino acids and polyamines analysis using FAA and IBCF as derivatizing reagents.

Reagent	Compounds analyzed	Limit of detection (LOD)	Limit of quantitation (LOQ)	Caliberation range	Retention time (min)	Detection	Reference
BSTFA	16	0.01-4.24 µmol/l	0.02-7.07 µmol/l	0.1-133 µmol/l	17	GC-MS	21
Propyl chloroformate	18	0.5-1.0 µmol/l	0.9-3.0 µmol/l	0.3-1072 µmol/l	20	GC-MS	28
FAA+ECF	19	0.1-0.3 µg/ml	0.31-0.91 µg/ml	1-10 µg/ml	10	GC-FID	34
FAA+IBCF	28	0.06-0.2 µg/ml	0.18-0.61 µg/ml	1-10 µg/ml	11	GC-FID	Present

Table 2: Comparative GC analysis of amino acids and polyamines by reported procedures.

The linearity of the calibration curves were checked for all the 28 compounds separated by recording average peak height/peak area (n=4) verses concentration and linear relationship was observed within 1-10 µg/ml for each of the compound with coefficient of determination within 0.9912-0.9999 using 6 calibrators. The limits of detection (LOD) measured as signal to noise ratio 3:1 were calculated within 60-200 ng/ml, corresponding to 6-20 pg injected on the column with split ratio 10:1. The limit of quantitation (LOQ) measured as signal to noise ratio (10:1) was calculated 0.18-0.61 µg/ml. The linear regression equations for the amino acids were calculated and are summarized in Table 1. The repeatability of the derivatization, separation and quantitation was examined inter (n=5) and intra (n=5) day by the same operator at the concentration 5 µg/ml in terms of retention time and peak height/peak area and RSD were within 0.96-3.33 % and 2.29-5.25 % respectively. The accuracy of the analysis of test mixtures (n=4) of amino acids within the calibration range was checked and the relative error was obtained within ± 02-03 %.

de Paiva et al have reported GC separation of 16 amino acids as N, O-bis (trimethylsilyl) trifluoroacetamide (BSTFA) derivatives within 17 min [21]. Kasper et al. separated 18 L-amino acids by GC as propyl chloroformate derivative with run time of 20 min [28]. The present work reports separation of 28 compounds with 11 min (Table 2).

Now comparing the GC using FAA and ECF as derivatizing reagent [34], it was observed an increase in the retention time with an improvement in the FID sensitivity for the analysis of amino acids using FAA and IBCF as derivatizing reagent due to an increase in carbon numbers. A change in elution order was also observed when using FAA and IBCF as compared to FAA and ECF as derivatizing reagent, may be due to the effect of bulky isobutyl instead of ethyl group (Table 2).

Sample analysis

The developed method was applied for the analysis of the compounds from acid hydrolyzed human skin samples. Skin samples collected from 10 healthy volunteers (H) and 8 each of Pv, P, L, and E patients were analyzed. The peak identification was made based on comparing the retention time of the standards. 23 compounds comprising 20 amino acids and three polyamines (Put, Spd, Spm) were identified from all the samples analyzed (Figure 2). The results of analysis are summarized in Table 3. The average values with standard deviation and range (minimum- maximum values are summarized in

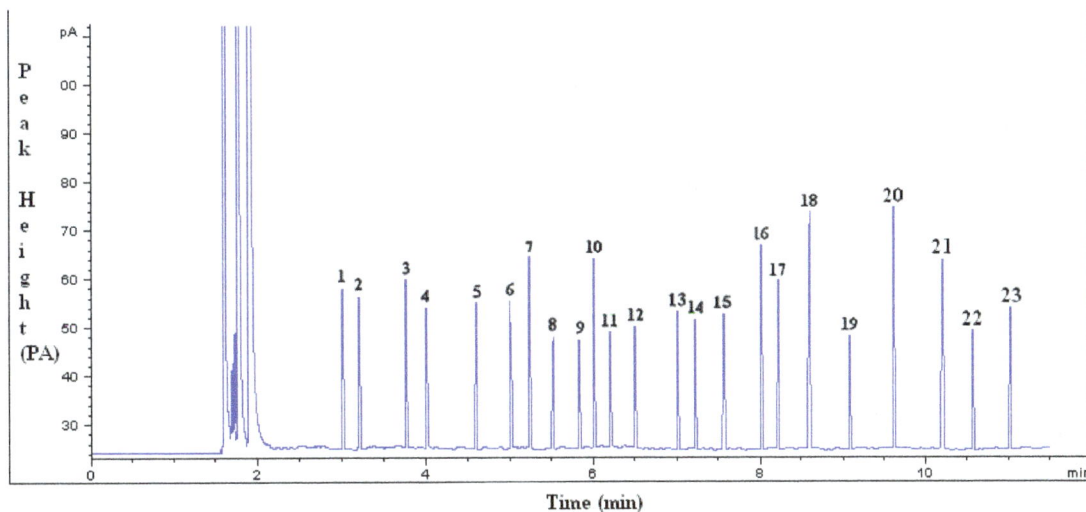

Figure 2: Quantitative response of amino acids and polyamines from leishmania patient using FAA and IBCF as derivatizing reagents.
1 Gly, 2 Ala, 3 Put, 4 Ser, 5 Pro, 6 Val, 7 Thr, 8 Leu, 9 Ile, 10 Cys, 11 Asn, 12 Asp, 13 Lys, 14 Gln, 15 Glu, 16 Met, 17 Spd, 18 His, 19 Phe, 20 Arg, 21 Tyr, 22 Trp, 23 Spm. GC conditions as fig. 1

	Gly	Ala	Put	Ser	Pro	Val	Thr	Leu	Ile	Cys	Asn	Asp	Lys	Gln	Glu	Met	Spd	His	Phe	Arg	Tyr	Trp	Spm
H1	65 (1.6)	65.1 (1.5)	45 (1.6)	92.4 (0.8)	77.9 (1.4)	21.5 (0.7)	37.3 (1.4)	54.7 (0.9)	55.6 (0.8)	42.4 (1.7)	34 (1.4)	27 (1.5)	45.3 (1.3)	33.8 (1.3)	29.5 (0.8)	36.8 (1.0)	39.6 (0.7)	58.1 (1.3)	30.3 (1.6)	48.3 (1.3)	36.7 (1.6)	19.3 (1.4)	35.1 (1.5)
H2	69.6 (0.9)	69.8 (1.4)	48.7 (1.3)	95.3 (1.4)	77 (1.5)	20.4 (1.2)	36.6 (1.6)	52.7 (0.9)	52.4 (1.4)	46.8 (1.3)	29.5 (0.8)	21.8 (1.5)	43.3 (0.8)	29.5 (0.7)	25.7 (1.4)	35.7 (1.5)	43.8 (0.9)	59.4 (1.3)	31.7 (1.0)	46.3 (1.8)	40.7 (1.8)	23.1 (1.6)	37.7 (2.0)
H3	66.3 (0.9)	66.5 (0.8)	53.4 (0.9)	94.5 (1.2)	76.6 (0.8)	23.2 (0.9)	31.9 (1.5)	53.4 (0.8)	50.5 (1.1)	40.7 (1.8)	35.3 (1.3)	19 (1.3)	42.2 (1.0)	30.8 (0.9)	26.8 (1.3)	40.6 (1.0)	46.5 (0.8)	55.5 (1.2)	27.7 (0.8)	44.3 (1.2)	39.1 (1.6)	26.1 (1.7)	41.5 (0.8)
H4	68.7 (1.0)	68.9 (1.6)	46.9 (1.2)	97.7 (1.3)	81.1 (1.5)	18.8 (1.4)	34.4 (1.1)	56.5 (0.9)	56.9 (1.5)	44.5 (0.8)	32.4 (0.9)	25.1 (0.9)	42.6 (0.9)	26.7 (1.2)	23.1 (1.4)	41.2 (1.4)	43.3 (0.8)	57 (1.7)	28.4 (0.7)	53 (1.5)	44.6 (1.1)	22.2 (1.5)	37.7 (1.2)
H5	70.2 (1.0)	70.4 (0.9)	45.3 (1.3)	93 (1.4)	83 (1.3)	21.2 (1.4)	36.5 (1.0)	51.5 (0.9)	54.3 (1.4)	41.5 (0.8)	34 (1.6)	27 (1.6)	44.9 (1.5)	28.1 (1.3)	24.9 (1.5)	40.3 (1.3)	40.4 (1.2)	58.9 (1.3)	24.1 (1.4)	48.3 (1.3)	41.1 (1.5)	20.3 (1.6)	35.3 (1.3)
H6	66.9 (1.4)	67.1 (1.1)	53.2 (1.3)	96.7 (1.6)	75.9 (1.4)	23.1 (1.4)	32.6 (1.1)	54.7 (0.9)	54.7 (0.9)	44.5 (0.7)	35.3 (1.3)	19 (1.4)	46.8 (1.3)	33.3 (0.7)	29.9 (1.5)	37.9 (1.2)	46.4 (0.8)	58.4 (1.0)	30.5 (0.9)	50.3 (0.9)	43.3 (0.8)	26.1 (1.2)	40.8 (1.3)
H7	65.9 (1.7)	66 (1.4)	47.5 (0.7)	95.6 (1.5)	78.1 (1.3)	18.9 (1.6)	37.3 (2.0)	52.3 (0.9)	53 (1.5)	43.4 (0.9)	29.1 (1.7)	26.3 (1.3)	40.2 (1.4)	23.4 (0.8)	27.5 (1.1)	41.7 (1.4)	42.2 (1.7)	61.2 (1.3)	26.9 (1.4)	47.1 (1.5)	38.5 (1.1)	22.2 (1.5)	37 (1.6)
H8	68.1 (1.1)	68.3 (0.9)	48.2 (1.1)	93.8 (1.5)	76.5 (0.7)	20.6 (0.9)	34.9 (1.4)	56.5 (0.9)	51.3 (1.0)	41.1 (1.5)	31.8 (0.9)	24.4 (0.8)	43.3 (0.8)	27.4 (0.9)	22.8 (0.9)	35.2 (1.2)	43.1 (1.5)	54.4 (0.7)	31.7 (1.0)	49.3 (0.8)	44.4 (1.3)	22.9 (1.4)	38.3 (2.1)
H9	69.7 (1.2)	70 (1.7)	46.9 (0.4)	91.3 (0.9)	78.1 (1.3)	22.8 (0.9)	36.5 (1.1)	51.2 (1.6)	55.1 (1.5)	43.7 (1.4)	35.1 (1.5)	21.1 (1.5)	45.5 (0.9)	31.3 (0.9)	28.8 (1.2)	36.5 (1.0)	41.5 (0.9)	56.4 (1.1)	29.5 (0.7)	48 (1.3)	41.7 (1.4)	21.4 (1.4)	37 (1.4)
H10	67.2 (1.3)	66.6 (0.7)	42.9 (1.9)	92.7 (1.0)	77.3 (1.6)	22.4 (1.0)	32.2 (1.1)	52.4 (0.7)	53.4 (1.4)	39.2 (1.0)	28.1 (1.6)	20.2 (1.3)	37.4 (0.8)	24.5 (0.7)	22 (1.4)	37.6 (1.0)	37.5 (0.9)	52.9 (1.6)	23 (1.4)	43.3 (0.8)	35.1 (1.5)	18.7 (1.5)	33.4 (0.7)
H10a	66.8 (1.4)	66.2 (1.5)	41.8 (1.5)	93.2 (1.4)	76.8 (1.2)	21.5 (0.7)	31.8 (0.9)	52.5 (0.9)	53.1 (1.3)	38.3 (2.1)	26.8 (0.4)	20.5 (1.3)	36.3 (1.2)	23.3 (0.8)	21.7 (1.3)	36.6 (1.3)	36.5 (1.0)	51.8 (1.2)	22.9 (1.7)	42 (1.7)	34.2 (1.6)	17.8 (1.5)	32.6 (1.1)
MEAN	67.8	67.9	47.8	94.3	78.2	21.3	35.0	53.6	53.7	42.8	32.5	23.1	43.2	28.9	26.1	38.4	42.4	57.2	28.4	47.8	40.5	22.2	37.4
SD	1.78	1.86	3.45	2.03	2.22	1.62	2.14	1.93	1.98	2.24	2.74	3.23	2.78	3.50	2.87	2.40	2.85	2.49	3.01	2.81	3.18	2.51	2.48
PV1	59.3 (0.9)	60.4 (1.2)	64.4 (1.2)	78.4 (0.9)	69 (1.4)	34 (1.4)	52.8 (1.2)	44 (1.4)	46.5 (0.9)	55.7 (1.5)	66.4 (1.3)	43.5 (0.8)	62.8 (1.8)	21.1 (1.5)	47.7 (1.4)	52.1 (1.4)	57.2 (1.4)	66.2 (0.9)	20.6 (0.9)	61.6 (1.4)	51.4 (1.1)	34.5 (0.8)	50.8 (0.9)
PV2	61.3 (1.1)	59.8 (1.3)	64.1 (1.4)	79.2 (1.6)	66.1 (1.6)	34.5 (0.8)	49.7 (0.9)	42.4 (1.7)	48.5 (1.7)	58.3 (0.9)	69.3 (0.7)	45.1 (1.3)	60.5 (0.8)	18.8 (1.4)	45.1 (1.3)	54.3 (1.4)	57.5 (1.1)	68.2 (1.6)	23.5 (1.8)	62.9 (1.7)	52.6 (0.9)	35.3 (1.3)	50.5 (1.1)
PV3	59.1 (1.5)	57.7 (1.4)	64.4 (0.7)	77.1 (1.2)	73.9 (1.8)	33.7 (1.3)	51.8 (1.2)	41.5 (0.8)	45.8 (2.2)	56.6 (0.8)	65.4 (0.6)	40.6 (0.8)	61.9 (1.3)	22.7 (1.8)	48.4 (0.7)	51.6 (0.9)	56.6 (0.8)	67.1 (1.1)	21.5 (0.7)	59.8 (1.3)	51.6 (0.6)	34.6 (0.8)	50.8 (0.9)
PV4	58.7 (1.7)	61 (1.2)	61.7 (0.7)	79.4 (0.8)	71.2 (1.8)	31.7 (1.0)	52.6 (0.9)	42.5 (1.9)	46.8 (1.3)	56.6 (1.4)	66.8 (1.4)	44.9 (1.5)	59.7 (1.4)	18.4 (0.8)	46.6 (1.1)	53 (1.5)	55.3 (1.3)	65.9 (1.7)	20.4 (1.2)	62.6 (0.6)	49.4 (0.8)	33.1 (1.7)	48.6 (1.0)
PV5	60.7 (1.2)	59.1 (1.5)	65.5 (0.9)	77.9 (1.4)	67.1 (1.1)	34.9 (1.4)	51.3 (1.1)	43.5 (1.9)	45.2 (1.3)	59.3 (0.9)	64.5 (0.8)	42.3 (1.3)	61 (1.2)	20.4 (1.2)	49.3 (0.8)	51.2 (1.6)	56.7 (1.6)	67.3 (1.5)	23 (1.3)	61.3 (1.1)	52.2 (1.5)	35 (1.3)	51.7 (1.5)
PV6	61.9 (1.3)	62.4 (0.9)	59.9 (1.6)	80.8 (1.1)	72.6 (1.1)	36.1 (1.4)	49.9 (1.5)	45.6 (0.9)	49.1 (1.5)	61.5 (1.3)	68.5 (0.9)	47.7 (1.4)	65.4 (0.6)	22.4 (1.2)	45.5 (0.9)	54.5 (1.2)	53.7 (0.9)	69.9 (1.4)	22.2 (1.5)	59.3 (0.9)	50.6 (0.9)	31.8 (0.9)	47.2 (1.4)

PV7	59.1 (1.5)	58.5 (1.3)	63 (1.9)	76.6 (0.8)	68.6 (0.8)	33.7 (1.3)	55.2 (1.2)	42.5 (0.9)	46.5 (1.1)	58.3 (0.9)	71.1 (1.4)	39.9 (1.5)	60.3 (0.9)	24.7 (1.2)	51.7 (1.0)	56.4 (1.1)	57 (1.4)	69.3 (0.7)	25.5 (0.7)	58.8 (1.6)	46.8 (1.8)	33.9 (1.8)	49.6 (1.7)
PV8	58.5 (1.2)	60.1 (1.9)	65.7 (0.6)	79.4 (0.8)	72.6 (0.8)	32.6 (1.1)	53 (1.4)	43.6 (0.9)	47.8 (1.8)	55.7 (1.1)	67.4 (1.5)	45.8 (2.2)	63.4 (0.9)	20.8 (1.2)	46.9 (1.2)	52.7 (1.2)	59 (1.5)	66.6 (1.5)	19.5 (1.2)	61.4 (1.5)	49.6 (0.6)	35.8 (1.6)	51.8 (1.2)
PV8a	58.7 (1.7)	60 (1.3)	64.7 (1.6)	79.3 (1.4)	72.5 (0.7)	31.7 (1.0)	52.4 (1.4)	44.1 (1.6)	47.5 (0.7)	54.6 (1.1)	66.8 (1.4)	45.7 (1.4)	62.3 (1.0)	20 (1.5)	46.4 (0.8)	51.4 (1.1)	58.7 (1.6)	66.2 (0.9)	18.9 (1.6)	60.5 (1.1)	48.8 (1.4)	34.6 (0.8)	51 (1.3)
MEAN	59.8	59.9	63.6	78.6	70.1	33.9	52.0	43.2	47.0	57.8	67.4	43.7	61.9	21.2	47.7	53.2	56.6	67.6	22.0	61.0	50.5	34.3	50.1
SD	1.29	1.47	1.97	1.38	2.84	1.35	1.79	1.26	1.34	2.00	2.15	2.67	1.91	2.08	2.15	1.74	1.59	1.45	1.95	1.51	1.89	1.29	1.58
P1	54.1 (1.1)	54.5 (1.2)	68.4 (1.9)	66.8 (1.4)	63.1 (1.2)	26.7 (1.2)	74.6 (0.9)	36.8 (1.2)	40.8 (1.3)	35.2 (0.9)	56.3 (0.6)	64.2 (1.4)	37 (1.6)	60.4 (1.2)	69.5 (0.7)	30.8 (0.9)	60.8 (1.8)	45.9 (1.6)	49.5 (0.7)	42.8 (1.5)	32.5 (0.8)	38 (2.2)	54 (1.5)
P2	51 (1.3)	55.9 (1.6)	66.6 (1.5)	63.9 (1.4)	61.9 (1.4)	28.7 (1.1)	77.1 (1.3)	35.8 (1.6)	41.8 (1.5)	30.7 (0.9)	54.9 (1.1)	62.5 (0.8)	33.6 (1.1)	58 (1.3)	66.7 (1.5)	32.6 (1.1)	58.8 (1.6)	42.6 (0.9)	51.8 (1.4)	40.2 (0.9)	29.7 (0.7)	36.1 (1.4)	52.5 (1.0)
P3	52.3 (0.9)	54.1 (1.1)	67.5 (1.2)	64.7 (1.2)	61.4 (1.5)	25.2 (1.4)	73.9 (1.8)	36.3 (1.2)	39.5 (1.1)	33 (0.3)	55.5 (1.2)	63.2 (1.5)	35.8 (0.9)	57.1 (1.6)	72.2 (1.3)	28.9 (1.2)	60.3 (0.9)	43.9 (1.3)	53.2 (1.3)	41.5 (0.8)	30.7 (0.9)	36.9 (1.3)	53.3 (1.7)
P4	53.4 (1.4)	58.5 (1.2)	69.1 (1.3)	62.6 (0.6)	66.5 (0.8)	27.1 (1.2)	76.3 (0.9)	34.5 (0.8)	41.2 (1.2)	31.6 (0.8)	52.4 (1.4)	66.7 (0.7)	34.5 (0.8)	56.4 (1.1)	66.9 (1.4)	32.3 (1.6)	60.3 (1.3)	46.5 (0.9)	50.9 (1.3)	42.5 (1.2)	32.1 (1.4)	37.8 (0.7)	54.6 (1.1)
P5	52.1 (1.6)	54.8 (1.7)	70 (1.7)	66.2 (1.5)	66.4 (1.3)	28.2 (1.5)	79 (1.2)	35.5 (1.0)	39.9 (1.5)	32.8 (1.2)	50.1 (1.2)	64.2 (1.5)	37.4 (0.8)	59.3 (0.9)	68.2 (1.6)	30.4 (1.0)	62.1 (0.9)	45.3 (1.3)	50 (1.5)	42 (1.6)	31.5 (1.0)	39.5 (0.9)	55.3 (1.3)
P6	50.7 (1.4)	57.7 (1.4)	65.9 (1.7)	65.7 (0.6)	62.3 (1.0)	25.2 (1.4)	74.7 (1.4)	37.2 (1.6)	42.1 (1.4)	34.7 (1.3)	53.8 (1.7)	62 (1.3)	36.5 (1.0)	62 (1.4)	74.1 (1.2)	29.6 (0.8)	59.5 (0.9)	44 (1.3)	52 (1.5)	40.2 (0.9)	29.5 (0.7)	36 (1.5)	52 (1.4)
P7	53.2 (1.3)	53.5 (0.9)	66.9 (1.4)	68.6 (0.8)	63.7 (1.1)	29.4 (1.0)	73.9 (1.8)	33.7 (1.3)	41.2 (1.4)	37.3 (1.4)	57.5 (1.1)	66 (1.4)	39.6 (0.7)	58.7 (1.7)	66.7 (1.2)	34 (1.4)	59.8 (1.3)	47.4 (0.7)	53.2 (1.4)	44.2 (1.6)	34.3 (1.5)	37.1 (1.3)	52.8 (1.7)
P8	52.5 (0.8)	56 (1.2)	70.2 (1.4)	66.5 (0.8)	61.6 (1.4)	27 (1.3)	76.1 (1.4)	35.6 (1.2)	39.6 (0.9)	31 (1.6)	52.6 (0.9)	64.9 (2.3)	37.6 (0.7)	59.8 (1.3)	71.3 (1.1)	31.9 (1.4)	63.7 (1.3)	45.3 (0.9)	50.2 (1.0)	43.7 (1.4)	33.5 (1.1)	40 (1.5)	55.4 (1.9)
P8a	52.4 (1.4)	56.5 (0.9)	69.9 (1.4)	66.1 (1.3)	61 (1.2)	26 (1.3)	75.4 (1.2)	35.3 (1.3)	38.8 (0.9)	30.6 (0.9)	51.3 (1.1)	64.7 (1.2)	36.7 (1.6)	58.7 (1.6)	71 (1.3)	31.2 (1.2)	62.7 (1.5)	45.1 (1.3)	49.7 (0.9)	42.7 (1.4)	32.2 (1.1)	49.4 (0.8)	54 (1.2)
MEAN	52.4	55.6	68.1	65.6	63.4	27.2	75.7	35.7	40.8	33.3	54.1	64.2	36.5	59.0	69.5	31.3	60.6	45.1	51.4	42.1	31.7	37.7	53.7
SD	1.16	1.76	1.61	1.86	2.05	1.53	1.78	1.15	1.00	2.28	2.40	1.63	1.88	1.82	2.82	1.69	1.55	1.55	1.43	1.47	1.72	1.47	1.29
L1	44.3 (1.2)	48.3 (1.3)	62.3 (1.0)	46.2 (1.3)	61.1 (1.4)	41.6 (1.0)	64.7 (1.2)	33 (0.3)	34.2 (1.6)	73.2 (1.2)	43.7 (1.4)	54.5 (1.2)	53 (1.5)	43.7 (1.4)	60.1 (1.9)	65.3 (1.4)	55.6 (0.8)	75.3 (0.8)	39.9 (1.5)	78.4 (0.9)	75 (1.7)	32.9 (1.5)	48.5 (1.7)
L2	47.8 (1.8)	45.9 (1.6)	57 (1.7)	48.5 (1.7)	55.1 (1.5)	43.2 (1.5)	62.7 (0.8)	31 (1.6)	30.7 (0.9)	74.2 (1.4)	48.7 (1.5)	53.1 (1.2)	55.5 (1.2)	46.2 (1.3)	55.2 (0.9)	64.2 (1.4)	50.3 (0.9)	76.5 (0.7)	37.5 (0.9)	79.9 (1.5)	76.4 (0.8)	29.5 (0.7)	45.1 (1.4)
L3	46.2 (1.3)	48.8 (1.4)	59.7 (1.4)	45.8 (2.2)	59.9 (1.6)	39.9 (1.5)	64.2 (1.4)	32.7 (0.9)	32.9 (1.5)	71.7 (1.3)	45.6 (1.3)	49.6 (0.6)	54.4 (1.2)	47.1 (1.5)	57.6 (1.1)	62.7 (0.8)	54 (1.4)	73 (1.4)	41.3 (0.7)	77.7 (0.8)	78.2 (1.5)	31.7 (1.0)	47.9 (1.1)
L4	47.2 (1.0)	46.8 (1.3)	63.5 (1.5)	47.2 (1.4)	58.5 (1.2)	41.9 (1.5)	61.8 (1.4)	33.6 (1.1)	30.8 (0.9)	74.4 (1.2)	43.9 (1.3)	54.7 (0.9)	52.2 (1.6)	43.9 (1.3)	60.3 (0.9)	65.2 (1.3)	56.2 (1.5)	75.3 (0.8)	39.2 (1.0)	81.2 (1.1)	76.8 (1.2)	34.2 (1.6)	49.6 (0.6)
L5	45.3 (1.3)	46.4 (0.8)	55.4 (1.9)	49.3 (1.3)	54.2 (0.4)	40.7 (1.6)	62.8 (1.2)	31.2 (1.2)	32 (1.4)	72.7 (1.0)	46.8 (1.3)	51 (1.3)	54.6 (1.5)	50 (1.5)	59 (1.5)	62.8 (1.2)	50.9 (1.6)	73.9 (1.8)	40.8 (2.0)	79.7 (1.5)	75 (1.6)	28.7 (1.4)	43.5 (1.4)
L6	48.8 (1.4)	50.4 (0.9)	57.6 (1.1)	45.8 (2.2)	60.7 (1.2)	44.5 (0.8)	65.9 (1.7)	35.4 (1.2)	34.2 (1.3)	76.8 (1.4)	49.7 (0.9)	51.5 (0.9)	55.9 (1.6)	48 (1.3)	56.1 (1.4)	63.5 (1.5)	50.9 (1.3)	76.1 (1.4)	38 (2.2)	78.6 (0.9)	78.2 (1.5)	29.8 (1.7)	46.2 (1.3)
L7	46.5 (0.9)	49.5 (0.7)	62.6 (1.2)	50.1 (1.6)	58.2 (1.2)	42.9 (1.9)	61.3 (1.1)	33.1 (1.7)	34.7 (1.3)	74.6 (0.9)	44.6 (1.1)	57.8 (1.7)	50.6 (0.9)	44.4 (1.7)	53 (1.5)	60.6 (1.0)	54.8 (1.7)	77.5 (1.0)	42.7 (1.4)	75.9 (1.4)	80.6 (0.9)	32.8 (1.2)	48.5 (1.7)
L8	47.6 (0.9)	46.4 (0.8)	61 (1.2)	47.7 (1.4)	59.2 (1.5)	40.5 (0.8)	65.1 (1.5)	30.8 (0.9)	31 (1.1)	71.7 (1.2)	48.1 (1.2)	55 (1.5)	53 (1.4)	45.1 (1.3)	60.1 (1.5)	65.5 (0.8)	53.8 (1.7)	75.3 (1.2)	39.4 (1.6)	81.4 (0.9)	75 (1.7)	31.8 (0.9)	47.9 (1.7)
L8a	47.3 (1.4)	45.9 (1.6)	60.2 (1.2)	47.3 (1.4)	59.4 (1.3)	40 (1.5)	65.7 (0.6)	29.8 (1.1)	30.7 (0.9)	71.2 (1.8)	47.2 (1.4)	53.9 (1.0)	52.8 (1.3)	44.6 (1.1)	59.4 (1.3)	65.1 (1.5)	52.7 (0.9)	75 (1.7)	38.2 (1.4)	80.4 (1.2)	74.3 (1.3)	31.3 (0.7)	47.7 (1.4)
MEAN	46.7	47.8	59.9	47.6	58.4	41.9	63.6	32.6	32.6	73.7	46.4	53.4	53.7	46.1	57.7	63.7	53.3	75.4	39.9	79.1	76.9	31.4	47.1
SD	1.45	1.67	2.96	1.63	2.51	1.56	1.65	1.56	1.66	1.71	2.29	2.64	1.78	2.21	2.70	1.68	2.31	1.42	1.72	1.84	2.01	1.92	2.03
E1	41.8 (0.9)	42.7 (1.4)	43.3 (0.8)	39.5 (0.8)	48.9 (1.3)	17.3 (0.9)	42.6 (0.9)	28.9 (1.2)	28.3 (1.1)	26 (1.4)	25.4 (1.4)	36.4 (0.8)	28.3 (1.1)	75.9 (1.4)	38.1 (1.2)	21.7 (1.3)	38.6 (0.7)	36.7 (1.6)	64 (1.4)	31.5 (0.8)	24.9 (1.5)	19.1 (1.3)	34.4 (1.1)
E2	40.4 (1.2)	40.2 (0.9)	40.8 (1.3)	36.9 (1.3)	46.9 (1.2)	15.3 (1.3)	41.4 (0.9)	27.7 (0.8)	25.7 (1.4)	23.7 (1.5)	26.8 (0.4)	31.2 (1.2)	24.8 (1.6)	74.1 (1.2)	34.8 (0.9)	18.7 (1.5)	36.7 (0.8)	33.8 (1.3)	62.1 (0.9)	34.2 (1.6)	21.7 (1.3)	16.9 (1.1)	31.5 (1.0)
E3	39.5 (0.8)	40.9 (1.3)	42.8 (1.5)	38.2 (1.4)	45.1 (1.3)	16.2 (1.3)	44 (1.3)	29.6 (1.0)	27.5 (1.1)	27.1 (1.2)	23.9 (1.1)	35.2 (1.2)	23.9 (1.1)	78.5 (1.0)	35.6 (1.2)	20 (1.8)	37.8 (1.6)	35 (1.4)	65.2 (1.3)	32.9 (1.5)	22.5 (0.8)	18 (1.6)	32.5 (1.3)
E4	40.1 (1.4)	42.9 (1.9)	44.4 (1.7)	40.6 (0.8)	48.1 (1.2)	14.6 (0.9)	41.1 (1.6)	27.1 (1.2)	26.5 (1.2)	25.6 (0.7)	27.7 (0.8)	34.5 (0.9)	27 (1.6)	79.2 (1.6)	34.5 (0.8)	17.9 (1.3)	40.7 (1.0)	33 (0.3)	62.4 (0.9)	34.7 (1.3)	24.9 (1.5)	20 (2.0)	35.1 (1.5)
E5	41.3 (0.8)	41.1 (1.5)	42.4 (1.7)	39.2 (1.6)	45.4 (1.2)	17.5 (0.7)	42.4 (1.7)	28.8 (1.2)	25 (1.6)	24.5 (0.7)	25.2 (1.4)	36.4 (0.8)	28.7 (1.1)	76.1 (1.4)	37.4 (0.8)	20.9 (1.4)	38 (2.2)	36 (1.5)	64.9 (2.3)	32.2 (1.1)	23.1 (1.5)	18.8 (1.4)	33.1 (1.7)

E6	37.9 (1.2)	39.7 (1.3)	43.3 (0.8)	36.9 (1.3)	47.4 (0.7)	18.9 (1.6)	45.5 (0.9)	30.2 (1.4)	28.8 (1.2)	29.2 (0.9)	29.1 (1.7)	30.8 (0.9)	23.7 (1.5)	78.1 (1.3)	40.5 (0.8)	22.9 (1.7)	37.7 (2.0)	38 (1.5)	66.9 (1.4)	36.3 (1.2)	26.3 (1.3)	18.1 (1.5)	33 (1.1)
E7	38.8 (0.9)	38.8 (0.9)	39.7 (1.3)	34.3 (1.5)	47.5 (0.7)	16.7 (1.6)	44.5 (0.8)	27.3 (1.7)	26.3 (1.3)	26.2 (1.6)	23.9 (1.1)	29.8 (1.1)	25.6 (0.7)	74.9 (1.3)	37 (1.6)	17.9 (1.5)	35.1 (1.5)	35.4 (1.2)	62.1 (1.4)	33.5 (1.1)	21.5 (0.7)	16.2 (1.7)	31.8 (1.4)
E8	40.1 (1.4)	41.7 (1.4)	46.6 (1.1)	41.4 (0.9)	42.8 (1.5)	17.1 (1.0)	41.8 (0.9)	28.4 (1.0)	28 (1.3)	23.5 (1.2)	26.8 (1.3)	34.3 (1.5)	26.8 (1.3)	79 (1.5)	33.4 (0.7)	19.6 (1.7)	40.2 (1.4)	33.5 (1.4)	63.5 (1.5)	34.5 (0.8)	24.9 (1.5)	21 (1.6)	36.6 (1.6)
E8a	40.4 (0.7)	41.5 (0.8)	45.6 (1.3)	40.4 (0.7)	42.4 (1.7)	16.2 (1.3)	41.2 (1.4)	29 (1.5)	27.6 (1.0)	22.2 (1.5)	26 (1.4)	34.4 (1.1)	26.3 (1.3)	78.3 (0.8)	33.2 (1.2)	18.6 (1.5)	39 (1.3)	33.2 (1.2)	62.8 (1.8)	33.1 (1.7)	23.7 (1.5)	20.3 (0.8)	36.1 (1.4)
MEAN	40.0	41.0	42.9	38.4	46.5	16.7	42.9	28.5	27.0	25.7	26.1	33.6	26.1	77.0	36.4	20.0	38.1	35.2	63.9	33.7	23.7	18.5	33.5
SD	1.27	1.42	2.10	2.30	1.97	1.34	1.58	1.10	1.34	1.88	1.83	2.61	1.91	1.99	2.29	1.80	1.82	1.71	1.72	1.53	1.76	1.57	1.74

H Healthy volunteers, Pv Pemphigus vulgaris patients, P Psorisis patients, L Leishmania patients, E Eczema patients, a   spiked samples
The values given in parenthesis represents standard deviation

Table 3: Amino acids and polyamines contents (μg/g) in skin samples of healthy volunteers, pemphigus vulgaris, psoriasis, leishmania and eczema patients.

	H (n=10)	PV (n=8)	P (n=8)	L (n=8)	E (n=8)
Gly	65-70.2	58.5-61.9	50.7-54.1	44.3-48.8	37.9-41.8
	67.8 (1.78)	59.8 (1.29)	52.4 (1.16)	46.7 (1.45)	40 (1.27)
Ala	65.1-70.4	57.7-62.4	53.5-58.5	45.9-50.4	38.8-42.9
	67.9 (1.86)	59.9 (1.47)	55.6 (1.76)	47.8 (1.67)	41 (1.42)
Put	42.4-53.4	59.9-65.7	65.9-70.2	55.4-63.5	39.7-46.6
	47.8 (3.45)	63.6 (1.97)	68.1 (1.61)	59.9 (2.96)	42.9 (2.1)
Ser	91.3-97.7	76.6-80.8	62.6-68.6	45.8-50.1	34.3-41.4
	94.3 (2.03)	78.6 (1.38)	65.6 (1.86)	47.6 (1.63)	38.4 (2.3)
Pro	75.9-83	66.1-73.9	61.4-66.5	54.2-61.1	42.8-48.9
	78.2 (2.22)	70.1 (2.84)	63.4 (2.05)	58.4 (2.51)	46.5 (1.97)
Val	18.8-23.2	31.7-36.1	25.2-29.4	39.9-44.5	14.6-18.9
	21.3 (1.62)	33.9 (1.35)	27.2 (1.53)	41.9 (1.56)	16.7 (1.34)
Thr	31.9-37.3	49.7-55.2	73.9-79	61.3-65.9	41.1-45.5
	35 (2.14)	52 (1.79)	75.7 (1.78)	63.6 (1.65)	42.9 (1.58)
Leu	51.2-56.5	41.5-45.6	33.7-37.2	30.8-35.4	27.1-30.2
	53.6 (1.93)	43.2 (1.26)	35.7 (1.15)	32.6 (1.56)	28.5 (1.1)
Ile	50.5-56.9	45.2-49.1	39.5-42.1	30.7-34.7	25-28.8
	53.7 (1.98)	47 (1.34)	40.8 (1.0)	32.6 (1.66)	27 (1.34)
Cys	29.2-46.8	55.7-61.5	30.7-37.3	71.7-76.8	23.5-29.2
	42.8 (2.24)	57.8 (2)	33.3 (2.28)	73.7 (1.71)	25.7 (1.88)
Asn	28.1-35.3	64.5-71.1	50.1-57.5	43.7-49.7	23.9-29.1
	32.5 (2.74)	67.4 (2.15)	54.1 (2.4)	46.4 (2.29)	26.1 (1.83)
Asp	19-27	39.9-47.7	62-66.7	49.6-57.8	29.8-36.4
	23.1 (3.23)	43.7 (2.67)	64.2 (1.63)	53.4 (2.64)	33.6 (2.61)
Lys	37.4-46.8	59.7-65.4	33.6-39.6	50.6-55.9	23.7-28.7
	43.2 (2.78)	61.9 (1.91)	36.5 (1.88)	53.7 (1.78)	26.1 (1.91)
Gln	23.4-33.8	18.4-24.7	56.4-62	47.7-50	74-79.2
	28.9 (3.5)	21.2 (2.08)	59 (1.82)	46.1 (2.21)	77 (1.99)
Glu	22-29.9	45.1-51.7	66.7-74.1	53-60.3	33.4-40.5
	26.1 (2.87)	47.7 (2.15)	69.5 (2.82)	57.7 (2.7)	36.4 (2.29)
Met	35.2-41.7	51.2-56.4	28.9-34	60.6-65.5	17.9-22.9
	38.4 (2.4)	53.2 (1.74)	31.3 (1.69)	63.7 (1.68)	20 (1.8)
Spd	37.5-46.5	53.7-59	58.8-63.7	50.3-56.2	35.1-40.7
	42.4 (2.85)	56.6 (1.59)	60.6 (1.55)	53.3 (2.31)	38.1 (1.82)
His	52.9-61.2	65.9-69.9	42.6-47.4	73-77.5	33-38
	57.2 (2.49)	67.6 (1.45)	45.1 (1.55)	75.4 (1.42)	35.2 (1.71)
Phe	23-31.7	19.5-25.5	49.5-53.2	37.5-42.7	62.1-66.9

	28.4 (3.01)	22 (1.95)	51.4 (1.43)	39.9 (1.72)	63.9 (1.72)
Arg	43.3-53	58.8-62.9	40.2-44.2	75.9-81.4	31.5-36.3
	47.8 (2.81)	61 (1.51)	42.1 (1.47)	79.1 (1.84)	33.7 (1.53)
Tyr	35.1-44.6	46.8-52.6	29.5-33.5	75-80.6	21.5-26.3
	40.5 (3.18)	50.5 (1.89)	31.7 (1.72)	76.9 (2.01)	23.7 (1.76)
Trp	18.7-26.1	31.8-35.8	36-40	28.7-34.2	16.9-21
	22.2 (2.51)	34.3 (1.29)	37.7 (1.74)	31.4 (1.92)	18.5 (1.57)
Spm	33.4-41.5	47.2-51.8	52-55.4	43.5-49.6	31.5-36.6
	37.4 (2.48)	50.1 (1.58)	53.7 (1.29)	47.1 (2.03)	33.5 (1.74)

H Healthy volunteers, Pv Pemphigus vulgaris patients, P Psoriasis patients, L Leishmania patients, E Eczema patients, n Number of samples analysed

Table 4: Quantitative analysis of amino acids and polyamines in skin samples with range (µg/g), mean values (standard deviation).

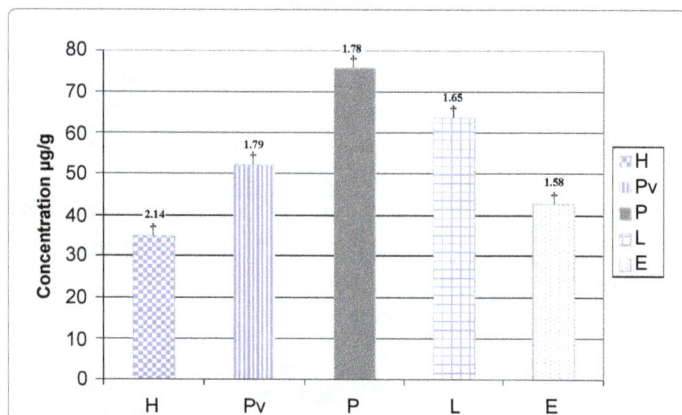

Figure 3: The variation in the average concentration µg/g with standard deviation for threonine in H=healthy volunteers (n=10), Pv=pemphigus vulgaris (n=8), P=psoriasis (n=8), L=leishmaniasis (n=8) and E=eczema (n=8) concentration as table 3.

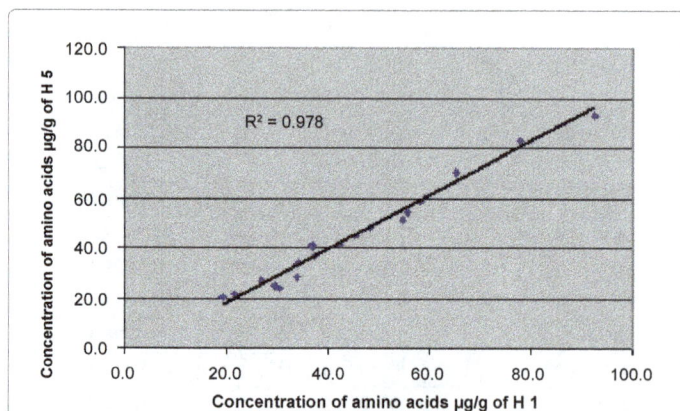

Figure 4a: Coefficient of determination (r²) of 20 amino acids and 03 polyamines for healthy volunteers 1 and 5. The concentration of amino acids and polyamines as table 2. GC conditions as fig 1.

and Ile were higher in healthy volunteers and minimum in eczema with the decreasing order: H > Pv > P > L > E. Val was observed higher in L with following decreasing order: L > Pv > P > H > E. Among hydroxy amino acids Thr and Ser, the highest concentration of Thr was obtained in P with decreasing order: P >L > Pv >E > H and Ser was highest in H with a sequence H > Pv >P > L > E. Among sulphur contaning amino acids Met, and Cys the highest concentration for both was in L with decreasing order L >Pv > H > P > E. Among acedic amino acids Glu and Asp and amide Asn the highest concentration of Glu and Asp was observed in P with decreasing order: P > L > Pv > E > H. Where as in the case of Asn the order was some what different Pv > P > L > H > E. The other amino acids such as aromatic (Phe) and basic (Lys) were also examined. Phe indicated the concentration of amino acids in the sequence E > P > L > H > Pv and Lys in Pv > L > H > P > E while polyamines Put, Spd, Spm indicated the concentration in the following decreasing order P > Pv > L > H > E (Figure 3).

The concentration of amino acids and polyamines were observed in the range of µg/g H 18.7-97.7, Pv 18.4-80.8, P 25.2-79, L 28.7-81.4 and E 14.6-79.2. The average results (n=10) for H indicated following decreasing order of concentration: Ser > Pro > Ala > Gly > His > Ile > Leu > Put > Arg > Lys > Cys > Spd > Tyr > Met > Spm > Thr > Asn > Gln > Phe > Glu > Asp > Trp > Val. The concentration order was somewhat different in the affected skin samples. Pv indicated Ser > Pro > His > Asn > Put > Lys > Arg > Ala > Gly > Cys > Spd > Met > Thr > Tyr > Spm > Glu > Ile > Asp > Leu > Trp > Val > Phe > Gln. P indicated Thr > Glu > Put > Ser > Asp > Pro > Spd > Gln > Ala > Asn > Spm > Gly > Phe > His > Arg > Ile > Trp > Lys > Leu > Cys > Tyr > Met > Val. L showed the following order Arg > Tyr > His > Cys > Met > Thr > Put > Pro > Glu > Lys > Asp > Spd > Ala > Ser > Spm > Gly > Asn > Gln > Val > Phe > Ile > Leu > Trp. Finally E patients indicated amino acids concentration in the decreasing order: Gln > Phe > Pro > Put > Thr > Ala > Gly > Ser > Spd > Glu > His > Arg > Asp > Spm > Leu > Ile > Asn > Lys > Cys > Tyr > Met > Trp > Val.

The coefficient of determination (r²) among the healthy volunteers for 23 compounds concentration µg/g were examined and r² were observed within 0.9659-0.9878, but when the average concentration values of healthy volunteers (n=10) on the x-axis were plotted against the affected skin samples of Pv, P, L and E on Y-axis (n=8), the values of r² were obtained 0.5162, 0.0364, 0.0045 and 0.0094 respectively. The results indicated that the amino acids from healthy volunteers agreed with each other, but they differed significantly from the affected skin samples from Pv, P, L and E (Figure 4 a, b).

Similarly the results of analysis of patients with the same disease agreed with each other with r² within 0.9621-0.9938, but differed from different diseases like Pv-P, Pv-L, Pv-E, P-L, P-E, and L-E with r² 0.0265, 0.1879, 0.0943, 0.0005, 0.3489 and 0.0067 respectively and supports the

Table 4. A sample from each of the group H, Pv, P, L and E was also analyzed by standard addition technique and the results obtained agreed with direct calibration with recovery within 95-102 % with RSDs within 0.94-7.78 %.

To understand these results, the amino acids were grouped by chemical features and differences between the skin samples were analyzed.

The aliphatic amino acids Gly, Ala, Val, Leu and Ile present in significant amount in all the skin samples were examined. Gly, Ala, Leu

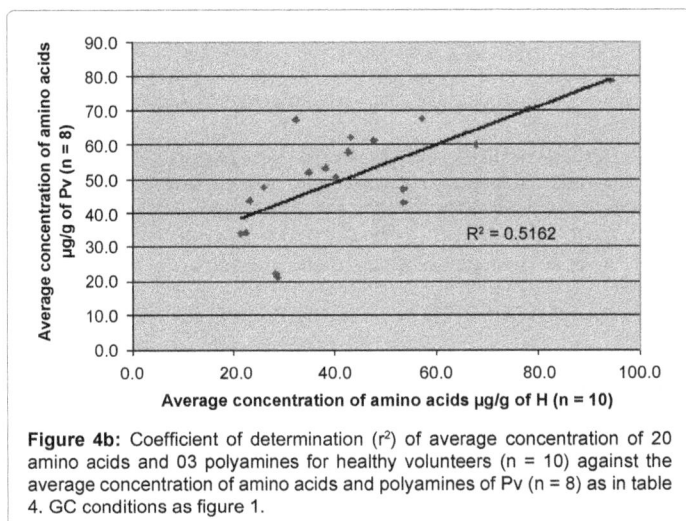

Figure 4b: Coefficient of determination (r^2) of average concentration of 20 amino acids and 03 polyamines for healthy volunteers (n = 10) against the average concentration of amino acids and polyamines of Pv (n = 8) as in table 4. GC conditions as figure 1.

hypothesis that the diseases Pv, P, L and E alter the protein molecules differently in the keratinocytes of the skin.

Conclusions

An analytical procedure has been developed for GC determination of 20 amino acids, 5 polyamines and 3 biological active compounds using double derivatization with FAA and IBCF with improved sensitivity. The method is applied for the analysis of amino acids and polyamines from acid hydrolyzed human skin samples from the patients of pemphigus vulgaris, Psoriasis, leishmaniatic, eczema and healthy volunteers. A variation in amino acids contents among the patients and healthy volunteers is noted.

Acknowledgement

We like to acknowledge Pakistan Science Foundation Islamabad for the grant of research project PSF/Res/S-SU/Chem (439). We also like to thank Professor Dr. Abdul Manan Bhutto chairman of department of Dermatology Shaheed Mohatarma Benazeer Bhutto Medical University Larkana for his generous help in identification and collection of samples of the patients suffering from skin diseases.

References

1. de Veer SJ, Furio L, Harris JM, Hovnanian A (2014) Proteases: common culprits in human skin disorders. Trends Mol Med 20: 166-178.

2. Robertson JH, Wheatley DN (1979) Pools and protein synthesis in mammalian cells. Biochem J 178: 699-709.

3. Meyer W, Poehling HM, Neurand K (1991) Intraepidermal distribution of free amino acids in porcine skin. J Dermatol Sci 2: 383-392.

4. Avino P, Campanella L, Russo MV (2001) High-performance liquid chromatography intercomparative study for amino acid analysis in two tissues by PITC- and OPA-derivatizations. Anal Lett 34: 867–882.

5. Nefiodov LI, Uglyanica KN, Smirrov VY, Doroshenko YM, Fomin KA, et al. (1996) Amino acids and their derivatives in tumour tissue from patients with breast cancer treated with Ukrain, Part VI. Drugs Exp Clin Res 22: 159–161.

6. Sánchez-Jiménez F, Ruiz-Pérez MV, Urdiales JL, Medina MA (2013) Pharmacological potential of biogenic amine-polyamine interactions beyond neurotransmission. Br J Pharmacol 170: 4-16.

7. Moinard C, Cynober L, de Bandt JP (2005) Polyamines: metabolism and implications in human diseases. Clin Nutr 24: 184-197.

8. Salato VK, Hacker-Foegen MK, Lazarova Z, Fairley JA, Lin MS (2005) Role of intramolecular epitope spreading in pemphigus vulgaris. Clin Immunol 116: 54-64.

9. Ding X, Diaz LA, Fairley JA, Giudice GJ, Liu Z (1999) The anti-desmoglein 1 autoantibodies in pemphigus vulgaris sera are pathogenic. J Invest Dermatol 112: 739-743.

10. Maintz L, Benfadal S, Allam JP, Hagemann T, Fimmers R, et al. (2006) Evidence for a reduced histamine degradation capacity in a subgroup of patients with atopic eczema. J Allergy Clin Immunol 117: 1106-1112.

11. Eyerich K, Pennino D, Scarponi C, Foerster S, Nasorri F, et al. (2009) IL-17 in atopic eczema: linking allergen-specific adaptive and microbial-triggered innate immune response. J Allergy Clin Immunol 123: 59-66.

12. Kuru T, Jirata D, Genetu A, Barr S, Mengistu Y, et al. (2007) Leishmania aethiopica: identification and characterization of cathepsin L-like cysteine protease genes. Exp Parasitol 115: 283-290.

13. Warren KS (1993) Immunoglobulin G Subclass Responses of Children during Infection with Onchocerca volvulus. Immunology and Molecular Biology of Parasitic Infection, third ed. Blackwell Scientific Publication, Oxford, UK .

14. Rafati S, Nakhaee A, Taheri T, Ghashghaii A, Salmanian AH, et al. (2003) Expression of cysteine proteinase type I and II of Leishmania infantum and their recognition by sera during canine and human visceral leishmaniasis. Exp Parasitol 103: 143-151.

15. Ortonne JP (1999) Recent developments in the understanding of the pathogenesis of psoriasis. Br J Dermatol 140 Suppl 54: 1-7.

16. Vanizor Kural B, Orem A, CimÅŸit G, Yandi YE, Calapoglu M (2003) Evaluation of the atherogenic tendency of lipids and lipoprotein content and their relationships with oxidant-antioxidant system in patients with psoriasis. Clin Chim Acta 328: 71-82.

17. Del Campo CP, Garde-Cerdan T, Sanchez AM, Maggi L, Carmona M, et al. (2009) Determination of free amino acids and ammonium ion in saffron (Crocus sativus L.) from different geographical origins. Food Chemistry 114: 1542-1548.

18. Liming W, Jinhui Z, Xiaofeng X, Yi L, Jing Z (2009) Fast determination of 26 amino acids and their content changes in royal jelly during storage using ultra-performance liquid chromatography. J Food Composit and Anal 22: 242-249.

19. Gwatidzo L, Botha BM, McCrindle RI (2013) Determination of amino acid contents of manketti seeds (Schinziophyton rautanenii) by pre-column derivatisation with 6-aminoquinolyl-N-hydroxysuccinimidyl carbamate and RP-HPLC. Food Chem 141: 2163-2169.

20. Chen G, Li J, Sun Z, Zhang S, Li G, et al. (2014) Rapid and sensitive ultrasonic-assisted derivatization micro extraction (UDME) technique for bitter taste-free amino acids (FAA) study by HPLC-FLD. Food Chem 143: 97-105.

21. de Paiva MJ, Menezes HC, Christo PP, Resende RR, Cardeal Zde L (2013) An alternative derivatization method for the analysis of amino acids in cerebrospinal fluid by gas chromatography-mass spectrometry. J Chromatogr B Analyt Technol Biomed Life Sci 931: 97-102.

22. Chang PL, Chiu TC, Chang HT (2006) Stacking, derivatization, and separation by capillary electrophoresis of amino acids from cerebrospinal fluids. Electrophoresis 27: 1922-1931.

23. Jones BN, Paabo S, Stein S (1981) Amino Acid Analysis and Enzymatic Sequence Determination of Peptides by an Improved o-Phthaldialdehyde Precolumn Labeling Procedure. J Liq Chromatogr 4: 565-586.

24. Heinrikson RL, Meredith SC (1984) Amino acid analysis by reverse-phase high-performance liquid chromatography: precolumn derivatization with phenylisothiocyanate. Anal Biochem 136: 65-74.

25. Wiedmeier VT, Porterfield SP, Hendrich CE (1982) Quantitation of Dns-amino acids from body tissues and fluids using high-performance liquid chromatography. J Chromatogr 231: 410-417.

26. Einarsson S, Josefeson B, Lagerkvist S (1983) Determination of amino acids with 9-fluorenylmethyl chloroformate and reversed-phase high-performance liquid chromatography. J Chromatogr A 282: 609-618.

27. Schwarz EL, Roberts WL, Pasquali M (2005) Analysis of plasma amino acids by HPLC with photodiode array and fluorescence detection. Clin Chim Acta 354: 83-90.

28. Kaspar H, Dettmer K, Chan Q, Daniels S, Nimkar S, et al. (2009) Urinary amino acid analysis: a comparison of iTRAQ-LC-MS/MS, GC-MS, and amino acid analyzer. J Chromatogr B Analyt Technol Biomed Life Sci 877: 1838-1846.

29. Visser WF, Verhoeven-Duif NM, Ophoff R, Bakker S, Klomp LW, et al. (2011) A sensitive and simple ultra-high-performance-liquid chromatography-tandem mass spectrometry based method for the quantification of D-amino acids in body fluids. J Chromatogr A 1218: 7130-7136.

30. You J, Ming Y, Shi Y, Zhao X, Suo Y, et al. (2005) Development of a sensitive fluorescent derivatization reagent ,2-benzo-,4-dihydrocarbazole-9-ethyl chloroformate (BCEOC) and its application for determination of amino acids from seeds and bryophyte plants using high-performance liquid chromatography with fluorescence detection and identification with electrospray ionization mass spectrometry. Talanta 68: 448-458.

31. Kaspar H, Dettmer K, Gronwald W, Oefner PJ (2008) Automated GC-MS analysis of free amino acids in biological fluids. J Chromatogr B Analyt Technol Biomed Life Sci 870: 222-232.

32. Zahradníckova H, Hartvich P, Simek P, Husek P (2008) Gas chromatographic analysis of amino acid enantiomers in Carbetocin peptide hydrolysates after fast derivatization with pentafluoropropyl chloroformate. Amino Acids 35: 445-450.

33. Pätzold R, Brückner H (2006) Gas chromatographic determination and mechanism of formation of D-amino acids occurring in fermented and roasted cocoa beans, cocoa powder, chocolate and cocoa shell. Amino Acids 31: 63-72.

34. Majidano AS, Khuhawar MY (2011) GC Analysis of Amino Acids Using Trifluoroacetylacetone and Ethyl Chloroformate as Derivatizing Reagents in Skin Samples of Psoriatic and Arsenicosis Patients. Chromatographia 73: 701–708.

35. Awan MA, Fleet I, Thomas PCL (2008) Determination of biogenic diamines with a vaporisation derivatisation approach using solid-phase microextraction gas chromatography–mass spectrometry. Food Chemistry 111: 462-468.

36. Sobolevsky TG, Revelsky AI, Revelsky IA, Miller B, Oriedo V (2004) Simultaneous determination of fatty, dicarboxylic and amino acids based on derivatization with isobutyl chloroformate followed by gas chromatography—positive ion chemical ionization mass spectrometry. J Chromatogr B 800: 101-107.

37. Husek P (1991) Rapid derivatization and gas chromatographic determination of amino acids. J Chromatogr 552: 289-299.

6

Anion Exchange Chromatography for Purification of Antigen B of Cystic Echinococcosis

Jeyathilakan N[1]*, Abdul Basith S[1], John L[1], Chandran NDJ[2] and Dhinakarraj G[3]

[1]Department of Veterinary Parasitology, Madras Veterinary College, Tamil Nadu Veterinary and Animal Sciences University, Chennai- India
[2]Department of Veterinary Microbiology, Madras Veterinary College, Tamil Nadu Veterinary and Animal Sciences University, Chennai- India
[3]Department of Animal Biotechnology, Madras Veterinary College, Tamil Nadu Veterinary and Animal Sciences University, Chennai- India

Abstract

Cystic echinococcosis or hydatidosis causes severe economic loss and public health problem to both human beings and livestock in many temperate and tropical areas of the world. Immuno diagnosis has been found to be useful not only in primary diagnosis but also for follow up of patients after surgical or pharmacological treatment in man. Various immune diagnostic tests for hydatidosis in man and animals have been attempted using hydatid cyst fluid antigens with varied sensitivity and specificity. However these assays using crude hydatid antigens have been non-specific due to cross reaction with *Cysticercus*, *Coenurus* and other helminthic infections. In order to overcome these difficulties various novel tests using purified antigens are essential for confirmative diagnosis of hydatidosis in man and animals. One of the subunit of antigen B, 8 kDa protein has been shown to be hydatid specific. Various types of chromatography have been used for purification of Hydatid antigens. However, Anion exchange chromatography had been used in this study to get more antigen B 8 kDa than other chromatographic methods.

Keywords: Hydatidosis; Antigen B; Anion exchange chromatography

Introduction

Cystic Echinococcosis

Cystic echinococciosis (CE) or hydatidosis is a chronic zoonotic parasitic infection in man and animals, with a cosmopolitan distribution, caused by the dog tapeworm *Echinococcus granulosus*. People become infected when they ingest eggs passed from tapeworms in dogs. This disease is of utmost public health problem in human beings and requires expensive and prolonged medical treatment with surgical intervention. The global annual monetary loss due to hydatidosis in man accounts for US$ 193,529,740.The most tangible economic effects in animals of this are the loss of offal from food animals. The body weight of infected animal will be 1 per cent less than uninfected animals. The global annual livestock production loss due to CE is estimated to be US$ 141, 605,195 [1].

Diagnosis of CE can provide substantial improvements in the quality of the management and treatment of disease. In livestock, infection with hydatid cyst is asymptomatic and diagnosis is made usually at necropsy. Immunodiagnosis has been found to be useful not only in primary diagnosis but also for follow up of patients after surgical or pharmacological treatment in man. However assays using crude hydatid antigens have been non-specific due to cross reaction with *Cysticercus*, *Coenurus* and other helminthic infections. In order to overcome these difficulties various novel tests using purified antigens are essential for confirmative diagnosis of hydatidosis in man and animals [2].

Purification of hydatid antigens by chromatography

Immunodiagnostic tests using crude hydatid cyst antigen are far from satisfactory. Thus, purification of hydatid antigens is mandatory to remove host components and other cross reactive proteins. Various types of chromatography have been used for purification of Hydatid antigens.

Sephadex G 200 gel filtration chromatography had been commonly used to purify the antigens of sheep hydatid fluid. Sephadex G-200 chromatography (15 × 50 cm column) used for purification of hydatid cyst fluid and membrane antigen. The elution profile of hydatid fluid revealed two peaks, while 3 distinct peaks were observed in case of hydatid membrane [3]. However Sephadex G-200 chromatography of sheep hydatid fluid gave three main peaks [4]. Six major peaks were got in elution profile of protoscolices aqueous soluble proteins on Sephadex G-200 column of 2 × 90 cm size [5].

A simple two step procedure involving Sephadex G 200, Sepharose 4B gel filtration chromatography was used to purify the lipoprotein antigens of sheep hydatid fluid [6]. Elution profiles showed two larger peaks designated as antigen A and B and two smaller peaks. Ovine liver *Echinococcus granulosus* hydatid cyst fluid was purified 24 and 30 fold by Sephadex G-200 and DEAE(Di- Ethyl Amino- Ethyl) - Sephadex chromatography in 25 and 31 cm column [7]. One major and 2 minor peaks were obtained in the sephadex G-200 column chromatography. Buffalo hydatid cyst fluid was fractionated in nine peaks by molecular exclusion chromatography using Sephadex G-150, 90 × 1.5 cm column. The third peak alone reacted with treated and untreated anti buffalo hydatid cyst fluid sera suggesting parasite derived antigen [8].

Affinity chromatographic purification may be a useful first step in production of species specific immuno diagnostic antigens for hydatid infections. Affinity chromatography of crude hydatid cyst fluid using CNBr (Cyanogen Bromide) activated sepharose 4B column was carried out by various workers [9]. Con (Concavalin) -A sepharose column was also used for preparation of fraction rich in glycoprotein from crude hydatid cyst fluid [10].

---

*Corresponding author: Jeyathilakan N, Professor and Head, Department of Veterinary Parasitology, VCRI, Orathanadu, Thanjavur, Tamil Nadu, India
E-mail: drjthilakan@yahoo.com

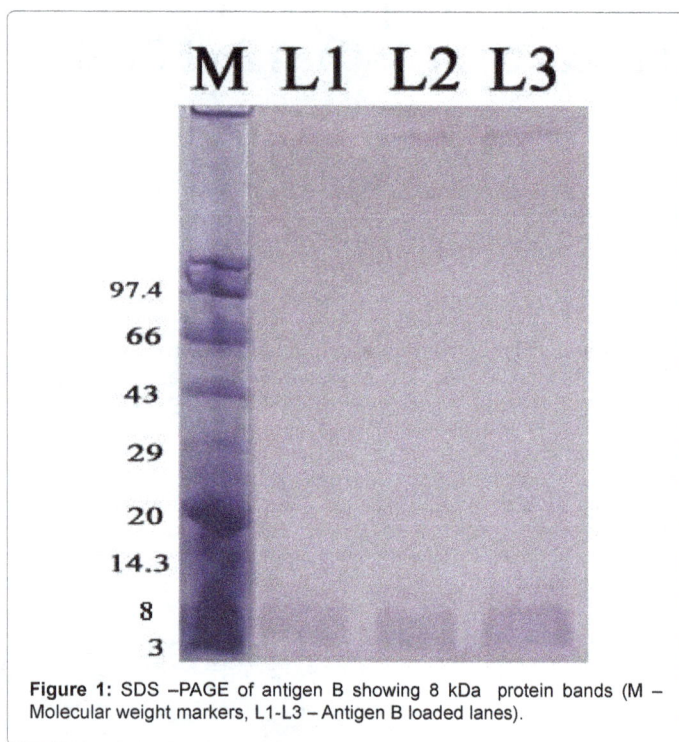

Figure 1: SDS –PAGE of antigen B showing 8 kDa protein bands (M – Molecular weight markers, L1-L3 – Antigen B loaded lanes).

Anion exchange chromatography

Anion exchange chromatography had been used to get more antigen B than other chromatographic methods [11].

Experimental

Collection of hydatid cysts

The hydatid cysts were collected from sheep slaughtered at Corporation Slaughter House in Perambur and Department of Meat Science and Technology, Madras Veterinary College. Chennai, India.

Anion exchange Chromatography

The immunodominant antigen B was prepared from hydatid cyst fluid by Anion exchange chromatography using DEAE- sepharose fast flow (Sigma, USA) by following the standard protocols [12,13].

Reagents

Application buffer pH 7.4.

20 mM sodium phosphate 1.2 g

200 mM sodium chloride 5.84 gm

TGDW (Triple glass distilled water) – 500 ml

Elution buffer pH 7.4.

20 mM sodium phosphate 1.2 g

500 mM sodium chloride 14.6 g

TGDW 500 ml

Conductivity buffer pH 7.4.

20 mM sodium phosphate 1.2 g

2 M sodium chloride 58.0 g

TGDW 500 ml

Preparation of hydatid fluid antigen

The hydatid cysts were thoroughly washed in distilled water to remove the adhering dirt and clotted blood. The fluid was aspirated slowly using 20 ml syringe. The aspirated fluid was pooled together and kept in a glass beaker for settling of brood capsules, protoscolices and dead tissues. The supernatant was collected and clarified by centrifugation at 10,000 rpm for 30 minutes to remove other sediments. The conductivity of hydatid fluid was adjusted with conductivity buffer as equal to that of application buffer. The hydatid cyst fluid was loaded bit by bit into the column.

Preparation of Anion exchange column

DEAE sepharose fast flow (Sigma, USA) was slowly packed to a 5 × 2.5 cm size column (Bio-rad, USA). The column was equilibrated with application buffer. Typically 1.5 liters of the hydatid cyst fluid supernatant were loaded in the column. The flow rate was adjusted to 3 ml / minute and the chromatography was undertaken at 4°C. The column was washed with 5 column volumes of application buffer. The bound antigen fractions were eluted with elution buffer. The antigen B was eluted in 7, 8, 9 and 10th fractions. These fractions were pooled together and concentrated using PEG 6000. The protein concentration of antigen B was estimated by BCA method [14]. The protein content was 0.987 mg / ml. SDS-PAGE analysis of DEAE Sepharose fast flow anion exchange chromatography fractions revealed the antigen B protein bands at 8 kDa (Figure 1).

Conclusions

Hydatid cyst fluid (HCF) is a complex mixture of glycol lipoproteins, carbohydrates and salts. Crude HCF has a high sensitivity, ranging typically from 75 per cent to 95 percent. However, its specificity is often unsatisfactory and cross-reactivity with sera from patients infected with other cestode (89 per cent), nematode (39 percent) and trematode (30 per cent) species is commonly observed. Hence, the crude HCF is specifically recommended for mass serological screening and it has now become more frequent to purify components such as the lipoproteins antigen B and antigen 5, the most relevant components of HCF for diagnostic purposes. Antigen B 8 kDa may have the opportunity to accumulate in the cyst fluid after being secreted by the parasite in such a way that the protein has the chance to aggregate into a form that is more immunogenic before the antigen gains contact with the host immune system [15]. Various authors have used different chromatographic protocols to isolate antigen B from hydatid cyst fluid, [16], however, the quantity of antigen available from the above methods was scanty. Therefore the Anion exchange chromatography using DEAE Sepharose fast flow was followed and it resulted in production of a large quantity of antigen B. The 8 kDa fraction can be separated from antigen B by SDS-PAGE method.

References

1. Budke CM, Deplazes P, Torgerson PR (2006) Global socioeconomic impact of cystic echinococcosis. Emerg Infect Dis 12: 296-303.

2. Shepherd JC, McManus DP (1987) Specific and cross-reactive antigens of *Echinococcus granulosus* hydatid cyst fluid. Mol Biochem Parasitol 25: 143-154.

3. Dev Sarma MK, Deka DK, Rahman H, Sarma S, Borkakoty MR (2002) Purification and partial characterization of cystic stages of *Echinococcus granulosus* and *Taenia hydatigena*. J Vet Parasitol 16: 63-64.

4. Pozzuoli R, Musiani P, Arru E, Piantelli M, Mazzarella R (1972) *Echinococcus granulosus* : Isolation and characterization of sheep hydatid fluid antigens. Exp Parasitol 32: 45-55.

5. Ahmad G, Nizami WA, Saifullah MK (2001) Analysis of potential antigens of protoscolices isolated from pulmonary and hepatic hydatid cysts of *Bubalus bubalis*. Comp Immunol Microbiol Infect Dis 24: 91-101.

6. Oriol R, Williams J.F, Perez Esandi M.V, Oriol C (1971) Purification of lipoprotein antigens of *Echinococcus granulosus* from sheep hydatid fluid. Am J Trop Med Hyg 20: 569-574.

7. Vessal M, Bambaea-Row N (1994) Partial purification and comparison of the kinetic properties of ovine liver *Echinococcus granulosus* hydatid cyst fluid malate dehydrogenase and the cytoplasmic enzyme from the host liver. Comp Biochem Physiol Biochem Mol Biol 107: 447-451.

8. Bandyopadhyay S, Singh BP (2000) Antibody response in secondary hydatidosis by ELISA using affinity purified buffalo hydatid cyst fluid antigen. J Vet Parasitol 14: 17-20.

9. Craig PS, Rickard MD (1981) Studies on the specific Immunodiagnosis of larval cestodes infections of cattle and sheep using antigens purified by affinity chromatography in an enzyme linked immunosorbent assay (ELISA). Int J Parasitol 11: 441-449.

10. Sbihi Y, Janssen D, Osuna A (1996) Serologic recognition of hydatid cyst antigens using different purification methods. Diagn Microbiol Infect Dis 24: 205-211.

11. Gonzalez G, Nieto A, Fernandez C, Orn A, Wernstedet C, et al. (1996) Two different 8 kDa monomers are involved in the oligomeric organization of the native *Echinococcus granulosus* antigen B. Parasite Immunol 18: 587-596.

12. Bollag DM, Edelstein SJ, Rozycki MD (Eds) 1996 Protein methods (2nd Edn). Wiley-Liss, New York: 231-269.

13. Rosenberg JM (1996) Protein analysis and Purification – Benchtop techniques, Birkhauser: 283 -296.

14. Smith PK, Krohn RI, Hermanson GT, Mallia AK, Gartner FH, et al. (1985) Measurement of protein using bicinchoninic acid. Anal Biochem 150: 76-85.

15. Mamuti W, Sako Y, Nakao M, Xiao N, Nakaya K, et al. (2006) Recent advances in characterisation of *Echinococcus* antigen B. Parasitol Int 55: S57-S62.

16. Kanwar JR, Kanwar R (1994) Purification and partial immunochemical characterization of a low molecular mass, diagnostic *Echinococcus granulosus* immunogen for sheep hydatidosis. FEMS Immunol Med Microbiol 9: 101-107.

7

HPTLC Method for Shanzhiside Esters: Simultaneous Quantitative Analysis of Barlerin, Acetylbarlerin and Shanzhiside Methyl Ester in *Barleria* Species

Preet Kawal Kaur[1,2]*, Karan Vasisht[2] and Maninder Karan[2]

[1]*IEC School of Pharmacy, IEC University, Kalujhanda, Solan, Baddi, Himachal Pradesh, India*
[2]*University Institute of Pharmaceutical Sciences–UGC Centre for Advanced Studies, Panjab University, Chandigarh, India*

Abstract

A simple, selective, precise, robust, rapid and reliable high-performance thin layer chromatographic method of analysis for simultaneous determination of acetylbarlerin, barlerin and shanzhiside methyl ester (the major iridoids of *Barleria*) was developed and validated. The three iridoid markers were chromatographed on aluminium base TLC plates precoated with silica gel $60F_{254}$ using chloroform-ethylacetate- methanol-acetic acid (3.0:3.0:3.0:1.0, v/v/v/v) as mobile phase having pH of 5.01. The compounds were quantified by quantitative analysis in absorbance mode at 233 nm. The system gave compact spots for acetylbarlerin, barlerin and shanzhiside methyl ester with optimum resolution (R_f 0.71, 0.61 and 0.50 respectively) in single development. The linear regression analysis data with 95 % confidence limits by the Software package for Statistical Analysis (SPSS software, version 16) for the calibration plots for acetylbarlerin, barlerin and shanzhiside methyl ester showed good linear relationship with r^2=0.997, 0.995 and 0.992 in the concentration range of 1.42-4.95, 0.28-1.67 and 0.60-3.60 µg/spot respectively for the three markers. The mean value of slope and intercept were 0.98 ± 0.013 and -203.14 ± 42.7 for acetylbarlerin, 2.29 ± 0.04 and 249.52 ± 40.5 for barlerin and 0.96 ± 0.021 and 53.82 ± 49.89 for shanzhiside methyl ester respectively. The method was validated for precision, recovery, repeatability and robustness as per the International Conference on Harmonisation (ICH) guidelines. The limit of detection and limit of quantification were 0.07, 0.21; 0.05, 0.15 and 0.05, 0.15 µg/spot respectively for acetylbarlerin, barlerin and shanzhiside methyl ester respectively. Statistical analysis showed the method to be repeatable and selective for the estimation of the three iridoid markers. Since the proposed mobile phase effectively resolves acetylbarlerin, barlerin and shanzhiside methyl ester, the developed method can be successfully applied for the identification and simultaneous quantification of these markers. The method will be of particular use for crude drug/herbal extracts quality testing etc in especially resource constrained countries and laboratories of Asia and Africa where this plant is known to grow in abundance.

Keywords: Iridoids; Shanzhiside esters; Barleria; HPTLC

Introduction

Barleria is a reputed plant of Ayurveda and enjoys high status for its versatile use in several ailments including inflammation [1,2]. The genus *Barleria* of family Acanthaceae comprises of small shrubs or under shrubs, which are distributed in warmer parts of the world. It has about 300 species, of which nearly 32 are reported to occur in India [3] and many are grown as low hedge plants in gardens [4,5]. The important medicinal species are *B. buxifolia* Linn., *B. courtallica* Nees., *B. cristata* Linn., *B. longifolia* Linn., *B. prionitis* Linn., *B. lupulina* Lindl. and *B. strigosa* Willd [4,6].

The well documented traditional uses of *Barleria* employs whole plant for various ailments like catarrhal affections of children which are accompanied by fever and much phlegm, leaves to relieve toothache, roots for boils and glandular swellings, bark for cough and anascara and decoction of the plant in dropsy and as an anti-inflammatory [1,6-14]. Some of the important Ayurvedic formulations of *Barleria* are Sahacaradi taila, Nilikadya taila, Astavarga kvatha curna and Rasnarandadi kvatha curna [14]. The genus *Barleria* has gained importance in recent years for the treatment of various diseases such as liver disorders, diabetes, neurological disorders, immunodeficiency, inflammation, ulcers, HSV-2 viral diseases, etc. [6,7,11].

The genus is reported to contain iridoids, anthraquinones, flavonoids, sterols, and fatty acids [14]. Iridoids constitute the major class of compounds isolated from *Barleria* and important bioactive iridoids are acetylbarlerin, barlerin and shanzhiside methyl ester [14]. Besides *Barleria*, these iridoids are also reported to be present

in *Gentian* [15], *Melampyrum* [15,16], *Morinda* [17], *Phlomis* [15,18], *Stilbe* [15] and *Valerian* [15,19] species etc. Principally, iridoids are cyclopentan-[c]-pyran monoterpenoids [15]. There is a continuing interest in iridoids as many of them have shown a host of biological and pharmacological activities like cardiovascular, antihepatotoxic, choleretic, hypoglycemic, analgesic, anti-inflammatory, antimutagenic, antispasmodic, antitumor, antiviral, immunomodulator and purgative properties [17,19,20]. The broad diversity of biological activities exhibited by the iridoids has always interested the chemists to explore new methods of isolation and determination. Various techniques for e.g. HPLC [21,22], capillary HPLC [23]/HPLC [24,25] coupled with Photodiode Array Detector and Electron Spray Ionisation Mass Spectrometric (DAD-ESI-MS), densitometric determination [26] and micellar electrokinetic capillary chromatography [27] are employed to analyse various groups of iridoids. After a bibliographic search, it was found that HPLC is the most common reported method for the

---

*Corresponding author: Preet Kawal Kaur, Associate Professor and Head, IEC School of Pharmacy, IEC University, Kalujhanda, Solan, Baddi, Himachal Pradesh, India, E-mail: preetkawalpu@gmail.com

analysis of shanzhiside esters [21,22,24]. To our knowledge, three reports are known where analysis of *B. prionitis* has been done by TLC densitometry using the solvent system chloroform-methanol (8:2, v/v) [28-30]. The first report was from our laboratory and this solvent system was developed for the estimation of single iridoid marker barlerin and the method was not validated. Subsequently in another publication 10-50 times high per cent content of barlerin (10.03 ± 1.69) along with exceptionally high content of shanzhiside methyl ester (21.55 ± 2.40 per cent) has been reported employing this solvent system. This indicated that the earlier reported method is not specific when more than one marker is desired to be quantified simultaneously. Therefore, in the present study, a battery of solvent systems were tested with an aim (i) to resolve the underlying spots interfering with the markers, (ii) to validate the developed method as per ICH guidelines for the drug quality assurance and (iii) to obtain the most accurate values of per cent content of the three iridoid markers. Hence in this study, a validated HPTLC method was developed employing a unique combination of solvents and was used as a rapid method of analysis for the simultaneous determination of shanzhiside iridoids in *Barleria* species. The proposed method was successfully used for the simultaneous estimation of acetylbarlerin, barlerin and shanzhiside methyl ester in at least 5 different *Barleria* species and varieties *viz.*, *B. prionitis* Linn., *B. lupulina* Lindl, *B. cristata* var. *dichotoma*, *B. cristata* (pink flower variety) and *B. cristata* (blue flower variety) without any other component interfering in the analysis. It is expected that the developed method could easily and successfully be applied with precise results for the estimation of shanzhiside esters in other *Barleria* species/related plant species, herbal extracts, formulations etc.

Experimental

Materials

Standard acetylbarlerin (purity 99.04%), barlerin (purity 99.31%) and shanzhiside methyl ester (purity 99.83%) were isolated in the laboratory from *B. cristata* Linn. whole plant, procured during February to March 2009 from the Medicinal Plants Garden of University Institute of Pharmaceutical Sciences, Panjab University, Chandigarh, using column chromatography, PTLC and flash chromatography as the preferred isolation techniques [31]. The authenticity of the samples was duly confirmed by National Institute of Science Communication and Information Resources (NISCAIR), New Delhi (Ref. NISCAIR/RHMD/Consult/-2008-09/1127/158) and voucher specimens of the same have been deposited in the Museum-cum-Herbarium of University Institute of Pharmaceutical Sciences-Centre of Advanced Study, Panjab University, Chandigarh, India, under the voucher numbers 1460, 1461,1462, 1463 and 1464 for *B. prionitis*, *B. cristata* var. *dichotoma*

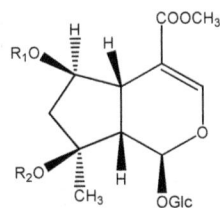

	R_1	R_2
Acetylbarlerin	Ac	Ac
Barlerin	H	Ac
Shanzhiside methyl ester	H	H

Figure 1: Structures of shanzhiside esters.

(white flower variety), *B. cristata* (pink flower variety), *B. cristata* (blue flower variety) and *B. lupulina* respectively. *B. prionitis* was collected from August to September 2008 from the Medicinal Plants Garden of University Institute of Pharmaceutical Sciences, Panjab University, Chandigarh. All the chemicals and reagents used were of analytical grade procured from E. Merck chemicals, India.

Isolation and characterization of shanzhiside esters

The three iridoids (Figure 1) were isolated from the whole plant of *B. cristata* using various chromatographic methods and were duly characterized with the help of different spectroscopic techniques like UV, IR, ^{1}H-NMR, ^{13}C-NMR and ESI-MS. These were identified as acetylbarlerin, barlerin and shanzhiside methyl ester.

Characterization of acetylbarlerin: $UV_{\lambda max}$ = 233 nm; ^{1}H NMR (400 MHz, CD_3OD) (δ) : 1.45 (H-10) [s], 1.92 (OCOOCH$_3$) [s], 1.95 (OCOOCH$_3$) [s], 1.99 (H-7) [d, J = 5.52 Hz], 2.03 (H-7) [d, J = 5.56 Hz], 2.84 (H-9) [dd, J=8.56 and 3.44 Hz], 3.09 (H-5) [dd, J = 7.98 and 1.12 Hz], 3.30 - 3.60 (H-2' - H-6') [m], 3.58 (COOCH$_3$) [s], 4.77 (H-I ') [s], 5.16 (H-6) [d, J= 5.56 Hz], 5.74 (H-I) [d, J = 3.48 Hz], 7.41 (H-3) [d, J = 1.48 Hz]; ^{13}C NMR (75 MHz, CD_3OD) (δ) : 93.96 (C_1), 153.1 1 (C_2), 107.10 (C_4), 38.52 (C_5), 77.55 (C_6), 43.5.9 (C_7), 88.27 (C_8), 48.87 (C_9), 29.39 (C_{10}), 20.88 (CH$_3$), 20.43 (CH$_3$), 50.58 (CH$_3$O), 167.17(CO), 171.62 (CO), 171.00 (CO), 98.84 (C_1'), 73.27 (C_2'), 76.52 (C_3'), 76.96 (C_5'), 70.23 (C_4') and 61.55 (C_6').

Characterization of barlerin: $UV_{\lambda max}$ = 233 nm; ^{1}H NMR (400 MHz, CD$_3$OD) (δ) : 1 .41 (H-10) [s], 1.90 (H-7) [d, J = 5.36 Hz], 1.93 (H-7) [d, J = 5.4 Hz], 2.08 (OCOOCH$_3$) [s], 2.09 (H-9) [dd, J = 15.92 and 2.9 Hz], 3.08 (H-5) [dd, J = 8.5 and 0.92 Hz], 3.20 - 4.42 (H-2' - H-6') [m], 3.62 (COOCH$_3$) [s], 3.80(H-6) [dd, J = 12.08 and 6.04 Hz], 4.8 (H-1') [s], 5.82 (H-I) [d, J = 2.32 Hz], 7.35 (H-3) [d, J = 1.36 Hz]; ^{13}C - NMR (75 MHz, CD$_3$OD) (δ) : 94.29 (C_1), 152.40 (C_2), 108.37 (C_4), 40.85 (C_5), 74.53 (C_6), 46.29 (C_7), 88.42 (C_8), 48.49 (C_9), 29.41 (C_{10}), 20.8 (CH$_3$), 50.57 (CH$_3$O), 167.75 (CO), 172.02 (CO), 98.88 (C_1'), 73.24 (C_2'), 76.54 (C_3'), 76.88 (C_5'), 70.18 (C_4') and 61.47 (C_6').

Characterization of shanzhiside methyl ester: $UV_{\lambda max}$ = 233 nm; m.p. = 98-100ºC; ^{1}H NMR (400 MHz, CD$_3$OD) (δ) : 1.27 (H-10) [s], 1.84 (H-7) [dd, J = 7.24 and 6 Hz], 2.12 (11-7) [dd, J = 6.84 and 6.4 Hz], 2.63 (H-9) [dd, J = 7.64 and 2.54 Hz], 3.01 (H-5) [dd, J = 7.04 and 3.06 Hz], 3.20 - 3.66 (H-2' - H-6') [m], 3.75 (COOCH$_3$) [s], 4.05 (H-6) [m], 4.97 (H-I') [s], 5.59 (H-I) [d, J = 2.68 Hz], 7.42 (H-3) [s]; ^{13}C - NMR (75 MHz, CD$_3$OD) (δ) : 93.44 (C_1), 151 .44 (C_2), 110.04 (C_4), 39.97 (C_5), 76.06 (C_6), 47.83 (C_7), δ 76.95 (C_8), 50.34 (C_9), 23.31 (C_{10}), 50.55 (CH$_3$O), 168.42 (CO), 98.41 (C_1'), δ 73.23 (C_2'), 76.56 (C_3'), 76.95 (C_5'), 70.22 (C_4') and 61.43 (C_6').

HPTLC instrumentation

The samples were applied as spots keeping a distance of 10 mm between the spots, on precoated silica gel G aluminium plate 60F$_{254}$ (20 cm × 10 cm, 0.2 mm thickness; Cat. no. 1.05554.0007, E. Merck, Darmstadt Germany, Ltd.) using Linomat 5 (Camag, Switzerland). The plates were prewashed by methanol and activated at 60ºC for 5 min prior to chromatography. The mobile phase consisted of chloroform-ethyl acetate- methanol-acetic acid (3.0:3.0:3.0:1.0, v/v/v/v) and 20 ml of mobile phase was used per chromatography. Linear ascending development was carried out in 20 cm × 20 cm twin trough glass chamber (Camag, Muttenz, Switzerland) saturated with mobile phase. The optimum chamber saturation time for mobile phase was 10 min at room temperature (25ºC ± 2) and a relative humidity of 60% ± 5. The development distance

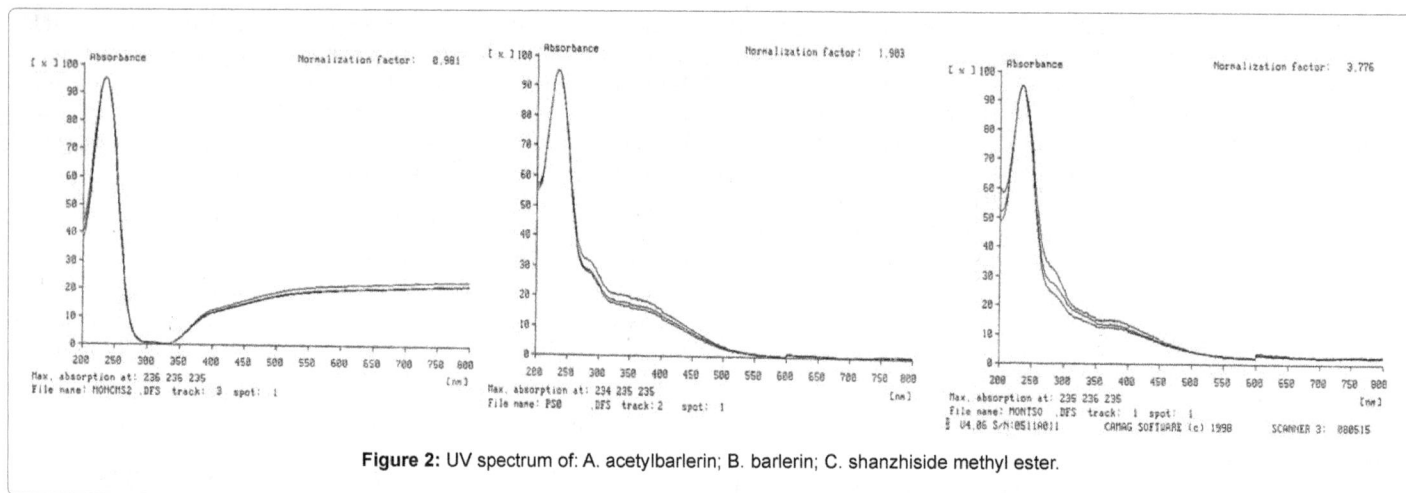

Figure 2: UV spectrum of: A. acetylbarlerin; B. barlerin; C. shanzhiside methyl ester.

was kept at 80 mm. Subsequent to the development, TLC plates were dried in a current of air using an air dryer. Densitometry scanning was performed on Camag TLC scanner 3 in the reflectance absorbance mode at 233 nm with slit dimension of 6 × 0.45 mm and operated by WINCATS software (Camag, version 4.06). Concentrations of the compounds chromatographed were determined from the intensity of diffusely reflected light and evaluation was done *via* peak areas with linear regression. The spectra of three iridoid markers developed in chloroform-ethyl acetate-methanol-acetic acid (3.0:3.0:3.0:1.0, v/v/v/v, pH - 5.01) was taken at peak start, peak apex and peak end of respective spot as a reference for comparison in the test samples and is shown in Figure 2.

Calibration curves of acetylbarlerin, barlerin and shanzhiside methyl ester

The standard solutions of acetylbarlerin (14.15 mg/10 ml), barlerin (5.55 mg/10 ml) and shanzhiside methyl ester (12 mg/10 ml) were prepared in methanol in a 10 ml volumetric flask to give concentration of acetylbarlerin (1.42 µg/µl), barlerin (0.56 µg/µl) and shanzhiside methyl ester (1.20 µg/µl). The standard solutions were spotted on the TLC plate in triplicate in increasing volumes of 1.0 to 3.5 µl (acetylbarlerin) and 0.5 to 3.0 µl (barlerin and shanzhiside methyl ester), in increments of 0.5 µl, to obtain final concentration range of 1.42-4.95 µg/spot for acetylbarlerin, 0.28-1.67 µg/spot for barlerin and 0.60-3.60 µg/spot for shanzhiside methyl ester. Each concentration was spotted three times on the TLC plate.

Method validation

Validation of the developed HPTLC method was carried out as per the International Conference on Harmonization (ICH, 1996) guidelines for specificity, sensitivity, accuracy, precision, repeatability and robustness.

Specificity: The specificity of the developed method was established by analysing various extracts and partitioned fractions of mother extract of different *Barleria* species *viz*., *B. cristata*, *B. prionitis*, and *B. lupulina* containing three bioactive iridoid markers namely acetylbarlerin, barlerin and shanzhiside methyl ester. The spots of the three iridoids in extracts and fractions were confirmed by comparing the relative front (R_f) values and UV spectrum in test samples with spots of the standard markers at three different levels of peak start, peak apex and peak end position.

Sensitivity: The values of limit of detection (LOD) and limit of

quantification (LOQ) were ascertained using diluted and known concentrations of all the three markers in replicate and comparing the results with a blank sample of methanol spotted six times following same method as mentioned in HPTLC instrumentation. Sensitivity of the method was determined with respect to limit of detection (LOD) and limit of quantification (LOQ). Series of concentration of acetylbarlerin (0.02-2.83 µg/spot), barlerin (0.03-1.11 µg/spot) and shanzhiside methyl ester (0.03-1.20 µg/spot) were applied on plate and analysed to determine LOD and LOQ.

Linearity: Linearity was determined using six concentration levels with calibration curves plotted over a wide concentration range of 1.42-4.95 µg/spot for acetylbarlerin, 0.28-1.67 µg/spot for barlerin and 0.60-3.60 µg/spot for shanzhiside methyl ester, respectively. The calibration curves were developed by plotting peak area versus concentration (n=6) with the help of WINCATS software.

Accuracy: Accuracy of the method was evaluated by carrying the recovery study at three levels. The recovery experiments were performed by spiking the sample of whole plant of *B. prionitis* with three different amounts of each marker at 50%, 100% and 150% of their pre-analysed content. The experiment was conducted in triplicate.

Precision: Precision of the developed method was evaluated in terms of intra-day and inter-day precision. Intra-day precision was established by analysing analytical concentrations of acetylbarlerin (2.12 µg/spot), barlerin (0.56 µg/spot) and shanzhiside methyl ester (1.20 µg/spot) in triplicate, for three times on the same day (Instrumental and manual precision). Inter-day precision was determined by analyzing three times measurement of same concentrations of acetylbarlerin, barlerin and shanzhiside methyl ester over a period of three days (n=3) for both system and manual precision. The intra-day and inter-day variation for the three iridoids were expressed in terms of the mean and % RSD (Relative Standard Deviation) values of the peak areas.

Repeatability: Repeatability of measurement of peak area was determined by analysing the spot of acetylbarlerin (2.12 µg/spot), barlerin (0.56 µg/spot) and shanzhiside methyl ester (1.20 µg/spot) of the calibration curve, nine times without changing the position of the plate.

Robustness: Small changes in mobile phase composition, its volume, chamber saturation time and slight change in the solvent migration distance were introduced to study their effect on the results. Robustness of the method was determined from triplicate measurement at an analytical concentration of 2.12 µg/spot of acetylbarlerin, 0.56 µg/

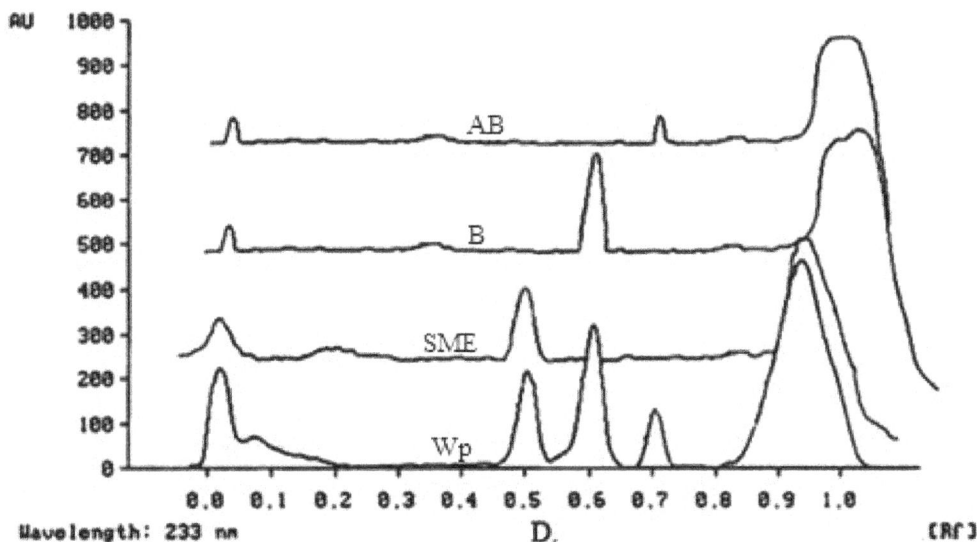

Figure 3: TLC densitometric scan of methanolic extract of *B. prionitis* using chloroform-ethyl acetate-methanol-acetic acid (3.0:3.0:3.0:1.0, v/v/v/v); Wp = whole plant; AB = acetylbarlerin; B = barlerin; SME = shanzhiside methyl ester.

spot of barlerin and 1.20 µg/spot of shanzhiside methyl ester and was represented as mean and % RSD of peak area.

Statistical analysis: The linear regression analysis data was statistically analysed using Software package for Statistical Analysis (SPSS software, version 16) and the mean and % RSD using Microsoft Excel 2010.

Analysis of shanzhiside esters in *Barleria* species/varieties

The test solutions were prepared by extracting accurately weighed 5 g of coarsely powdered material of whole plant, leaf, stem and root of *B. prionitis*, *B. cristata* var. *dichotoma*, *B. cristata* (blue flower variety), *B. cristata* (pink flower variety) and *B. lupulina* with methanol for 4 h using Soxhlet apparatus. The extracts were filtered and concentrated under reduced pressure. The final volume was adjusted to 50 ml with methanol. However, in species with too low content of iridoids, the quantities were further concentrated accordingly.

Another 5 g of coarsely powdered material of whole plant of *B. prionitis* and *B. cristata* var. *dichotoma* was similarly extracted with 50 ml methanol for 4 h in Soxhlet, filtered and the solvent was removed under reduced pressure. The residue was suspended in 15 ml of distilled water and successively partitioned with hexane, chloroform, ethyl acetate and saturated butanol (each 7 ml × 3). The three washings of each solvent were combined, filtered and concentrated under reduced pressure, and the final volume of residue was adjusted to 5 ml with the respective solvent. The solutions were either diluted or concentrated further and an appropriate quantity of each test solution was applied in triplicate on a precoated TLC plate and the plate was developed in solvent system chloroform-ethyl acetate-methanol-acetic acid (3.0:3.0:3.0:1.0, v/v/v/v, pH - 5.01) up to a distance of 80 mm and scanned following the same procedure as used for the preparation of the standard plot. The average AUC of the peak corresponding to acetylbarlerin, barlerin and shanzhiside methyl ester was noted for each test sample and their concentration was calculated from the respective standard plots.

Results and Discussion

High-performance thin layer chromatography (HPTLC) is increasingly gaining popularity as one of the accepted techniques of analysis of pharmaceuticals and herbal drugs. With the advancements in instrumentation of this technique, it is now possible to accomplish more precise and reliable results comparable to HPLC. A validated HPTLC method of analysis was developed for simultaneous estimation of three iridoids, the shanzhiside esters in *Barleria* species. The TLC procedure was first optimized using a range of solvent systems with varying compositions of chloroform-methanol; ethyl acetate-methanol-formic acid; ethyl acetate-methanol-acetic acid; chloroform-ethyl acetate-methanol-acetic acid etc. The reported solvent system of chloroform-methanol (8:2, v/v) gave variable results ranging from a low barlerin content of 0.18% to extremely high content of 10.03%. This clearly indicated the possibility of interference from other components of extract in the analysis as the exceedingly high values could be the result of other underlying spots. Most of the listed solvent systems gave either poor or no resolution of the three markers. The optimum solvent system that gave compact spots with most selective resolution was chloroform-ethyl acetate-methanol-acetic acid (3.0:3.0:3.0:1.0, v/v/v/v, pH - 5.01) with chamber saturation of 10 min at 25°C and solvent migration distance of 80 mm. Further, the desired resolution was obtained with single development having significantly different R_f values of 0.50, 0.61 and 0.71 for shanzhiside methyl ester, barlerin and acetylbarlerin and none of the desired components had any underlying or interfering spots (Figure 3). The optimum resolution was obtained using unmodified silica layer as stationary phase on precoated plates and slit-scanning densitometry with UV-Visible light as the detection technique. The iridoid markers namely acetylbarlerin, barlerin and shanzhiside methyl ester were examined directly on the stationary layer. It did not involve any pretreatment or derivatization and optical densitometric scanning was used for the *in situ* measurement of compounds directly on the layer.

Under the experimental conditions employed, the minimum amount of markers that could be detected was found to be 0.07 µg for acetylbarlerin and 0.05 µg for both barlerin and shanzhiside methyl ester; the lowest quantifiable amount of markers was 0.21 µg/spot for acetylbarlerin, 0.15 µg/spot for barlerin and shanzhiside methyl ester (Table 1). The method was confirmed to be particularly specific as

Parameter	Acetylbarlerin	Barlerin	Shanzhiside methyl ester
Limit of detection (LOD) (µg ± SD)	0.07 ± 3.10	0.05 ± 2.69	0.05 ± 2.08
Limit of quantification (LOQ) (µg ± SD)	0.21 ± 1.5	0.15 ± 2.8	0.15 ± 3.1
Linearity range (µg/spot)	1.42 – 4.95	0.28 – 1.67	0.60 – 3.60
r^2	0.997	0.995	0.992
Slope ± SD	0.98 ± 0.013	2.29 ± 0.04	0.96 ± 0.021
95 % Confidence interval of slope	0.956 - 1.009	2.209 - 2.368	0.917 - 1.007
Intercept ± SD	-203.14 ± 42.7	249.52 ± 40.5	53.82 ± 49.89
95 % Confidence interval of intercept	-293.60- (-112.68)	163.73 - 335.32	-51.94 - 159.58

Table 1: Limit of detection, quantification and linear regression data for the calibration curves of shanzhiside esters.

Original marker content in the sample (µg)	Level of spiking in %	Theoretical concentration after spiking (µg ± SD)	Total amount of drug analysed in spiked sample (µg ± SD)	% Average recovery ± SD	% RSD
Acetylbarlerin (0.35)	50	0.53 ± 23.8	0.50 ± 7.3	95 ± 3.10	3.25
	100	0.70 ± 15.1	0.68 ± 27.0	97 ± 4.02	4.15
	150	0.88 ± 4.5	0.84 ± 39.3	96 ± 4.46	4.65
Barlerin (0.81)	50	1.21 ± 19	1.19 ± 14	99 ± 2.62	2.65
	100	1.62 ± 29	1.59 ± 55	98 ± 2.36	2.41
	150	2.02 ± 37	1.99 ± 32	99 ± 1.04	1.05
Shanzhiside methyl ester (1.36)	50	2.04 ± 22	2.01 ± 20	99 ± 1.29	1.30
	100	2.72 ± 28	2.66 ± 115	98 ± 3.77	3.86
	150	3.39 ± 81	3.35 ± 146	99 ± 1.95	1.98

Table 2: Recovery studies.

Parameter	Average amount detected (µg ± SD), % RSD		
	Acetylbarlerin	Barlerin	Shanzhiside methyl ester
Reproducibility (*n*=9)	2.12 ± 11, 0.51	0.56 ± 11.0, 1.98	1.20 ± 2, 0.15
Manual precision Intra-day Inter-day	2.12 ± 33, 1.56 2.13 ± 6, 0.29	0.55 ± 17.0, 3.03 0.55 ± 9.7, 1.75	1.20 ± 20, 2.04 1.201 ± 14, 1.15
Instrumental precision Intra-day Inter-day	2.12 ± 40, 1.89 2.12 ± 31, 1.47	0.56 ± 13.8, 2.49 0.56 ± 5.91, 1.06	1.19 ± 28, 2.34 1.20 ± 28, 2.37

Table 3: Reproducibility and precision studies.

no interference of matrix was observed for any of the markers. The method was also linear in a concentration range of 1.42-4.95 µg/spot for acetylbarlerin, 0.28-1.67 µg/spot for barlerin and 0.60-3.60 µg/spot for shanzhiside methyl ester (n=6), with respect to peak area. The linear regression data shown in (Table 1) using SPSS software with 95% confidence limits revealed a good linear relationship over the concentration range studied demonstrating its suitability for analysis. Results of accuracy from recovery studies as shown in (Table 2), showed the method accuracy in the desired range. The results of reproducibilty, intra-day and inter-day precision are shown in (Table 3) and robustness of the method for acetylbarlerin, barlerin and shanzhiside methyl ester is shown in (Table 4) respectively.

Analysis of the selected iridoids in Barleria species/varieties - Barleria should be italicised

The quantitative analysis of the three selected markers, *viz.*, acetylbarlerin, barlerin and shanzhiside methyl ester was done in various *Barleria* species and extracts using the technique of TLC-densitometry. Comparative chromatograms (Figure 4) of various species and varieties of *Barleria* were prepared and analyzed for the per cent content of three selected iridoids. The details are given in (Table 5). The estimation of the three iridoids was accomplished in all the parts and partitioned fractions for two abundantly growing species of *Barleria* i.e. *B. prionitis* and *B. cristata* var. *dichotoma*. In case of *B. lupulina*, the quantitative estimation was restricted only to different parts due to limited quantity of the available material.

There was no interference in analysis of the three shanzhiside esters from any other components present in the matrix. The three iridoids resolve in the chromatogram at significantly different R_f values. The highest content of acetylbarlerin was found to be 3.82%, w/w in leaf of *B. lupulina*, that of barlerin was 0.97%, w/w in stem of *B. prionitis* and shanzhiside methyl ester was 2.62% w/w in leaf of *B. prionitis*. The iridoids were found to be absent in two varieties of *B. cristata* (pink and blue flower variety) as indicated by flat chromatogram in (Figure 4). Statistical evaluation of the results was performed with respect to accuracy and precision using SPSS software (version 16) at 95% confidence limit. The low % RSD value indicated the suitability of this method for routine analysis of shanzhiside methyl esters.

The HPTLC method for simultaneous analysis of three shanzhiside esters has been developed and validated. As of now, HPLC is the most common reported method for the analysis of these esters. The two reports on HPTLC technique employ estimation of single or two markers, but the methods are not validated. Moreover the findings of Ghule et al. [29] report very high content of barlerin and shanzhiside methyl ester indicating interference from other components and hence cannot be considered for analysis. The HPTLC method in present study may provide a suitable alternative for the reported HPLC technique by providing fast and reliable analysis. The method will be of special interest in resource constrained countries and laboratories of Asia and Africa, where large number of species and varieties of this plant are known to grow in abundance.

Parameter	Amount spotted (µg)	Average amount detected (µg ± SD)	% RSD
(A) Acetylbarlerin			
Mobile phase composition 3:3.1:2.9:1		2.13 ± 51	2.39
Mobile phase composition 3:3:3:1		2.12 ± 43	2.01
Mobile phase composition 3.2:2.8:3:1		2.12 ± 51	2.40
Mobile phase volume 8 ml		2.12 ± 54	2.53
Mobile phase volume 10 ml		2.12 ± 43	2.01
Mobile phase volume 12 ml		2.13 ± 39	1.86
Chamber saturation time: 5 min		2.12 ± 57	2.69
Chamber saturation time: 10 min	2.12	2.12 ± 43	2.01
Chamber saturation time: 15 min		2.12 ± 55	2.59
Solvent migration distance: 78 mm		2.12 ± 53	2.51
Solvent migration distance: 80 mm		2.12 ± 43	2.01
Solvent migration distance: 82 mm		2.12 ± 39	1.84
(B) Barlerin			
Mobile phase composition 3:3.1:2.9:1		0.56 ± 15.1	2.72
Mobile phase composition 3:3:3:1		0.56 ± 9.4	1.69
Mobile phase composition 3.2:2.8:3:1		0.55 ± 9.6	1.74
Mobile phase volume 8 ml		0.56 ± 7.9	1.43
Mobile phase volume 10 ml		0.55 ± 9.4	1.69
Mobile phase volume 12 ml		0.56 ± 15.5	2.80
Chamber saturation time: 5 min		0.55 ± 11.6	2.11
Chamber saturation time: 10 min	0.56	0.56 ± 9.4	1.69
Chamber saturation time: 15 min		0.56 ± 14.2	2.56
Solvent migration distance: 78 mm		0.56 ± 11.1	2.00
Solvent migration distance: 80 mm		0.55 ± 9.4	1.69
Solvent migration distance: 82 mm		0.55 ± 10.8	1.95
(C) Shanzhiside methyl ester			
Mobile phase composition 3:3.1:2.9:1		1.19 ± 32	2.71
Mobile phase composition 3:3:3:1		1.20 ± 18	1.54
Mobile phase composition 3.2:2.8:3:1		1.19 ± 38	3.14
Mobile phase volume 8 ml		1.20 ± 38	3.16
Mobile phase volume 10 ml		1.20 ± 18	1.54
Mobile phase volume 12 ml		1.19 ± 26	2.16
Chamber saturation time: 5 min		1.20 ± 29	2.38
Chamber saturation time: 10 min	1.20	1.20 ± 18	1.54
Chamber saturation time: 15 min		1.19 ± 39	3.21
Solvent migration distance: 78 mm		1.20 ± 34	2.82
Solvent migration distance: 80 mm		1.20 ± 18	1.54
Solvent migration distance: 82 mm		1.19 ± 37	3.11

Table 4: Robustness of the method.

Conclusion

The HPTLC method was developed for the simultaneous estimation of three shanzhiside esters (iridoid markers) *viz.*, acetylbarlerin, barlerin and shanzhiside methyl ester. The developed method is simple, sensitive, precise, robust, specific, accurate and reliable for the determination of shanzhiside methyl ester iridoids. Statistical analysis indicates that the method is repeatable and selective for the simultaneous analysis of these three biologically active iridoids in different species/varieties of *Barleria*. The method was validated to ensure that it fits well the intended purpose and meet the strict regulatory requirements for analysis of herbal drugs. It is expected that the proposed method would also be useful for comparing and differentiating more *Barleria* or related plant species especially in drug discovery programmes. Interestingly, the method can detect and quantify very low concentration of iridoids to high per cent content and can be extended to study the degradation of shanzhiside esters under different stress conditions as per recommendation in ICH guidelines. The proposed method will definitely facilitate the drug quality assurance in such laboratories/ countries where more sophisticated and costly analytical equipments are insufficiently available especially those of Asia and Africa where this plant is known to grow in abundance.

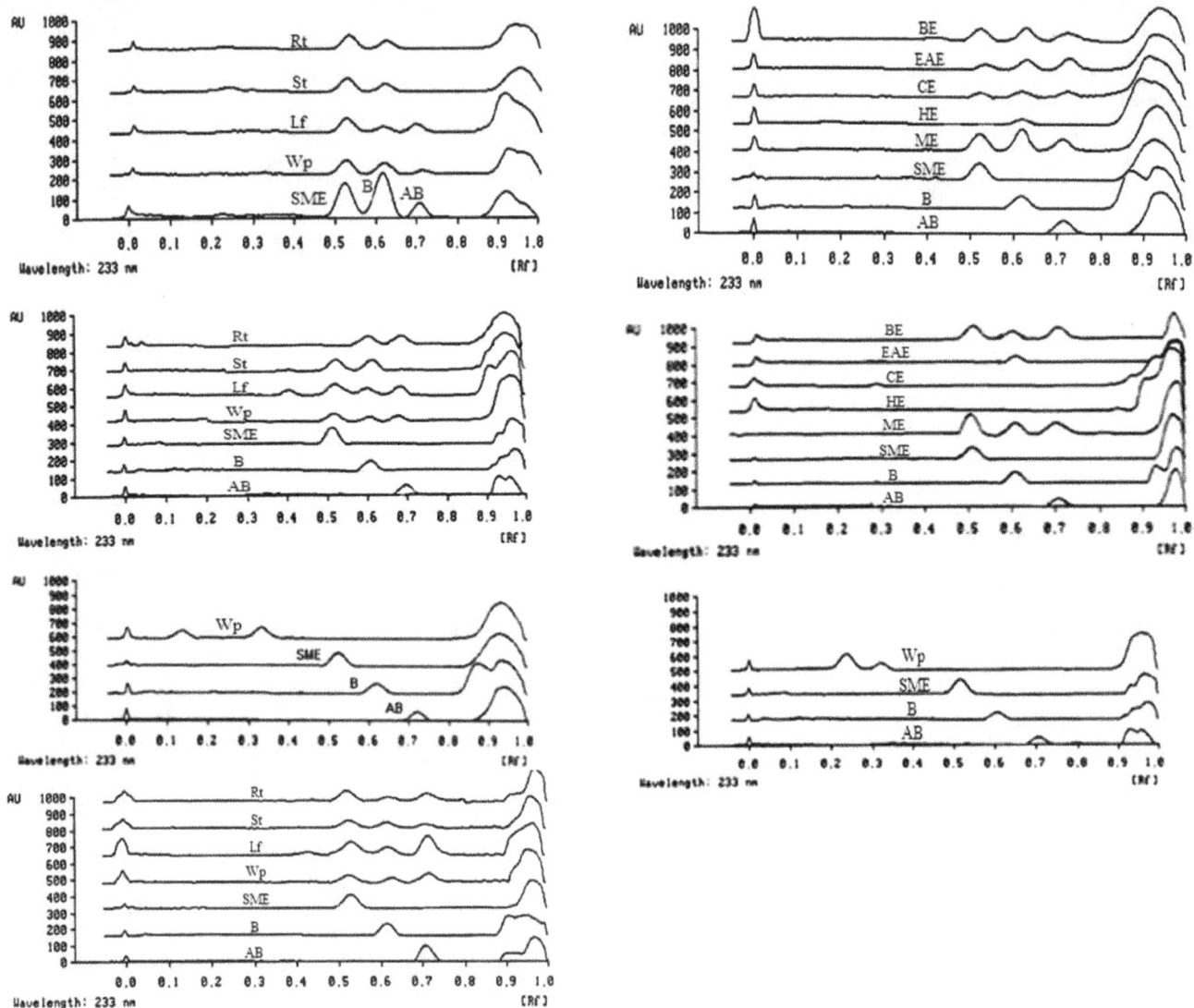

Figure 4: TLC densitometric chromatograms of shanzhiside methyl esters in different *Barleria* species/varieties: **A.** plant parts of *B. prionitis*; **B.** various fractions of methanolic extract of *B. prionitis*; **C.** plant parts of *B. cristata* var. *dichotoma*; **D.** various fractions of methanolic extract of *B. cristata* var. *dichotoma*; **E.** whole plant of *B. cristata* (pink flower variety); **F.** whole plant of *B. cristata* (blue flower variety); **G.** plant parts of *B. lupulina*; **AB:** acetylbarlerin; **B:** barlerin; **SME:** shanzhiside methyl ester; **Wp:** whole plant; **Lf:** leaf; **St:** stem; **Rt:** root; **ME:** mother extract; **HE:** hexane soluble fraction of mother extract; **CE:** chloroform soluble fraction of mother extract; **EAE:** ethyl acetate soluble fraction of mother extract; **BE:** butanol soluble fraction of mother extract.

S. No.	Sample	Acetylbarlerin	Barlerin	Shanzhiside methyl ester
1.	**B. prionitis**			
	Whole plant	0.35	0.81	1.36
	Leaf	0.95	0.25	2.62
	Stem	ND	0.97	1.46
	Root	ND	0.96	1.35
	Hexane soluble fraction	ND	0.02	ND
	Chloroform soluble fraction	0.28	0.08	0.11
	Ethyl acetate soluble fraction	1.28	1.10	1.00
	Butanol soluble fraction	1.68	2.82	3.92
2.	**B. cristata var. dichotoma**			
	Whole plant	0.26	0.18	0.52
	Leaf	0.46	0.18	0.70
	Stem	ND	0.23	0.88
	Root	0.16	0.14	ND
	Hexane soluble fraction	ND	ND	ND
	Chloroform soluble fraction	ND	ND	ND
	Ethyl acetate soluble fraction	ND	0.21	ND
	Butanol soluble fraction	0.89	0.59	1.28

3.	*B. cristata* (Pink flower variety) Whole plant	ND	ND	ND
4.	*B. cristata* (Blue flower variety) Whole plant	ND	ND	ND
5.	*B. lupulina* Whole plant Leaf Stem Root	2.20 3.82 0.43 0.72	0.64 0.73 0.59 0.45	0.91 1.30 0.80 1.03

ND: not detected

Table 5: Per cent content of shanzhiside esters in different *Barleria* species/varieties.

Conflict of Interest

There are no conflicts of interest.

Acknowledgement

One of the authors, Preet Kawal is grateful to University Grants Commission, New Delhi for the award of fellowship under Research Fellowship for Meritorious Students scheme.

References

1. Pandey G (2004) Saireyaka. In: Dravyaguna Vijnana : Materia Medica Vegetable Drugs. Krishnadas Ayurvedic series, Chaukhambla Orientalia, Varanasi.

2. Sharma PV (1998) Barleria prionitis. In: Dhanvantari-Nighantu. Jaikrishnadas series, Chaukhambla Orientalia, Varanasi.

3. Shendage SM, Yadav SR (2010) Revision of the genus, Barleria (Acanthaceae) in India. Rheedea 20: 81-130.

4. Anonymous (1948) Barleria prionitis. In: The Wealth of India, A Dictionary of Indian Raw Materials and Industrial Products. Council of Scientific and Industrial Research, New Delhi.

5. Hooker JD (1885) Barleria prionitis. In: The Flora of British India. The authority of the secretary of state for India in council. L. Reeve and Co. Ltd.

6. Chopra RN, Nayar SL, Chopra IC (1956) Barleria prionitis. In: Glossary of Indian Medicinal Plants. Council of Scientific and Industrial Research, New Delhi.

7. Anonymous (2004) Barleria. In: Indian Medicinal Plants. Indian Council of Medical Research, New Delhi.

8. Geeta M, Wahi AK (1997) Identity of root of *Barleria prionitis* Linn. Indian J Nat Prod 13: 14-16.

9. Shastry JLN (2005) Barleria prionitis. In: Dravyaguna Vijnana. Study of Essential Medicinal Plants in Ayurveda. Chaukhambha Orientalia, Varanasi.

10. Suba V, Murugesan T, Kumaravelrajan R, Mandal SC, Saha BP (2005) Antiinflammatory, analgesic and antiperoxidative efficacy of Barleria lupulina Lindl. extract. Phytother Res 19: 695-699.

11. Kirtikar KR, Basu BD (1956) Barleria prionitis. In: Bishen Singh, Mahendra Pal Singh (eds) Indian Medicinal Plants. (2ndedn), Council of Scientific and Industrial Research, New Delhi.

12. Singh VK, Ali ZA (1989) Folk medicines of Aligarh (Uttar Pradesh), India. Fitoterapia 64: 483-489.

13. Borthakur SK, Nath K, Gogoi P (1996) Herbal remedies of the Nepalese of Assam. Fitoterapia LXVII: 231.

14. Anonymous (2002) Sahacara. In: (1stedn), The Ayurvedic Pharmacopoeia of India. Department of Indian System of Medicine and Homeopathy, Government of India, Ministry of Health and Family Welfare, New Delhi.

15. El-Naggar LJ, Beal JL (1980) Iridoids. A review. J Nat Prod 43: 649-707.

16. Boros CA, Stermitz FR (1990) Iridoids - An updated review Part I. J Nat Prod 53: 1055-1147.

17. Dinda B, Chowdhury DR, Mohanta BC (2009) Naturally occurring iridoids, secoiridoids and their bioactivity. An updated review, part 3. Chem Pharm Bull (Tokyo) 57: 765-796.

18. Dinda B, Debnath S, Harigaya Y (2007) Naturally occurring iridoids. A review, part 1. Chem Pharm Bull (Tokyo) 55: 159-222.

19. Dinda B, Debnath S, Harigaya Y (2007) Naturally occurring secoiridoids and bioactivity of naturally occurring iridoids and secoiridoids. A review, part 2. Chem Pharm Bull (Tokyo) 55: 689-728.

20. Dinda B, Debnath S, Banik R (2011) Naturally occurring iridoids and secoiridoids. An updated review, part 4. Chem Pharm Bull (Tokyo) 59: 803-833.

21. Premjet D, Premjet S, Lelono RAA, Tachibana S (2010) Callus induction and determination of iridoid glycosides from Barleria prionitis Linn leaf explants, Aust. J Basic Appl Sci 4: 4461-4467.

22. Singh B, Chandan BK, Prabhakar A, Taneja SC, Singh J, et al. (2005) Chemistry and hepatoprotective activity of an active fraction from Barleria prionitis Linn. in experimental animals. Phytother Res 19: 391-404.

23. Cheng J, Sony Y, Li P (2007) Capillary high-performance liquid chromatography with mass spectrophotometry for simulataneous determination of major flavonoids, iridoid glycosides and saponins in Flos Lonicerae. J Chromatogr A 1157: 217-226.

24. Yalcin FN, Kaya D, Calis T, Ersoz T, Palaska E (2008) Determination of iridoid glycoside from four Turkish Lamium species by HPLC-ESI/MS. Turk J Chem 32: 457-467.

25. Li C, Xue X, Zhou D, Zhang F, Xu Q, et al. (2008) Analysis of iridoid glucosides in Hedyotis diffusa by high-performance liquid chromatography/electrospray ionization tandem mass spectrometry. J Pharm Biomed Anal 48: 205-211.

26. Krzek J, Janeezko Z, Walusiak O, Podolak I (2002) Densitometric determination of aucubin in syrups in the presence of other iridoids - An approach to standardization J Planar Chrom 15: 196-199.

27. Suomi J, Wiedmer SK, Jussila M, Riekkola ML (2002) Analysis of eleven iridoid glycosides by micellar electrokinetic capillary chromatography (MECC) and screening of plant samples by partial filling (MECC)-electrospray ionization mass spectrometry. J Chromatogr A 970: 287-296.

28. Anonymous (2006) Barleria prionitis. In: Quality Standards of Indian Medicinal Plants. Indian Council of Medical Research New Delhi.

29. Ghule BV, Yeole PG (2012) In vitro and in vivo immunomodulatory activities of iridoids fraction from Barleria prionitis Linn. J Ethnopharmacol 141: 424-431.

30. Ghule BV (2012) Validated HPTLC Method for simultaneous determination of Shanzhiside methyl ester and barlerin in B prionitis Linn, J Planar Chromatogr 25: 426-432.

31. Karan M, Kaur I, Chopra D, Vasisht K (2010) Anti-inflammatory activity of the iridoid glycosides. Indian Patent Application no 2182/DEL/2008.

8

Environmental Fate of Two Psychiatric Drugs, Diazepam and Sertraline: Phototransformation and Investigation of their Photoproducts in Natural Waters

Jakimska A[1]*, Śliwka Kaszyńska M[2], Nagórski P[1], Kot Wasik A[1] and Namieśnik J[1]

[1]*Department of Analytical Chemistry, Faculty of Chemistry, Gdańsk University of Technology (GUT), G. Narutowicza Street 11/12, 80-233 Gdańsk, Poland*
[2]*Department of Organic Chemistry, Faculty of Chemistry, Gdańsk University of Technology (GUT), G. Narutowicza Street 11/12, 80-233 Gdańsk, Poland*

Abstract

Experimental studies were conducted to investigate the photodegradation of diazepam and sertraline, two of the most frequently used psychiatric drugs, induced by xenon lamp irradiation, which overlap the sunlight spectra, and natural sunlight. Degradation kinetics was established indicating the occurrence of autocatalytic reactions. The application of liquid chromatography coupled to quadrupole time-of-flight mass spectrometry allowed for accurate mass measurements and proper fragmentation of target compounds enabling structure elucidation of various photoproducts of diazepam and sertraline formed during the forced photolysis. As a result, phototransformation pathways were proposed including the compounds and routes of degradation described here for the first time. In the final step, the presence of parent and identified compounds was examined in various environmental water samples. Although, the analytes were non-detected it was assumed that they may undergo different process in the environment such as adsorption, dilution, advection or rapid photodegradation of all compounds.

Keywords: Pharmaceuticals; Photolysis; Aqueous environment; Transformation products; Degradation kinetics

Abbreviations: DIZ: Diazepam; LC-HRMS: Liquid Chromatography-High Resolution Mass Spectrometry; MeOH: Methanol; PhAC: Pharmaceutically Active Compound; SER: Sertraline; SPE: Solid Phase Extraction; SSRI: Serotonin Reuptake Inhibitor; TP: Transformation Product; UHPLC-QTOF-MS: Ultra-High Pressure Liquid Chromatography–Quadrupole Time-of-Flight-Mass Spectrometry; WTP: Water Treatment Plant; WWTP: Waste Water Treatment Plant

Introduction

Pharmaceutically active compounds (PhACs) are considered as emerging contaminants, which are not regulated, and have raised a great concern in recent years. The major contributors of these pollutants into the water bodies are wastewater treatment plants (WWTPs), where PhACs are not effectively removed during conventional treatment process [1-3]. It is not unambiguously stated what adverse effects PhACs can cause once they reach natural waters. However, it is suspected that such phenomena as increased bacterial drug resistance [4] genotoxicity [5] or endocrine disruption [6,7] origin from chronic exposure to low doses of PhACs present in aquatic environment.

While various techniques of wastewater treatment or drinking water disinfection are applied, many transformation products (TPs) may be formed. TPs can have the same impact on human health and the environment as the parent pharmaceutical compound; however, they can also be more or less toxic depending on the transformation pathway (possessing or losing the toxicophores). It is also feasible, that the transformation product will receive 'new' toxicophore, thus the toxicity of the compound will increase [8]. In this context, it is important to investigate the transformation pathway of various pharmaceuticals in order to track their level and estimate their influence on the environment and living organisms.

Many processes such as sorption, biodegradation, volatilization and photolysis may contribute to PhACs degradation [9]. Due to the pseudo-persistence of many drugs, biochemical reactions have a limited share in the removal of such pollutants [10,11]. It is suspected that in the superficial layer of water bodies photolysis is a main process responsible for the transformation of PhACs [12,13]. In direct photolysis the pollutant, which contain aromatic rings, heteroatom and other functional groups in the structure, can absorb a photon leading to its degradation. In indirect photolysis, the degradation occurs via reaction with active molecules e.g. •OH, $^1O_2$, ROO•, $^3DOM*$, e^-_{aq} (generated by photosensitizers such as dissolved organic matter) [9,14,15].

The present work focuses on the photodegradation of two selected psychiatric drugs, sertraline and diazepam. Sertraline is an anti-depressant classified as a second generation serotonin reuptake inhibitor (SSRIs). It is also widely used in the treatment of depression, panic disorders, social phobia, obesity and obsessive-compulsive disorders [16-18]. Diazepam is the most frequently prescribed benzodiazepine used in the treatment of anxiety, insomnia, seizures and alcohol withdrawal, and is commonly known as Valium [19]. The use of psychiatric medications and anti-depressants is rapidly increasing and they become the most widely distributed psychiatric drugs in the world. Therefore, these types of pharmaceuticals are of a great concern since their occurrence in the aquatic environment has been reported in various studies [20-23].

Considering that residues of these drugs have been found in both water bodies and aquatic organisms, the aim of this work was to perform forced degradation under the xenon lamp irradiation in

*Corresponding author: Jakimska A, Department of Analytical Chemistry, Faculty of Chemistry, Gdańsk University of Technology (GUT), G. Narutowicza Street 11/12, 80-233 Gdańsk, Poland, E-mail: anik.jakimus@interia.pl

order to identify photoproducts of diazepam and sertraline in water samples. For this purpose liquid chromatography - high resolution mass spectrometry (LC-HRMS) was applied due to its accurate mass measurements, possibility to obtain the isotope pattern and the fragmentation pathway of selected compound. Due to these advantages of this technique the presented results are of a high accuracy and certainty. Furthermore, degradation kinetics in various matrices have been investigated. In the final step different water samples were analyzed for the presence of the parent compounds and their photoproducts. To the best of our knowledge this is the most extensive study on diazepam and sertraline photo-degradation induced by xenon lamp,which simulates natural process occurring in the environment. The performed experiments enabled proposing photodegradation pathways and certain transformations are reported here for the first time.

Materials and Methods

Sample collection and preparation

Influent and effluent water samples from municipal wastewater treatment plant (WWTP) were collected in Gdańsk, Poland. The wastewater treatment in the facility was based on activated sludge process with nitrification and partial denitrification. Treated and untreated water samples from drinking water treatment plant (WTP) were collected in Gdańsk, Poland. The facility employed partial chlorination and ozonation during water treatment processes. River water samples were collected from two rivers (Radunia and Kacza) exposed to anthropogenic activity located in Pomerania area. The average pH were 6.9 for wastewater influent, 7.5 for wastewater effluent, 7.4 for untreated water, 8.2 for treated water and 7.9 for river water. All kinds of samples for the analysis of the presence of DIZ, SER and their photoproducts were collected in 5 different months (from February to June) and were stored in the dark at 4°C prior to the analysis.

Water samples were prepared using solid phase extraction (SPE) with the application of Oasis HLB cartridges (500 mg, 6 mL; Waters). Briefly, SPE cartridges were washed with MeOH and ultrapure water. Then, a certain volume of water sample containing 1 mL 0.1 M EDTA/100 mL sample was loaded on the sorbent at a flow rate of 5 mL/min. After the sorbent was dried, the analytes were eluted with 6mL of MeOH. The extracts were evaporated to dryness in a stream of nitrogen and dry residues were redissolved in 1 mL of MeOH/H$_2$O (1:9, v/v). 5 µL of the final extract was injected to UHPLC-QTOF-MS system. Detailed procedure has been previously described [24].

Reagents

Analytical standards of diazepam (DIZ) and sertraline (SER) were purchased from LGC Standards (Łomianki, Poland) and Sigma-Aldrich (Poznań, Poland). LC-MS grade methanol (MeOH) was purchased from Sigma-Aldrich (Poznań, Poland). Formic acid was from Merck (Warsaw, Poland). Ultrapure water was obtained from a HLP5 system (Hydrolab, Poland). Individual stock solutions of DIZ and SER at concentration level of 1 g/L were prepared in methanol and stored at -80°C.

Irradiation experiments

Degradation kinetic studies of DIZ and SER (at concentration level of 1 mg/L) were investigated in two types of conditions: (1) under simulated solar radiation using xenon lamp and (2) under natural solar radiation. Xe-lamp emits concentrated and constant light spectrum which range overlaps the range of the solar irradiation, and allows performing the experiment independently from the external factor, e.g. weather conditions. While exposing the samples on the solar irradiation it is possible to simulate conditions similar to those in the natural environment.

Solar irradiation experiments were performed for eight different samples: wastewater influent and effluent, untreated and treated water, river water, methanol, ultrapure water for pH 3 and pH 10. The extracts were analyzed by UHPLC-QTOF-MS system every two weeks until the decrease of the concentration below 10% of the initial value.

Xenon lamp irradiation experiments of river water samples containing DIZ or SER (separate experiments) were performed using a small-scale system consisting of a cylindrical photoreactor (25 mL), cooling system, external light system and optical filters. The temperature of the samples during the experiment was maintained at 10°C. The equipment was previously applied for identification of phototransformation products of other pharmaceuticals in river water samples [24,25] and by Górska et al. [26] for photodegradation of phenol.

Identification of photodegradation products in river water samples was performed during the Xe-lamp exposure in order to observe only the photoproducts formed during the irradiation and reducing the possibility ofthe occurrence of other processes (e.g. hydrolysis or biodegradation)to a minimum.

In case of the samples exposed to Xe-lamp light, around 500 µL of a sample was collected during the experiment every 0.5 h (SER) or 1 h (DIZ) until the total degradation of the parent compound. The irradiation by xenon lamp lasted for 4 h (SER) and 18 h (DIZ) that is, until the decrease of the parent analyte concentration below 5%. Each extract was filtered through a 0.45µm nylon syringe filter (Rockwood, USA) and analyzed immediately by UHPLC-QTOF-MS system for further data evaluation.

Analytical methods

The UHPLC-QTOF-MS analysis were carried out using an Agilent 1290 UHPLC system coupled to a hybrid quadrupole time-of-flight (QTOF) mass spectrometer (Agilent 6540 Series Accurate Mass QTOF-MS) with Dual ESI interface operated in positive ion mode. Detailed chromatographic conditions and mass spectrometer parameters were previously described in our publication [24]. Briefly, the separation of analytes was performed on XTerra MS C18 column (30 mm×2.1 mm; 3.5µm) maintained at 22°C. The mobile phase A was ultrapure water with 0.05% FA and B was MeOH and the flow rate was set to 0.4 ml/min. The analysis were carried out in gradient elution (starting from 95% A to 0% A in 15 min, back to 95% A in 1 min and kept at 95% A for 5 min) and the injection volume was 5µL. The QTOF-MS conditions were as follows: sheath gas temperature 400°C at the flow rate of 12 L/min, capillary voltage 4000 V, nebulizer pressure 20 psig, drying gas 10 L/min, gas temperature 325°C, skimmer voltage 45 V, octopole RF peak 750 V and fragment or voltage 100 V. Analysis were performed in two different modes, MS/MS or TargetMS/MS with various collision energies (10, 20 and 30 V) and the masses were scanned from 50 to 1000 m/z. The instrument was working in the 4GHz high-resolution mode with the acquisition rate of 1.5 spectra/s. Acquisition data were processed using Agilent Mass Hunter Workstation software.

Results and Discussion

Degradation kinetics

Degradation kinetics were estimated for two types of experiments: (1) during the exposure to xenon lamp radiation, and (2) during the exposure to solar radiation in order to simulate natural processes occurring in the environment. Both experiments lasted until the decrease of DIZ and SER concentration below 10% of the initial value. Based on the values of the coefficient of determination (R^2) the degradation kinetics in all samples followed a pseudo-first-order kinetic and the results are presented in Table 1. The time-based pseudo-first-order rate constant (k) was calculated according to the Eq. (1):

$$ln\frac{C_t}{C_0} = -kt \qquad (1)$$

where C_t is the concentration of the analyte at the specific irradiation time t and C_0 is the initial concentration of the compound.

Half-life $t_{1/2}$ was calculated using Eq. (2) which was derived by transforming Eq. (1) and replacing C_t with $C_0/2$:

$$t_{1/2} = \frac{ln2}{k} \qquad (2)$$

The photodegradation rate constants k and the half-lives $t_{1/2}$ for xenon lamp experiment were 0.001 min^{-1} and 9 h for DIZ, and 0.009 min^{-1} and 1.3 h for SER.

The photodegradation kinetics of DIZ and SER induced by solar radiation was investigated in eight different matrices: wastewater influent and effluent, untreated and treated water, river water, methanol, ultrapure water for pH 3 and pH 10 in order to estimate the influence of the type of the matrix and the sample pH on the degradation rate of selected compounds.

Autocatalytic degradation reactions of DIZ and SER have been observed in certain matrices. In such case the initial concentration of the compounds is constant (so-called. delay time) until the specific degradation product is formed in a certain amount that triggers the degradation process running consecutively as the first-order kinetic reaction. An example of such autocatalytic reaction is presented in Figure 1. However, autocatalytic reactions did not occur for all analyzed matrices and was strictly dependent on the compound, the conditions and complexity of the sample composition. The autocatalytic effect of diazepam was observed in all samples, except for wastewater effluent which matrix has a very complex composition, and water of acidic pH (~3), which in this case, significantly accelerates the degradation process. For sertraline this effect was observed in samples with matrices containing a small amount of interfering substances (with a simpler composition), which means that the formation of desired transformation products occurs later than in case of other samples and significantly affects the initiation of the sertraline degradation. The half-life times $t_{1/2}$ presented in Table 1 are a sum of 'delay time' (given in square brackets) and real degradation time. As it can be noticed, the 'delay time' is quite long for most of the matrices were it was observed and after it was finished the degradation proceeded rapidly comparatively to the conditions were autocatalysis was not observed.

The photodegradation rate constants k varied from 0.007 to 0.158 day^{-1} for DIZ and from 0.005 to 0.140 day^{-1} for SER. The half-lives $t_{1/2}$ were in range from 4.4 (pH 3) to 102.1 days (wastewater influent) for DIZ and from 4.9 (wastewater influent) to 129 days (methanol) for SER. These results indicate various behaviors of these two psychiatric drugs in considered matrices which may be related to their structures and the presence of different functional groups.

Photodegradation of selected pharmaceuticals

Identification of photodegradation products of DIZ and SER was performed during the exposition of the water extract to Xe-lamp light. The advantage of this approach was independence of various external factors which may negatively affect the experiment. Furthermore, xenon lamp emits the light in the range 250-1000 nm which overlaps the spectrum of the light reaching the Earth. Both direct and indirect photolysis may occur simultaneously due to the constant radiation provided by the lamp.

In order to identify photoproducts of DIZ and SER, samples collected at various time intervals were analyzed by UHPLC-QTOF-MS. Preliminary identification of the probable photoproducts was based on accurate mass measurements of selected (pseudo)molecular ions with a mass error below 5 ppm. Further accurate MS/MS measurements (mass error below 10 ppm) were performed in order to obtain characteristic fragmentation ions necessary to elucidate the structures of suspected photoproducts and the transformation pathways of diazepam and sertraline.

The irradiation of psychiatric drugs solutions resulted in the formation of several chromatographic peaks corresponding to 9 and 7 new compounds for DIZ and SER, respectively (Figure 2a,b). Detailed results are presented in the next sections.

Identification of photoproducts of DIZ: Phototransformation of the widely used benzodiazepine pharmaceuticals diazepam under simulated sunlight in water has been investigated. A total of nine photoproducts, including benzophenones, acridinones and quinazolinones or quinazolines have been identified. All photoproducts, with only two exceptions, have lower molecular weight than the parent compound and photo-induced dimerization was not observed. On the

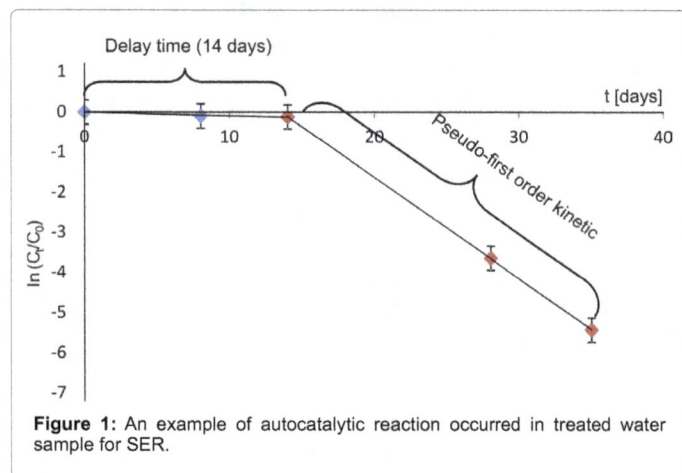

Figure 1: An example of autocatalytic reaction occurred in treated water sample for SER.

Matrix	Solar irradiation			
	DIZ		SER	
	k (day^{-1})	$t_{1/2}$(day)	k (day^{-1})	$t_{1/2}$(day)
Wastewater influent	0.018	38.3	0.100	6.9
Wastewater effluent	0.007[a]	102.1 [a][85][b]	0.140	4.9
River water	0.007[a]	94.5 [a][85][b]	0.100	6.9
Untreated water	0.007[a]	94.5 [a][85][b]	0.067	10.3
Treated water	0.007[a]	94.4 [a][85][b]	0.041[a]	16.8 [a][14][b]
Methanol	0.009[a]	72.9	0.005[a]	129.0 [a][120][b]
Ultrapure water pH 3	0.158	4.4	0.005[a]	127.5 [a][120][b]
Ultrapure water pH 10	0.009[a]	73.5 [a][49][b]	0.071	9.7
	Xe-lamp irradiation			
River water	k (min^{-1})	$t_{1/2}$(min)	k (min^{-1})	$t_{1/2}$(min)
	0.001	540.2	0.009	76.5

[a] autocatalytic reaction
[b] 'delay time' duration

Table 1. Photodegradation rate constants (k) and half-lives ($t_{1/2}$) of DIZ and SER in various matrices exposed to solar or Xe-lamp irradiation.

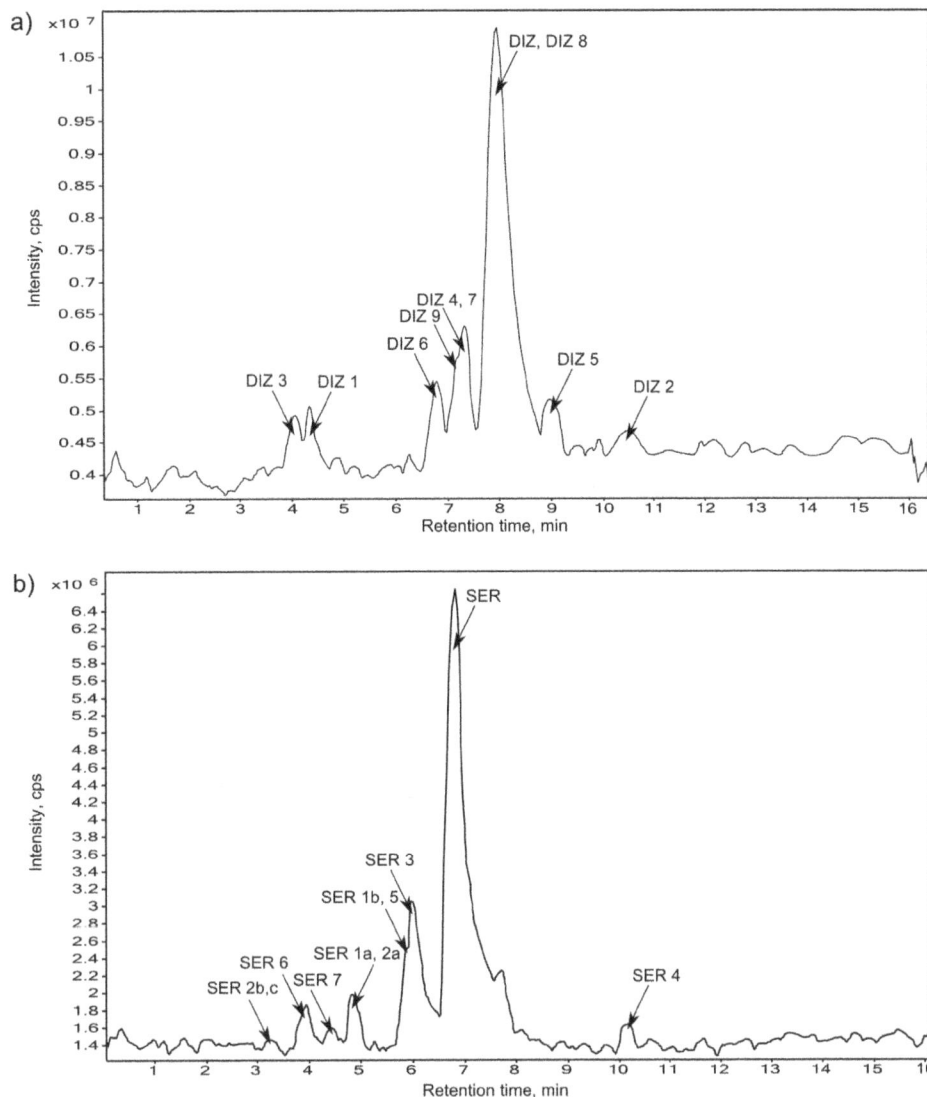

Figure 2: Chromatograms obtained during the photodegradation experiments for DIZ (a) and SER (b).

basis of the identified photoproducts (Table 2) the degradation pathway of diazepam has been proposed, and schematically presented in Figure 3.

Xe-lamp irradiation of Diazepam solution caused after 20 minutes formation of two by-products of its degradation (DIZ 1 and DIZ 2). The photoproduct with the m/z 303.0880 may correspond to mono-hydroxylated derivative of diazepam molecule followed by addition of two hydrogen atoms (reduction of imine fragment). The product DIZ 1 was completely degraded in 180 minutes by photolytic irradiation. The structure of this compound was confirmed by the MS/MS spectrum obtained during the fragmentation of DIZ 1 (Figure 4). There are three characteristic fragment ions of m/z 285.0770, 239.0379 and 167.0368 which correspond to the loss of specific groups as -H_2O (formation of diazepam protonated molecule), -$C_2H_6O_2$ and -$C_8H_8O_2$, respectively. The formation of this photoproduct was not reported in the literature before.

The second product DIZ 2 (m/z 246.0690) was identified as the 5-chloro-2-methylaminobenzophenone by comparison with the mass spectra of an authentic standard of 5-chloro-2-

methylaminobenzophenone known in the literature and reported by West et al. [19]. This compound has also been identified during the exposure of methanolic solutions of diazepam to UV radiation [27]. The structure of this compound was elucidated on the basis of fragmentation pattern obtain during the LC-QTOF-MS analysis in MS/MS mode, which follows by the formation of radicals and it was previously noticed by West et. al. [19].

Continued irradiation (40 min) of Diazepam water solution revealed formation of another degradation product (DIZ 3, m/z 273.0777). Based on its characteristic mass spectra it has been identified as (6-chloro-4-phenyl-1,2-dihydroquinazolin-2-yl)methanol-photoproduct already known in the literature [19]. Photodegradation product DIZ 4 with m/z 289.0736, appeared after 150 minutes of Xe-lamp irradiation of Diazepam. Its formation probably resulted from the opening of the diazepinone seven-membered ring followed by a rearrangement into [4-chloro-2-(phenyl carbonoimidoyl) phenyl] methylcarbamic acid as. This photoproduct is also known in the literature and recent has been reported by West et al.[19].

Name		Structure	Retention time, min	Molecular formula	[M+H]$^+$theoret.	[M+H]$^+$experim.	Error, ppm	Fragment ions m/z (relative intensity)
Diazepam (DIZ)		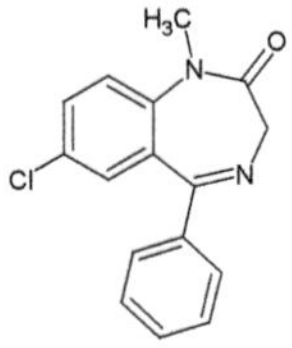	8.6	$C_{16}H_{13}ClN_2O$	285.0789	285.0791	-0.7	285 (100), 154 (25), 257 (19), 193 (17), 222 (17), 228 (15), 182 (5)
DEGRADATION PRODUCTS	1	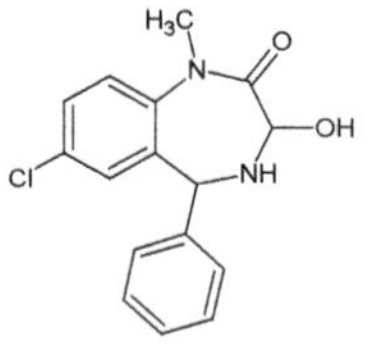	4.5	$C_{16}H_{15}ClN_2O_2$	303.0895	303.0880	4.9	167 (100), 239 (72), 285 (40), 303 (36)
	2	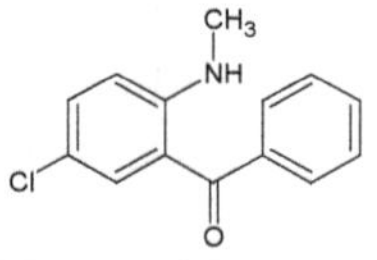	10.3	$C_{14}H_{12}ClNO$	246.0680	246.0690	-4.1	246 (100), 195 (50), 228 (42), 177 (40), 149 (37)
	3	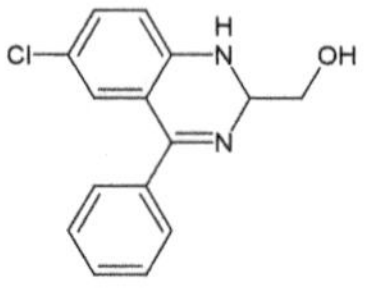	4.2	$C_{15}H_{13}ClN_2O$	273.0789	273.0777	4.4	158 (100), 273 (80), 196 (13), 228 (12)
	4	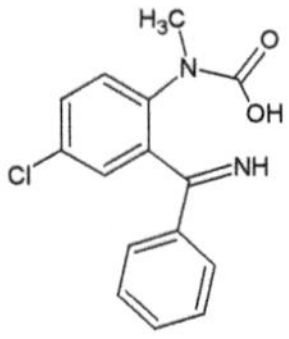	7.6	$C_{15}H_{13}ClN_2O_2$	289.0738	289.0736	0.7	289 (100), 156 (27), 260 (19), 223 (17), 230 (17), 193 (15)
	5	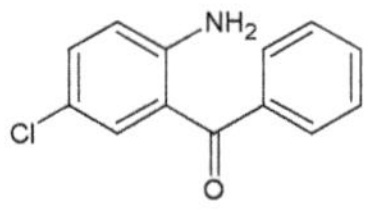	8.9	$C_{13}H_{10}ClNO$	232.0524	232.0533	-3.9	232 (100), 129(48), 158 (44), 117 (40) 214 (36), 191 (32), 154 (28), 141 (28)
	6	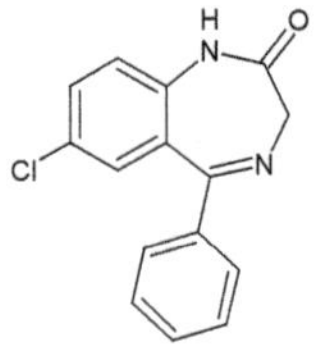	6.9	$C_{15}H_{11}ClN_2O$	271.0632	271.0645	-4.8	271 (100), 140 (88), 165 (56), 208 (50), 243 (18), 193 (13), 226 (9)
	7	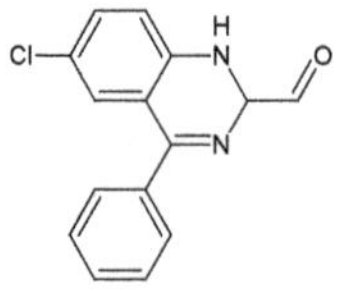	7.6	$C_{15}H_{11}ClN_2O$	271.0632	271.0638	-2.2	
	8	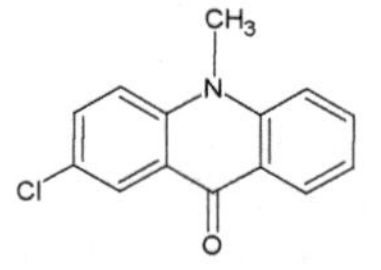	8.1	$C_{14}H_{10}ClNO$	244.0524	244.0535	-4.5	153 (100), 209 (90), 177 (52)
	9	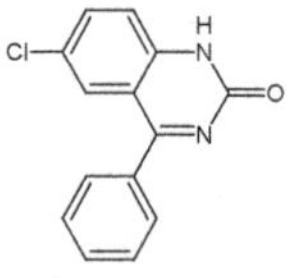	7.4	$C_{14}H_9ClN_2O$	257.0476	257.0488	-4.7	207 (100), 124 (530, 223 (20)

Table 2. Degradation products of DIZ identified during the irradiation of river water samples by Xe-lamp light.

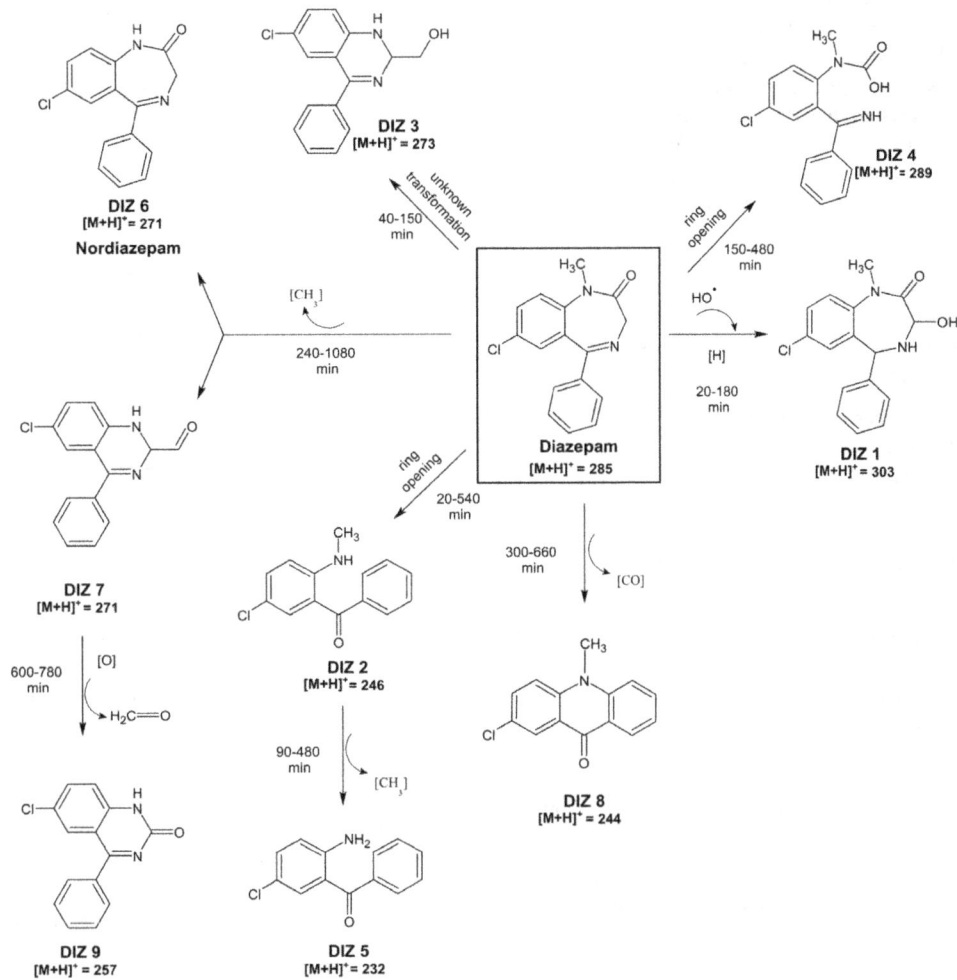

Figure 3: Proposed photodegradation pathway of DIZ.

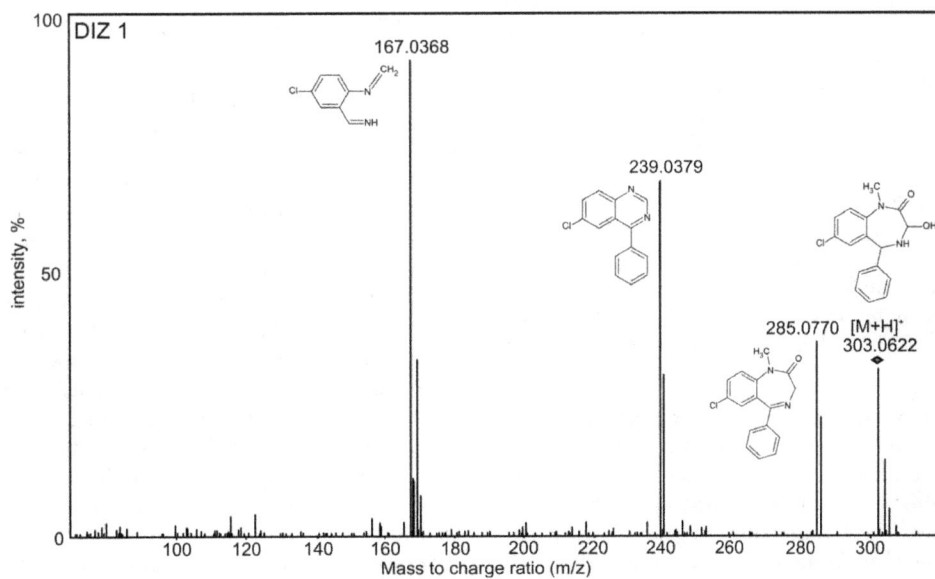

Figure 4: MS/MS fragmentation spectra of photoproduct DIZ 1 identified by UHPLC-QTOF-MS.

Xe-lamp irradiation of diazepam solution caused after 90 minutes formation of another by-product of its degradation (DIZ 5). This product (m/z 232.0533) is 15 mass units smaller than the compound DIZ 2 suggesting demethylation of aminomethyl group (elimination of methyl group). The two next products (DIZ 6 and DIZ 7, m/z 271.0645) of diazepam degradation have been identified after 240 minutes of exposure to light. The retention times of these products were different, whereas their mass spectra are identical. Both products' mass spectra with the most important fragment m/z 270 showed that their mass is 14 units lower, suggesting loss of methylene group from diazepam molecule. The structure of DIZ 6 may correspond to the nordiazepam molecule which is the human metabolite of the diazepam, whereas the structure of DIZ 7, based on fragmentation ions and the reported data [19], was tentatively identified as 6-chloro-4-phenyl-1,2-dihydroquinazoline-2-carbaldehyde.

Continued irradiation of diazepam water solution (300 min) revealed formation of another degradation product (DIZ 8, m/z 244.0535). This compound was assumed to be the result of the opening of the diazepinone seven-membered ring followed by a decarbonylation and rearrangement into a highly stabilized six-membered ring - 2-chloro-10-methylacridin-9(10H)-one. Note, that this first photodegradation product (DIZ 8) is already known in the literature [19].

The appearance of compound DIZ 9 (m/z 257.0488) after 600 minutes of Xe-lamp exposure shows that the byproduct DIZ 7 is probably oxidized during irradiation and converted into 6-chloro-4-phenylquinazolin-2(1H)-one. Formation of this product is confirmed based on the fragmentation ion (m/z 223.0885) which may correspond to phenylquinazolin-2(1H)-one after elimination of chlorine. This compound is already known in the literature as the product of photodegradation of oxazepam [27].

The irradiation of DIZ was terminated after 18 h when no signal from DIZ was registered and other transformation products were not observed. Further examples of tentative compound identifications for diazepam and its photoproducts DIZ 3 – DIZ 9 based on their mass spectra together with proposed structures of fragment ions during obtained UHPLC-QTOF-MS analysis are provided in the electronic supplementary information (Figures S1–S8).

Photodegradation products of SER: The identified compounds during the photodegradation experiment are summarized in Table 3 and the proposed transformation pathway of sertraline during the photolysis process is presented in Figure 5. As it can be noticed, the process starts by hydroxyl radical attack at the aromatic rings of sertraline molecule, that is much easier then hydroxylation of non-aromatic ring, forming two hypothetical isomers SER 1a and SER 1b (m/z 322.0760). Unfortunately, the obtained information during the UHPLC-QTOF-MS analysis was not sufficient to establish the exact position of the hydroxyl substituent at the aryl ring. Similar products have been reported in the literature [18], however, the suggested hydroxylation occurred in non-aromatic ring which was confirmed during the fragmentation of SER 1a,b at various collision energies no ions (Figure S10).

Continued irradiation of the sertraline solution revealed after 20 minutes formation of three new degradation products (SER 2a–c, SER 3 and SER 4). The compounds SER 2 a-c (m/z 288.1145) were suspected to be formed as a result of the mono-dechlorination of the SER 1a-b molecules. These compounds appear in two peaks indicating the formation of all these three compounds after the same time of irradiation. Furthermore, different fragmentation patterns were observed for products SER 2b,c (t_r=3.3 min) and SER 2a (t_r=4.9 min)

(Figure 6a,b) allowing proper structure elucidation of these analytes. Although SER 2b,c was observed by Pliego et al. [28] during photo-Fenton oxidation, further transformation of SER 2a-c was not observed. The exact structure of these products (specific hydroxylation of aryl ring) could not be defined on the basis of fragmentation data.

The formation of the SER 3 molecule with the m/z 272.1205 suggests that it must be the product of mono-dechlorination of the parent sertraline molecule. Simultaneously the creation of SER 4 (m/z value 291.0340) was observed. The structure of this compound may correspond to the oxidation of the parent molecule of sertraline to the ketone, followed by the elimination of CH_3NH_2 fragment. This degradation product (SER 4) has already been known in the literature as the product formed after photodegrative destruction as reported by Shen et al. [29].

Further exposition of the sertraline solution to Xe-lamp light caused formation of another degradation product (SER 5, m/z 304.0658) after 40 minutes. This product may correspond to dehydrogenated derivative of the parent molecule. SER 6 (254.1544), appeared after 50 minutes of the irradiation and may corresponds to the full dechlorination of SER 1a-b photoproduct. Shen et al. suggested [29] transformation pathway of sertraline via dehydrogenation to SER 5 followed by hydrolysis to SER 4. However, our study showed that the simultaneous exposition to Xe-lamp light lead to the formation of sertraline ketone (SER 4) after 20 min of irradiation, and compounds SER 5 appears after 40 min. These results prove that this transformation is not possible when using Xe-lamp, and these photoproducts are formed via different mechanisms.

The final compound detected in this solution (SER 7, m/z value 270.1491) was probably formed by hydroxylation of the SER 6 photoproduct, which was formed from compound SER 1a,b. This transformation chain was not reported in the literature before.

Irradiation of sertraline solution by the xenon lamp has been terminated after 240 minutes. The signal originated from sertraline, SER 1a-b, SER 2b-c, SER 3 and SER 7 has been still recorded, whereas the presence of degradation products SER 2a, SER 4, SER 5 and SER 6 was not detected in the examined solution.

Mass spectra obtained during UHPLC-QTOF-MS analysis for sertraline and photoproducts SER 3 – SER 7 together with proposed structures of fragment ions are provided in the electronic supplementary information (Figures S9–S15).

Dynamics of the transformation products generated during photolysis: The continuous irradiation of a sample containing diazepam let to the formation of nine main photoproducts. As presented in Figure 7a, the first two compounds, DIZ 1 and DIZ 2, are created after 20 min of irradiation in large amount (80% and 20%, respectively, of total relative amount of all 9 photoproducts). Their concentrations slowly decrease proportionally to the irradiation time. DIZ 2 is transformed to DIZ 5, which is degraded after 8 h of exposition, while DIZ 1 is decomposed to unknown compounds. DIZ 3, DIZ 5, DIZ 8 and DIZ 9 are unstable photoproducts formed at low concentration levels, which after continuous irradiation are degraded to unidentified compounds. DIZ 6 and DIZ 7 are two isomers which are formed slowly from 4 h of exposition to Xe-lamp light. Even after the experiment was terminated (after 18 h) they are still present in the sample and the increase of their concentration is noticeable. Although, it was established that DIZ 7 is transformed to DIZ 9, it can be easily observed that product DIZ 7 is more stable and dominant compound. The obtained results are not fully consistent with the results obtained by West et al. [19]. As far as DIZ 2 converts to DIZ 5 during the exposure to Xe-lamp light, the other

Name		Structure	Retention time, min	Molecular formula	[M+H]⁺ theoret.	[M+H]⁺ experim.	Error, ppm	Fragment ions m/z (relative intensity)
Sertraline (SER)			6.7	$C_{17}H_{17}Cl_2N$	306.0811	306.0815	-1.3	159 (100), 275 (55), 129 (45)
DEGRADATION PRODUCTS	1a		4.8	$C_{17}H_{17}Cl_2NO$	322.0760	322.0762	-0.6	238 (100), 273 (86), 211 (48)
	1b		5.8		322.0760	322.0765	-1.6	
	2a		4.9	$C_{17}H_{18}ClNO$	288.1150	288.1136	4.9	111 (100), 101 (96), 211 (93), 129 (73)
	2b		3.3		288.1150	288.1146	1.4	204 (100), 239 (47)
	2c		3.3					
	3		6.0	$C_{17}H_{18}ClN$	272.1201	272.1193	2.9	125 (100), 241 (68), 129 (45)
	4		10.3	$C_{16}H_{12}Cl_2O$	291.0338	291.0325	4.5	117 (100), 145 (50), 238 (45)
	5		5.8	$C_{17}H_{15}Cl_2N$	304.0654	304.0664	-3.3	129 (100), 287 (86), 263 (48)
	6		3.7	$C_{17}H_{19}NO$	254.1539	254.1528	4.3	239 (100), 101 (84), 133 (37), 223 (29)
	7		4.6	$C_{17}H_{19}NO_2$	270.1489	270.1478	4.1	111 (100), 129 (58), 67 (32), 211 (17), 183 (11), 239 (7)

Table 3 Degradation products of SER identified during the irradiation of river water samples by Xe-lamp light.

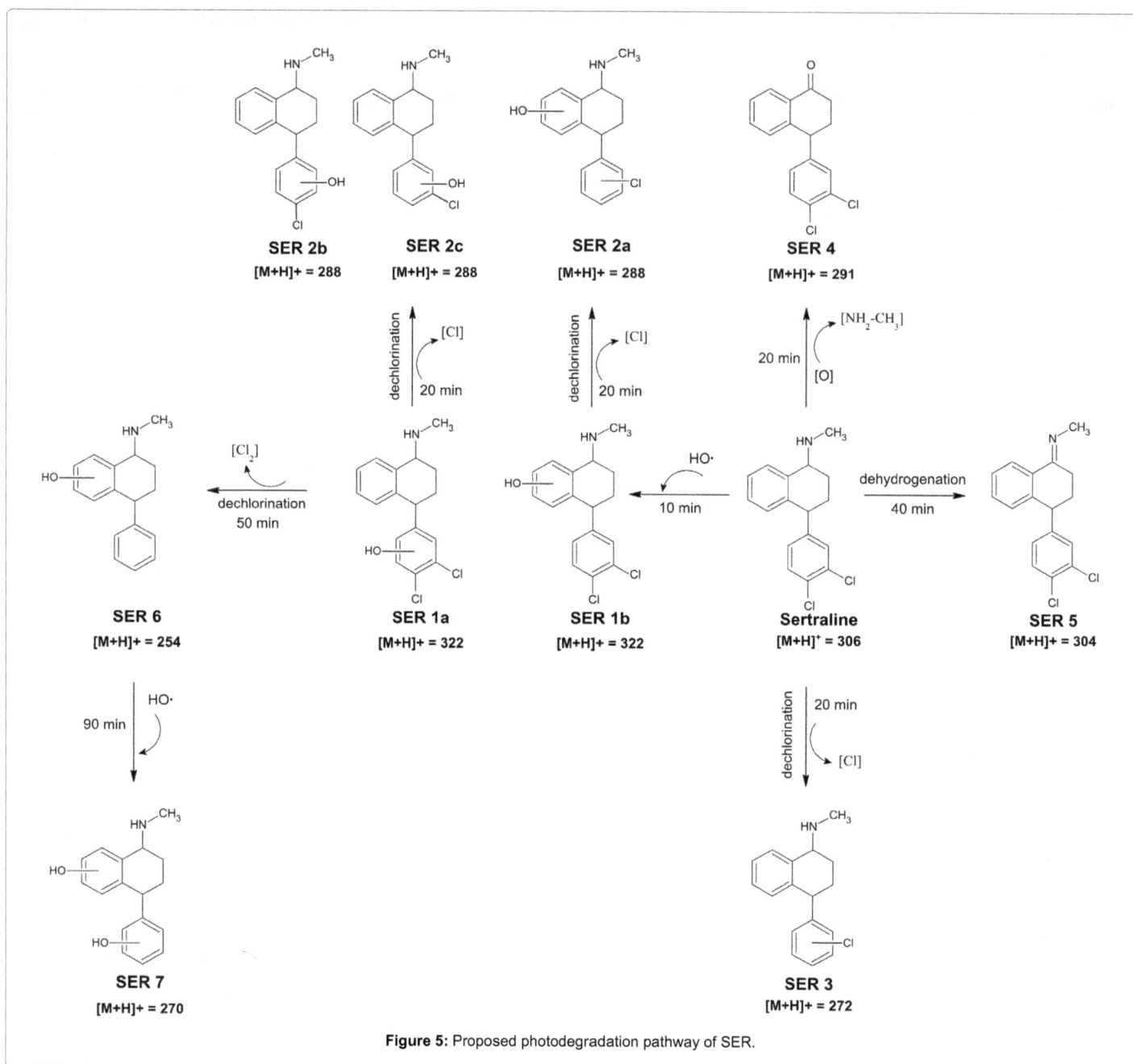

Figure 5: Proposed photodegradation pathway of SER.

compounds are formed in different ways form the parent molecule of diazepam, not necessarily from its metabolites. However, the possibility of transformation of diazepam to nordiazepam, known as human metabolite of DIZ, was observed and described here for the first time.

Photolysis of sertraline generated tenanalytes including three isomers of SER 1 and SER 2, however, SER 2a, SER 4 and SER 6 were considered as unstable products since they were created in small amounts and were decomposed while continuing the exposure to Xe-lamp light (Figure 7b). It was established that product SER 6 is transformed to SER 7 what is observed in a decrease of the amount of compound SER 6 whilst SER 7 concentration is increasing. Compound SER 5 is created spontaneously after 40 min of irradiation, however, it is slowly degrading during the continuation of the experiment. One of the isomers, SER 1a, and products SER 3 are formed in the highest

amounts and are present in the solution even after the experiment was terminated together with SER 1b, SER 2b,c and SER 7. As it was established product SER 1 is transformed into SER 6 and later to SER 7, which is confirmed by the decrease of the amount of one compound whilst the other concentration is increasing. This transformation pathway was reported here for the first time.

The experiment proved the formation of several degradation products of diazepam and sertraline, some of them appeared persistent and certain transformations are reported here for the first time. It leads to the conclusion that tracing the occurrence of parent and transformed compounds in the aqueous environment as well as in WWTPs is of a great importance and necessary to perform.

Occurrence of photoproducts in environmental waters: The presence of DIZ, SER and their photoproducts was investigated in

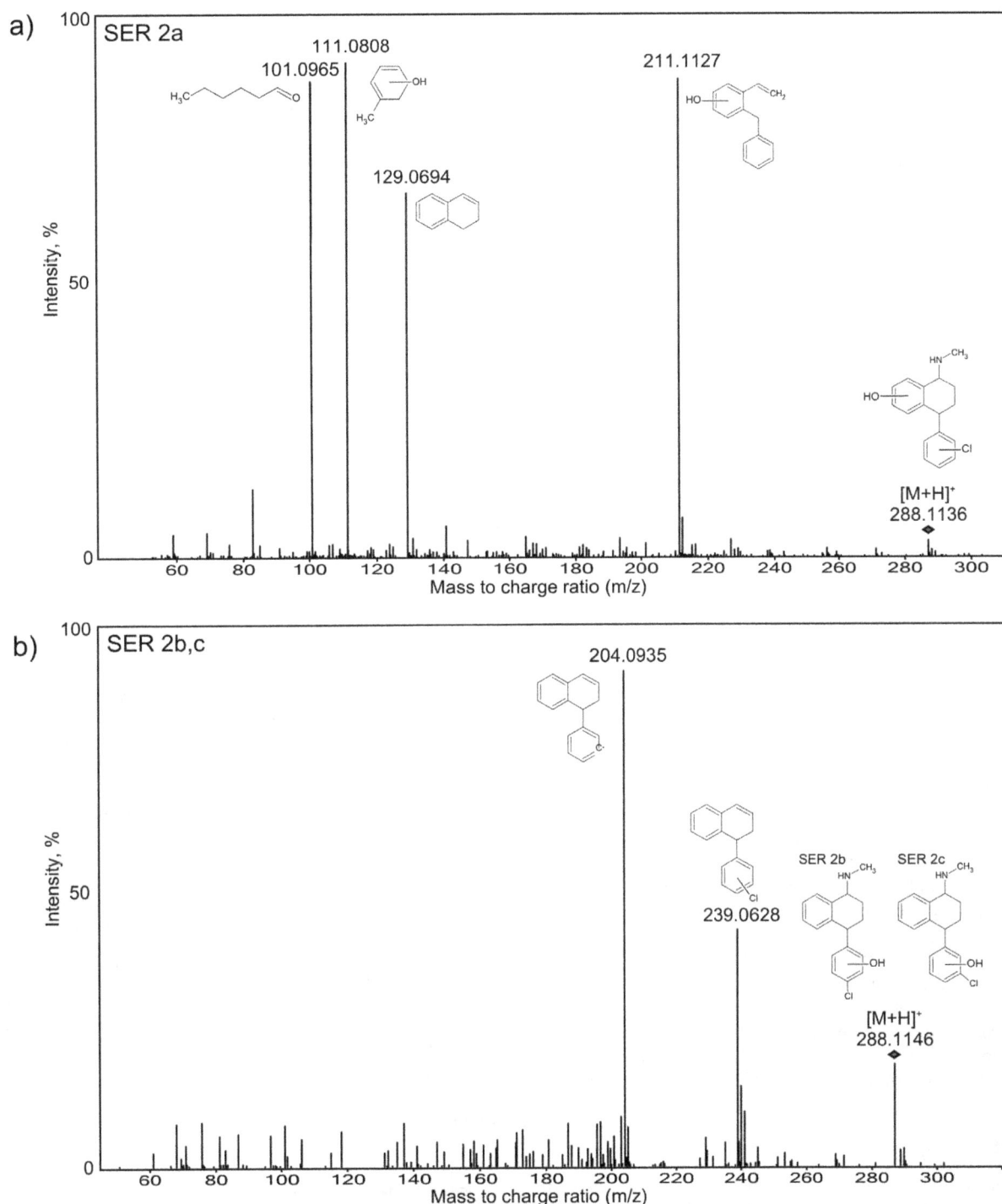

Figure 6: MS/MS fragmentation spectra of photoproduct SER 2a (a) and SER 2b,c (b) identified by UHPLC-QTOF-MS.

various water samples: influent and effluent wastewater, treated and untreated water, and river water from Pomerania area (Northern Poland) collected in 5 different months (from February to June). Samples were subjected to SPE procedure described in section 'Sample collection and preparation' and the recoveries of the extraction for parent compounds are presented in supplementary materials in Table S1. In order to estimate the actual contamination of the aqueous environment samples were analyzed by UHPLC-QTOF-MS in TargetMS/MS mode. Several

parameters were used for proper identification of target compounds, such as accurate mass measurements of (pseudo)molecular ions (< 5 ppm) and at least two fragment ions (< 10 ppm), relative intensity of fragment ions (deviation < 5 %), and compliance of retention times(deviation < 2%).

The obtained results showed that neither parent compounds (DIZ and SER) nor their photoproducts were detected in all analyzed samples.

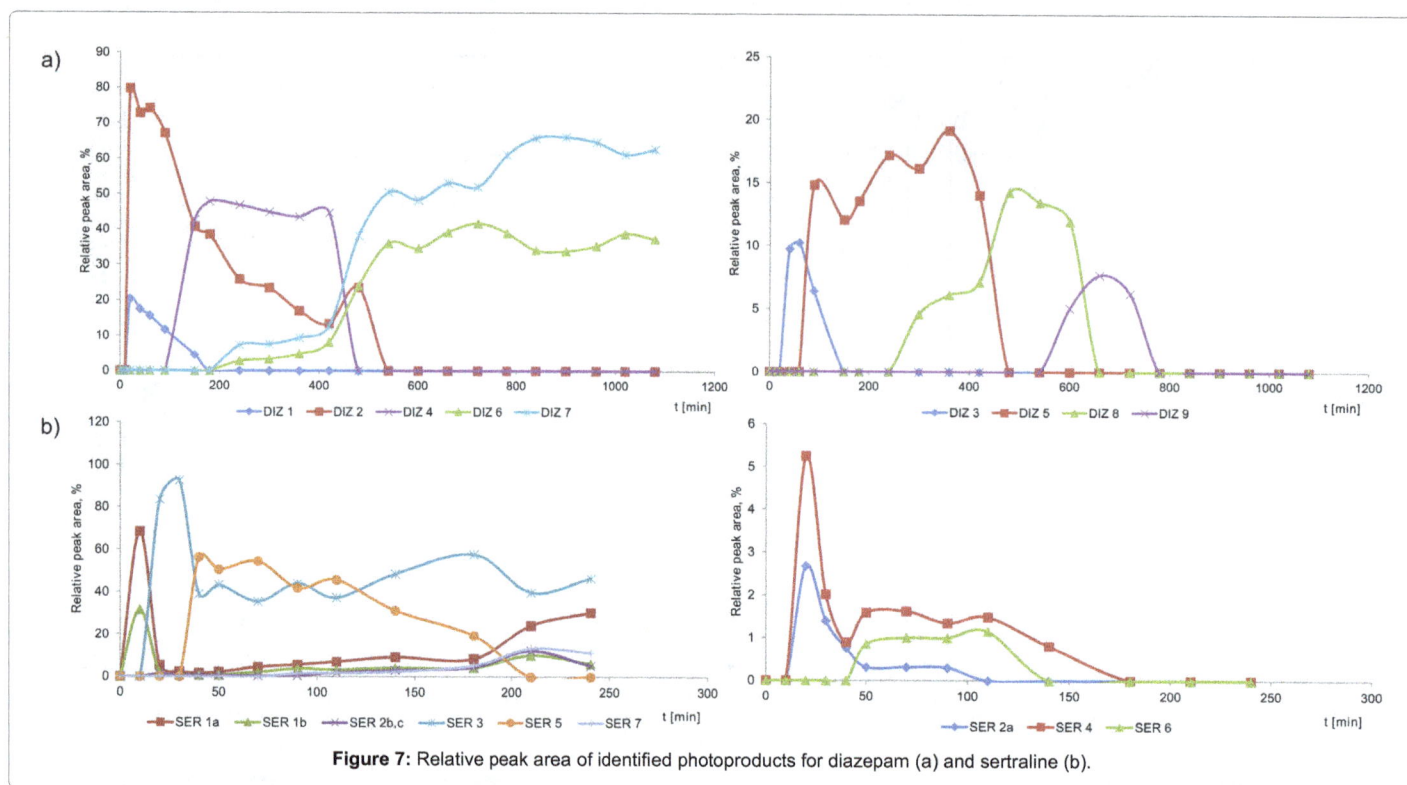

Figure 7: Relative peak area of identified photoproducts for diazepam (a) and sertraline (b).

The main reason of non-detection of sertraline and its photoproducts is probability of adsorptions to organic matter during wastewater treatment while applying activated sludge processes [18,30,31]. Diazepam and its transformed compounds are probably removed from the water samples via sorption to sedimenting particles, dilution, advection or rapid photodegradation in superficial water layer [32,33]. It is possible that in natural water samples, in the presence of microorganism and dissolved organic matter, photodegradation processes are more rapid than in laboratory scale leading to total decomposition of parent and transformed compounds causing their non-detection. However, it is also probable that these compounds may infiltrate from stored sludge to groundwater resulting in its contamination. In order to determine the actual pollution of the environment it would be valuable to determine the parent compound together with its degradation products in the sewage sludge, where such pharmaceuticals can be adsorbed. Therefore, further analysis should include determination of both parent and transformed compounds in various matrices, liquid and solid, in order to track the fate of pharmaceuticals in different parts of the aquatic environment.

Conclusion

In this work, a photodegradation of two psychiatric drugs, diazepam and sertraline was investigated with the application of hyphenated technique, UHPLC-QTOF-MS, which provided highly accurate and certain results. The degradation kinetic was investigated for two types of conditions, (1) while exposing to xenon lamp light, and (2) while exposing to natural sunlight in order to simulate process occurring in the environment. The degradation kinetic was pseudo-first order in all cases, however, autocatalytic reactions were observed in case of the second experiment for various natural water samples.

The degradation pathways of DIZ and SER induced by xenon lamp irradiation were proposed, whereas, particular photoproducts and transformations were described here for the first time. Nine

photoproducts of DIZ and seven transformed compounds of SER were identified in total, and their presence was investigated in various water samples. However, none of the compounds were detected, probably due to the adsorption of these pharmaceuticals on sewage sludge, sediment or other components of the environment. According to the authors knowledge this is the most accurate and extensive research on photodegradation of diazepam and sertraline with simulation of natural environmental conditions.

Acknowledgement

This scientific work (as a research project no. 2012/05/N/ST4/01991) has been financially supported within the framework of a grant awarded by the Polish National Science Centre (in the years 2013-2014). Anna Jakimska has been financially supported for the preparation of a doctoral dissertation within the funding of a doctoral scholarship awarded by Polish National Science Centre on the basis of the decisionnumber2013/08/T/ST4/00637. The authors also acknowledge Professor Adriana Zaleska from Department of Chemical Technology (Faculty of Chemistry, Gdańsk University of Technology) for providing access to xenon lamp. Additionally, the authors acknowledge Mr. Andrzej Szwacki for providing water and wastewater samples for this research.

References

1. Gros M, Petrovic M, Ginebreda A, Barcelo D (2010) Removal of pharmaceuticals during wastewater treatment and environmental risk assessment using hazard indexes. Environ Int 36: 15-26.

2. Homem V, Santos L (2011) Degradation and removal methods of antibiotics from aqueous matrices--a review. J Environ Manage 92: 2304-2347.

3. Jelic A, Gros M, Ginebreda A, Cespedes-Sánchez R, Ventura F, et al. (2011) Occurrence, partition and removal of pharmaceuticals in sewage water and sludge during wastewater treatment. Water Res 45: 1165-1176.

4. Mondragon VA, Llamas-Perez DF, Gonzalez-Guzman GE, Marquez-Gonzalez AR, Padilla-Noriega R, et al. (2011) Identification of Enterococcus faecalis bacteria resistant to heavy metals and antibiotics in surface waters of the Mololoa River in Tepic, Nayarit, Mexico. Environ Monitor Assess 183: 329-340.

5. Ragugnetti M, Adams ML, Guimaraes ATB, Sponchiado G, Vasconcelos ECd, et al. (2011) Ibuprofen Genotoxicity in Aquatic Environment: An Experimental Model Using Oreochromis niloticus. Water Air Soil Pollut 218: 361-364.

6. Milnes MR, Bermudez DS, Bryan TA, Edwards TM, Gunderson MP, et al. (2006) Contaminant-induced feminization and demasculinization of nonmammalian vertebrate males in aquatic environments. Environ Res 100: 3-17.

7. Rodríguez EM, Medesani DA, Fingerman M (2007) Endocrine disruption in crustaceans due to pollutants: a review. Comp Biochem Physiol A Mol Integr Physiol 146: 661-671.

8. de Jongh CM, Kooij PJ, de Voogt P, ter Laak TL (2012) Screening and human health risk assessment of pharmaceuticals and their transformation products in Dutch surface waters and drinking water. Sci Total Environ 427-428: 70-77.

9. Dong MM, Trenholm R, Rosario-Ortiz FL (2015) Photochemical degradation of atenolol, carbamazepine, meprobamate, phenytoin and primidone in wastewater effluents. J Hazard Mater 282: 216-223.

10. Fatta-Kassinos D, Vasquez MI, Kümmerer K (2011) Transformation products of pharmaceuticals in surface waters and wastewater formed during photolysis and advanced oxidation processes – Degradation, elucidation of byproducts and assessment of their biological potency. Chemosph 85: 693-709.

11. Michael I, Vasquez MI, Hapeshi E, Haddad T, Baginska E, et al. (2014) Metabolites and transformation products of pharmaceuticals in the aquatic environment as contaminants of emerging concern. In: Lambropoulou DA, Nollet LML (Eds) Advanced Mass Spectrometry-based techniques for the identification and structure elucidation of transformation products of emerging contaminants: Wiley.

12. Boreen AL, Arnold WA, McNeil K (2003) Photodegradation of pharmaceutical in the aquatic environment: A review. Aquat Sci 65: 320-341.

13. Lin AY, Reinhard M (2005) Photodegradation of common environmental pharmaceuticals and estrogens in river water. Environ Toxicol Chem 24: 1303-1309.

14. Alapi T, Dombi A (2007) Comparative study of the UV and UV/VUV photolysis of phenol in aqueous solution. J Photchem Photobiol A 188: 409-418.

15. Andreozzi R, Raffaele M, Nicklas P (2003) Pharmaceuticals in STP effluents and their solar photodegradation in aquatic environment. Chemosphere 50: 1319-1330.

16. DeVane CL, Liston HL, Markowitz JS (2002) Clinical pharmacokinetics of sertraline. Clin Pharmacokinet 41: 1247-1266.

17. MacQueen G, Born L, Steiner M (2001) The selective serotonin reuptake inhibitor sertraline: its profile and use in psychiatric disorders. CNS Drug Rev 7: 1-24.

18. Li H, Sumarah MW, Topp E (2013) Persistence and dissipation pathways of the antidepressant sertraline in agricultural soils. Sci Total Environ 452-453: 296-301.

19. West CE, Rowland SJ (2012) Aqueous phototransformation of diazepam and related human metabolites under simulated sunlight. Environ Sci Technol 46: 4749-4756.

20. Brooks BW, Chambliss CK, Stanley JK, Ramirez A, Banks KE, et al. (2005) Determination of select antidepressants in fish from an effluent-dominated stream. Environ Toxicol Chem 24: 464-469.

21. Gros M, Rodríguez-Mozaz S, Barceló D (2012) Fast and comprehensive multi-residue analysis of a broad range of human and veterinary pharmaceuticals and some of their metabolites in surface and treated waters by ultra-high-performance liquid chromatography coupled to quadrupole-linear ion trap tandem mass spectrometry. J Chromatogr A 1248: 104-121.

22. Vasskog T, Berger U, Samuelsen PJ, Kallenborn R, Jensen E (2006) Selective serotonin reuptake inhibitors in sewage influents and effluents from Tromsø, Norway. J Chromatogr A 1115: 187-195.

23. Metcalfe CD, Chu S, Judt C, Li H, Oakes KD, et al. (2010) Antidepressants and their metabolites in municipal wastewater, and downstream exposure in an urban watershed. Environ Toxicol Chem 29: 79-89.

24. Jakimska A, Sliwka-Kaszynska M, Reszczynska J, Namiesnik J, Kot-Wasik A (2014) Elucidation of transformation pathway of ketoprofen, ibuprofen, and furosemide in surface water and their occurrence in the aqueous environment using UHPLC-QTOF-MS. Anal Bioanal Chem 406: 3667-3680.

25. Jakimska A, Śliwka-Kaszyńska M, Nagórski P, Namieśnik J, Kot-Wasik A (2014) Phototransformation of amlodipine: degradation kinetics and identification of its photoproducts. PLoS One 9: e109206.

26. Górska P, Zaleska A, Hupka J (2009) Photodegradation of phenol by UV/TiO2 and Vis/N,C-TiO2 processes: Comparative mechanistic and kinetic studies. Sep Purif Technol 68: 90-96.

27. Cornelissen PJG, Beijersbergen-van-Henegouwen GMJ, Gerritsma KW (1978) Photochemical decomposition of 1,4-benzodiazepines. Diazepam. Int J Pharm 1: 173-181.

28. Pliego G, Xekoukoulotakis N, Venieri D, Zazo JA, Casas JA, et al. (2014) Complete degradation of the persistent anti-depressant sertraline in aqueous solution by solar photo-Fenton oxidation. J Chem Technol Biotechnol 89: 814-818.

29. Shen LQ, Beach ES, Xiang Y, Tshudy DJ, Khanina N, et al. (2011) Rapid, biomimetic degradation in water of the persistent drug sertraline by TAML catalysts and hydrogen peroxide. Environ Sci Technol 45: 7882-7887.

30. Styrishave B, Halling-Sørensen B, Ingerslev F (2011) Environmental risk assessment of three selective serotonin reuptake inhibitors in the aquatic environment: a case study including a cocktail scenario. Environ Toxicol Chem 30: 254-261.

31. Hörsing M1, Ledin A, Grabic R, Fick J, Tysklind M, et al. (2011) Determination of sorption of seventy-five pharmaceuticals in sewage sludge. Water Res 45: 4470-4482.

32. Straub JO, Kummerer K (2008) Deterministic and probabilistic environmental risk assessment for diazepam. In Pharmaceuticals in the Environment: Sources, Fate, Effects and Risks: Springer-Verlag Berlin Heidelberg. 343-383.

33. Jelic A, Gros M, Petrovic M, Ginebreda A, Barcelo D (2012) Occurrence and Elimination of Pharmaceuticals During Conventional Wastewater Treatment: Springer-Verlag Berlin Heidelberg.

9

Liquid Chromatographic Assay for the Analysis of Atazanavir in Rat Plasma after Oral Administration: Application to a Pharmacokinetic Study

Gurinder Singh, Roopa S. Pai * and Sanual Mustafa

Department of Pharmaceutics, Faculty of Pharmacy, Al-Ameen College of Pharmacy, Bangalore, Karnataka, India

Abstract

A new reverse phase liquid chromatographic method for the investigation of atazanavir in rat plasma was developed after oral administration. The chromatographic separation was achieved on Phenomenex C_{18} column (250 mm × 4.6 mm I.D., 5 μm), under isocratic conditions using UV detection at 249 nm. The optimized mobile phase consisted of a mixture of potassium dihydrogen phosphate (pH 3.5) and acetonitrile (58: 42, v/v) at a flow rate of 1ml/min. The system was found to produce sharp and well-resolved peak for atazanavir with retention time of 5.78 min. The linear regression analysis for the calibration curves showed a good linear correlation over the concentration range 0.050–2.0 μg/ml, with determination coefficients, R^2, exceeding 0.9989. The limits of detection (LOD) and quantitation (LOQ) were found to be 0.004 µg/ml and 0.012 µg/ml, respectively. The method was successfully applied for the pharmacokinetic in rats. Atazanavir concentration in plasma reached (C_{max}) was 0.087 μg/ml at 3 h after oral administration of 7.2 mg/kg/rat. The AUC_{0-12} was 0.4812 µg/ml*h and the apparent plasma half-life ($t_{1/2}$) was 7.5 h. This method was found to be suitable for examining atazanavir concentration in rats, after oral administration of atazanavir in a single dose.

Keywords: Atazanavir; High-performance liquid chromatography; Pharmacokinetic study

Introduction

Atazanavir (ATV) belongs to the HIV protease inhibitor (PI) class of the antiretrovirals (ARVs), which have played an important role in lowering the morbidity and mortality of HIV/AIDS [1]. The compound selectively inhibits the virus-specific processing of viral Gag and Gag-Pol polyproteins in HIV-1 infected cells, thus preventing formation of mature virions. Atazanavir is distinguished from other protease inhibitors in that it can be given once-daily and has lesser effects on the patient's lipid profile [2]. Moreover, several cytochrome P450 isoenzymes can be inhibited by atazanavir [3]. The pharmacokinetics of ATV allows for once daily dosing [4], which may improve patient compliance.

There are several methods described in the literature for the quantitative analysis of atazanavir in plasma, either alone [5-7] or in combination with other ARVs [7-10]. Some authors reported the use of mass spectrometry for detection [10], which is not routinely available in all laboratories. Furthermore, all the reported methods had run times exceeding 15 min, even those for the analysis of atazanavir alone.

HPLC-UV detection method was developed for atazanavir with LOQ of 0.044 μg/ml [11]. A validated HPLC method for the estimation of atazanavir was not appropriate for detection of low atazanavir concentration. In this method, the LOQ value was found to be 0.090 μg/ml and retention time was more than 8.3 min [12]. The purpose of the present study was to develop and validate a simple and time-saving RP-HPLC method with UV detection for the investigation of atazanavir after oral administration to Wistar rats. The method was validated according to Food and Drug Administration (FDA) and International Conference on Harmonization (ICH) guidelines with respect to linearity, precision, accuracy, and specificity and stability studies [13,14]. Indinavir (0.250 μg/ml) was used as an internal standard (IS).

Experimental

Chemicals and reagents

Atazanavir and Indinavir (99.8% w/w and 98.7% w/w, respectively,

High performance liquid chromatography, HPLC) were provided ex-gratis by M/s Hetero Labs, Hyderabad, India. HPLC grade acetonitrile was purchased from SD fine-chem limited (Mumbai, India). Deionized water used in all the experiments was passed through a Milli-Q water purification system (18.2 MΩ/cm) Millipore (Bangalore, Karnataka, India).

Instrumentation

The HPLC (Shimadzu, Kyoto, Japan) instrument was equipped with binary pump and SPD- 20AVPUV detector. Sample injection was done by Rheodyne injector with a 50 μL loop and a computer running Varian workstation version 6.42 software for data acquisition and processing. The chromatographic analysis was carried out on Phenomenex C18 column (250 mm × 4.6 mm I.D., 5 μm).

Chromatographic conditions

The mobile phase was composed of potassium dihydrogen phosphate (pH 3.5) and acetonitrile in a ratio of (58: 42, v/v) run under isocratic elution and pumped at a flow rate of 1 ml/min. The column was thermostated at 30°C. Under these conditions the run time was less than 8 min.

Preparation of Calibration Curve (CC) and Quality Control Samples (QC)

Eight-point calibration curve (CC) was prepared by serial dilution

*Corresponding author: Roopa S. Pai, Professor, Faculty of Pharmacy, Department of Pharmaceutics, Al-Ameen College of Pharmacy, Bangalore 560027, Karnataka, India, E-mail: roopaspai@yahoo. com, gurindersingh181@gmail.com

of atazanavir stock solution (100 $\mu g/ml$) in the range of 0.050, 0.1, 0.5, 1.0, 1.5, and 2.0 $\mu g/ml$ that were obtained by measuring the required amount of 100 $\mu g/ml$ working standard solution, mixed with a sufficient quantity of mobile phase and making up to 10 ml. Similarly, six standard solutions were prepared by serial dilution of atazanavir stock solution (10 $\mu g/ml$) in the range of 0.005, 0.010, 0.020, 0.030, 0.040 and 0.050 $\mu g/ml$. Six standard solutions were obtained from the 10 $\mu g/ml$ working standard solution, in order to determine the LOD and LOQ of the method. Calibration standards were prepared daily by spiking 100 μL of blank plasma with 10μl of the appropriate working solution resulting in concentrations of 0.050, 0.1, 0.5, 1.0, 1.5, and 2.0$\mu g/ml$ and 0.005, 0.010, 0.020, 0.030, 0.040, and 0.050 $\mu g/ml$, respectively, of atazanavir. Stock solution (0.250 $\mu g/ml$) of IS was prepared in methanol and stored at −20°C. The solutions were stable for one day when stored at room temperature (20–25°C). The stock and standard solutions were prepared on a daily basis and stored in the dark at about 5°C. All solutions were used on the day they were prepared.

Quality control (QC) samples (low quality control (LQC), 0.1 $\mu g/ml$; medium quality control (MQC), 1.0 $\mu g/ml$; high quality control (HQC), 2.0 $\mu g/ml$; limit of quantification (LOQ), 0.009 $\mu g/ml$ were prepared by spiking 0.1ml aliquot of blank plasma with 10 μl of spiking solution of drug as well as the IS. All solutions were stored in the refrigerator at 4.0 ± 2.0°C. The bulk spiked CC and QC samples were stored at −20°C and brought to room temperature before use.

Sample preparation

To a 50 μl of rat plasma, 10 μl of IS and 100 μl of atazanavir were added and the mixture was incubated at 37°C for 1 h. Atazanavir was then extracted using 30 μl of acetonitrile followed by vortexing for 2 min. After vortexing, the samples were subjected to centrifuge at 12,000×g for 10 min. The supernatant was decanted into a china dish and evaporated to dryness at room temperature. This was further reconstituted with 100 μl of mobile phase and vortexed for 30 s and 20 μl was injected into an HPLC system. Atazanavir was detected at a wavelength of 249 nm.

Method validation study

The developed method was validated as per ICH guidelines using atazanavir with respect to the following parameters: accuracy, precision, LOD, LOQ, specificity, stability, and system suitability.

Linearity

For testing linearity, five calibration standards were prepared in the concentration range of 0.05–2.0 $\mu g/ml$ (0.050, 0.1, 0.5, 1.0, 1.5 and 2.0 $\mu g/ml$). Standard curve was achieved by plotting peak area against concentration, and the evaluation of linearity was completed by linear regression analysis using least square method.

Limit of detection and limit of quantitation

Normally, limit of detection (LOD) and limit of quantitation (LOQ) are estimated at a signal to noise ratio of 3:1 and 10:1, respectively [13]. LOD and LOQ were determined based on the response and slope of a specific calibration curve obtained from six standard solutions (0.005, 0.010, 0.020, 0.030, 0.040, and 0.050 $\mu g/ml$) that were in proximity of these limit concentration values.

Selectivity, specificity and linearity

Selectivity was verified by analyzing the blank plasma from rats to test interference at the analyte retention times. By employing the proposed extraction procedure each blank plasma sample was tested

and then compared with the results of plasma samples spiked with atazanavir ($n=6$) in calibration standard to ensure no interference of atazanavir from plasma. Spiked plasma samples that contained increasing concentrations of atazanavir from 0.050 to 2.0 $\mu g/ml$ were analyzed according to the procedure described above. The linearity was detected by calculating the correlation coefficient (r) of the curves by means of least-squared linear regression method. All calibration curves of atazanavir were constructed prior to the experiments with correlation coefficient of ($r^2 > 0.9989$).

Accuracy

The accuracy of the assay method was evaluated in triplicate at three different concentration levels (0.1, 1.0, and 2.0 $\mu g/ml$), and the percentage recoveries were calculated.

Precision: The precision is usually reported as the percent relative standard deviation (%RSD) of a set of responses. Precision was represented into two categories, namely, repeatability (intraday precision) and intermediate precision (interday precision).

Repeatability or intraday precision: Repeatability was tested by analyzing six determinations at three different concentrations, namely, low, medium, and high within the linearity range.

Intermediate or interday precision: The inter-day variability of this method was assessed over three days at three low, medium, and high concentrations of atazanavir standard in replicates of six.

Pharmacokinetic study in Rats: The pharmacokinetic studies were carried out in healthy male Wistar rats (200– 250 g), and the animals were fasted overnight before dosing with free access to water. The animals were acclimatized to laboratory conditions over the week before experiments and fed with standard rat diet, under controlled conditions of a 12:12 h light : dark cycle, with a temperature of 22 ± 3°C and a relative humidity of 50 ± 5% RH. The experimental protocol was approved by the Institutional Animal Ethical Committee (AACP/IAEC/Jun-2012-02).

Twelve rats were randomly separated into two groups (six animals each group). The grouping of animals was as follows:

Group I: Control normal rats (received saline solution)

Group II: Administered with pure drug (as solution) (7.2 mg/kg/rat) [15]

At regular time intervals 0, 0.5, 1, 2, 3, 4, 6, 8, 10 and 12h samples of blood were withdrawn (100 μl) from the retro-orbital plexus by microcapillary technique under light ether anesthesia into heparinized microcentrifuge tubes (50 units heparin/ml of blood). Plasma was separated by centrifugation at 12,000×g for 15 min and analyzed by the following method. Plasma samples were deproteinated with 1ml of acetonitrile, vortexed for 2 min, and centrifuged at 12,000×g for 10 min. The supernatant was decanted into a China dish and evaporated to dryness at room temperature. This was further reconstituted with 100 μl of mobile phase and vortexed for 30 s and 20 μl was injected into an HPLC system. Atazanavir was detected at a wavelength of 249 nm.

Results and Discussion

Method development and optimization of HPLC-UV conditions

A liquid chromatographic method for the estimation of atazanavir in rat plasma has been developed and validated according to the principles of Good Laboratory Practices. An appropriate wavelength

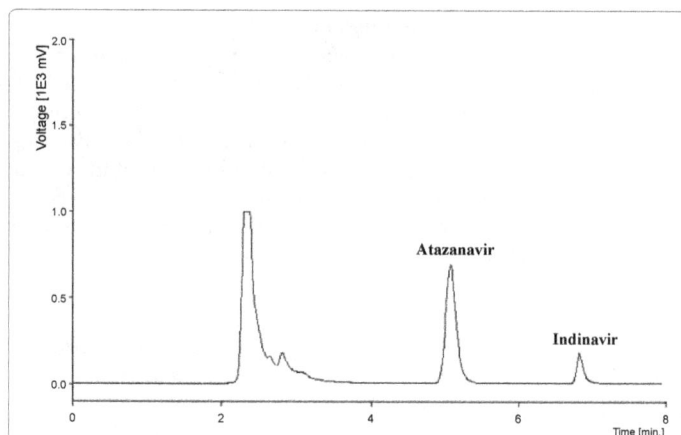

Figure 1: At optimized conditions: (a) chromatographic profile of the plasma spiked with atazanavir (2 μg/ml) in the presence of the I.S (0.250 μg/ml).

Concentration (μg/ml)	Observed concentration (μg/ml)	% Precision	% Accuracy
Intra-day			
0.1	0.098 ± 0.002	2.00	98.00
1.0	0.988 ± 0.016	1.61	98.80
2.0	2.105 ± 0.011	0.54	105.25
Inter-day			
0.1	0.094 ± 0.003	3.19	94.00
1.0	0.981 ± 0.027	2.75	98.10
2.0	1.996 ± 0.068	3.40	99.80

Table 1: Intra-day and inter-day precision and accuracy of atazanavir in rat plasma (n = 6).

Sample condition	Spiked concentration (μg/ml)	Mean determined concentration (μg/ml)	Accuracy (%)
Bench-top stability[a]	0.1	0.096	96.00
	1.0	0.97	97.00
	2.0	1.99	99.50
Freeze–thaw stability[b]	0.1	0.098	98.00
	1.0	0.99	99.00
	2.0	2.07	103.5
One-week stability[c]	0.1	0.092	92.00
	1.0	0.89	89.00
	2.0	2.38	95.20
[a]Exposed at ambient temperature (25 °C) for 4h.			
[b]After three freeze–thaw cycles			
[c]Stored at −16 °C.			

Table 2: Stability of atazanavir in rat plasma (n = 6)

was important for good sensitivity. Atazanavir has a special conjugation structure which leads to strong UV absorption at the wavelength of 249 nm. Therefore, the detection wavelength was set at 249 nm. It was necessary to use an IS in extraction techniques and HPLC method to compensate for extraction variation, efficiency, and analytical errors. Indinavir was adopted as the IS in this study for the reasons that it is structurally similar to atazanavir and its behavioural characteristics and properties conform to the chemical requirement for IS in HPLC. In addition, indinavir is commercially available in high purity, and it is stable and nonreactive with sample or mobile phase. Meanwhile, it also has good response at the detection wavelength of 249 nm.

The mixture of potassium dihydrogen phosphate (pH 3.5) and acetonitrile (58: 42, v/v) at a flow rate of 1ml/min could achieve the

above purpose that was found to be optimum and provided adequate peak separation, with less tailing and resulted in good resolution among all the other combinations tested which was finally adopted as the mobile phase.

Limit of detection and limit of quantitation

Concentrations of LOD and LOQ were found to be 0.003 μg/ml and 0.012 μg/ml, respectively.

Specificity

Specificity is expressed as the capability of a method to distinguish the analyte from all potentially intrusive substance [16]. The specificity of the method was scrutinized by blank plasma detection, peak purity, and spiking blank plasma with pure standard compounds. Blank plasma had no interference, when atazanavir and the IS were eluted. At optimized conditions, the separation of atazanavir and indinavir was completed within 8 min (Figure 1).

Linearity

Each sample was analyzed in replicates of six to verify the reproducibility of detector response at each concentration level. The detector responses were found to be linear over the concentration range from 0.050 to 2.0 μg/ml. The regression equation for the graph is $y = 0.7986x - 0.6198$, and the correlation coefficient R^2 is 0.9989 showing excellent correlation between the area and the concentration.

Precision

The percentage relative standard deviation (%RSD) of the area of atazanavir during intraday study was found to be less than 5 and for interday study was found to be less than 6.5, which indicated a good precision of the method (Table 1). Intra-day and inter-day precision (% R.S.D) of the methods were lower than 4% and were within the acceptable limits to meet the guidelines for bioanalytical method validation which is considered to be ≤ 15% [17,18].

Accuracy

The quantitative recovery of atazanavir achieved ranged from 94.0 to 105.25% with a low %RSD value. The results of the recovery experiments done at three concentration levels and the %RSD values are given in Table 1.

Stability

Bench-top stability was investigated to ensure that atazanavir was not degraded in plasma samples at room temperature for a time period to cover the sample preparation. It was measured by divulging the QC samples to ambient laboratory conditions for 10 h. Freeze-thaw stability was measured over three cycles. Because of the need for occasional delayed injection of extraction samples, the stability of reconstituted samples was assessed at ambient temperature for 12 h. The freezer storage stability of atazanavir rat plasma at -20°C was evaluated by assaying QC samples at the beginning and one week later. All stability QC samples were analyzed in six replicates. The results indicated that atazanavir had an acceptable stability under those conditions (Table 2).

Application of the assay

The validated method was successfully applied to investigate the content of atazanavir in in vivo, after administered orally to rats. Oral administration of atazanavir in the present study resulted in a sharp C_{max} of 0.087 μg/ml at 3 h after which the plasma concentration declined rapidly, indicating a rapid absorption of atazanavir. The area under the

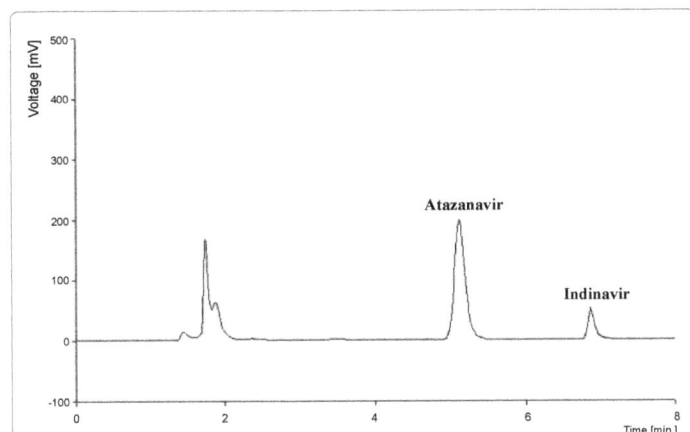

Figure 2: Chromatogram of plasma sample collected from rats 3 h after oral administration of atazanavir.

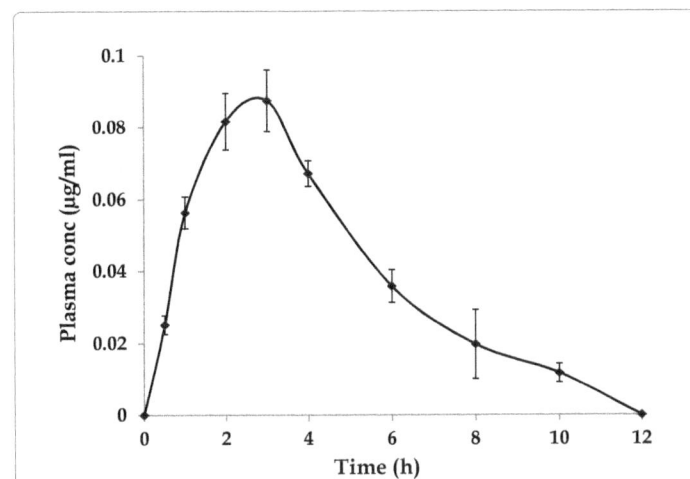

Figure 3: Plasma concentration-time curve of atazanavir after being orally administered male Wistar rats at dose of 7.2mg/kg/rat ($n = 6$, mean ± S.D).

Pharmacokinetic parameters	Atazanavir
C_{max} (µg/ml)	0.0874 ± 0.032
t_{max} (h)	3.2 ± 0.014
$t_{1/2}$ (h)	7.5 ± 0.190
AUC_{0-12} (µg/ml*h)	0.481 ± 0.023
K_e (1/h)	0.2921 ± 0.005
MRT	4.736 ± 0.189

Data presented as mean ± standard deviation (n=6)

Table 3: Pharmacokinetic parameters of atazanavir at a dose of 7.2 mg/kg/rat.

concentration versus time curve was $0.4812\ \mu g/ml\ ^\cdot h$. The representative chromatogram of a plasma sample, which was collected from Wistar rats 3 h following oral administration of atazanavir as portrays in Figure 2. The plasma profile of atazanavir is shown in Figure 3. The results substantiate the suitability of the developed method for determining atazanavir concentration in plasma after oral administration. The pharmacokinetic data of atazanavir after oral administration in rats is shown in Table 3.

Conclusion

A specific, linear, accurate, reliable, and reproducible new method of atazanavir in rat plasma was developed and fully validated over the range 0.050-2.0 µg/ml with LOQ of 0.012 µg/ml. The method was successfully applied to measure the drug concentration in plasma after oral administration to rats. Reproducible high recovery of atazanavir was achieved. Because of its highly satisfactory sensitivity, accuracy, linearity, and specificity, this HPLC methodology could thus be an appropriate tool for further determination of atazanavir in plasma samples in the pharmacokinetic studies.

Conflict of Interest

The authors confirm that this article content has no conflicts of interest.

Acknowledgement

The authors would like to thank Prof. B.G Shivananda for his advice and support to carry out this research work. The authors gratefully acknowledge financial support and granting research fellowship (45/38/2011/Nan-BMS) from ICMR (Indian Council of Medical Research, Govt of India, New Delhi). Authors are also grateful to M/s Hetero Labs, Hyderabad, India for providing the gift sample of atazanavir and indivnavir.

References

1. Harrison TS, Scott LJ (2005) Atazanavir-a review of its use in the management of HIV infection. Drugs 65: 2309-2336.

2. Chitturi SR, Somannavar YS, Peruri BG, Nallapati S, Sharma HK, et al. (2011) Gradient RP-HPLC method for the determination of potential impurities in atazanavir sulfate. J Pharm Biomed Anal 55: 31-47.

3. Cateau E, Tournier N, Dupuis A, Moalc GL, Venisse N (2005) Determination of atazanavir in human plasma using solid-phase extraction and high-performance liquid chromatography. J Pharm Biomed Anal 39: 791-795.

4. Singh G, Pai RS (2013) High-Performance Liquid Chromatographic Method for Analysis of Emtricitabine in Rat Plasma: Method Development, Validation and Application to a Pharmacokinetic Study. ISRN Chromatography Article ID 329072.

5. Cattaneo D, Maggiolo F, Ripamonti D, Perico N (2008) Determination of atazanavir in human plasma by high-performance liquid chromatography with UV detection. J Chromatogr Sci 46: 485-489.

6. Loregian A, Pagni S, Ballarin E, Sinigalia E, Parisi SG, et al. (2006) Simple determination of the HIV protease inhibitor atazanavir in human plasma by high-performance liquid chromatography with UV detection. J Pharm Biomed Anal 42: 500-505.

7. Elens L, Veriter S, Fazio VD, Vanbinst R, Boesmans D, et al. (2009) Quantification of 8 HIV-protease inhibitors and 2 nonnucleoside reverse transcriptase inhibitors by ultra-performance liquid chromatography with diode array detection. Clin Chem 55: 170-174.

8. Weller DR, Brundage RC, Balfour HH, Vezina HE (2007) An isocratic liquid chromatography method for determining HIV non-nucleoside reverse transcriptase inhibitor and protease inhibitor concentrations in human plasma. J Chromatogr B Biomed Sci Appl 848: 369-373.

9. Notari S, Bocedi A, Ippolito G, Narciso P, Pucillo LP, et al. (2006) Simultaneous determination of 16 anti-HIV drugs in human plasma by high-performance liquid chromatography. J Chromatogr B Biomed Sci Appl 831: 258-266.

10. Avolio AD, Siccardi M, Sciandra M, Baietto L, Bonora S, et al. (2007) HPLC-MS method for the simultaneous quantification of the new HIV protease inhibitor darunavir, and 11 other antiretroviral agents in plasma of HIV-infected patients. J Chromatogr B Biomed Sci 859: 234-240.

11. Sparidans RW, Dost F, Crommentuyn KML, Huitema ADR, Schellens JHM, et al. (2006) Liquid chromatographic assay for the protease inhibitor atazanavir in plasma. Biomed Chromatogr 20: 72-76.

12. Muller AC, Kanfer I (2010) An efficient HPLC method for the quantitative determination of atazanavir in human plasma suitable for bioequivalence and pharmacokinetic studies in healthy human subjects. J Pharm Biomed Anal 53: 113-118.

13. Food and Drug Administration (2001) Guidance for Industry: Bioanalytical Method Validation, US Department of Health and Human Services, FDA, Center for Drug Evaluation and Research, Rockville,Md, USA.

14. ICH Harmonised Tripartite Guideline (2005) Validation of Analytical Procedures: Methodology, Q2 (R1), International Conference on Harmonisation of Technical Requirements for Registrations of Pharmaceuticals for Human Use, ICH, Geneva, Switzerland.

15. Fukushima K, Terasaka S, Haraya K, Kodera S, Seki Y, et al. (2007) Pharmaceutical approach to HIV protease inhibitor atazanavir for bioavailability enhancement based on solid dispersion system. Biol Pharm Bull 30: 733-738.

16. Singh G, Pai RS, Pandit V (2012) Development and validation of a HPLC method for the determination of trans-resveratrol in spiked human plasma. J Adv Pharm Tech Res 3: 130-135.

17. Singh G, Pai RS, Pandit V (2014) *In vivo* pharmacokinetic applicability of a simple and validated HPLC method for orally administered trans-resveratrol loaded polymeric nanoparticles to rats. J Pharm Investig 44: 69-78.

18. Singh G, Pai RS (2014) Optimization (central composite design) and validation of HPLC method for investigation of emtricitabine loaded poly(lactic-co-glycolic acid) nanoparticles: *in vitro* drug release and in *vivo* pharmacokinetic studies. The Scientific World Journal.

10

Development and Validation of a Liquid Chromatography Method with Electrochemical Detection for Hydroxyurea Quantification in Human Plasma and Aqueous Solutions

Anar Rodriguez[1], Delphine Beukens[1], Nicole Debouge[2], Béatrice Gulbis[2] and Frédéric Cotton[1,2]*

[1]*Laboratory of Biological and Medical Chemistry, Faculty of Pharmacy, Université Libre de Bruxelles (ULB), Boulevard du Triomphe, 1050 Brussels, Belgium*
[2]*Department of Clinical Chemistry, Erasme Hospital, Université Libre de Bruxelles, Route de Lennik 808 1070 Brussels, Belgium*

Abstract

Hydroxyurea is the unique drug having demonstrated a significant efficacy in sickle cell disease treatment. We developed a liquid chromatography method with electrochemical detection for hydroxyurea analysis in plasma and aqueous solutions. Analytical goals included an analytical range from 2 to 50 mg/L, a total imprecision lower than 15% and a total error lower than 30%. After protein precipitation with acetonitrile, the separation was performed on a C18 Atlantis T3 column and eluted with sodium acetate 25 mM, acetonitrile 2.5%, pH 6.5. Thioacetamide was used as internal standard. The method was linear for drug concentrations ranging from 0.5 to 50 mg/L and recovery was comprised between 100 and 120%. The intra-day precision was lower than 6.0% and between-day precision was lower than 11%. The detection limit was 0.18 and 0.63 mg/L for aqueous solution and plasma, respectively and the quantification limit was 1.0 and 1.2 mg/L for aqueous solution and plasma, respectively. No interference from urea was observed. The liquid chromatography method developed can be used for pharmacokinetic studies in plasma and other biological samples such as saliva or urine. It requires low sample volumes and a simple pre-treatment and it allows a direct measure of non-derivatized hydroxyurea.

Keywords: Hydroxyurea; Hydroxycarbamide; Thioacetamide; Liquid chromatography; Electrochemical detection; Sickle cell disease

Introduction

Hydroxycarbamide, better known as hydroxyurea, is a cytostatic agent used in the treatment of myeloproliferative disorders such as essential thrombocythemia and myelofibrosis [1,2]. It is also the most active drug in sickle cell disease, a genetic disorder due to the mutation of the 6th codon of the β-globin gene, leading to the synthesis of an abnormal hemoglobin (Hb S); sickle cell disease is characterized by chronic hemolytic anaemia and vaso-occlusive crises [3]. The main beneficial effect of hydroxyurea is to increase cellular levels of fetal hemoglobin (Hb F), which reduces Hb S polymerization [4]. Others mechanisms were demonstrated or suggested such as reduced expression of adhesion molecules, increased nitric oxide production, cation transport changes and myelosuppressive effects [5,6].

Despite hydroxyurea therapy has shown clinical improvement for sicke cell patients [7-9], differences in response i.e., the increase in Hb F levels, are observed and accurate predictors of hydroxyurea efficacy do not currently exist [10]. When hydroxyurea is administrated orally at a dose of 20 mg/kg, plasma concentrations normally reach a peak within 2 h (T_{max}) with a mean maximal concentration of 26 mg/L (C_{max}). Nevertheless, two phenotypes have recently been described: a "fast" (T_{max} of 15 or 30 min) and a "slow" one (T_{max} of 60 or 120 min) [10]. Pharmacokinetics studies are therefore required to explain these differences and to establish prediction factor.

Several methods have been developed to quantify hydroxyurea in biological samples: colorimetric techniques [10], liquid chromatography coupled with UV spectrophotometry [11,12], GC-MS [13,14], reversed phase liquid chromatography coupled with mass spectrometry [15]. Colorimetric methods require sample volumes of 250-500 µL and are insensitive. GC-MS needs a preliminary derivatization of hydroxyurea.

Here we describe the development and validation of a simple liquid chromatography method with electrochemical detection to quantify hydroxyurea in aqueous solutions and plasma.

Material and Methods

Chemicals and pre-analytical treatment

A stock solution of hydroxyurea (Sigma Aldrich, Steinheim, Germany) at 1000 mg/L was prepared and kept at -20°C. Aqueous and plasma standards were prepared by dilution of the stock solution in distilled water or human plasma pool.

Thiourea, methylurea, 2-thiouracil and thioacetamide (Sigma Aldrich, Steinheim, Germany) were tested as internal standard. The solutions were prepared at a final concentration of 100 mg/L containing 3 g/L albumin (Behring Institut, Marburg, Germany).

10 µL of samples or standards were added with 10 µL of internal standard solution and 100 µL acetonitrile (Biosolve, Dieuze, France). After centrifugation at 4000 rpm during 10 minutes, 80 µL of supernatant were recovered and either diluted to 400 µL with mobile phase or dried under nitrogen and reconstituted in 400 µL of mobile phase.

Two mobile phases were tested. The first one was described by Pujari and collaborators [16] and was composed of 0.2 M perchlorate and methanol 95/5 (V/V) (Sigma Aldrich, Steinheim, Germany). The

*Corresponding author: Frédéric Cotton, Department of Clinical Chemistry, Erasme Hospital, Université Libre de Bruxelles, 808 Route de Lennik - 1070 Brussels, Belgium, E-mail: fcotton@ulb.ac.be

second one, reported by Jong and collaborators [17] was composed of 25 mM sodium acetate (Merck, Darmstadt, Germany), 120 mg/L NaCl (VWR International, Leuven, Belgium), 0.1% diethylamine (Sigma Aldrich, Steinheim, Germany) and 2.5 M acetonitrile 2.5% (Biosolve, Dieuze, France), pH 6.5.

Liquid chromatography

The chromatography was performed on a Waters instrument (Waters 717 plus autosampler with a Waters 515 pump) connected to a Waters 2465 electrochemical detector using a carbon working electrode and an Ag/AgCl reference electrode. Elution was carried out on an Atlantis T3 column (5 μm, 250×4.6 mm). Two potentials (560 and 610 mV) were applied to the working electrode. Data were collected at rate of 5 points / second. Five μL of sample were injected and eluted at a flow rate of 0.5 mL/min.

Calibration

Calibration curve was performed correlating HU-enriched plasma or HU aqueous solution concentrations and height ratio between HU and internal standard. Five points are used: 2; 5; 10; 20; 30 and 50 mg/l. Two injections were performed and analyzed. Regression lines were built by linear regression with $1/x^2$ weighting, using EMPOWER2 software.

Method validation

Linearity: Aqueous and plasma standards at different concentrations (0.5, 1, 2, 5, 10, 20, 30, 50, 100 mg/L of hydroxyurea) were analyzed in duplicate. Regression lines corresponding to concentrations ranging from 0.5 to 100 mg/L, 0.5 to 50 mg/L, 0.5 to 30 mg/L, and 1 to 50 mg/L of hydroxyurea were calculated and slopes were compared by the Student t-test. A statistical difference between slopes was considered as a proof of non linearity.

Limit of detection: The mean hydroxyurea concentration corresponding to 10 determinations of hydroxyurea-free plasma or aqueous solutions plus two times the standard deviation was considered as the limit of detection.

Limit of quantification: Plasma and aqueous standards at different concentrations (0.5, 1.0, 1.2, 1.4, 1.6, 1.8, 2.0, 3.0; 4.0 and 5.0 mg/L of hydroxyurea) were analyzed in triplicate. The limit of quantification

was the lowest concentration associated with a total error of less than 30%.

Accuracy: The relative recovery of drug added to plasma was calculated at 4 levels in duplicate. The method was considered as accurate at concentrations where the recovery was comprised between 80 and 100%.

Precision: For intra-day precision, aqueous and plasma standards containing 10 mg/L of hydroxyurea, were analyzed 20 times during the same day. For inter-day precision, standards containing 10 mg/L of hydroxyurea were analyzed 20 times on different days over a period of 2 months. Mean and coefficient of variation were calculated.

Specificity: The influence of urea, tested at a concentration of 100 mg/L, was assessed in the regular chromatographic conditions.

Results and Discussion

Optimization of extraction and liquid chromatography conditions

Hydroxyurea is a very hydrophilic compound and solvent extraction from plasma is not possible. Therefore, the samples were treated by protein precipitation with acetonitrile. An additional step of evaporation did not enhance chromatographic results and was not retained for the final protocol.

Thiourea, methylurea, 2-thiouracil and thioacetamide were tested as internal standard. Only the last one provided the required criteria of stability, detection and separation from hydroxyurea.

A mobile phase containing sodium acetate was selected for the final method, with a detection voltage of 610 mV. A voltage of 560 mV was associated with lower signal stability. Symmetrical hydroxyurea peaks were detected at a retention time of 5.8 minutes. The detector response for thioacetamide was not stable in aqueous solutions but well in plasma. To overcome this problem, solutions were enriched with human albumin at a concentration of 3 g/L and were renewed each day.

Method validation

Hydroxyurea and the internal standard were separated within 12 minutes without interference from endogenous compounds (Figure 1

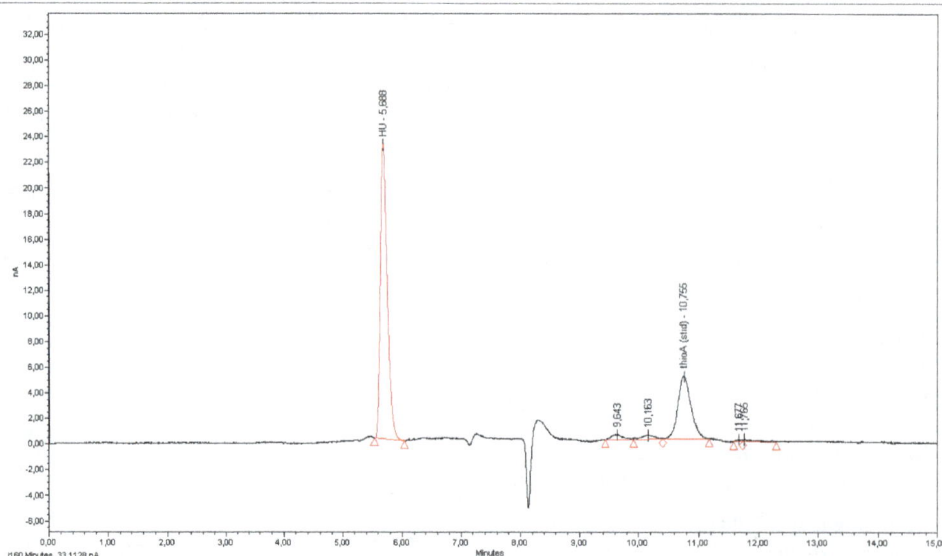

Figure 1: Chromatogram of hydroxyurea aqueous solution (10 mgL) enriched with internal standard.

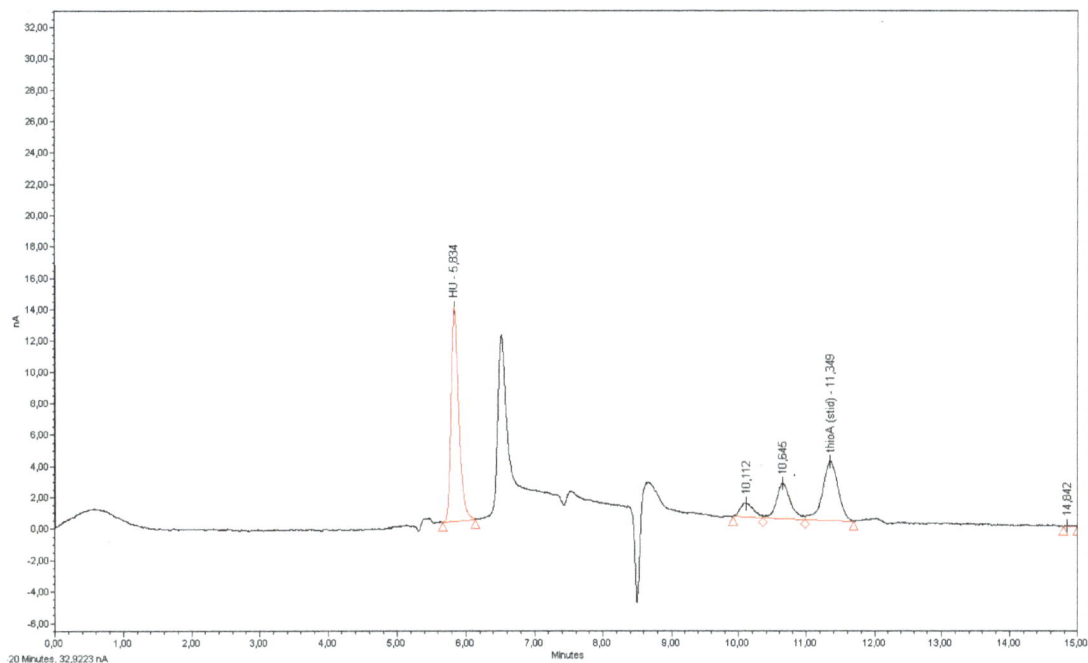

Figure 2: Chromatogram of hydroxyurea plasma solution (10 mgL) enriched with internal standard.

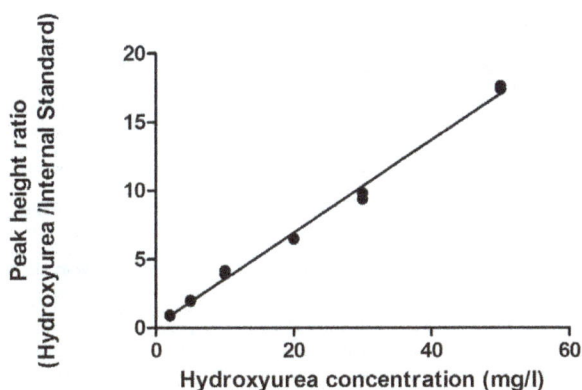

Figure 3: Typical plasma calibration curve.

Theoretical concentration (mg/l)	Calculated concentration (mg/l)	% Recovery
2.0	2.4	118
5.0	6.0	120
20.0	20.0	100
30.0	31.8	106

Table 1: Accuracy and recovery values for concentrations: 2; 5; 20 and 30 mg/l of hydroxyurea standards based in enriched plasma.

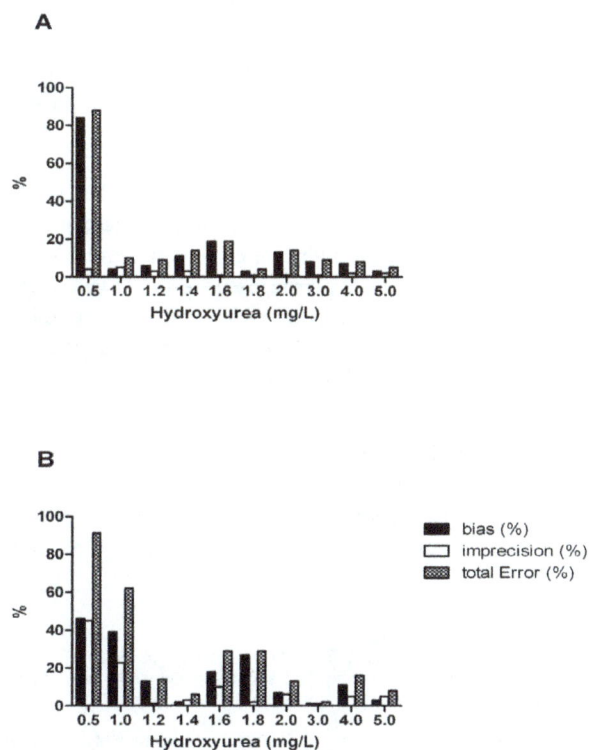

Figure 4: Estimation of the limit of quantification (A) In water (B) In plasma.

and 2). The column efficiency, the resolution and the peak symmetry were satisfactory. A typical calibration line is shown in Figure 3.

The method displayed good analytical performances with a linear range of 0.5-50 mg/L. Accuracy and recovery values for each concentration tested are presented in Table 1. All were lower or equal to 120%. The intra-day precision was 10.3% for aqueous and plasma standards and the inter-day precision was 5.1% and 6.0% for aqueous and plasma standards, respectively.

The limit of detection was 0.18 and 0.63 mg/L for aqueous and plasma solutions, respectively, and the limit of quantification was 1.0 mg/L for aqueous and 1.2 mg/L for plasma standards, in agreement with the analytical goals (Figure 4).

No interfering peak was observed when the urea solution was

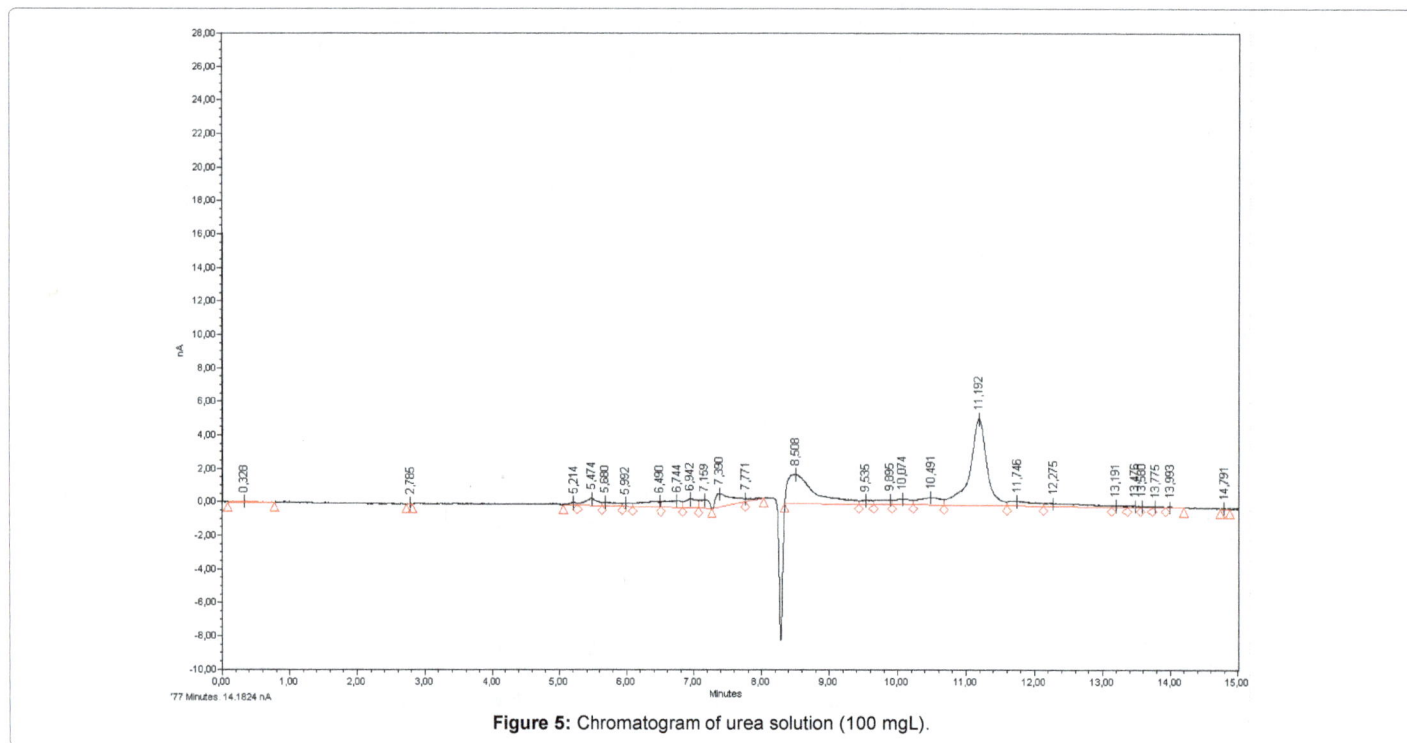

Figure 5: Chromatogram of urea solution (100 mgL).

analysed, reflecting a good specificity regarding this endogenous compound (Figure 5).

These data confirmed that the method was reliable and reproducible for the measurement of all targeted values expected in pharmacokinetic studies, both in plasma and saliva.

Conclusions

The accurate measurement of hydroxyurea in plasma is critical for a better understanding of hydroxyurea response and toxicity and to adapt the patient dose for a personalised treatment with improved clinical benefits. In different situations such as anemia or childhood, the sampling of saliva in place of blood is beneficial. Therefore, a method suitable for saliva analysis is needed too.

In this report, we describe a liquid chromatography method with electrochemical detection that permits the quantification of hydroxyurea in plasma and in aqueous solutions. Moreover, compared to previously published methods, this one presents several advantages like a low sample volume, a simple pretreatment and the direct measure of non-derivatized hydroxyurea.

References

1. Simunović M, Perković I, Zorc B, Ester K, Kralj M, et al. (2009) Urea and carbamate derivatives of primaquine: synthesis, cytostatic and antioxidant activities. Bioorg Med Chem 17: 5605-5613.

2. Gunnar Engström K, Löfvenberg E (1998) Treatment of Myeloproliferative Disorders WithHydroxyurea: Effects on Red Blood Cell Geometry and Deformability. Blood 91: 3986.

3. Ballas SK (2002) Sickle cell anaemia: progress in pathogenesis and treatment. Drugs 62: 1143-1172.

4. Levensburger JD, Pestina TI, Ware RE, Boyd KL, Persons DA (2010) Hydroxyurea therapy requires HbF induction for clinical benefit in a sickle cell mouse model. Haematologica 95: 1599-1603.

5. Adragna NC, Fonseca P, Lauf PK (1994) Hydroxyurea affects cell morphology,

cation transport, and red blood cell adhesion in cultured vascular endothelial cells. Blood 83: 553-560.

6. Halsey C, Roberts IAG (2003) The role of hydroxyurea in sickle cell disease. Brit J Haematol 120: 177-186.

7. Gulbis B, Haberman D, Dufour D, Christophe C, Vermylen C, et al. (2005) Hydroxyurea for sickle cell disease in children and for prevention of cerebrovascular events: the Belgian experience. Blood 105: 2685-2690

8. De Montalembert M (2008) Hydroxyurea treatment in patients affected with sickle cell anemia: efficacy and safety. Transfus Clin Biol 15: 34-38.

9. Ware RE (2010) How I use hydroxyurea to treat young patients with sickle cell anemia. Blood 115: 5300-5311

10. Ware RE, Despotovic JM, Mortier NA, Flanagan JM, He J (2011) Pharmacokinetics, pharmacodynamics, and pharmacogenetics of hydroxyurea treatment for children with sickle cell anemia. Blood 118: 4985-4991.

11. Bachir D, Hulin A, Huet E, Habibi A, Nzouakou R, et al. (2007) Plasma and urine hydroxyurea levels might be useful in the management of adult sickle cell disease. Hemoglobin 31: 417-425.

12. Sassi H, Bachir D, Habibi A, Astier A, Galactéros F, et al. (2010) No effect of CYP450 and P-glycoprotein on hydroxyurea in vitro metabolism. Fundam Clin Pharmacol 24: 83-90.

13. James H, Nahavandi M, Wyche MQ, Taylor RE (2006) Quantitative analysis of trimethylsilyl derivative of hydroxyurea in plasma by gas chromatography-mass spectrometry. J Chromatogr B Analyt Technol Biomed Life Sci 831: 42-47.

14. Kettani T, Cotton F, Gulbis B, Ferster A, Kumps A (2009) Plasma hydroxyurea determined by gas chromatography-mass spectrometry. J Chromatogr B Analyt Technol Biomed Life Sci 877: 446-450.

15. Sreedevi V, Kumar PR, Thatavarti R (2011) LC-MS Method Development and validation for the estimation of Felodipine in human plasma and Stability studies of freeze thaw analyte. IJPSR 2: 65-73.

16. Pujari MP, Barrientos A, Muggia FM, Koda RT (1997) Determination of hydroxyurea in plasma and peritoneal fluid by high-performance liquid chromatography using electrochemical detection. J Chtomatogr B Biomed Sci Appl 694: 185-191.

17. Jong YJ, Hsu HO, Wu HL, Kou HS, Wu SM (2003) Analysis of hydroxyurea in human plasma by high performance liquid chromatography with electrochemical detection. Anal Chim Acta 488: 223-230.

11

An Improved HPLC Method for Quantification of Metanephrine with Coulometric Detection

Pedro AQ[1#], Soares RF[1#], Oppolzer D[1], Santos FM[1], Rocha LA[1], Gonçalves AM[1], Bonifacio MJ[2], Queiroz JA[1], Gallardo E[1] and Passarinha LA[1*]

[1]CICS-UBI-Centro de Investigacaoem Ciencias da Saude, Universidade da Beira Interior, 6201-506 Covilha, Portugal
[2]Departamento de Investigacao e Desenvolvimento, BIAL 4745-457 Sao Mamedo Coronado, Portugal
[#]Authors contributed equally to this work

Abstract

A rapid and straightforward analytical method, based on the use of RP-HPLC with coulometric detection, was developed and validated for the quantification of metanephrine, an O-methylated product in catechol-O-methyltransferase enzymatic assays. The isocratic separation was achieved on a reverse column with a mobile phase consisting of 0.1 M sodium dihydrogen phosphate, 0.024 M citric acid monohydrate, 0.5 mM sodium octyl sulphate and 9% acetonitrile (%v/v). The method was found to be linear between 0.25 and 15 nmol/mL with a determination coefficient of 0.9997 for metanephrine. Intra-and interday precision and accuracy were in conformity with the criteria accepted in bioanalytical method validation and the LOD and LLOQ were 0.25 nmol/mL. The main focus of the developed method is the lower LLOQ achieved that can have important implications in laboratory research for COMT activity determinations, in particular for the methionine 108/158 variant obtained either from native or recombinant extracts. Another major advantage of the present method is the shorter run times on automated chromatographic systems that allow the analysis of several samples in a short time.

In addition, metanephrine was stable in the samples for at least 24 h at room temperature, for at least 24 h in HPLC system injector and for at least three freeze/thaw cycles. The developed method demonstrated higher sensitivity, precision, accuracy, stability, and linearity when compared with the methods previously described. Finally, a catechol-O-methyltransferase activity assay, resulting in an O-methylated reaction product, was used to evaluate the method applicability.

Keywords: HPLC; Coulometric detection; Metanephrine; Catechol-O-methyltransferase.

Introduction

Normetanephrine (NMN) and metanephrine (MN), O-methylated metabolites of norepinephrine and epinephrine respectively, are produced by the actions of catechol-O-methyltransferase (COMT) (EC 2.1.1.6), an enzyme largely confined to extraneuronal tissues [1], that needs accurate and selective measurements, not only for clinical diagnosis but also for pathological studies of several diseases. In fact, as the catechol-metabolizing system that comprises COMT has a potential pathophysiological and pathogenic significance in several disorders [2,3], it becomes important to the determination of COMT biological activity and determination of metanephrine with lower detection limits and improved sensitivity.

MN quantification, after chemical or enzymatic hydrolysis of the conjugated forms, is still important for diagnosing neural crest tumors, particularly in differentiating between pheochromocytoma and hypertension [4]. Several analytical methods, in particular, HPLC, have been proposed for the analysis of catecholamines and their metabolites (O-methylated reaction products) in biological fluids. Several detecting techniques are depicted in bibliography, namely MS [5], UV spectrophotometry [6], Fluorometry [7], RIA [8] and chemical luminescence [9].

Liquid chromatography coupled with electrochemical detection (LC-ECD) has provided a new tool to evaluate the levels of these compounds in urine [10], plasma [11] and COMT assays [12] and is considered a reliable technique for catecholamine assay [13]. Typically, catecholamines and their metabolites can be separated by RP HPLC systems with ion-pairing reagents or by ion-exchange HPLC and detected by their reversible oxidation by amperometric [11,12] or coulometric carbon-based working electrodes [14,15]. Analysis of electroactive compounds, like catecholamines, in biological samples/

extracts by HPLC with electrochemical detectors in coulometric mode has gained more interest in the last years. As a matter of fact, ECD has been improved since the appearance of the flow-through electrodes (coulometric detectors) that can react with near of 100% of the electroactive components of the analyte. Recently, developed coulometric sensors provide selectivity, identification, and resolution of compounds when used as detectors in HPLC systems. Coulometric detectors claim a better performance for sensitivity and selectivity than the classic amperometric detector, making the coulometric detection an advanced tool capable of addressing the analytical complexity of biological samples/extracts. In fact, coulometric detection in HPLC ECD, a technique in which all of the analyte in the column effluent is oxidized or reduced at the surface of an electrode at constant potential, offer certain advantages over amperometric detection, in which only a few percent of the analyteis converted [16]. A major concernin amperometric HPLC ECD is the decrease of electrode response in time, which is often attributed to the reduction of the active area by adsorption phenomena. On the other hand, when the electrode area of a coulometric detector is large enough, the conversion efficiency may still be 100% despite some loss of active area, so that the response is unaffected [16]. Coulometric detection has been used in detection of

***Corresponding author:** Passarinha LA, Health Sciences Research Centre - University of Beira Interior (CICS-UBI), Av. Infante D. Henrique, 6200-506, Covilha, Portugal, E-mail: lpassariha@fcsaude.ubi.pt

clinical important substances in Plasma [17], Saliva [18], Urine [19], including catecholamine detection [20].

The sensitivity limitations imposed by the large areas of coulometric (>99% efficient) detectors and the theoretical advantages of 100% detection compared to the 2-5% detection achieved in amperometric sensors, have been recognized with detection limits inside the femtogram (10-15) region [15]. In coulometric detection, 100% oxidation efficiency at one electrode of the analyte and their reduction at a second electrode allow a great increase in selectivity and sensitivity for compounds with reversible oxidation over non-reversible compounds, such as MN, a typical O-methylated product of COMT in the presence of epinephrine. Furthermore, COMT has been described as an important drug target in Parkinson's disease [2], emphasizing the importance of the development of methods with improved LLOQ that allow the quantification of lower MN levels and, consequently, the determination of low enzyme activity levels.

In this work, we describe a novel method using HPLC with coulometric detection, with adequate sensitivity for reliable measurement of MN in biological bacteria lysate extracts for the first time, by calibration procedures, including intra- and interday precision and stability. Moreover, this method was fully validated in a wide concentration range with lower limit of quantification. We evaluated the method applicability through in vitro membrane COMT activity assays, using epinephrine as substrate, resulting in an O-methylated reaction product such as MN. The method reported here can be applied to clinical MN assessment on biological fluids (plasma, urine) after suitable extraction procedures due to the high sensitivity and speed shown, associated to a greater selectivity of the coulometric detectors.

Experimental

Reagents and standards

Analytical standards of MN (DL-Metanephrine Hydrochloride), citric acid monohydrate and 1-Octanesulfonic acid were purchased from Sigma Aldrich (Steinheim, Germany). Acetonitrile (HPLC grade) and sodium dihydrogen phosphate were obtained from Fisher Scientific (Leicestershire, UK) and USB Corporation (Ohio, USA), respectively. Perchloric acid was purchased from Panreac (Barcelona, Spain). Deionized water was obtained from a Millipore purification system in our research laboratory. Stock solution at 25 nmol/mL of MN was prepared in 0.2 M perchloric acid. Standard solutions were obtained by diluting stock solution with the same solution of perchloric acid. All these buffers were stored and protected from light at 4°C and were stable for at least three months.

Instrumentation

Chromatographic analysis was performed using a HPLC model Agilent 1260 system (Agilent, Santa Clara, California, USA) equipped with an autosampler and quaternary pump coupled to an ESA Coulochem III (Milford, Massachussets, USA) coulometric detector. Chromatographic separation was achieved on an analytical column Zorbax 300SB C_{18} RP analytical column (250×4.6 mm i.d. 5 μm) (Agilent, Santa Clara, California, USA). The mobile phase (0.1 M sodium dihydrogen phosphate, 0.024 M citric acid monohydrate, 0.5 mM sodium octyl sulphate and 9% acetonitrile, v/v), pH 2.9, was filtered under vacuum (0.2 μM hydrophilic polypropylene filter), degassed in ultrasonic bath before use. Under this procedure the mobile phase was maintained stable during 2 weeks. Column effluent was monitored with an electrochemical detector in the coulometric mode, which was equipped with a 5011 high sensitivity dual electrode

analytical cell (electrodes I and II) using a procedure of oxidation/reduction (analytical cell #1: +410 mV; analytical cell #2: -350 mV). The high surface area of electrodes I and II results in a 100% reaction of the electroactive compound (MN). The method sensitivity was set at 1 μA and the flow rate applied was 1mL/min. Column temperature was optimized to 30°C. The chromatograms were obtained by monitoring the reduction signal of the working electrode II. The retention time was around 8.8 min for MN.

Standards preparation

A stock solution of MN (25 nmol/mL) was prepared by dissolving the appropriate salt in 0.2 M perchloric acid. The initial solution was diluted in 0.2 M perchloric acid in order to obtain MN standard solutions at several concentrations (0.25 to 15 nmol/mL). The standard samples were agitated by rotation/inversion movements for 2 min and injected into the HPLC-ECD system according to the experimental conditions.

Validation procedure

The procedure was validated in terms of selectivity, linearity, intra- and inter-day precision, accuracy, and stability. Calibration data was generated by spiking samples and the calibration curve was established between 0.25 and 15 nmol/mL (eight calibrators evenly distributed). Five calibration curves were prepared, and the criteria for acceptance included a R^2 value of at least 0.99, and the calibrators accuracy within a ± 15% interval, except at the lower LLOQ (LLOQ), for which ± 20% was accepted. The LLOQ was defined as the lowest amount of analyte that presented a signal-to-noise ratio of at least 5 and could be measured with adequate precision and accuracy (coefficient of variation of less than 20% and an accuracy of ± 20%). Intraday precision was characterized in terms of RSD (%) by analyzing sets of 5 MN samples at five different concentrations (0.25, 2, 4, 8 and 12 nmol/mL) over a 5-day period. Interday precision was assessed at eight concentrations (0.25, 0.5, 1, 2, 4, 8, 12, 15 nmol/mL) over a 5-day period. Accuracy was evaluated in terms of mean relative error between the measured and the spiked concentrations for the calibrators and also in the intra- and interday precision assays; the limits of acceptable variability were set at 15% for all concentrations, except at the LLOQ, for which 20% was accepted. Processed sample-stability, short-term stability and freeze/thaw stability were studied (n=3) at two concentration levels (3 and 10 nmol/mL).

MN samples were subjected to different storage conditions, and the obtained results were compared to those achieved after analysis of freshly prepared samples. The compound was considered stable under the tested conditions if the coefficients of variation between the two sets of samples were less than 15%.

Membrane-bound catechol-O-methyltransferase expression and enzymatic assay

Recombinant human membrane-bound catechol-O-methyltransferase (hMBCOMT) biosynthesis was carried out according to the procedure previously described by Pedro et al. [3]. Cells were grown in a mineral salt medium supplemented with soytone 1% (w/v) for 50 hrs at 30°C and 120 rpm. The experiments of activity were designed to evaluate the methylation efficiency of recombinant hMBCOMT, by measuring the amount of MN, using epinephrine as substrate, as previously described by Passarinha et al. [12]. In hMBCOMT activity assay, a 2000 μg/mL aliquot of the membrane extract after suitable solubilization, was incubated in 5 mM sodium phosphate buffer (pH 7.8) containing 0.2 μM $MgCl_2$, 2 mM EGTA, 250 μM SAM e and 1 mM epinephrine in a total sample volume of 1 mL. Reactions were carried out at 37°C during 15 min and were stopped by incubation in ice following the addition of 2M 200 μl perchloric acid. The supernatants were centrifuged, filtered and subsequently injected into the HPLC system.

Results and Discussion

The methodology presented in this work concerns the development and validation of an HPLC method with coulometric detection for assessment of MN, allowing the determination of the biological activity of recombinant COMT as well as native COMT extracted from animal tissues or cell lines. This method was validated using authentic samples obtained from a bioprocess intended to synthesize the membrane-bound isoform of COMT [3]. On the other hand, in humans, the activity of COMT can be distributed into three classes with high, intermediate and low activity groups in which such difference is correlated with a functional polymorphism at codon 108/158 (SCOMT/MBCOMT) involving a methionine/valine substitution in the polypeptide chain [2]. While the metionine 108/158 variant is associated with low enzymatic activity [2], the development of HPLC methods with improved LLOQ and sensitivity capable of measuring lower quantities of MN seems to be of great importance not only to monitor bioprocesses by measuring the biological activity of the target recombinant COMT but also to measure the biological activity of low activity native COMT extracted from animal tissues or cell lines in different pathophysiological and pathogenic conditions.

Chromatographic methodology

The chromatographic conditions were chosen to allow selectivity on the basis of the ionic and hydrophobic characteristics of MN.

Chromatographic parameters such as temperature and mobile phase (data not shown) were investigated in order to obtain a higher detection of MN within an acceptable time span. Cell potentials were optimized according to preliminary experiments (data not shown) and ESA Biosciences, Inc indications. The C18 stationary phase has been successfully employed for the separation and quantification of catecholamines and its metabolites, as previously shown by Passarinha et al. [12], Hollenbach and collaborators [21] and Unceta [22]. Therefore and based on these results, the C_{18} was applied as the stationary phase in this work.

In order to obtain better electrochemical oxidation efficiency, the larger electrode surface area coulometric cell, compared to the amperometric, results in a 99% reaction of our electroactive compound-MN. While the electrochemical behavior of MN is reversible at carbon electrodes, we used the oxidation/reduction mode in the coulometric detector because it plays an important role in improving both selectivity and sensitivity of our analysis. The electrode 1 (E1-oxidation channel) was applied only to modify the molecules and reduce contaminants in the sample before the detection on electrode 2 (E2-reduction channel). In the E1, MN was oxidized generating oxidation products which were reduced at E2. In this way, monitoring of the channel E2 current eliminates a great deal of the interferences and improves the signal/noise ratio, emphasizing the chromatographic response for MN (see Figures 1 and 2).

Method Validation

The method was validated in a 5-day validation protocol. The validation parameters included linearity and LLOQ, intra-and interday precision and accuracy, and stability were performed according to the

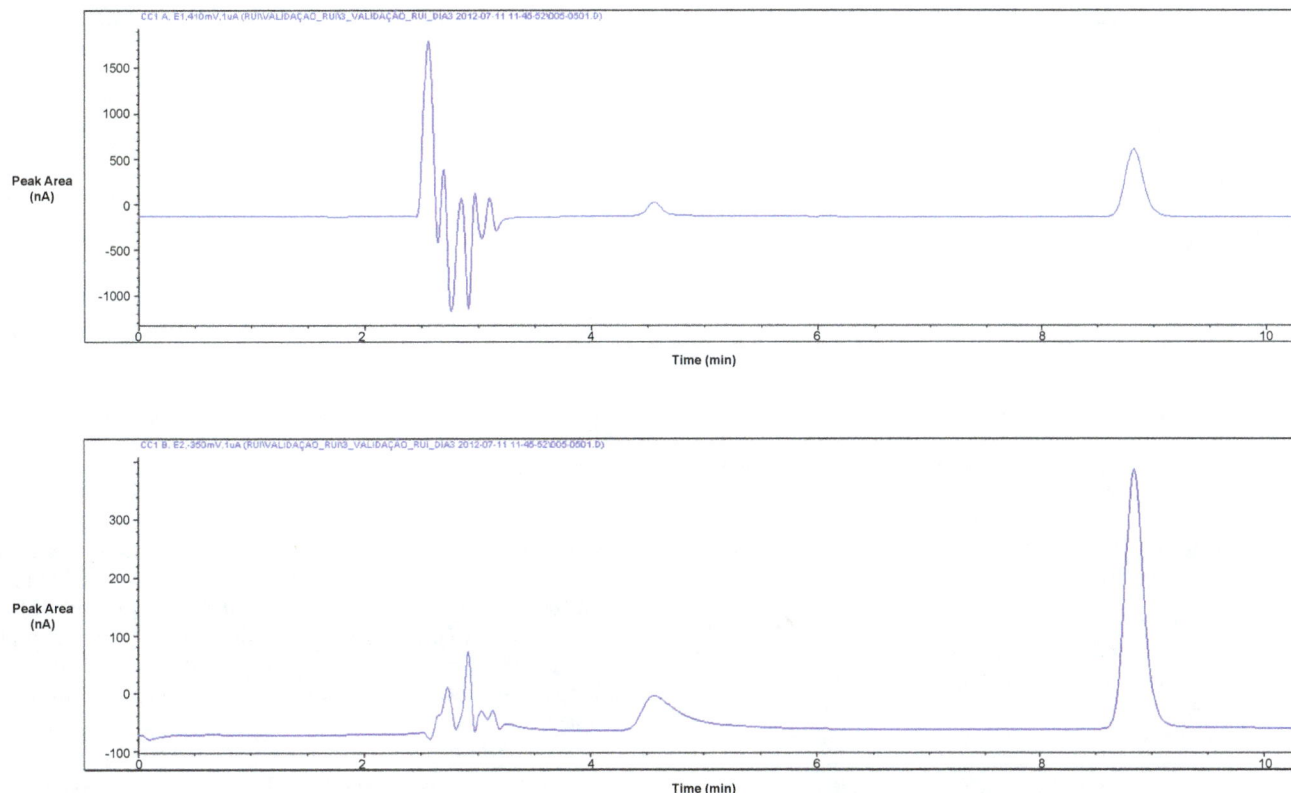

Figure 1: Chromatogram of MN at 2 nmol*mL^{-1}; (a) E1-Oxidation, (b) E2-Reduction (down); retention time (MN) of 8.873 min.

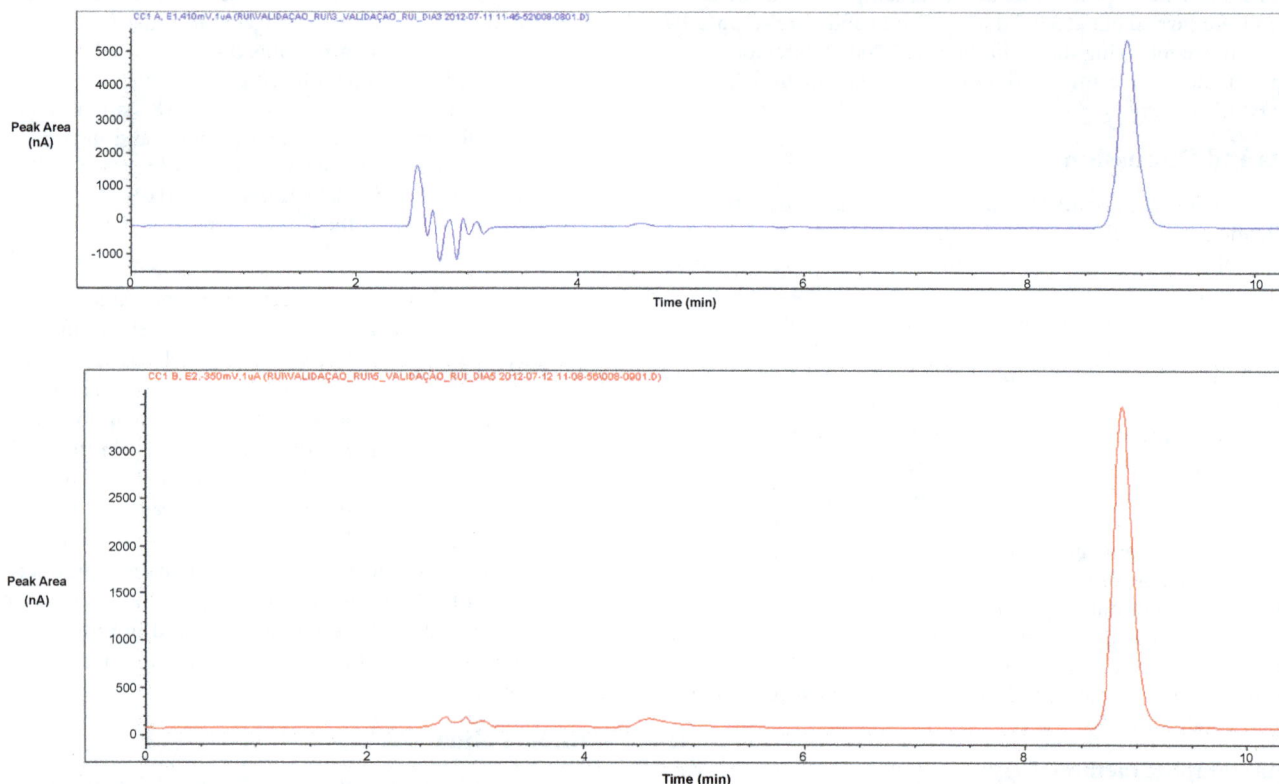

Figure 2: Chromatogram of MN at 15 nmol*mL^{-1}; (a) E1-Oxidation, (b) E2-Reduction (down); retention time (MN) of 8.873 min.

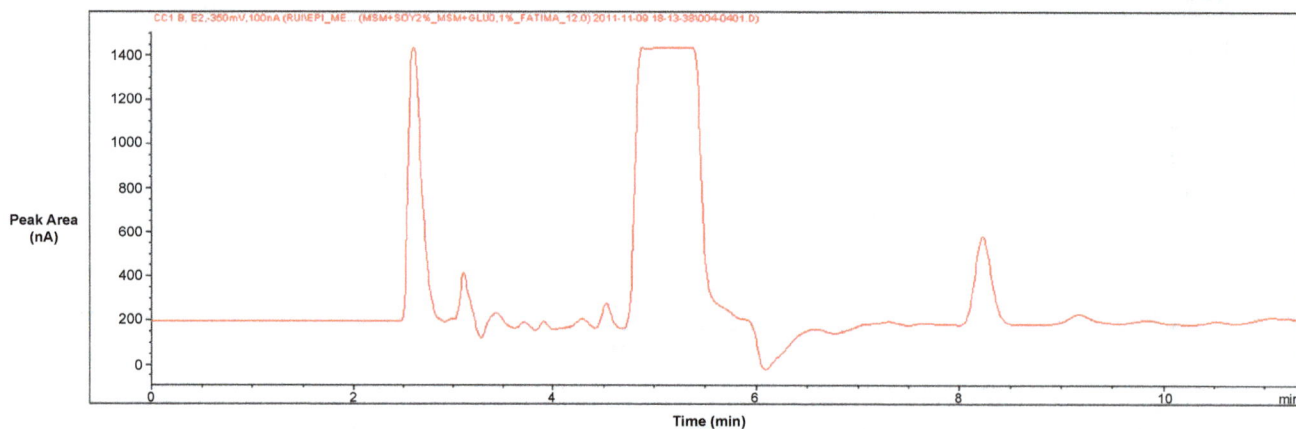

Figure 3: Chromatogram of a biological extract sample (Retention time (MN) of 8.220 min).

guidelines principles of the Food and Drug Administration [23] and International Conference on Harmonization [24].

As shown in Figure 4, the HPLC chromatogram of a fermentation extract sample demonstrates that the method provides an excellent resolution and selectivity between the compounds of interest, allowing the samples to be directly injected without pretreatment. In addition, the different components of the bacteria extract sample were injected as a "blank analysis" in order to evaluate possible interferents at the respective retention times of the analyte but none was observed.

The linearity of the method for MN quantification was established

between 0.25 and 15 nmol/mL with eight evenly distributed calibrators and the following values were obtained (mean values ± SD): 2943.4 ± 24.2664 for slope (m), 126.3828 ± 76.0743 for intercept (b) and a R^2 of 0.9997 ± 0.0001. Each calibration level was in quintuplicate and determination coefficients higher than 0.99 were obtained. The calibration curves were obtained by plotting the peak area ratio between the analyte versus analyte concentration. The calculated concentration of each calibrator had to be within a ± 15% interval of target except for the LLOQ, where ± 20% was accepted. In addition, to each calibration curve, two control samples were prepared in triplicate (n=15) at the concentration 3 and 10 n nmol/mL, the low quality control (LQC)

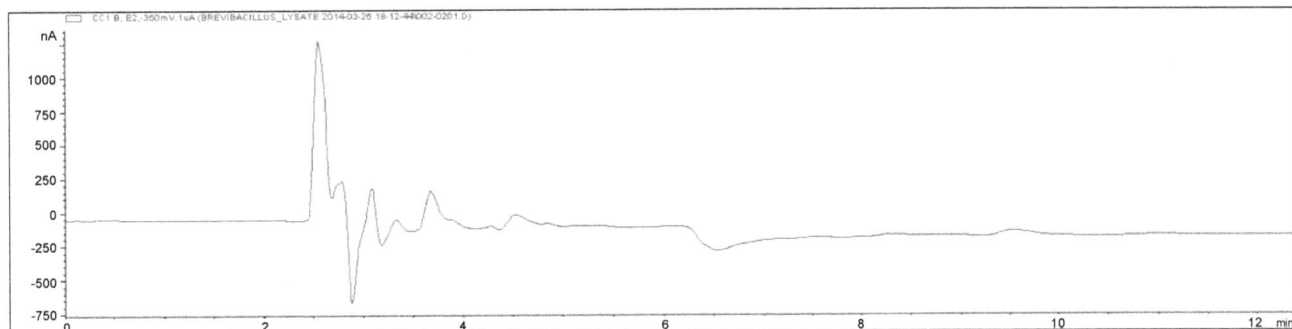

Figure 4: Representative chromatogram of a *Brevibacillus choshinensis* lysate extract sample (blank).

Spiked	Measured concentration		CV (%)		RE (%)	
	Intraday (n=6)	Interday (n=5)	Intraday (n=6)	Interday (n=5)	Intraday (n=6)	Interday (n=5)
0.25	0.28 ± 0.01	0.24 ± 0.02	4.36	8.78	11.55	-4.13
0.5	-	0.51 ± 0.02	-	4.52	-	1.42
1	-	1.01 ± 0.04	-	4.16	-	1.33
2	1.98 ± 0.02	1.98 ± 0.04	1.03	1.96	-0.85	-0.87
4	3.94 ± 0.03	3.94 ± 0.07	0.74	1.67	-1.57	-1.56
8	7.94 ± 0.03	8.10 ± 0.07	0.33	0.87	-0.79	1.22
12	11.81 ± 0.05	12.03 ± 0.14	0.47	1.18	-1.60	0.28
15	-	14.94 ± 0.13	-	0.88	-	-0.40

All concentrations in nmol*mL^{-1}; CV, coefficient of variation; RE, relative error [(measured concentration−spiked concentration)/spiked concentration×100].

Table 1: Intra-day and Inter-day precision and accuracy (n=5).

and high quality control (HQC) respectively. The LLOQ was defined as the smallest concentration of analyte (MN) that could be measured reproducibly and accurately (coefficient of variation less than 20% and calculated concentration within a ±20% interval from the target level) and was established at 0.25 nmol/mL. The LOD, was not systematically evaluated, and was defined to be the same as the LLOQ for practical reasons.

On the other hand, since the coulometric detector presents a high sensitivity, this parameter was set at 1 μA, allowing the detection of the MN standards with higher concentration without channel saturation and not compromising the detection of low standards. Moreover, following this methodology, a retention time around 8.8 min for MN was obtained, faster and more stable hands-on sample than reported by Gamache et al. [25] and Lenders [11] using coulometric detection and better precision/accuracy and lower retention time than reported by Passarinha et al. [12] and Parker and collaborators [26] with amperometric detection. In addition, our method showed a better efficiency and sensitivity, being able to detect MN concentrations of 0.25 nmol/mL with an excellent signal-to-noise ratio, better than reported by amperometric detectors [12], making the coulometric detection an advantageous method to MN quantification.

Other parameters evaluated for this method were the intraday, interday and intermediate precisions. In what concerns to the intraday precision and accuracy (relative error, %) for MN, it was determined by analysis of six independent replicates at five concentrations across the dynamic range of the assay: 0.25; 2; 4; 8 and 12 nmol/mL. The obtained CVs were typically below 5% at all concentrations, while relative errors were within a ±12% interval (see Table 1).

In addition, the interday precision and trueness were evaluated at eight concentrations within a 5-day period. The analysis of the interday precision and trueness yielded CVs generally lower than 9% at all concentration levels, while trueness was within a ±5% interval (see Table 1).

Another parameter evaluated in this work was the intermediate precision (combined intra- and interday) that was determined using the QC samples (LQC and HQC) that were prepared and analysed simultaneously with the calibration curves on 5 different days (15 measurements for each concentration). The CVs were typically below 2% at all concentrations, while accuracy was within ±6% of the nominal concentration.

Finally, the stability of metanephrine was assessed using samples spiked at the above-mentioned QC concentrations (n=3). In order to study the stability of the processed samples, the samples were left standing at room temperature and in the HPLC autosampler for 24 hrs. Those samples were compared to freshly prepared samples, and the obtained coefficients of variation were less than 10% for all compounds, meaning that MN are stable in the samples for at least 24 h at room temperature. Furthermore, the freeze/thaw stability was evaluated as follows: the MN QC samples were spiked at the intended concentrations and stored at −20°C for 1hrs and after this period, the samples were thawed single-handedly at room temperature. Following the third freeze/thaw cycle, the samples were analyzed and the obtained peak areas were compared to those obtained by analysis of freshly prepared samples. The MN was found to be stable for at least 3 freeze/thaw cycles (the obtained CVs were less than 10% for all compounds) with all values lying within ±15% and, hence, were considered acceptable (see Table 2). Also, the target analyte shows stability at room and injector temperature (4°C) up to 24 hrs, which is in accordance with previous findings [27]. Extended storage is possible at -20°C, because at least 3 freeze-thaw cycles had no influence on MN detection.

Method application with biological samples

After validation, the described procedure was applied in three

Spiked	Stability at room temperature			Stability of injector			Freeze stability		
	Measured	CV (%)	RE (%)	Measured	CV (%)	RE (%)	Measured	CV (%)	RE (%)
3	2.63 ± 0.28	9.83	-12.49	2.72 ± 0.13	2.72	-9.45	2.63 ± 0.26	9.81	-12.48
10	9.45 ± 0.93	9.89	-5.53	9.70 ± 0.58	9.70	-3.01	9.51 ± 0.84	8.81	-4.86

All concentrations in nmol*mL^{-1}; CV, coefficient of variation; RE, relative error [(measured concentration-spiked concentration/spiked concentration)×100]

Table 2: Stability data (n=3).

biological samples obtained from a MB-COMT enzymatic activity assay. A typical chromatogram of one of these biological samples is shown in Figure 3. Moreover, as MN is susceptible to oxidation due to the presence of the catechol structure, the presence of a reducing agent in standard solutions (perchloric acid 0.2 M) usually guarantees a higher stability of the analyte avoiding the oxidative degradation products of MN, which strongly influence the accuracy of the analysis [12]. Therefore, as can be seen in the chromatogram of a biological sample (see Figure 3), is possible to observe extra peaks corresponding to perchloric acid and epinephrine (used as substrate in MB-COMT enzymatic assay). However, these peaks don't interfere with MN (reaction product of MB-COMT enzymatic assay) in terms of magnitude and shape of peaks. In addition, all the compounds of the sample, namely MN, give neat and resolved chromatographic peaks as well as a very good selectivity to our interest compound. The mean concentration of MN present in the 3 biological samples was 1.499 ± 0.0135 nmol/mL. Finally, in order to evaluate possible interferents that could overlapping at the respective retention time of MN, a blank (bacteria lysate extract deproteinized with percloric acid) was injected into the HPLC system with the same operatory conditions (Figure 4). As we can observed for the retention time range of 8 to 9 min none components from the bacteria extract sample can be visualized, demonstrating that the analytical method develop is appropriated for this biological matrices.

Conclusions

A simple procedure employing a HPLC system with coulometric detector was developed and fully validated for the qualitative and quantitative determination of MNs from MB-COMT recombinant *Brevibacillu schosinensis* extracts. The proposed HPLC method with ECD, which employs an innovative coulometric detection in MB-COMT enzymatic assays, is suitable for the analysis of the main endogenous catecholamines, namely epinephrine and its main metabolite MN.

The optimization of the chromatographic conditions led to a more rapid, efficient and sensitive method, allowing the clear determination of MNs in MB-COMT enzymatic assays. The HPLC-ECD method developed in this work, in comparison with previously reported methods, namely the metanephrine detection quantification by HPLC coupled with amperometric detection [12], presents shorter chromatographic runs (10 minutes in contrast with 20 minutes). In addition, this method also presents a very good resolution to detect MN and the greater sensitivity associated with the coulometric detector, in contrast to other existing methods. Finally, another major advantage of this method is the improvement on the signal/noise ratio achieved that allowed us to lower the LLOQ to metanephrine from 1 nmol/mL [12] to 0.25 nmol/mL^{-1}, what can be of extremely importance in determining low COMT activity levels.

The method sensitivity, linearity, limits and intra- and interday precision and accuracy were adequate, allowing MN detection even at very low concentrations. Therefore, this procedure may be useful both in research and in routine analysis, more specifically in the quantification of low metanephrine concentrations associated with

the COMT biological activity obtained either from recombinant or native sources. In conclusion, the accurate determination of lower metanephrine levels within a short time analysis provided by this method may have important implications in COMT research and, consequently, in the development of new and more effective COMT inhibitors for application in Parkinson's disease.

Acknowledgements

A.Q. Pedro acknowledges a doctoral fellowship (SFRH/BD/81222/2011) from Fundaçãopara a Ciência eTecnologia within the scope of QREN-POPH-Advanced Formation programs co-funded by Fundo Social Europeu and MEC. This work was partially funded by Fundação da Ciência e Tecnologia I.P. (PIDDAC) and Fundo Europeu de Desenvolvimento Regional-FEDER funds through Programa Operacional Factores de Competitividade (POFC)-COMPETE: FCOMP-01-0124-FEDER-027563 and by National Funds through FCT-Fundaçãopara a Ciência eTecnologia within the scope of Project "EXPL/BBB478/BQB/0960/2012". Also, Santos F.M. and Gonçalves A.M. acknowledges a fellowship from the project "Technologies for purification and controlled release of biopharmaceuticals to be applied in age-related disease" (CENTRO-07-ST24_FEDER-002014), funded by ProgramaOperacional Regional do Centro2007-2013 QREN (Programa "Mais Centro"). The authors also acknowledge the MscFilipa Correia, Luís Martins and Guilherme Espirito Santo for technical support on HPLC-ECD.

References

1. Chan E, Ho P (2000) High performance liquid chromatography/atmospheric pressure chemical ionization mass spectrometric method for the analysis of catecholamines and metanephrines in human urine. Rapid Commun Mass Spectrom 14: 1959-1964.

2. Bonifacio MJ, Soares-da-Silva P (2007) Catechol-O-methyltransferase and its inhibitors in Parkinson's disease. CNS Drug Rev 13: 352-379.

3. Pedro AQ, Bonifacio MJ, Queiroz JA, Maia CJ, Passarinha LA (2011) A novel prokaryotic expression system for biosynthesis of recombinant human membrane-bound catechol-O-methyltransferase. J Biotechnol 156: 141-146.

4. Rosano T, Swift T, Hayes L (1991) Advances in catecholamine and metabolite measurements for diagnosis of pheochromocytoma. Clin Chem 37: 1854-1867.

5. Middelkoop C, Dekker G, Kraayenbrink A, Popp-Snijders C (1993) Platelet-poor plasma serotonin in normal and preeclamptic pregnancy. Clin Chem 39: 1675-1678.

6. Camanas RMV, Mallols JMS, Lapasió JRT, Ramis-Ramos G (1995) Analysis of pharmaceutical preparations containing catecholamines by micellar liquid chromatography with spectrophotometric detection. Analyst 120: 1767-1772.

7. Lizuka H, Ishige T, Ohta Y, Yajima T (1999) Simultaneous determination of 5-hydroxyndoles and catecholamines by HPLC with fluorometric precolumn derivatization. Adv Exp Med Biol 467: 821-826.

8. Engbaek F, Voldby B (1982) Radioimmunoassay of serotonin (5-hydroxytryptamine) in cerebrospinal fluid, plasma and serum. Clin Chem 28: 624-628.

9. Takezawa K, Tsunoda M, Murayama K, Santa T, Imai, K (2000) Full papers–Separation science-automatic semi-microcolumn liquid chromatographic determination of catecholamines in rat plasma utilizing peroxy oxalate chemiluminescence reaction. Analyst 125: 293-296.

10. Volin P (1992) Determination of urinary normetanephrine, metanephrine and 3-methoxytyramine by high-performance liquid chromatography with electrochemical detection: comparison between automated column-switching and manual dual-column sample purification methods. J Chromatogr B 578: 165-174.

11. Lenders J, Eisenhofer G, Armando I, Keiser H, Goldstein D, et al. (1993) Determination of metanephrines in plasma by liquid chromatography with electrochemical detection. Clin Chem 39: 97-103.

12. Passarinha LA, Bonifacio MJ, Queiroz JA (2006) The effect of temperature on

the analysis of metanephrine for catechol-O-methyltransferase activity assay by HPLC with electrochemical detection. Biomed Chromatogr 20: 937-944.

13. Kagedal B, Goldstein DS (1988) Catecholamines and their metabolites. J Chromatogr 429: 177-233.

14. Achili G, Perego C, Ponzio F (1985) Application of the dual-cell coulometric detector: a method for assaying monoamines and their metabolites. Anal Biochem 148: 1-9.

15. Matson W, Langlais P, Volicer L, Gamache P, Bird E, et al. (1984) n-Electrode three-dimensional liquid chromatography with electrochemical detection for determination of neurotransmitters. Clin Chem 30: 1477-1488.

16. Frei RW, Zech K (1988) Selective sample handling and detection in high-performance liquid chromatography part A. Journal of High Resolution Chromatography 11: 613.

17. Saracino M, Raggi M (2010) Analysis of soy isoflavone plasma levels using HPLC with coulometric detection in postmenopausal women. J Pharmaceut Biomed 53: 682-687.

18. Saracino MA, de Palma A, Boncompagni G, Raggi MA (2010) Analysis of risperidone and its metabolite in plasma and saliva by LC with coulometric detection and a novel MEPS procedure. Talanta 81: 1547-1553.

19. Marklund M, Landberg R, Aman P, Kamal-Eldin A (2011) Comparison of gas chromatography-mass spectrometry and high-performance liquid chromatography with coulometric electrode array detection for determination of alkylresorcinol metabolites in human urine. J Chromatogr B 879: 647-651.

20. Sabbioni C, Saracino MA, Mandrioli R, Pinzauti S, Furlanetto S, et al. (2004) Simultaneous liquid chromatographic analysis of catecholamines and 4-hydroxy-3-methoxyphenylethylene glycol in human plasma: Comparison of amperometric and coulometric detection. J Chromatogr A 1032: 65-71.

21. Hollenbach E, Schulz C, Lehnert H (1998) Rapid and sensitive determination of catecholamines and the metabolyte 3-methoxy-3-hydroxyphen-ethyleneglycol using HPLC following novel extraction procedures. Life Sci 63: 737-750.

22. Unceta N, Rodriguez E, de Balugera ZG, Sampedro C, Goicolea MA, et al. (2001) Determination of Catecholamines and their metabolites in human plasma using liquid chromatography with coulometric multi-electrode cell-desing detection. Anal Chim Acta 444: 211-221.

23. http://www.fda.gov/downloads/drugs/guidancecompilance/regulatoryinformation/guidances/ucm070107.pdf.

24. http://www.ich.org/lob/media/media417.pdf

25. Gamache P, Kingery M, Acworth I (1993) Urinary metanephrine and normetanephrine determined without extraction by using liquid chromatography and coulometric array detection. Clin Chem 39: 1825-1830.

26. Parker NC, Levtzow C, Wright P, Woodard LL, Chapman JF (1986) Uniform chromatographic conditions for quantifying urinary catecholamines, metanephrines, vanillylmandelic acid, 5-hydroxyindoleacetic acid, by liquid chromatography with electrochemical detection. Clin Chem 32: 1473-1476.

27. de Jong WHA, Graham KS, Van der Molen JC, Links TP, Morris MR, et al. (2007) Plasma free metanephrine measurement using automated online solid-phase extraction HPLC tandem mass spectrometry. Clin Chem 53: 1684-1693.

Liquid Chromatography – Tandem Mass Spectrometry for Simultaneous Determination of Ticarcillin and Vancomycin in Presence of Degradation Products: Application to the Chemical Stability Monitoring of Ticarcillin-Vancomycin Solutions

Belissa E[1], Nino C[1], Bernard M[1], Henriet T[1], Sadou-Yaye H[2], Surget E[1], Boccadifuoco G[1], Yagoubi N[3] and Do B[1,3]*

[1]Assistance Publique–Hôpitaux de Paris, Agence Générale des Produits et Equipements de Santé, Département de contrôle qualité et de développement analytique, France
[2]Assistance Publique – Hôpitaux de Paris, Groupe hospitalier Pitié-Salpétrière, Département Pharmacie, France
[3]Paris-SudUniversity, Faculty of Pharmacy, EA 401 Groupe Matériaux et Santé, France

Abstract

Acute bacterial conjunctivitis and their complications are commonly treated by 6 mg/mL ticarcillin and 50 mg/mL vancomycin eyes drops, formulated separately in hospitals. A rapid, selective and specific LC-tandem MS method has been developed and validated for simultaneous quantification of ticarcillin and vancomycinin solution and in presence of degradation products. Atlantis® C18 Column T3 (250×4.6 mm I.D., 5 µm) and gradient mobile phase composed of 44 mM trifluoracetic acid/methanol allowed for good separation. Degradation products were separated on LC and detected with mass spectrometry in full scan mode. The method was suitable forthe stability study of ticarcillin-vancomycin pH 5.2 solutionsover an 8-day period storage at 5 ± 3°C and at room temperature (25 ± 2°C). The solutions remained stable up to 5 days at 5 ± 3°C. At room temperature, degradation of ticarcillin was accelerated in presence of vancomycin, while vancomycin stability was not further affected by ticarcillin. High-resolution mass spectrometry was used for the characterization of the degradation products.

Keywords: LC-MS/MS; Ticarcillin; Vancomycin; Stability; Degradation products

Introduction

Acute bacterial conjunctivitis is one of the most common eye diseases. It is often caused by *staphylococcus aureus, staphylococcus epidermitis, streptococcus pneumonia, streptococcus viridans, haemophilus influenza, pseudomonas aeruginosa* and gram-negative intestinal bacteria [1-4]. It often evolves spontaneously to remission but could deteriorate into ocular keratitis, corneal abscess or endophtalmitis. In such cases, topical or systemic antibiotic treatment is required, reducing contagiousness and the emergence of bacterial resistance [5]. One of these treatments involves association of a β-lactamineas ticarcillin and a glycopeptideas vancomycin [3,6-8]. Because of their well-known instability in aqueous solutions, these antibiotics are only available as powder for injectable preparations and not as long-term stable eye-drops drug products. Locally eye drops are commonly prepared at hospital for standard clinical use [9]. So far, fortified antibiotic is individually formulated at high concentrations, required for cornea penetration [10]. Ticarcillin at 6 mg/mL and vancomycin at 50 mg/mL are commonly prepared [10-12].

In order to help formulation combining the two antibiotics, the stability of drugs mixtures stored at 5 ± 3°C and at room temperature (25 ± 2°C) was investigated. That's why, an LC-tandem MS (MRM) method for the simultaneous determination of the two antibiotics in presence of degradation products was developed and validated. High-resolution mass spectrometry (HR-MS) was used for the characterization of the degradation products formed.

The literature survey reveals a great number of HPLCmethods for individual quantifications of ticarcillin and of vancomycin in body fluids [13,14] and in pharmaceutical formulations [15,16]. Ticarcillin as well as vancomycin have been successfully separated from their respective synthesis impurities [17,18]. LC-MS methods were also described for chirality study [19], but no one had dealt with the simultaneous quantification of the two antibiotics in prepared mixtures and in presence of degradation products. Data on the individually drugs stability are available [20,21] but still no clue about their behaviour in mixtures.

Experimental

Instruments

The HPLC system consisted of an Ultimate 3000 HPLC from Dionex® (Sunnyvale, CA, USA), which is composed of a binary pump, a C18 Column Atlantis® T3 (250×4.6 mm I.D., 5 µm) from Waters® (Milford, MA, USA) and a refrigerated autosampler with a 100 µL syringe. Gradient mobile phase was made up with methanol(A) and 44 mM trifluoroacetic acid (TFA) (B). The injection volume was set at 50 µL, the flow rate at 0.8 mL/min and the column temperature at 20°C. The mobile phase gradient program was set as follows: 5 min: A:B (20:80 v/v); 25 min: A:B (90:10 v/v) and 30 min: A:B (20:80 v/v).

ABSciex® (Framingham, MA, USA) Mass Spectrometer 3200QTRAP® system was fitted with an electrospray ionization (ESI) interface operating in positive ion mode. MS data were treated using

*Corresponding author: B Do, Université Paris-Descartes, UFR de Pharmacie, 4 Avenue de l'Observatoire, 75006 Paris
E-mail: bernard.do@parisdescartes.fr

Analyst Software® from ABSciex® (Framingham, MA, USA). Direct infusion of each antibiotic was realised to achieve common optimal MS and collision-induced dissociation (CID) parameters: collision energy=20 eV, capillary temperature=700°C, nitrogen as curtain gas=50 arbitrary units (a.u), air as sheath gas=50 a.u and as auxiliary gas=70 a.u. The degradation products were detected in full-scan mode within a mass/charge range of 50-800 uma.

OrbitrapVelos Pro LTQ Tune Plus 2.7.0.1103 SP1 (Thermo Fisher Scientific, CA, USA) was used in positive ion mode for accurate mass measurement as follows: the source voltage was set at 3.4 kV and the temperatures were fixed at 53°C (source) and 298°C (capillary). Data were treated with Xcalibur® software (version 2.2 SP 1.48).

Materials and reagents

Chemicals: Ticarcillin disodium salt from Sigma Aldrich® (Saint-Louis, MO, USA) and Vancomycin hydrochloride from Xellia® (Oslo, Norway) were used. A Milli-Q water purification system from Millipore (Bedford, MA, USA) was used to generate ultrapure water. Sodium hydroxide 1.0 M and sodium hydroxide 0.1 M were prepared from sodium hydroxide 5.0 M Combi-Titrisol® purchased from Merck laboratory® (Darmstadt, Germany). Sodium acetate buffer (pH 5.2; 0.305 M) was made from anhydrous acetic acid purchased from Merck® (Billerica, MA, USA)where pH was adjusted to 5.2 with sodium hydroxide 1.0 M. Methanol HPLC grade was obtained from Sigma Aldrich® (Saint-Louis, MO, USA) and TFA analysis grade from VWR international® (Leuven, Belgium).

Samples preparation: Ticarcillin, vancomycin and ticarcillin-vancomycin working solutions were prepared from the same stock solutions obtained by dissolution of appropriate amounts of ticarcillin disodium salt and of vancomycin hydrochloride in ultrapure water. Stock solutions of ticarcillin were diluted to 2.3 mg mL^{-1} of ticarcillin base into pH 5.2 acetate buffer. Solutions of vancomycin were set at a concentration of 16.7 mg mL^{-1} vancomycin base in the same solvent. Ticarcillin-vancomycin mixture solutions were prepared in pH 5.2 acetate buffer with the same concentrations in antibiotics as for the individual solutions. Each solution, prepared in triplicate, was allocated into sealed glass vials to be stored for 8 days at 5 ± 3°C or at 25 ± 2°C. Stability was monitored at days 0, 3, 5 and 8. In order to have a standard degraded solution, an aliquot of ticarcillin-vancomycin solution was kept until 2 months at 5 ± 3°C. Working solutions were systematically diluted at 1/250 in acetate buffer/ultrapure water (75:25 (v/v)) prior to analysis.

Validation of the LC-MS/MS method

Linearity, accuracy and limits of detection and quantification: Linearity over concentration ranges 2.3-9.2 µg mL^{-1} for ticarcillin base and 16.7-66.8 µgmL^{-1} for vancomycin base was established by plotting the peak areas versus the concentration of each analyte. Accuracy was also determined over these concentration ranges. LOQ was determined at signal to noise (S/N)10, by injecting a series of dilute solutions with known concentration.

Precision: The precision of the method was established by analysing the standard solutions at low (0.58 mg mL^{-1} ticarcillin and 4.2 mg mL^{-1} vancomycin) and high (2.3 mg mL^{-1} ticarcillin and 16.7 mg mL^{-1} vancomycin) concentration levels for each analyte. As mentioned above, solutions were diluted at 1/250 prior to analysis. To determine the intra-day precision of each standard, the same standard solutions were examined six times within one day. The inter-day precision was established by analysing each standard solution on three consecutive days. The precision of the assay was expressed by the RSD (%) of the replicate measurements.

Results and Discussion

Chromatographic performanceand MS² studies

HPLC-UV-MS² was used to specifically quantify the antibiotics and also to characterize the degradation products detected. HPLC allowed separating the active substances and their degradation products in a single run. Owing to their respective apparent pK as, the use of TFA led to the formation of ion pairs with vancomycin, whereas ticarcillin would be chromatographed upon its molecular form. Chemical structures of ticarcillin and vancomycin were presented in figure 1. To improve the resolution between related substances generated during stability studies, mobile phase was tuned in gradient mode. Vancomycin was satisfactorily separated from ticarcillin epimers [22] as is shown in Figure 3a. "Ticarcillin active pharmaceutical ingredient (API) is basically composed of 2 epimers [18]. The two peaks, which result from analysis of ticarcillin reagent, present identical mass spectra so that confirms that the product is composed of epimers and shows the capacity of the method to also resolve them from each other. Besides, the more the elution is delayed, the better arethe drugs separated from the degradation products generated in the 2-month storage solutions used for the LC development (Figure 3b, Table 1). The full-scan mass spectrum measured for the peak of each drug is comparable to that of the reference, suggesting lack of co-elution.

Parallel to LC development, 0.1 mg/mL vancomycin and ticarcillin solutions were individually infused into ESI+mass spectrometer in order to select the most relevant transitions for specific quantification of the antibiotics in MRM mode. The full MS² spectrum of vancomycin showed an abundant precursor ion at m/z=725 corresponding to ion [M+2H]$^{2+}$ (Figure 3a). Among the major fragmentation mechanisms detected, the loss of vancosamine molecule (transition m/z 725 (307) was considered specific of vancomycin. Indeed, the degradation products formed by change upon the vancosamine part could hardly produce such an MS² transition. Besides, the rupture-rearrangement process resulting in the separation of the two parts of the glycoside group yielded the m/z 144 ion, which accounts for the base-peak. Therefore, m/z 307 ion and m/z 144 ion were set as the confirmation ion and the quantitative ion for vancomycin determination, respectively. As a result, quantitative determinations were performed in multiple reactions monitoring (MRM) using the following transitions: (i) 725 (z=2) 307 (z=1) for confirmation and 725 (z=2) 144 (z=1) for quantitation (vancomycin) and (ii) 385 198 for confirmation and 385 315 for quantitation (ticarcillin).

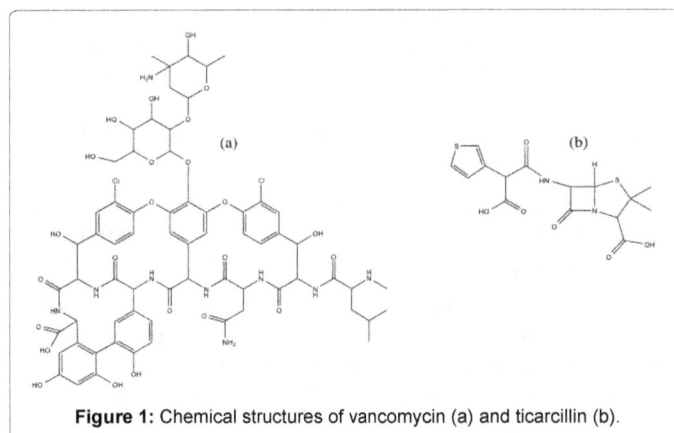

Figure 1: Chemical structures of vancomycin (a) and ticarcillin (b).

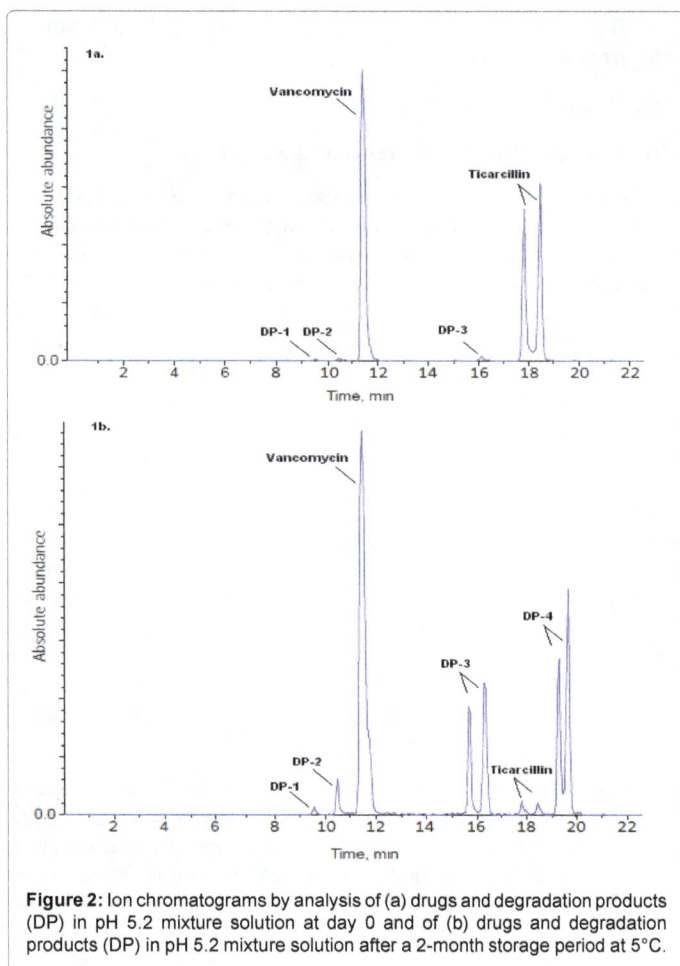

Figure 2: Ion chromatograms by analysis of (a) drugs and degradation products (DP) in pH 5.2 mixture solution at day 0 and of (b) drugs and degradation products (DP) in pH 5.2 mixture solution after a 2-month storage period at 5°C.

The full MS^2 spectrum of ticarcillinincludes the protonated ion as well as [M+Na]$^+$ ion commonly observed for β-lactam antibiotics [21]. Similarly to the other penicillins, [M+H-26]$^+$ ion (m/z=359) was formed by β-lactam ring opening due to hydration and by loss of CO_2 (Figure 3b). Another loss of CO_2 yielded m/z 315 ion, accounting for the base-peak. Instead of CO_2 departure, loss of a 161 Da moiety from m/z359 ion (affording m/z 198 ion) could be considered as quite specific to ticarcillin. As a result, m/z 315 ion and m/z 198 ion were set as the confirmation ion and the quantitative ion for ticarcillin determination, respectively.

Method validation

The responses of vancomycin and of ticarcillin were linear with respect to the concentration range 2.0-9.2 μg mL^{-1} for ticarcillin base and 13.2-66.8 μg mL^{-1} for vancomycin base, yielding a correlation coefficient R of 0.9923 and of 0.9984, respectively. Basically, as ticarcillin is composed of 2 epimers eluted at tR17.8 and tR18.4 min, the sum of the two areas accounted for its signal response.

LOD and LOQ, based on ratios S/N 3 and S/N 10 measurement, were found to be 0.6 and 1.7 μg mL^{-1}, respectively for ticarcillin base and 2.7 and 8.2 μg mL^{-1}, respectively for vancomycin base.

The method was found to be precise with % RSD intermediate precision <2.86 at the calibration low level and <1.67 at the calibration high level for ticarcillin. % RSD intermediate precision was <1.19 at the calibration low level and <1.34 at the calibration high level for vancomycin.

The method was also shown accurate with excellent recoveries and low relative errors over the calibration range tested (Figure 4).

Stability of individual and of mixture solutions fixed at pH 5.2

Degradation products: DP1 and DP2 are related to vancomycin as absent from the ticarcillin individual solutions (Table 1). DP3s and DP4s pertain to ticarcillin as not detected in the vancomycin idivdual solutions. The MS spectra of DP1 and of DP2 both exhibited a molecular mass with 1 Da greater than vancomycin's doubly protonated ion (Table 1). Accurate masses of DP1 and of DP2 are then consistent

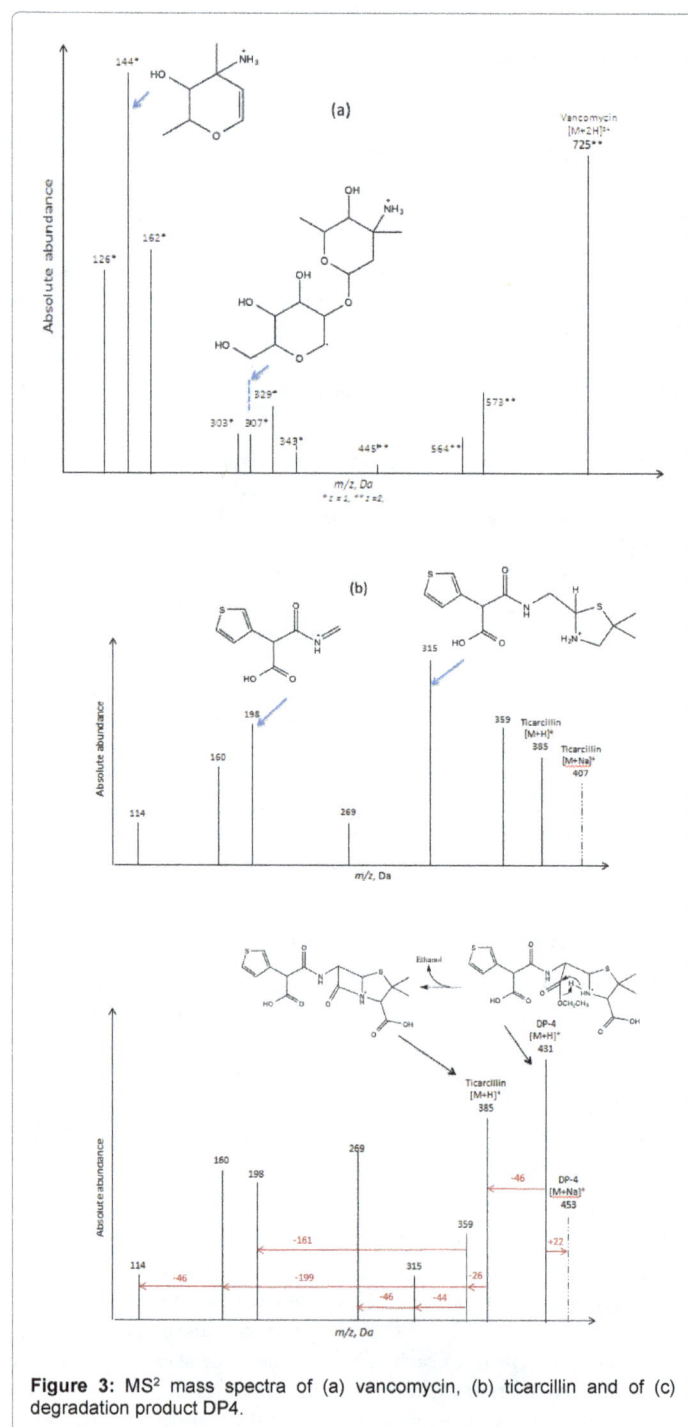

Figure 3: MS^2 mass spectra of (a) vancomycin, (b) ticarcillin and of (c) degradation product DP4.

	DP1	DP2	Vancomycin	DP3/epimer	Ticarcillin/epimer	DP4/epimer
RT (min)	9.6	10.4	11.6	15.8/16.3	17.8/18.4	19.2/19.7
Resolution factors	2.5	1.67	4.65	1.81	1.53	
$(M+nH)^{n+}$ m/z	725.3 (z=2)	725.3 (z=2)	724.7 (z=2)	403.0/403.0	385.0/385.0	431.0/431.0
Accurate masses (relative errors ppm)	725.2190 (-2.04)	725.2190 (-2.04)	724.7208 (-2.17)	403.0620 (-2.03)	385.0516 (-1.69)	431.0569 (-1.92)
Elemental composition	$C_{66}H_{76}Cl_2N_8O_{25}^{2+}$	$C_{66}H_{76}Cl_2N_9O_{25}^{2+}$	$C_{66}H_{77}Cl_2N_9O_{24}^{2+}$	$C_{15}H_{19}N_2O_7S_2^{+}$	$C_{15}H_{17}N_2O_6S_2^{+}$	$C_{16}H_{19}N_2O_8S_2^{+}$

Table 1: Retention times (RT), resolution factors (Rs) and MS data from ion chromatogram of drugs pH 5.2 antibiotics mixture solution after a 2-month storage period at 5°C. Accurate masses were measured by flow injection analysis and best possible elemental formula are proposed.

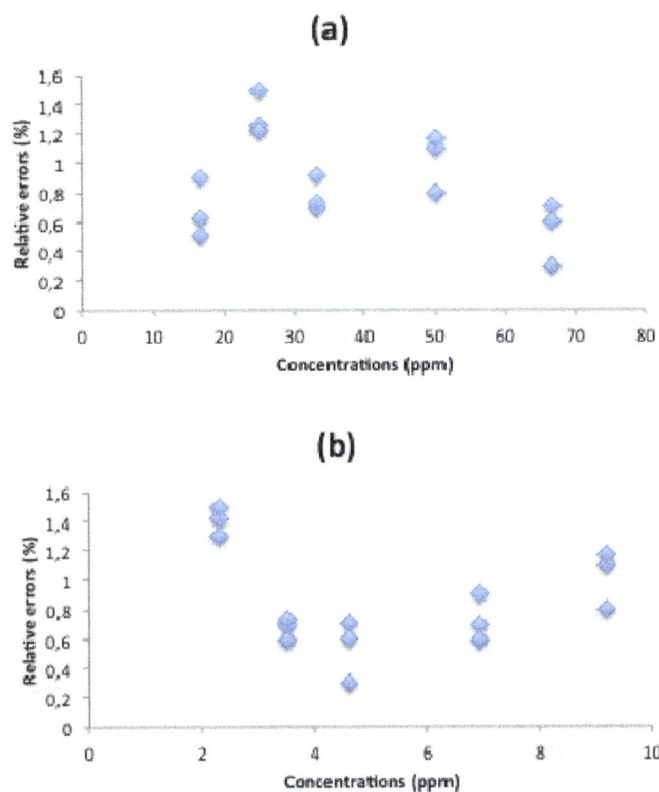

Figure 4: Accuracy profiles using the linear regression model (Acceptability limits at ± 2%).

with the elemental formula $C_{66}H_{76}Cl_2N_8O_{25}^{2+}$. The stability data have shown that DP1 was more likely a synthesis impurity than a degradant as it was already detected at T0 and at ainvariable concentration rate ≤ 0.3%. A derivative in which the asparagine residue is replaced by an aspartic acid residue has already been described as a synthesis by-product of vancomycin [23,24]. DP2 ion was assumed to bea crystalline degradation product (CPD1), a well-known degradation product of vancomycin [25-27]. Chemically, a vancomycin in pH 5.2 individual or mixture solution was quite stable over a 5-day storage period at 5°C as well as at room temperature (Figure 5). However, beyond a certain threshold, DP2 was shown to crystallize notably at day 5 at ambient temperature, based on HR-MS analysis of the particles isolated.

DP3s and DP4s are two groups of epimers formed from ticarcillin degradation. The epimers produced comparable mass spectra (Table 1). With a mass shift of 18 Da (Table 1), DP3slikely accounted for the two-penicilloïc acid derivative epimers, which is consistent with the

literature data and the elemental formuladeducted from the observed accurate mass (Table 1) [17,28-30]. DP3s increased in time parallel to a drop in ticarcillin concentration and even went through the roof with temperature or in presence of vancomycin. Unlike DP3s, DP4s were only detected in the 2-month storage solutions. There of yielded a protonated ion at m/z 431.0569, which might correspond to $C_{17}H_{23}N_2O_7S_2^{+}$ elemental formula and be consistent with penicilloic ethanolate ester. Its CID mass spectrum has confirmed the structure (Figure 3c). An internal rearrangement mechanism leading to ethanol loss gave rise to the formation of protonated ticarcillin, explaining the great similarity with the MS/MS spectrum of ticarcillin (Figure 3b and 3c).

Stability of the mixture: Basically, at the concentrations used, the mixtures containing ticarcillin and vancomycin precipitated as soon as the two antibiotics were mixed in solution at the aforementioned concentrations, likely by formation of poorly soluble ticarcillin-

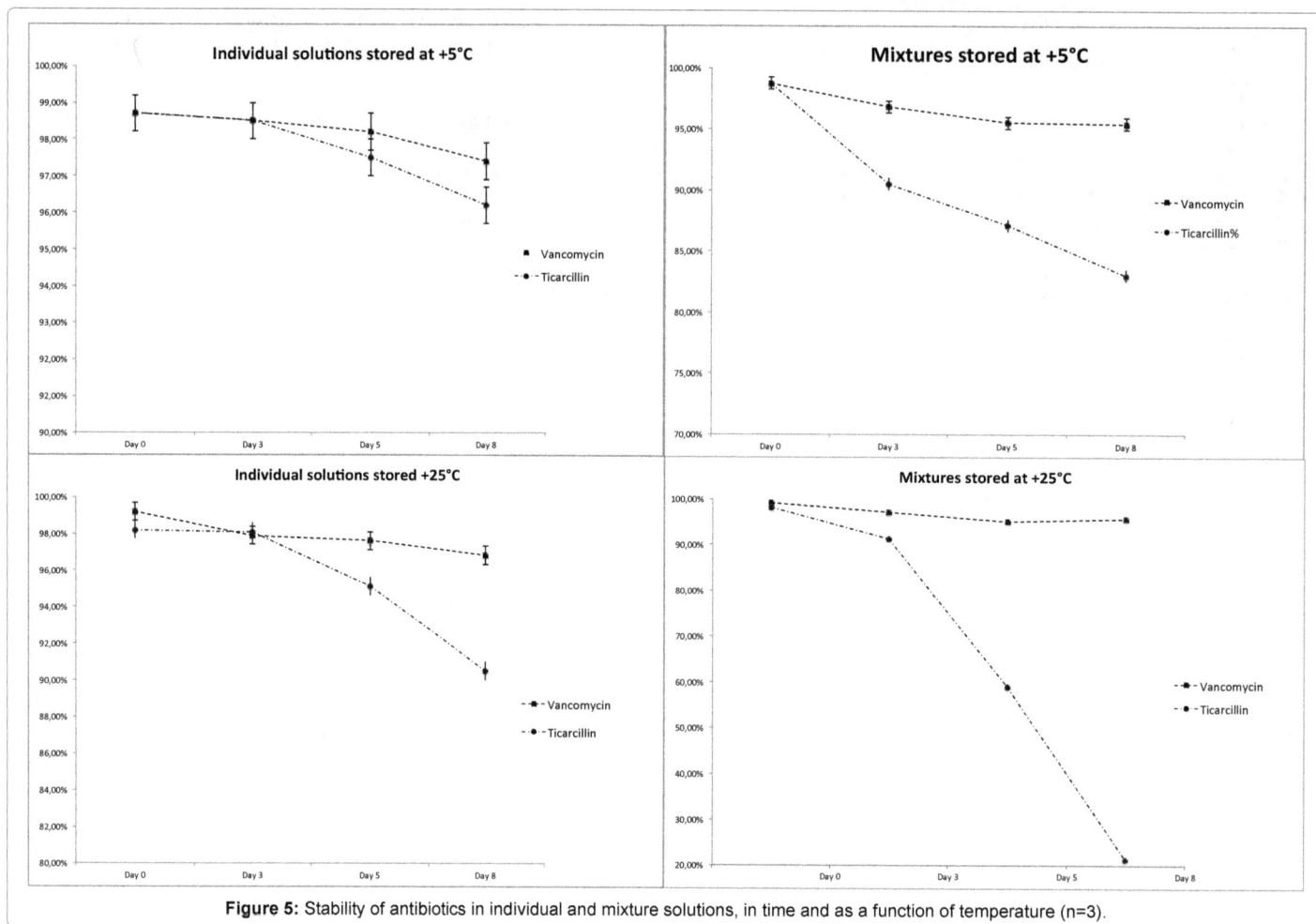

Figure 5: Stability of antibiotics in individual and mixture solutions, in time and as a function of temperature (n=3).

vancomycin ion-pairs. The precipitate was partly solubilised in 0.01 M sodium hydroxide ethanol and analysed by LC-MS. The sample was shown to exclusively contain ticarcillin and vancomycin. Based on this outcome, some solubilisation experiments were undertaken and the use of pH 5.2 acetate buffers had made co-solubilisation possible.

Unlike vancomcin, stability of ticarcillin in pH 5.2 individual solutions differed from that of ticarcillin in pH 5.2 mixture solutions (Figure 5). At day 8 and at 5°C, ticarcillin concentration dropped by 10% in presence of vancomycin, whereas its individual solutions still contained at least 96% of the drug. The phenomenon was much more significant at room temperaturewith sharp increase of DP3s. It seems that nucleophilic functions of vancomycin could catalyse ticarcillin hydrolysis [29-32].

At day 5 and similarly to what was previously described for the pH 5.2 individual vancomycin solutions, DP2 crystals were formed in the mixture solutions and thereof could initiate vancomycin-ticarcillin precipitation, actually progressively formed and visually detected fromday 5-6 in the solutions stored at 5°C and at room temperature.

Conclusion

A method for the simultaneous determination of ticarcillin and vancomycin in presence of degradation products has been developed and validated. It is based on the gradient LC separation performance and *MRMMS* detection. It has allowed the stability monitoring of the

pH 5.2 antibiotics individual and mixture solutions. Itwas shown that ticarcillin mixed with vancomycin was much less stable than ticarcillin alone in solution. Unlike ticarcillin, vancomycin stability was not affected by the presence of ticarcillin at the studied concentration levels. Vancomycin degradation could make the solutions physically unstable.

References

1. Chung CW, Cohen EJ (2000) Eye disorders: bacterial conjunctivitis. West J Med 173: 202-205.

2. Gigliotti F, Williams WT, Hayden FG, Hendley JO, Benjamin J, et al. (1981) Etiology of acute conjunctivitis in children. J Pediatr 98: 531-536.

3. Seal DV, Barrett SP, McGill JI (1982) Aetiology and treatment of acute bacterial infection of the external eye. Br J Ophthalmol 66: 357-360.

4. Weiss A, Brinser JH, Nazar-Stewart V (1993) Acute conjunctivitis in childhood. J Pediatr 122: 10-14.

5. Lohr JA (1993) Treatment of conjunctivitis in infants and children. Pediatr Ann 22: 359-364.

6. Block SL, Hedrick J, Tyler R, Smith A, Findlay R, et al. (2000) Increasing bacterial resistance in pediatric acute conjunctivitis (1997-1998). Antimicrob Agents Chemother 44: 1650-1654.

7. Buznach N, Dagan R, Greenberg D (2005) Clinical and bacterial characteristics of acute bacterial conjunctivitis in children in the antibiotic resistance era. Pediatr Infect Dis J 24: 823-828.

8. Høvding G (2008) Acute bacterial conjunctivitis. Acta Ophthalmol 86: 5-17.

9. Steinert RF (1991) Current therapy for bacterial keratitis and bacterial conjunctivitis. Am J Ophthalmol 112: 10S-14S.

10. Leeming JP (1999) Treatment of ocular infections with topical antibacterials. Clin Pharmacokinet 37: 351-360.

11. Deshpande AD, Baheti KG, Chatterjee NR (2004). Degradation of b-lactam antibiotics. Curr Sci 87: 1684-1695.

12. Lamm A, Gozlan I, Rotstein A, Avisar D (2009) Detection of amoxicillin-diketopiperazine-2', 5' in wastewater samples. J Environ Sci Health ATox Hazard Subst Environ Eng 44: 1512-1517.

13. Hosotsubo H (1989) Rapid and specific method for the determination of vancomycin in plasma by high-performance liquid chromatography on an aminopropyl column. J Chromatogr 487: 421-427.

14. Kwan RH, MacLeod SM, Spino M, Teare FW (1982) High-pressure liquid chromatographic assays for ticarcillin in serum and urine. J Pharm Sci 71: 1118-1121.

15. Tsou TL, Lee CW, Wang HJ, Cheng YC, Liu YT, et al. (2009) Development of a new HPLC method for the simultaneous determination of ticarcillin and clavulanic acid in pharmaceutical formulations. J AOAC Int 92: 1089-1094.

16. Enrique M, García-Montoya E, Miñarro M, Orriols A, Ticó JR, et al. (2008) Application of an experimental design for the optimization and validation of a new HPLC method for the determination of vancomycin in an extemporaneous ophthalmic solution. J Chromatogr Sci 46: 828-834.

17. European Pharmacopoeia 8.0, monograph 01/2008:0956.

18. European Pharmacopoeia 8.0, monograph 01/2008:1058.

19. Bakhtiar R, Tse FL (2000) High-throughput chiral liquid chromatography/tandem mass spectrometry. Rapid Commun Mass Spectrom 14: 1128-1135.

20. Carbonne M, Henne-Menetre H, May I (2012) Stability of pharmacy-preparedvancomycin and ceftazidin–fortified antibiotic eye drops and solutions in polypropylene, a review. Euro J Hosp Pharm 19: 152-153.

21. Zhang Y, Trissel LA (2001) Stability of piperacillin and ticarcillin in AutoDose infusion system bags. Ann Pharmacother 35: 1360-1363.

22. Blondeel S, Pelloquin A, Pointereau-Bellanger A, Thuillier A, Fernandez C (2005). Effects of freezing on stability of a fortified 5 mg/mL ticarcillin ophthalmic solution. Can J Hosp Pharm 58: 65-70.

23. Di Corcia A, Nazzari M (2002) Liquid chromatographic-mass spectrometric methods for analyzing antibiotic and antibacterial agents in animal food products. J Chromatogr A 974: 53-89.

24. European Directorate for the Quality of Medecines and HealthCare (EDQM), 2011. European Pharmacopeia 7th ed. [§2.7.2., Part A: 222-224].

25. Lancini GC, Cavalleri (1990). In Biochemistry of Peptide Antibiotics, Kleinhauf H, Von Döhren H Walterde Gruyter: Berlin, 159

26. Diana J, Visky D, Hoogmartens J, Van Schepdael A, Adams E (2006) Investigation of vancomycin and related substances by liquid chromatography/ion trap mass spectrometry. Rapid Commun Mass Spectrom 20: 685-693.

27. Barbault S, Aymard G, Feldman D, Pointereau-Bellanger A, Thuillier A (1999) Stability of vancomcycin eye drops. J Pharm Clin 18: 183-189.

28. Ghassempour A, Abdollahpour A, Tabar-Heydar K, Nabid MR, Mansouri S, et al. (2005) Crystalline degradation products of vancomycin as a new chiral stationary phase for liquid chromatography. Chromatographia 61: 151-155.

29. Somerville AL, Wright DH, Rotschafer JC (1999) Implications of vancomycin degradation products on therapeutic drug monitoring in patients with end-stage renal disease. Pharmacotherapy 19: 702-707.

30. Galvez O, Mullot JU, Mullot H, Simon L, Grippi R, et al. (2007) Etude de stabilité d'un collyre à 6 mg/mL de ticarcillin. Pharm Hosp 42: 171-176.

31. Lamm A, Gozlan I, Rotstein A, Avisar D (2009) Detection of amoxicillin-diketopiperazine-2', 5' in wastewater samples. J Environ Sci Health ATox Hazard Subst Environ Eng 44: 1512-1517.

32. Glew RH, Pavuk RA (1983) Stability of gentamicin, tobramycin and amikacin in combination with four beta-lactam antibiotics. Antimicrob Agents Chemother 24: 474-477.

13

Development of a Highly Sensitive, Fast and Efficient Screening Technique for the Detection of 2,3-Butanediol by Thin Layer Chromatography

Saran S[1], Yadav S[2] and Saxena RK[1]*

[1]Technology Based Incubator, University of Delhi South Campus, Biotech Centre, Benito Juarez Road, New Delhi-110021, USA
[2]Department of Microbiology, University of Delhi South Campus, Benito Juarez Road, New Delhi-110021, USA

Abstract

The interest in microbial production of 2,3-butanediol is based on the awareness that 2,3 butanediol is a promising bulk chemical due to its extensive industrial applications. Here, we report a novel method for the detection of 2,3-butanediol. The proposed method has a wide applicability to harness the commercial potential of microorganisms which produce 2,3-butanediol as the end product. Experimentally 32 bacterial strains were screened for 2,3-butanediol production. After 72 h the samples were spotted on thin-layer chromatography plates and ran in a solvent system comprising of hexane:ethyl-acetate:glacial acetic acid in the ratio of 70:30:1.5, followed by colour development using vanillin reagent. The appearance of a blue-colored spot of 2,3-butanediol with a retention factor (R_f) of 0.68 corresponds to the R_f value of the standard 2,3-butanediol which forms the basis for the selection of 2,3-butanediol producers. Apart from being a rapid detection system the proposed method is highly sensitive as it was able to detect a concentration as low as 1.0 mg/ml and its authenticity was reconfirmed by GC and GC-MS. Low cost of this method provides an effective support for 2,3-butanediol detection at any scale.

Keywords: 2,3-butanediol, Thin-layer chromatography, Rapid detection

Introduction

Continuous depletion of petroleum fuel-reserves is one of the prime concerns in this era. Despite of oil reserve availability, current estimates show that accessing them will become extremely difficult in a few decades [1,2]. Due to this, the bio-refinery systems that integrate biomass conversion processes and equipment to produce fuels, power, and chemicals from annually renewable resources are at the stage of worldwide development [3]. Many chemicals that could only be produced by chemical processes in the past can now have the potential to be generated biologically using renewable resources. Microbial production of 2,3-butanediol (2,3-BD) is one such example [4-6]. Interest in this bioprocess has increased remarkably because 2,3-BD has a large number of industrial applications, and microbial production will alleviate the dependence on oil supply.

2,3-BD can be produced efficiently *via* mixed acid fermentation with prokaryotes such as *Klebsiella pneumonia Klebsiella oxytoca Enterobacter aerogenes Serratia*, and *Bacillus polymyxa*. In these bacteria, pyruvate is first converted into α-acetolactate by acetolactate synthase. In anoxic state, α-acetolactate decarboxylase catalyzes the conversion of α-acetolactate into acetoin. 2,3-Butanediol is resulted from the reduction of acetoin by butanediol dehydrogenase. It exists in three stereoisomers, the dextro- (d-), levo- (l-) and meso-forms. The available reports reveal that quantitative estimation of 2,3-BD in fermentation broth is carried out only by two methods: high-performance liquid chromatography (HPLC) equipped with an Aminex HPX-87H column or by Alltech IOA-2000 organic acid column [7,8] and gas chromatography (GC) equipped with a glass column packed with Chromosorb 101 [9-11]. Although HPLC can detect 2,3-butanediol along with other by-products (1,4-propanediol, lactic acid, butyric acid, acetic acid, acetoin, 1,3-butanediol, ethanol) simultaneously; however, the process is expensive, time consuming, tedious and uneconomical and together with the limited column lifetime it fails to meet the criteria necessary to become a rapid screening procedure. Therefore it was of utmost important to have an efficient, authentic, reliable and inexpensive screening procedure to hunt for the most potent 2, 3-butanediol producer/s. Therefore, in the present investigation, the objective is to design a simple and fast protocol for 2,3-butanediol detection with wide applicability to harness the undoubted commercial potential of microorganisms to produce 2,3-butanediol as the major end product.

Materials and Methods

Materials

2, 3-butanediol, 1, 4-propanediol, 1, 3-butanediol, lactic acid, butyric acid, acetic acid, acetoin and ethanol were purchased from Sigma Chemicals (St. Louis, USA). TLC plates (Silica gel 60 F254) were purchased from E. Merck Ltd., Germany.

Fermentation conditions

Thirty two bacterial cultures, including different strains of *Klebsiella pneumoniae* were procured from the laboratory stock culture collection, Department of Microbiology, University of Delhi South Campus, New Delhi, India and screened for 2,3-butanediol production. The growth medium consisting of glucose 20 g/l, $(NH_4)_2HPO_4$ 5 g/l, $MgSO_4$ 0.3 g/l, KCl 1.0 g/l, pH 7.0 was used to grow and maintain the bacterial cultures.

The bacteria were grown in 250 ml Erlenmeyer flasks containing 50 ml production medium with the composition (g/L): glycerol, 25.0; Tryptone 10.0; $K_2HPO_4 \cdot 3H_2O$ 5.0; KH_2PO_4, 3.48; $MgCO_3$, 150 mM; $CaCl_2 \cdot 2H_2O$, 0.20; $CoCl_2 \cdot 6H_2O$, 0.004, $MgCl_2 \cdot 7H_2O$, 0.40 and $Na_2S \cdot 9H_2O$, 0.02 at pH 7.0 ± 0.2. Incubation was carried out at 37°C,

---

*Corresponding author: Saxena RK, Technology Based Incubator, University of Delhi South Campus, Biotech Centre, Benito Juarez Road, New Delhi-110021
E-mail: rks.micro2012@gmail.com

150 rpm for 72 h. The samples obtained after 72 h of fermentation were centrifuged at 10,000×g for 20 min to pellet the cells, followed by filtration using syringe filters (pore size 0.22 μm, Mdi filters, India). The pellet thus obtained was used for growth estimation and supernatant was used for 2, 3-butanediol estimation.

Thin-layer chromatography

One-dimensional chromatography on pre-activated silica gel TLC plates was carried out in saturated tanks. Samples were spotted 1.5 cm from the lower edge of the plate and at least 1.0 cm from the lateral border. The chromatograms were developed by the ascending technique with the desired mobile phase. The solvent front was drawn 10.0 cm from the application line.

Analytical methods

The concentration of 2,3-butanediol and other by-products was determined by the GC method described by [9].

GC–MS analysis

GC–MS analysis was performed with the Shimadzu QP-2010 Plus with Thermal Desorption System. Compounds were separated on a capillary column, the injector temperature was 260°C, and 1 μL samples were injected in split mode. The split ratio was 100; the total flow was 125.20 mL/min. The oven temperature was maintained at 40°C for 1

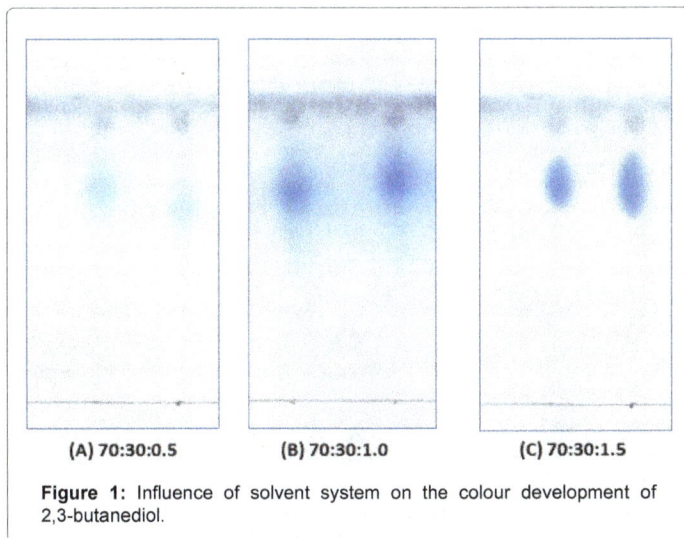

Figure 1: Influence of solvent system on the colour development of 2,3-butanediol.

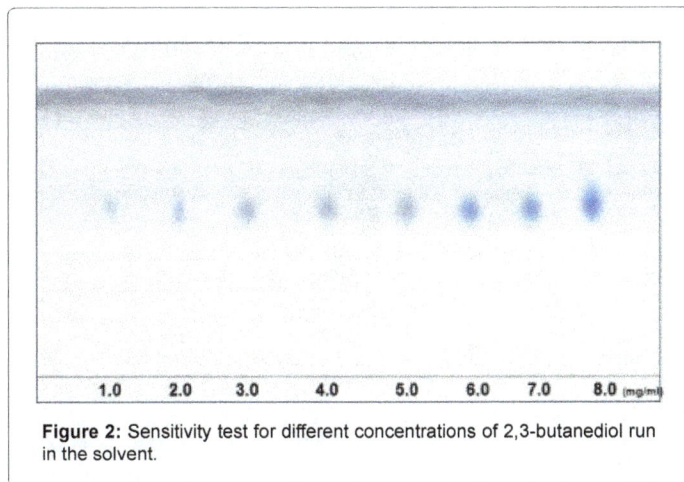

Figure 2: Sensitivity test for different concentrations of 2,3-butanediol run in the solvent.

min after injection, then programmed at 10°C min-1 to 200°C, which was held for 4 min, and then at 10°C min^{-1} to 200°C, which was held for 4 min. 2,3-Butanediol was identified by the use of standard database of mass spectra.

Results and Discussion

Selection of solvent system

The choice of best solvent system and the optimization of its composition are very important because the chromatographic separation is difficult to achieve. In the present investigation, 10 μl of standards of 2, 3-butanediol, 1, 4-propanediol, 1, 3-butanediol, lactic acid, butyric acid, acetic acid, acetoin and ethanol (50 mg/ml) were spotted individually and as a mixture on two TLC plates. These plates were then run in two solvent systems, separately, which were (a) chloroform:methanol 90:10 (b) hexane: ethyl acetate: glacial acetic acid 70:30:1. The solvents were allowed to run up to two-thirds of the plate height followed by its development using vanillin reagent. It was observed that with the solvent system 'b', except 2,3-butanediol, none of the compounds moved on to the TLC plates, however, a smear was observed with the solvent system 'a' and each standard spotted were not distinct and the mixture of these standards produced a smear due to overlapping. To further maximize the efficiency of the procedure and to enhance the resolution of 2,3-butanediol, different permutation combination of mobile phase were tested. Three different ratios, (A) 70:30:0.5, (B) 70:30:1.0, (C) 70:30:1.5, of the hexane: ethyl acetate: glacial acetic acid was used under the same conditions. It is clearly evident from the Figure 1 that with the ratio of (A) 70:30:0.5, no distinct resolution was achieved. Only one distinct spot corresponding to R$_f$ 0.68 was observed. With the ratio (B) 70:30:1.0, a bit cleaner plate was observed. Finally, with the ratio (C) 70:30:1.5, much clearer plates were obtained with distinct 2,3-butanediol spots. The above edge of this solvent system is that the procedure is specific for 2,3-butanediol and other by-products were not detected.

Test of sensitivity of the TLC method

A detection method is only reliable if it is able to detect even a minimal quantity of the desired product. Here, to examine the critical concentration of 2,3-butanediol required to give a visible spot on TLC plates, 10μl of different concentrations (1.0, 2.0, 3.0, 4.0, 5.0, 6.0, 7.0, 8.0 mg/ml) of 2,3-butanediol was spotted on TLC plates and run in hexane: ethyl acetate: glacial acetic acid (70:30:1.5) followed by a similar process of colour development. The results obtained are presented in Figure 2 which shows that even at 10 mg ml^{-1}, a distinct blue spot appeared. Therefore, it can be inferred that this process is highly sensitive and can be efficiently used for the screening of 2,3-butanediol producers.

Screening of 2,3-butanediol producers using TLC procedure

To authenticate the efficiency of the proposed TLC method, a total of 32 randomly selected bacterial cultures were used. For each culture, fermentation broth obtained at 72 h of incubation was used. Ten microliters of each of these culture supernatants (72 h) with the standard 2,3-butanediol as a reference was spotted and resolved on the TLC plates. The plates were developed under the same conditions described. The results of TLC are listed in (Table 1) which shows that out of 32 only 12 were 2,3-butanediol producers. To further authenticate the accuracy and efficiency of the proposed procedure, the samples were analysed by GC (Table 2). Among the 12 strains producing 2,3-butanediol, strain no. UDSC 6 was found to be a potent producer resulting in 2.97 g/L of 2,3-butanediol production as detected by GC. The results of prominent 2,3-butanediol producers along with a two negative ones are presented

Isolates No.	2,3-Butanediol production		Isolates No.	2,3-Butanediol production	
	24 hrs.	48 hrs.		24 hrs.	48 hrs.
UDSC 1	+	+	UDSC 17	+	+
UDSC 2	+	+	UDSC 18	-	-
UDSC 3	-	-	UDSC 19	+	+
UDSC 4	-	-	UDSC 20	-	-
UDSC 5	-	-	UDSC 21	-	-
UDSC 6	++	+++	UDSC 22	+	++
UDSC 7	+	++	UDSC 23	-	-
UDSC 8	+	+	UDSC 24	-	-
UDSC 9	-	-	UDSC 25	-	-
UDSC 10	-	-	UDSC 26	+	+
UDSC 11	-	-	UDSC 27	-	-
UDSC 12	+	++	UDSC 28	-	-
UDSC 13	-	-	UDSC 29	-	-
UDSC 14	-	-	UDSC 30	-	-
UDSC 15	-	-	UDSC 31	+	+
UDSC 16	+	+	UDSC 32	-	-

−blue spot not appeared, + Faded blue spot appeared, ++ Visible blue spot appeared, +++ Prominent dark blue spot appeared

Table 1: Analysis of the bacterial samples for 2,3-butanediol on TLC plates.

S.No	Isolates no.	48 hrs.	2,3-butanediol yield g/L
1.	UDSC 1	+	0.85
2.	UDSC 2	+	0.16
3.	UDSC 6	+++	2.97
4.	UDSC 7	++	2.68
5.	UDSC 8	+	0.73
6.	UDSC 12	++	1.60
7.	UDSC 16	+	0.56
8.	UDSC 17	+	0.14
9.	UDSC 19	+	0.65
10.	UDSC 22	++	1.26
11.	UDSC 26	+	0.89
12.	UDSC 31	+	0.94

+ Faded blue spot appeared, ++ Visible blue spot appeared, +++ Prominent dark blue spot appeared

Table 2: Evaluation of positive 2,3-butanediol producers on TLC plates and by GC method.

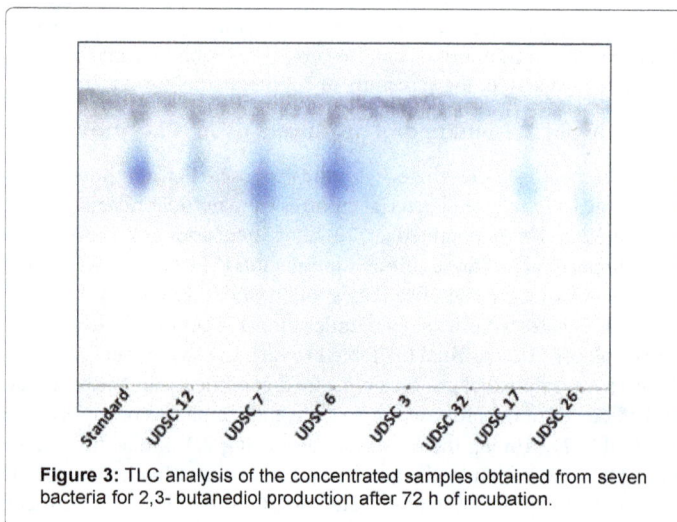

Figure 3: TLC analysis of the concentrated samples obtained from seven bacteria for 2,3-butanediol production after 72 h of incubation.

in Figure 3 which shows that in five of the culture filtrates (after 72 h), a blue coloured spot corresponding to the R_f value of standard 2,3-butanediol (0.68 ± 0.02) was observed. Two other cultures that were non 2,3-butanediol producers did not show any blue spot corresponding to 2,3-butanediol. Results obtained were coherent with the GC data. On the basis of the results, it can be concluded that the proposed method is very rapid and reliable and such a fast method has not been reported so far for screening of 2,3-butanediol producers. This TLC procedure will allow even hundreds of samples to be analysed rapidly as it requires only 15 min for each TLC run. Moreover, the method is economical as no special equipment or chemicals are required.

GC–MS analysis

2,3-butanediol scrapped from the TLC plates was confirmed by GC–MS in split mode. The mass spectrum showed it was 2,3-butanediol.

Conclusions

This method is simple, reliable and well suited for the routine laboratory assay of a large number of samples. Using this protocol, hundreds of samples can be analyzed in a day's time. Moreover, this method is economical as no special equipments or chemicals are required. Realizing the importance of 2,3-butanediol and its application in different industries, devising the methods for rapid screening of 2,3-butanediol producers is of great importance.

Acknowledgement

Authors sincerely acknowledge and thank the Department of Science and Technology (DST), Govt. of India for providing the financial support to carry out this work. The infrastructural support provided by Technology Based Incubator (TBI), South Campus is also highly acknowledged

References

1. Lin L, Cunshan, Z, Vittayapadung S, Xiangqian S, Mingdong D (2010) Opportunities and challenges for biodiesel fuel. Appl Energy 88: 1020-1031.

2. Li L, Chen C, Li K, Wang Y, Gao C (2014) Efficient Simultaneous Saccharification and Fermentation of Inulin to 2,3-Butanediol by thermophilic Bacillus licheniformis ATCC14580. Appl Environ Microbiol 80: 6458-6464.

3. Ji XJ, Huang H, Li S, Du J, Lian M (2008) Enhanced 2,3-butanediol production by altering the mixed acid fermentation pathway in Klebsiella oxytoca. Biotechnol Lett 30: 731-734.

4. Biebl H, Zeng AP, Menzel K, Deckwer WD (1998) Fermentation of glycerol to 1,3-propanediol and 2,3-butanediol by Klebsiella pneumoniae. Appl Microbiol Biotechnol 50: 24–29.

5. Perego P, Converti A, Del Borghi A, Canepa P (2000) 2,3-Butanediol production by Enterobacter aerogenes: selection of the optimal conditions and application to food industry residues. Bioprocess Eng 23: 613-620.

6. Ji XJ, Huang H, Li S, Du J, Lian M (2008) Enhanced 2,3-butanediol production by altering the mixed acid fermentation pathway in Klebsiella oxytoca. Biotechnol Lett 30: 731-734.

7. Kim SJ, Seo SO, Jin YS, Seo JH (2013) Production of 2,3-butanediol by engineered Saccharomyces cerevisiae. Bioresour Technol 146: 274-281.

8. Kopke M, Mihalcea C, Liew F, Tizard JH, Ali MS (2011) 2,3-Butanediol Production by Acetogenic Bacteria, an Alternative Route to Chemical Synthesis, Using Industrial Waste Gas. Appl Environ Microbial 77: 5467-5475.

9. Wang Y, Li L, Ma C, Gao C, Tao F, Ping Xu P (2013) Engineering of cofactor regeneration enhances (2S,3S)-2,3-butanediol production from diacetyl. Scientific Reports 3: 2643

10. Xiao Z, Wang X, Huang Y, Huo F, Zhu X (2012) Thermophilic fermentation of acetoin and 2,3-butanediol by a novel Geobacillus strain. Biotechnol Biofuels 5: 88.

11. Wang X, Lv M, Zhang L, Li K, Gao C (2003) Efficient bioconversion of 2,3-butanediol into acetoin using Gluconobacter oxydans DSM 2003. Biotechnol Biofuels 6: 155.

14

A Simple and Specific Method for Estimation of Lipoic Acid in Human Plasma by High Performance Liquid Chromatography

Ezhilarasi K[1], Sudha V[2], Geetha Ramachandran[2], Dhamodharan Umapathy[1], Rama Rajaram[3], Indira Padmalayam[4], Vijay Viswanathan[1] and Hemanth Kumar AK[2]*

[1]Department of Biochemistry and Molecular Genetics, Prof. M. Viswanathan Diabetes Research Centre and M.V. Hospital for Diabetes (WHO Collaborating Centre for Research, Education and Training in Diabetes), No. 4, West Madha Church Road, Royapuram, Chennai 600013, Tamil Nadu, India
[2]Department of Biochemistry, National Institute for Research in Tuberculosis (Indian Council of Medical Research), No. 1, Mayor Sathiyamoorthy Road, Chetpet, Chennai 600031, Tamil Nadu, India
[3]Department of Biochemistry, Central Leather Research Institute, Adyar, Chennai, Tamil Nadu. India
[4]Drug Discovery Division, Southern Research Institute, Birmingham, USA

Abstract

A rapid high performance liquid chromatographic method for determination of lipoic acid in human plasma was developed and validated. The method involved extraction of lipoic acid in ethanol consisting of 50 mM disodium hydrogen phosphate: acetonitrile: methanol in the ratio of 50:30:20. The separation was done using a C18 column (150 mm) and detection was carried out using UV detection at 201 nm. The assay was found to be linear in the range of 0.78-50 µg/ml with the correlation coefficient of 0.9998. Intra and inter-day variations were determined by processing each standard concentration in duplicate for five consecutive days. The average recovery of lipoic acid from plasma was 113%. The developed method demonstrates better sensitivity, precision, accuracy, stability and linearity when compared with the methods previously used. The method is simple and can be used for the determination of lipoic acid in basic research studies as well as in standard clinical laboratories.

Keywords: HPLC; Lipoic acid; Antioxidant; Free radicals; Co-factor; Therapeutic agent; Thioctic acid; Vitamin E; Vitamin C

Introduction

α- Lipoic acid (LA) (1, 2-dithiolane-3-valeric acid, 1, 2-dithiolane-3-pentanoic acid, 6, 8-dithiooctanic acids, thioctic acid) (Figure 1) is a crucial cofactor in the pyruvate dehydrogenase (PDH) and α-ketoglutarate dehydrogenase multienzyme complexes which are responsible for the production of acetyl-CoA in metabolic pathways [1]. It is also known as a metabolic antioxidant synthesized in human and animal cells and affords a protection to the cellular membrane by interacting with vitamin C and glutathione, which in turn recycles other antioxidants such as vitamin E to their reduced state and both oxidized and reduced forms of LA (Figure 2) reacts with reactive oxygen species and reactive nitrogen species, preventing damage due to oxidative stress in cells [2]. Many studies have proved its potential

therapy for chronic diseases associated with oxidative stress [3]. Because of its potent antioxidant properties, LA has been tested for its potential therapeutic effects in a variety of pathological conditions and used as an effective therapeutic agent in various diseases, including diabetes, mitochondrial cytopathies, cardiovascular diseases, hepatitis, cataract, radiation damage, HIV infections, also it plays a role in neurodegenerative disorders and neurovascular abnormalities associated with diabetic neuropathy and in delaying or inhibiting the development of neuropathy in diabetes [4-7].

Because of its extensive and multidisciplinary therapeutic use, a simple and accurate method for analysis of LA would be useful. Different analytical methods have been developed so far to determine the LA content in biological samples (tissue and plasma) various foodstuffs, both from animal and plant sources, in dietary supplements and in pharmaceutical preparations. Only few studies have been reported for the estimation of LA in the human body fluids in the minimum number of samples. New and rapid method should be developed to assay the LA in a large number of samples in various clinical stages to know the significance which should be economical and reproducible.

The methods so far reported by various studies include the use of the different extraction procedure to improve the recovery and of different

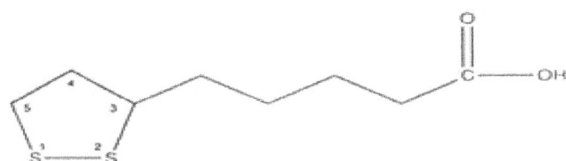

Figure 1: The chemical structure of LA (carbon 3 of the ring is the chiral center).

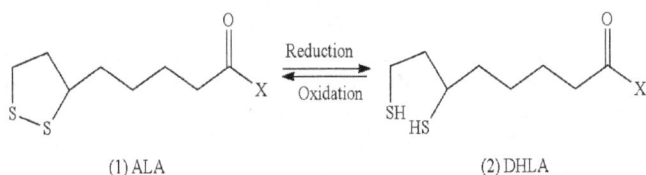

Figure 2: The oxidation and reduction structure of LA.

*Corresponding author: Hemanth Kumar AK, Department of Biochemistry and Clinical Pharmacology, National Institute for Research in Tuberculosis (Indian Council of Medical Research), Mayor Sathiyamoorthy Road, Chetpet, Chennai 600031, Tamil Nadu. India, E-mail: akhemanth20@gmail.com

analytical techniques to improve the sensitivity of measurements. However, most of these methods faced limitations either in terms of recoveries or sensitivities. This required the development of a rapid and reliable method to determine LA with better sensitivity in human plasma. Among the different available forms, free alpha-LA is of more interest due to its more pharmacological and therapeutic function.

The microbial assay, which was the first one to be reported for the determination of LA in human serum suffered from reproducibility [8,9]. Methods using liquid chromatography–mass spectrometry (LC–MS) [10,11] and gas chromatography–mass spectrometry (GC–MS) have been reported; however, these methods are quite expensive and unaffordable for many laboratories [12-14].

High performance liquid chromatographic (HPLC) is one of the most widely used analytical methods with different detection modes including ultraviolet [15], fluorescence [16,17] and electrochemical detection (ECD) [18-20]. Although HPLC with fluorescence detection was reliable for the determination of LA in biological fluids, the sample pre-treatment and intensive extraction involves derivatization which is tedious and cumbersome [12,16]. The HPLC method using an ECD with a dual gold- mercury electrode was found to be highly sensitive, but the electrodes lost sensitivity after 30–50 injections and had to be reconditioned [14,21].

However, there is still a need to improve the chromatographic conditions for better sensitivity, simple sample preparation and shorter run time. UV detector is a very common and primarily used detector for HPLC analysis. Since LA contains a carboxyl group it can be detected by using UV detector. Hence the study is aimed to develop and validate a simple, specific and reproducible method for quantitation of LA in human plasma by HPLC.

Experimental

Reagents and standards

Alpha-LA pure powder was purchased from Sigma Chemical Company, India; methanol, acetonitrile, chloroform, ethanol (HPLC grade), and disodium hydrogen phosphate (Na_2HPO_4) were purchased from Qualigens India. Deionized water was processed through a Milli-Q water purification system (Millipore, USA). Pooled human plasma was obtained from a Blood Bank in Chennai, India.

Instrumentation

Chromatographic analysis was performed using HPLC system (Shimadzu Corporation, Kyoto, Japan) consisted of two pumps (LC-10ATvp) and an auto sampler (SIL-HTA) with built in system controller (SCL-10Avp). Class VP-LC workstation was used for data collection and acquisition. The analytical column was a C18, 250×4.6 mm ID, 5μ particle size protected by a compatible guard column (Lichrospher 100 RP-18e, Merck, Germany). The mobile phase comprised of 50mM disodium hydrogen phosphate buffer (pH 2.7 adjusted with 1N HCl): acetonitrile: methanol in the ratio of 50:30:20. Prior to the preparation of mobile phase, the solvents were degassed separately using a Millipore vacuum pump with 0.45 μm filter paper. Sonication was done for 10 min to remove dissolved gas. The UV detector was set at 201 nm. The chromatogram was run for 5 minutes at a flow rate of 1 ml/min, with column oven temperature maintained at 40°C. Unknown concentrations were derived from linear regression analysis of the peak height of the analyte vs. concentration curve. The linearity was verified using estimates of correlation coefficient (r).

Preparation of standard solution

Stock standard solution of (1mg/ml) of LA was prepared by dissolving the pure powder in methanol. The working standards of LA were prepared in pooled human plasma. The concentrations of LA in the working standards were 0.78, 1.56, 3.125, 6.25, 12.5, 25 and 50 μg/ml.

Sample preparation

Three hundred microlitres of the blank/standard was mixed with 300 μl of ethanol and 1.8 ml of chloroform, vortexed for 2 min and centrifuged for 5 min at 2500 rpm. Six hundred microlitres of the supernatant was separated and evaporated in nitrogen evaporator for 10 min. The dried residue was reconstituted with 100 μl of the mobile phase and 50 μl of the sample was directly injected into the HPLC column.

Validation procedure

The accuracy and linearity of alpha-LA standards were evaluated by analyzing a set of standards ranging from 0.78 to 50 μg/ml. Unknown concentrations were derived from linear regression analysis of the peak height of the analyte vs. concentration curve. The linearity was verified using estimates of correlation coefficient (r). Inter day and between intra day variations were determined by processing each standard concentration in duplicate for five consecutive days.

Precision studies were carried out on the basis of injection repeatability of spiked plasma samples. In order to evaluate the precision of the method, three different concentrations of alpha-LA were prepared in pooled plasma and analyzed in duplicate on three consecutive days.

For the recovery experiment, known concentrations of LA were prepared in pooled human plasma samples spiked with 0.78 and 3.125 μg/ml of LA and assayed. The percentage of recovery was calculated by dividing sample differences with the added concentrations. Recovery experiments were carried out on three different occasions.

The sensitivity of the method was evaluated by quantifying the limits of quantification (LOQ) and detection (LOD) for alpha-LA. These values were estimated mathematically from the standard curve equations. The LOQ and LOD were obtained by multiplying the standard deviation (S.D.) of the Y-axis intercepts by 10 and 5 respectively.

Results and Discussion

Several HPLC methods have been described to measure LA levels in human plasma. Some of them are quite complex and lengthy because of the extraction procedures involved in the sample preparation. Although a spectrophotometric method has been reported very recently, the sensitivity of that method is quite low and requires a large volume of sample [22].

This method was optimized under different chromatographic conditions in order to validate various parameters. Several methods, such as liquid-liquid extraction, deproteinization using solvents like chloroform, ethanol and acetonitrile for sample preparation have been used [6,11,23,24]. Among the various solvents used as mobile phase, 50 mM Na_2HPO_4 (pH adjusted to 2.7): acetonitrile: methanol in the ratio of 50:30:20 produced a good resolution peak. While developing the method, the run time was kept at 8 min (Figure 3). For the selection of optimal flow rate, mobile phase was pumped at different flow rates in the range of 1–2 ml/min. Better separation of the target compound was achieved at 1 ml/min flow rate.

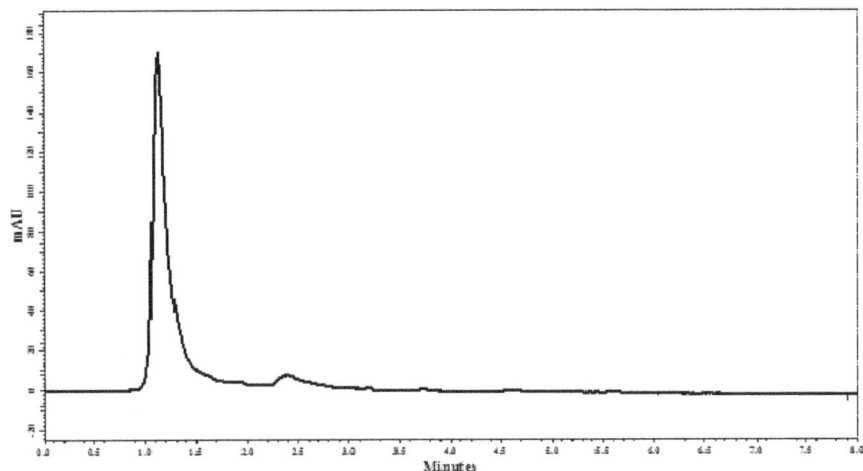

Figure 3: Blank plasma sample was run for 8 minutes.

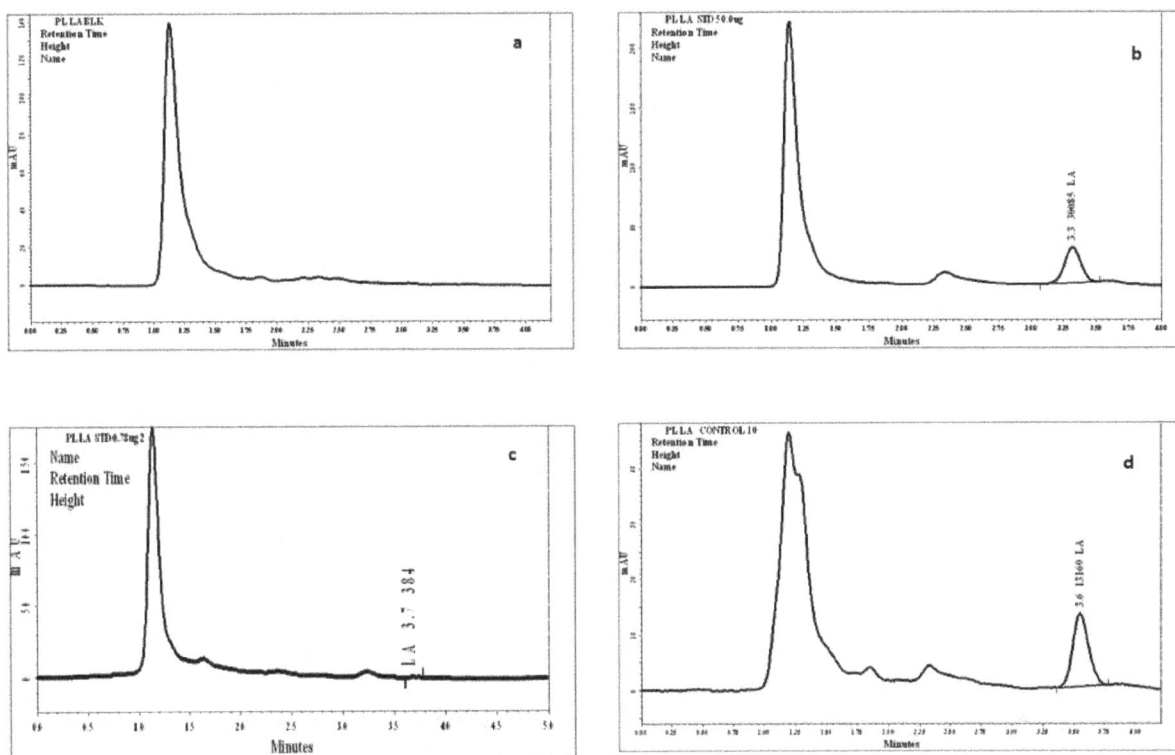

Figure 4: Chromatograms of LA a) Blank b) 50µg/ml c) 0.78 µg/ml d) Healthy volunteer.

In the present study, simple extraction procedure with ethanol and chloroform was standardized which resulted in better recovery of the analyte obtained as compared in the other solvents used. However the solvent used in the study provide satisfactory extraction efficiency. Under these chromatographic conditions, sufficient resolution was obtained with a good resolution peak at a retention time of 3.5 min. Temperature is an important parameter that greatly affects the chromatographic analysis. To study its effect on the analysis of LA, column oven temperature variations were evaluated in the range of 25–45°C. The sensitivity and retention time of analyte were greatly influenced by column oven temperature. The chromatographic peak LA

was found to be sharp at 40°C and resolution of the peak decreased with increasing temperature. The UV detector was used in the range of 201–401 nm. The detector at which the analyte showed optimal response was selected as the optimized range for the determination of LA. The UV detector at 201 nm showed an optimal response and it was used for measurement of LA content. This was performed using photodiode array (PDA), though LA can be detected using UV detector, as the detection wavelength falls within the range of UV detector. The method was validated by evaluating the linearity, precision, recovery, LOD and LOQ, specificity and sensitivity. Complete separation of the target peak was achieved in 5min. Plasma LA was separated as a discrete peak at a

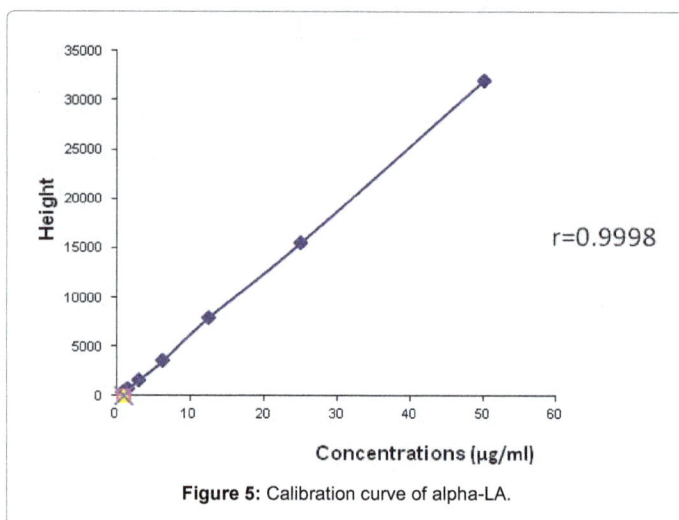

Figure 5: Calibration curve of alpha-LA.

Standard concentration (µg/ml)	Inter day (n=5) Mean height ± SD (% CV)	Intra day (n=5) Mean height ± SD (% CV)
0.78	305.8 ± 32.78 (10.7)	368 ± 39.76 (10.8)
1.56	564.8 ± 43.1 (7.6)	725.2 ± 72.5 (9.9)
3.125	1587.6 ± 174.4 (10.9)	1645.4 ± 177.7 (10.7)
6.25	3481.6 ± 299 (8.5)	3535 ± 356.9 (10.0)
12.5	7747.2 ± 499.9 (6.4)	7955.8 ± 321.5 (4.04)
25	15030.2 ± 855.7 (5.6)	15537.8 ± 1049.5 (6.75)
50	34021.8 ± 1702 (5.0)	32011.6 ± 1974.2 (6.16)

All concentrations in µg/ml. SD- standard deviation, CV- coefficient of variation

Table 1: Linearity and reproducibility of plasma alpha-LA standards.

Added (µg/ml)	Estimated (µg/ml) Mean ± SD	Recovery (%)
Baseline	1.39 ± 0.18	
0.78	1.82 ± 0.26	84%
Baseline	5.86 ± 0.38	
3.125	10.12 ± 1.06	113%

All concentrations in µg/ml, SD- standard deviation

Table 2: Recovery of alpha-LA in plasma.

Actual concentration (µg/ml)	Found concentration (µg/ml) Ratio ±SD (RSD %)
50	51.9 ± 0.55 (3.8)
6.25	5.87 ± 0.12 (6.08)
0.78	0.64 ± 0.02 (17.9)

All concentrations in µg/ml, SD- standard deviation

Table 3: Precision for plasma alpha-LA assay.

retention time of 3.5min. Representative chromatograms of blank, 50 µg/ml, 0.78 µg/ml and one healthy volunteer sample are shown in the Figure 4 a, b, c and d.

The peak for LA was observed linear at different concentration. The calibration curve of LA standard solutions constructed using seven concentrations shows good linearity in the range 0.78 to 50.0 µg/ml as shown in Figure 5. A linear relationship was observed between peak heights and the concentrations over this range with a correlation coefficient (r) of 0.9998.

The linearity, reproducibility (inter-day and intra-day) of plasma LA data is presented in Table 1. The inter-day and intra-day co-efficient variation (CV %) for standards containing 0.78-50 µg/ml ranged from 6.4to 10.9% and 4.04 to 10.8% respectively. The mean % variation

from the actual concentrations was 100.7%. Furthermore, recovery of LA at two concentrations (0.78 and 3.125 µg/ml) were 84 and 113% respectively (Table 2). The inter-day and intra-day relative standard deviation (RSD) for standards containing 0.78 to 50 µg/ml ranged from 3.8 to 17.9% (Table 3). The LOD and LOQ values were 0.39 and 0.78 µg/ ml respectively.

Conclusions

In summary, the HPLC-UV method described for the estimation of LA in human plasma is simple, sensitive, reproducible and precise with a recovery of 113%. The assay is less laborious, economical and less time consuming than other reported methods previously. Plasma concentration of LA was determined in healthy volunteers (n=10), the mean plasma LA concentration with standard deviation was 0.4570 ± 0.26 µg /ml. The validated HPLC-UV method may be applied to the measurement of LA levels in clinical practice. Further the LA concentrations can be confirmed by LC-MS quantification; thereby this method may be applied in future studies with shorter run time thereby large number of samples can be quantitated.

Acknowledgement

The authors thank Dr. Aleyamma Thomas and Dr. Soumya Swaminathan, of NIRT for their support and Dr. M. Parthiban and Dr. K. Satyavani of Prof M Viswanathan Diabetes Research Centre for their encouragement. The technical help rendered by Ms. V.V. Sandhya and Mr. S. Venkatesh is gratefully acknowledged.

References

1. Packer L, Witt EH, Tritschler HJ (1995) alpha-Lipoic acid as a biological antioxidant. Free RadicBiol Med 19: 227-250.

2. Biewenga GP, Haenen GR, Bast A (1997) The pharmacology of the antioxidant lipoic acid. Gen Pharmacol 29: 315-331.

3. Smith AR, Shenvi SV, Widlansky M, Suh JH, Hagen TM (2004) Lipoic acid as a potential therapy for chronic diseases associated with oxidative stress. Curr Med Chem 11: 1135-1146.

4. BalkisBudin S, Othman F, Louis SR, Abu Bakar M, Radzi M, et al. (2009) Effect of alpha lipoic acid on oxidative stress and vascular wall of diabetic rats. Rom J MorpholEmbryol 50: 23-30.

5. Abdin A, Sarhan NI (2011) Intervention of mitochondrial dysfunction- oxidative stress-dependent apoptosis as a possible neuroprotective mechanism of a-lipoic acid against rotenone- induced parkinsonism and L-dopa toxicity. Neurosci Res 71:387–395.

6. Trivedi RK, Kallem RR, Mamidi RN, Mullangi R, Srinivas NR (2004) Determination of lipoic acid in rat plasma by LC-MS/MS with electrospray ionization: assay development, validation and application to a pharamcokinetic study. Biomed Chromatogr 18: 681-686.

7. Jain SK, Lim G (2000) Lipoic acid decreases lipid peroxidation and protein glycosylation and increases (Na(+) + K(+))- and Ca(++)-ATPase activities in high glucose-treated human erythrocytes. Free RadicBiol Med 29: 1122-1128.

8. Shigeta Y, Hiraizumi G, Wada M, Oji K, Yoshida T (1961) Study on the serum level of thioctic acid in patients with various diseases. J Vitaminol 7: 48-52.

9. Natraj CV, Gandhi VM, Menon KKG (1984)Lipoic acid and diabetes: Effect of dihydrolipoic acid administration in diabetic rats and rabbits. JBiosci 6: 37-46.

10. Satoh S, Toyo'oka T, Fukushima T, Inagaki S (2007) Simultaneous determination of alpha-lipoic acid and its reduced form by high-performance liquid chromatography with fluorescence detection. J Chromatogr B AnalytTechnol Biomed Life Sci 854: 109-115.

11. Chen J, Jiang W, Cai J, Tao W, Gao X, et al. (2005) Quantification of lipoic acid in plasma by high-performance liquid chromatography-electrospray ionization mass spectrometry. J Chromatogr B AnalytTechnol Biomed Life Sci 824: 249-257.

12. Durrani AI, Schwartz H, Schmid W, Sontag G (2007) alpha-Lipoic acid in dietary supplements: development and comparison of HPLC-CEAD and HPLC-ESI-MS methods. J Pharm Biomed Anal 45: 694-699.

13. Haj-Yehia AI, Assaf P, Nassar T, Katzhendler J (2000) Determination of lipoic acid and dihydrolipoic acid in human plasma and urine by high-performance liquid chromatography with fluorimetric detection. J ChromatogrA 870: 381-388.

14. Durrani AI, Schwartz H, Schmid W, Sontag G (2007) alpha-Lipoic acid in dietary supplements: development and comparison of HPLC-CEAD and HPLC-ESI-MS methods. J Pharm Biomed Anal 45: 694-699.

15. Aboul-EneinV, Hoenen H (2005) Validated Method for Determination of a-Lipoic Acid in Dietary Supplement Tablets by Reversed Phase Liquid Chromatography. J LiqChromatogrRelatTechnol 27: 3029-3038.

16. Witt W, Rüstow B (1998) Determination of lipoic acid by precolumn derivatization with monobromobimane and reversed-phase high-performance liquid chromatography. J Chromatogr B Biomed SciAppl 705: 127-131.

17. Witt W, Rüstow B (1998) Determination of lipoic acid by precolumn derivatization with monobromobimane and reversed-phase high-performance liquid chromatography. J Chromatogr B Biomed SciAppl 705: 127-131.

18. Inoue T, Sudo M, Yoshida H, Todoroki K, Nohta H, et al. (2009) Liquid chromatographic determination of polythiols based on pre-column excimer fluorescence derivatization and its application to alpha-lipoic acid analysis.J Chromatogr A 1216: 7564-7569.

19. Teichert J, Preiss R (2002) High-performance liquid chromatographic assay for alpha-lipoic acid and five of its metabolites in human plasma and urine. J Chromatogr B AnalytTechnol Biomed Life Sci 769: 269-281.

20. Teichert J, Preiss R (1995) Determination of lipoic acid in human plasma by high-performance liquid chromatography with electrochemical detection. J Chromatogr B Biomed Appl 672: 277-281.

21. Sen CK, Roy S, Khanna S, Packer L (1999) Determination of oxidized and reduced lipoic acid using high-performance liquid chromatography and coulometric detection. Methods Enzymol 299: 239-246.

22. Han D, Handelman GJ, Packer L (1995) Analysis of reduced and oxidized lipoic acid in biological samples by high-performance liquid chromatography. Methods Enzymol 251: 315-325.

23. Deepthi K, Divakar TE, Bhanumurthy N, Pavankumar D (2012) Spectrophotometric methods for the determination of lipoic acid by using MBTH and ferric chloride. IJERT 7: 2278-2286.

24. Montero O, Ramírez M, Sánchez-Guijo A, González C (2012) Determination of lipoic acid, Trolox methyl ether and tocopherols in human plasma by liquid-chromatography and ion-trap tandem mass spectrometry. Biomed Chromatogr 26: 1228-1233.

15

HILIC Chromatography – An Insight on the Retention Mechanism

Aikaterini I Piteni, Maria G Kouskoura* and Catherine K Markopoulou

Laboratory of Pharmaceutical Analysis, Department of Pharmaceutical Technology, School of Pharmacy, Faculty of Health Sciences, Aristotle University of Thessaloniki, 54124 Thessaloniki, Greece

Abstract

Hydrophilic interaction chromatography (HILIC) could be characterized as a complex chromatographic system that involves multiple mechanisms. These are partitioning as well as polar and ionic interactions. Among several HILIC columns, ZIC-HILIC can be used to separate small organic ionic compounds. The presence of both positive and negative charge on the stationary phase may facilitate separations of both anionic and cationic analytes. Based on the Partial Least Squares methodology, an attempt to clarify the mechanism on this column revealed that the forces dominating are mainly determined by structural features. Consequently, the physicochemical properties which are related to the analytes' structure may heighten or attenuate the process. Ionic interactions are stronger for analytes containing moieties with basic properties since the interaction with the sulfonyl group is facilitated. The partition mechanism is prevailing for those analytes that are not sufficiently ionized at the experimental conditions (mobile phase pH 3 and 6.5) and for analytes that can create halogen bonds. Moreover, the stagnant water layer on the silica bed enhances the retention of water soluble compounds due to the increased hydrophilic interactions.

Keywords: ZIC-HILIC; Retention mechanism; PLS; Hydrophilic interaction

Introduction

Hydrophilic Interaction Chromatography (HILIC) is a term that was first suggested in 1990 by Alpert [1] who used the column for the separation of peptides, amino acids and other polar compounds. This methodology provides an alternative approach to separate small polar molecules efficiently using a stationary phase with increased polarity. Hydrophilic chromatography employs traditional polar stationary phases which are used in normal phase chromatography (NP-LC) and a mobile phase that is more suitable for reversed phase liquid chromatography (RP-LC) [2]. Moreover, it is suitable for highly hydrophilic and amphiphilic analytes that are too polar to be retained in RP-LC; still their charge is sufficient to allow the conditions of ion-exchange chromatography (IC) [3].

An important advantage of HLIC is that it can overcome the drawback of poor water solubility of many analytes that is often a problem in NP-LC. The most common eluent that is used in HILIC technique is acetonitrile, which is relatively polar. This has been the merit of HILIC towards NP-LC which uses mobile phases not friendly to the environment; still it seems to be a disadvantage in terms of green chemistry [4,5].

In terms of chromatography, it is known that analytes which are strongly hydrophilic but cannot attain charge in solution, do not seem to be retained on either type of stationary phases. On the contrary, in the case of highly hydrophilic compounds with functional groups allowing dipolar bonding, HILIC may retain molecules since it has been supplemented by the use of ion exchange chromatography [6].

Conceptually, HILIC could be considered as reversed phase chromatography, where a polar stationary phase has the ability to retain polar analytes which are eluted by mobile phase consisting of a mixture of an organic modifier and water.

Many researchers Buszewski [3], Bernal [7] contend that the mechanism of retention that prevails in HILIC chromatography is a difficult and complicated task that has not yet been clarified. In this vein, a large number of characteristics of numerous analytes have been employed so as to study their effect on the elution on a zwitterionic column (ZIC-HILIC) and explain the chromatographic behavior.

Based on the assertion that HILIC is partly ion exchange chromatography, stationary phases with zwitterionic functionalities (such as ZIC-HILIC) were intended for ion exchange separations [8,9] Still, the retention's mechanism complexity includes partition together with adsorption and hydrophobic interactions under specific conditions [10-12].

According to the theory of chromatography, it has been established that pH of the mobile phase is a tool that may control selectivity, since it determines the analyte's form. However, in reversed phase chromatography, molecules are not completely ionized in the experimental conditions and the presence of both ionized and neutral species leads in peak tailing or fronting [13]. For this instance, it is preferred to use a mobile phase where the aqueous phase pH is adjusted so as to induce ion formation. Based on the Henderson-Hasselbach equation [14]. The pH of the mobile phase will affect ionization of an analyte in accordance to the pK_a of the compound at a given pH value. This drawback has been surpassed using HILIC chromatography where the peak shape is improved [15].

The retention in HILIC is mainly affected by adjusting the eluent (e.g., the fraction of organic modifier), the type and concentration of the buffer and the pH value [7]. But in the case that there is high diversity in the structure of the analytes, the retention may depend on several additional factors.

The aim of this work is to investigate the mechanism of retention on a ZIC-HILIC column. Such an attempt was made through the study of a large number of analytes that belong in different groups of compounds (mainly drugs) at different pH conditions and different organic

*Corresponding author: Maria G Kouskoura, Laboratory of Pharmaceutical Analysis, Department of Pharmaceutical Technology, School of Pharmacy, Faculty of Health Sciences, Aristotle University of Thessaloniki, 54124 Thessaloniki, Greece E-mail: mkousk@gmail.com

modifier. The choice of ZIC-HILIC among several HILIC columns was made because it has been proved superior and supplementary to both NP-LC and RP-LC. Additionally this column can be conveniently coupled with many types of detectors such as UV and MS (especially in the ESI mode) [3].

Previous researches revealed that exploring the mechanism of retention of analytes on a stationary phase is feasible by the development of a model that takes into consideration as many parameters as possible [16,17]. There had been several studies using Quantitative Structure Retention Relationships (QSRR) [18] Principal Component Analysis (PCA) [19,20] or Multiple Linear Regression Analysis (MLR) [21] and Partial Least Squares (PLS) [22] on several columns.

According to the literature there had been studies aiming to explain the chromatographic behavior of specific categories of compounds using a limited number of descriptors [23,24]. However, there has not yet been an attempt to clarify the retention mechanism of analytes on a HILIC column using such techniques for a miscellaneous gathering of analytes using chemometric methodologies.

Therefore, this work focuses on explaining and modeling the retention behaviour with the aim of PLS. The advantage of PLS technique is that it can correlate significantly larger datasets compared to other methods [16]. The descriptors used in the current study include several physicochemical properties of the analytes along with their structural characteristics. Thus, within the results of the present study lies some perspective in the enlightenment of the complex retention mechanism on a ZIC-HILIC column.

Materials and Methods

Chemical and reagents

The analytes used in the experiments employ a miscellaneous group consisting of 136 compounds (Supplementary Information; Tables A and B) and analytical grade reagents. The compounds were dissolved in methanol at concentrations ranging from 5 μg/mL to 10 μg/mL depending on the solubility of each compound. 25 μL of the final dilutions were injected with a SIL-20AD auto-sampler.

The analytes were USP-grade and were obtained by i. Carlo Erbra (Milano, Italy), ii. Panreac Company (Controla, Thessaloniki, Greece), iii. Sigma-Aldrich (Life Science Chemilab), iv. Riedel-de-Haën (Life Science Chemilab), v. Fluka (Life Science Chemilab, authorized distributor in Greece), vi. Applichem (Bioline Scientific, Athens, Greece), vii. Acros-Organics (Life Science Chemilab), viii. Pierce Chemicals (Pierce Chemical Company, Rockford, Illinois), and ix. Boehringer Ingelheim (GmbH, Ingelheim am Rhein) companies.

Acetonitrile (ACN), methanol (MeOH) and water (H_2O) were gradient grade for liquid chromatography and were obtained from Panreac. The adjustment of the pH value of the mobile phase was made by the addition of aqueous solution of $NH_4OH > 25\%$ or HCl 37%, obtained by Fluka. Mobile phases were degassed by filtering through a Millipore HV 0.45 μm pore membrane filter.

Instrumentation

The experimental procedure was carried out on an HPLC (Shimadzu) instrument, equipped with two LC-20AD pumps, a SIL-10AD auto-sampler and a UV-diode array detector. LC Solution software was used to record and elaborate the chromatographic peaks. In order to ensure stable experimental conditions, the temperature was maintained at 30°C using a column oven (Shimadzu).

The stationary phase was a ZIC-HILIC column obtained from Merck Sequant, with dimensions 150 × 4.6 mm and particle size 5 μm.

Chromatographic characterization of the probes was studied on the basis of the pH difference of the mobile phase, as well as on the difference of the mobile phase organic modifier.

The solvent mixtures used as mobile phase were 40% $MeCN/H_2O$ adjusted at two pH values (3 and 6.5) using aqueous solutions of HCl and NH_3 respectively. The flow rate of the mobile phase was adjusted to 0.2 mL/min and the retention time was determined by the mean of at least duplicate measurements. The retention time (t_r) of naphthol was checked at frequent time intervals in order to control the stability of the chromatographic system.

Partial Least Squares (PLS)

The software used to develop and evaluate the models, is an independent modeling of class analogies Simca-P (Version 9; Umetrics, Upsala, Sweden). For this purpose, three datasets were separately designed and three PLS models were compiled, labelled as total. The models designed represent data at three different experimental conditions: (a) mobile phase 40% $MeCN/H_2O$ at pH 3, (b) mobile phase 40% $MeOH/H_2O$ at pH 3 and (c) mobile phase 70% $MeCN/H_2O$ at pH 6.5. Each dataset contained 79X variables and experimental data (Y variable) for 136 observations (analytes).

Despite the fact that some of the descriptors studied were proved to be of minor interest, the range of physicochemical properties recorded was wide. The study aimed to eliminate the most important factors affecting the retention among numerous physicochemical properties as well as structural features. The major division of the dataset is shown in Figure 1.

The structural characteristics are encountered within the constitutional descriptors. They were inserted in the dataset using integer numbers and zero, which were used to denote the presence, the multiplicity or the absence of a structural feature. The chosen observations represent a diverse gathering of compounds so as to take into account as many structural characteristics as possible. This way several conclusions on the retention mechanism could be drawn concerning either the total dataset or specific classes containing groups of similar compounds.

The structure of chemical compounds is the major factor determining their physicochemical properties. Thus, the descriptors employed for the implementation of the models include lipophilicity parameters, as well as topological, geometrical and electronic properties (Supplementary Information; Table C). The above data were derived by different databases such as ACD/Labs [24] and calculation platforms

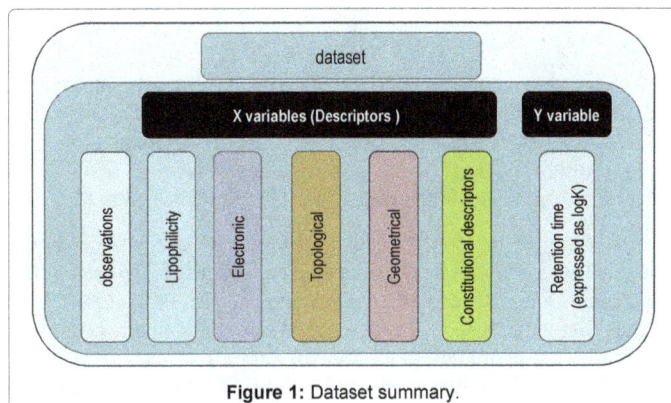

Figure 1: Dataset summary.

Dragon [25] Osiris [26], Hyperchem [27] Marvin [28]. All of them included drawing and displaying structures, as well as their geometry optimization which was made based on the Polak-Ribiere algorithm [29].

The distribution coefficient (logD) is directly related to the ionization of the compounds and had to be thoroughly studied. LogD indicates the distribution of the neutral or ionic form of a molecule in water against the non-ionized species in octanol. Based on this fact, it was necessary to calculate the corresponding logD values of the analytes studied and insert them as one of the variables in the dataset. Apart from the logD values, the ionization percentages of each analyte were also calculated and recorded (Supplementary Information; Table B).

Aside from the X variables which represent the aforesaid characteristics of the analytes a Y variable was also inserted. It refers to the logK which corresponds to the retention of each analyte on the stationary phase. The Y variable is the logK of the analytes, where K is the capacity factor which represents the relative speed of the analyte through the column and is expressed by the fraction $K=t_r-t_o/t_o$ (t_r is the retention time of the analytes and t_o is the solvent front).

The observations studied were divided in two large groups (Figure 2). The first consists of chemical compounds (including drugs) which cannot ionize at the experimental conditions or their ionization percentage (IP%) is lower than 45%.

This percentage has been shown to be the turning point for ionization; when ionization is lower, the effect on the retention is negligible [30]. The second group includes molecules that are affected by the pH of the mobile phase and the ionization degree (of their acidic or basic group) is higher than 45%. These groups were also studied as separate models.

The pK_a values of the basic or acidic groups of the analytes are determining their ionization, which will vary based on the pH of the mobile phase. Therefore, an overall and more detailed view for the retention behavior of the analytes on HILIC column, as well as a clarification of the effect of their ionization, was achieved through the overall and individual models studied.

Method validation: Cross-validation (CV) is a method leaving out part of the observations each time, and it was used to select the number of components (model dimensions) in the PLS algorithm. Based on the CV, the response values (Y) for the observations kept out were predicted by the model and then compared with the true values. The number of rounds used by Simca-P9 for model fitting was 7. Using the appropriate number of significant components, the total models were fit according to Haaland and Thomas [31] criteria. The results of the statistical data along with the validation results are given in Table 1 (and in Supplementary Information; Figure A-C) To affirm full statistical significance of the original estimates, the intercept limits were set to be $r^2<0.3$ and $q^2<0.05$ [32].

Table 1 shows the statistical coefficients for all models compiled. The first model for each mobile phase is characterized as total since it includes all observations. The other two models are characterized as group A for observations whose ionization degree is 0% ≤ (IP%) ≤ 45% and group B for observations with IP% >45% (as described in Figure 2).

According to the statistical data, it was obvious that the model developed when the mobile phase contained acetonitrile at pH 3 was the most reliable since it was statistically significant ($R^2=0.914$, $Q^2=0.792$). On the other hand, when the organic modifier of the mobile phase was replaced by methanol, although the correlation coefficient was good, the models appeared to lack predictive ability. It is worth mentioning

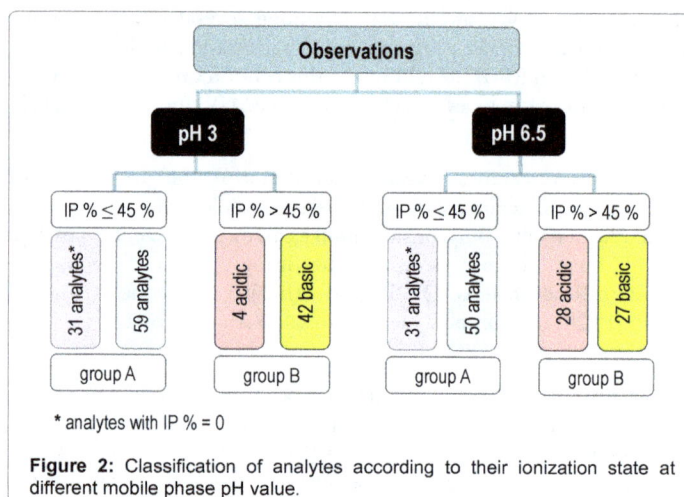

Figure 2: Classification of analytes according to their ionization state at different mobile phase pH value.

that the models at pH 6.5 also show poor predictive ability.

These results are justified due to the fact that the gathering of analytes was miscellaneous and they have many differences in structure and physicochemical properties. At pH 3 the behaviour has a more consistent pattern because the ionized analytes are those containing basic groups. On the other hand, at higher pH (6.5) there are compounds whose ionization occurs in both basic and acidic moieties in their structure and the interaction with the stationary phase is more complex. Finally, in the case of the model where the mobile phase contains MeOH at pH 3, the predictive ability is low because the mobile phase may also interact with the stationary phase shielding the charged functional groups. Therefore, methanol competes with the analytes for the active sites of the column and may interfere in the retention mechanism.

Results and Discussion

The commercially available HILIC columns are divided in three categories; these are neutral, charged (anionic, cationic) and zwitterionic [7]. The choice of the ZIC-HILIC stationary phase was made as it is the most widely used HILIC column, because it is applicable in many categories of analytes. The structure of this zwitterionic column (Figure 3) includes a quaternary ammonium and a sulfonic group attached on the silica bed.

The chromatographs obtained showed that the majority of the chromatographic peaks of the analytes were symmetrical with no distortion at all experimental conditions.

Initially, this research focused on the correlation of the retention time of an analyte with its ionization degree. In short, the pH (3 and 6.5) of the mobile phase determines the ionization degree of a molecule, its logD value and consequently its behavior (t_r) on a stationary phase.

A second part of the study has been an investigation of the factors affecting the retention on the HILIC column for all the observations regardless their ionization state. The effect of each variable (xk) on the Y variable in the model is labelled as VIP (Variable Importance in the Projection). VIP is the sum over all model dimensions of the contributions VIN (variable influence).

Moreover, the total PLS models were explored via the loadings' plots (w*c Ref. [1]) which display the correlation between the X variables and the Y scores in the first dimension (component) [32,33]. The loadings' plots provide the additional information of whether the effect of each variable is positive or negative on the logK.

Exploring the retention mechanism of the total models

The pH value of the mobile phase has been adjusted at pH 3 by the addition of HCl (37%). This choice has been made in order to avoid any other interferences of a buffer solution on the retention which would be caused by secondary mechanisms. The retention of the analytes at this pH value can be studied based on the different classes at which the analytes belong. First of all, the compounds that cannot be ionized are not retained on the stationary phase. Most of them are eluted along with the solvent front (~ 8-9 min). Moreover, there is a number of analytes that although bearing functional groups which are susceptible to ionization, the degree of ionization at the pH value of the mobile phase is not sufficient (lower than 45%). Consequently, the retention of the majority of them is also not affected and their elution is fast when the mobile phase consists of acetonitrile in water.

When the organic modifier of the mobile phase was replaced by methanol the majority of the analytes seem to be affected by this alteration and the retention is delayed. In the case of compounds that are not ionized, the retention time is slightly longer with no exceptions. Still in both cases, this delay in the retention time is short and its variance is limited.

For analytes whose ionization is lower than 45%, the retention is also slightly delayed; but there are a few compounds that have an odd behaviour. Some examples of these compounds are shown in Table 2. According to the information provided, it is obvious that even in the case of analytes that are not ionized (most of the compounds of Table 2 have IP << 45%; Supplementary Information; Table B), the retention is slightly higher when the mobile phase contains methanol. This can be explained as acetonitrile is an aprotic polar solvent which does not participate in the ionization of the analytes. Moreover, methanol can act as both hydrogen bond donor and acceptor and can compete for active polar sites on the HILIC surface. Therefore, if an analyte is more hydrophilic (even slightly) due to low ionization percentage it might be more retained on the stationary phase compared to its corresponding

retention time when the mobile phase consists of acetonitrile.

The study of the elution strength of the solvents used must be based on the ionization percentage of the analytes due to the differences occurring in the mechanism retention (Table 2). Therefore, in the case of molecules from the group A (Figure 2, IP<45%, Supplementary Information; Table B), the elution of the analytes is fast (sometimes along with the solvent front) while there are no significant difference when the organic modifier is not the same. The general pattern is that the retention time for all analytes is slightly increased when the mobile phase contains methanol against acetonitrile. Still, there are specific compounds that are retained longer with acetonitrile due to special structural characteristics or physicochemical properties. An example is clofoctol whose structure results in very high values at descriptors assigned to specific structural features such as C atoms on side chains. The aforesaid results in high value for free rotatable bonds (FRB) and the Balaban index (J3D) which is a topological index increasing when the number of branches on a molecule increases [34]. Probably, the complex structure, including many branches, makes the molecules bulky and in a way it is trapped within the structure of the stationary phase, even if the ionic interactions are negligible.

A similar behaviour is observed for analytes that are highly ionized (even in their acidic moieties). Obviously, there is a steric hindrance of quaternary amine group of the column and cannot develop ionic interactions with analytes that are negatively charged.

On the other hand, highly ionized basic groups exhibit a clear delay in their retention due to the ionic interactions with the sulfonic group. Actually, the eluting strength of the solvents is different for these analytes; water enhances the elution strength while acetonitrile causes a delay in the retention.

Apart from the conditions mentioned in the present study, the retention time of the analytes was also experimentally measured with a mobile phase containing 70% MeCN in water at pH 3. The model compiled for these data gave the same results in terms of retention behaviour and the effect of the descriptors studied. The only difference

	40% MeCN at pH 3			40% MeOH at pH 3			40% MeCN at pH 6.5		
	Total	Group A	Group B	Total	Group A	Group B	Total	Group A	Group B
Observations	136	90	46	136	90	46	136	81	55
Components	6	3	3	4	4	4	5	3	3
R^2	0.914	0.875	0.860	0.753	0.8	0.926	0.805	0.669	0.801
Q^2	0.792	0.818	0.662	0.577	0.632	0.493	0.584	0.518	0.599
Validation									
r^2	0.0, 0.22	0.0, 0.174	0.0, 0.314	0.0, 0.194	0.0, 0.0548	0.0, 0.306	0.0, 0.196	0.0, 0.19	0.0, 0.124
q^2	0.0, -0.435	0.0, -0.29	0.0, -0.148	0.0, -0.174	0.0, -0.0746	0.0, -0.161	0.0, -0.388	0.0, -0.113	0.0, -0.0846

Shows the statistical coefficients for all models compiled. The first model for each mobile phase is characterized as total since it includes all observations. The other two models are characterized as group A for observations whose ionization degree is 0% ≤ (IP%) ≤ 45% and group B for observations with IP%>45%.

Table 1: Statistical data for the compiled models.

	Retention Time				Retention Time		
Analyte	40% MeOH	40% MeCN	40% MeCN	Analyte	40% MeOH	40% MeCN	40% MeCN
p-nitrophenol	11.97	8.68	9.45	amitryptiline	37.99	24.35	189.98
p-nitroaniline	13.34	8.97	12.45	atropine	32.95	27.31	285.62
naphthol	14.70	8.78	8.41	desipramine	39.04	25.66	300.00
dichlorphenamide	22.96	10.26	8.96	dibucaine	31.82	26.17	132.27
clofoctol	18.00	16.63	9.21	dipyridamole	17.45	26.89	201.88
furosemide	15.10	8.44	8.21	ephedrine	31.17	33.38	300.30
Sulfadiazine	12.11	9.13	8.99	imipramine	39.04	24.79	244.53
Sulfabenzamide	13.84	8.70	8.51	nicergoline	9.31	25.87	122.47

Table 2: Comparison of retention times of specific analytes at methanol or acetonitrile at different mobile phases.

is that the retention time was increased for the majority of the analytes. More specifically, the increase was higher in the case of analytes containing ionisable groups but minor for the analytes that are either ionized at a lower extend or lack such moieties.

Factors affecting the retention mechanism

The compiled models of correlation along with the VIP table (Supplementary Information; Table D) and the loadings' column plots (Figures 4-6), brought about not only the impact of the descriptors on the retention time in priority order, but also whether the effect of each descriptor was positive or negative.

- **The effect of logD**

It is known that the logD value is related to the distribution of the ionic form of a molecule in water against the non-ionized species in octanol. Among the analytes studied, there are compounds that lack both acidic and basic groups and their logD value coincides with their partition coefficient (logP).

On the other hand there are compounds that have different logD and their elution is usually delayed. As it is shown in the loadings' plots (Figures 4-6), this delay is longer when the increase in the logD is higher which means that the ionization degree is higher (IP%). The phenomenon is even more intense in the case that the ionization exceeds 45%. The above is imprinted in the VIP (Variable Importance) (Supplementary Information; Table D) and its effect on the retention

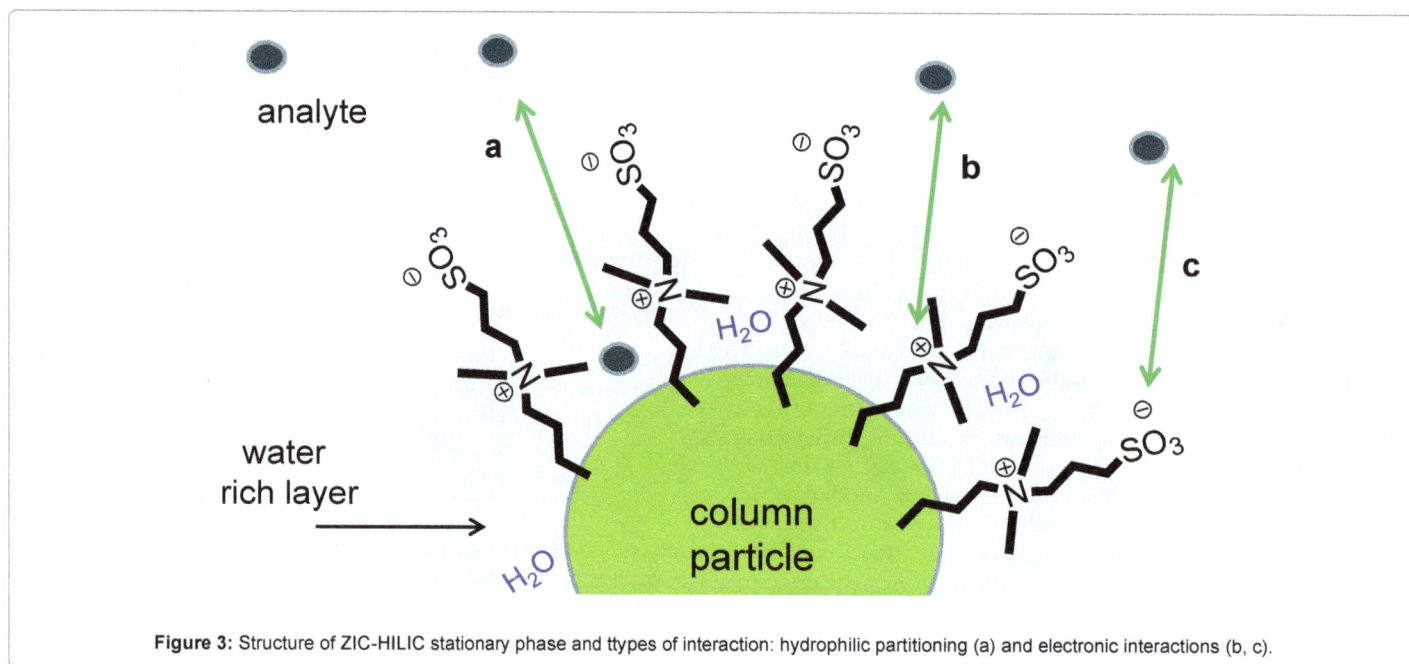

Figure 3: Structure of ZIC-HILIC stationary phase and ttypes of interaction: hydrophilic partitioning (a) and electronic interactions (b, c).

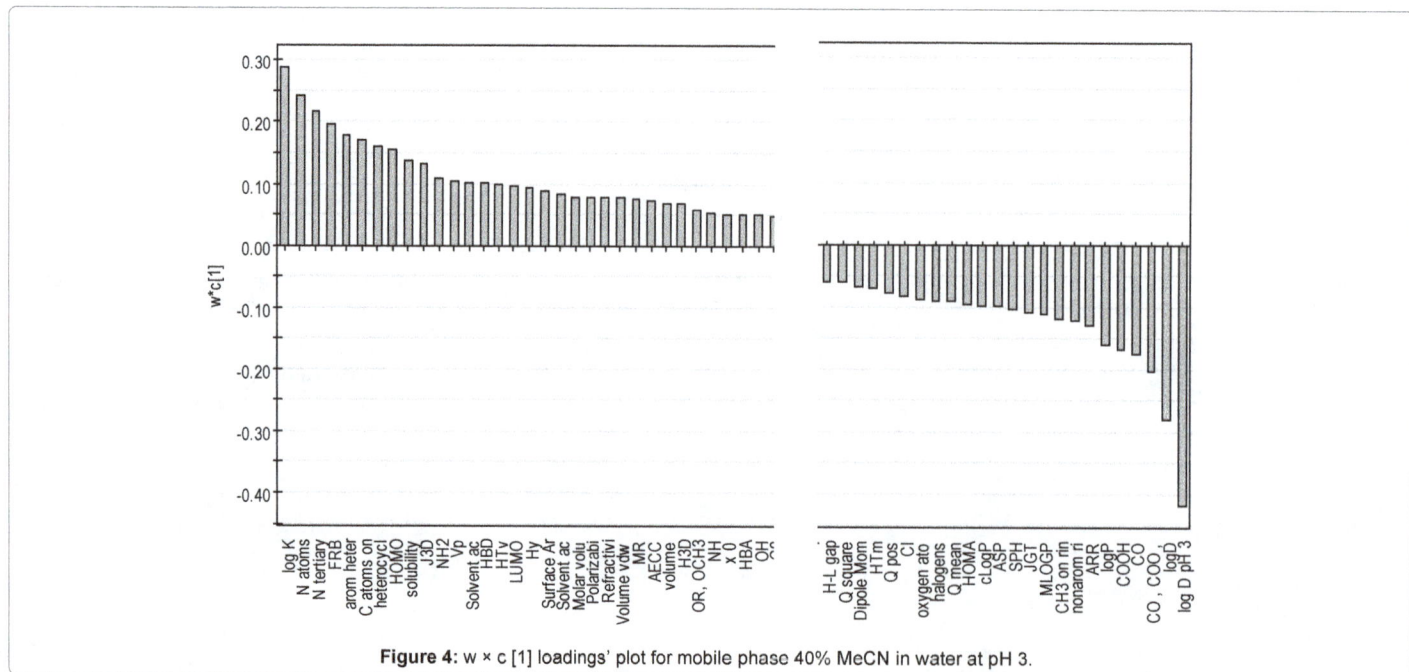

Figure 4: w × c [1] loadings' plot for mobile phase 40% MeCN in water at pH 3.

behaviour is similar for both mobile phases, i.e., regardless the organic modifier (methanol or acetonitrile). At pH 6.5, the effect of logD is similar, although the impact of this descriptor (VIP value) is relatively lower compared to the effect observed at pH 3. This can be explained because the groups ionized at pH 3 are mainly functional groups with basic properties which interact with the sulfonyl group of the stationary phase which is more active. On the other hand, at pH 6.5 more analytes are in their ionic form, but at least for half of them the ionized group has acidic properties and the ionic interaction with the ammonium group of the stationary phase is partly hindered.

- **The effect of the analytes' structure**

The presence of nitrogen atoms (N tertiary, NH, NH$_2$ etc.) as well as the presence of functional groups containing oxygen atoms (such as carbonyl, ester, carboxylic groups) are also important for the retention of the analytes. These descriptors have the same effect when the pH value is 3 at both mobile phases (MeOH or MeCN) as shown in the figures illustrating the loadings' plots of the total models (Figures 4 and 5) as well as at pH 6.5 (Figure 6).

More specifically, the presence of N atoms (e.g., N atoms, N tertiary, N heterocyclic) cause an increase in the retention time, because these groups are affected by the pH value of the mobile phase. At pH 3 the percentage of ionic micro species of analytes containing N atoms is higher because they increase the basicity of such molecules. Consequently, ionic interaction with the stationary phase (Figure 3c) is extended and the polar compounds will be delayed due to this mechanism, which is prevailing. Many of the analytes studied have functional groups with basic properties which are charged and therefore will be retained longer.

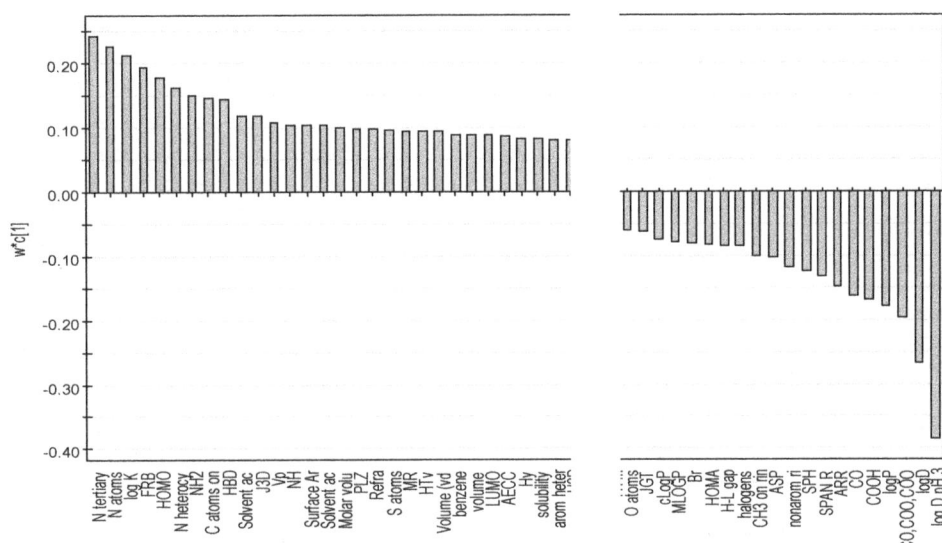

Figure 5: w × c [1] loadings' plot for mobile phase 40% MeOH in water at pH 3.

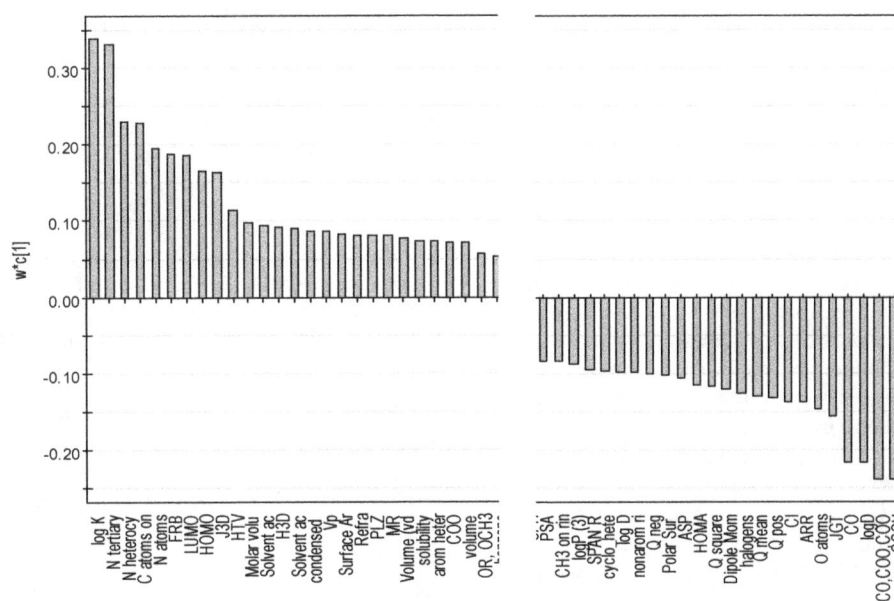

Figure 6: w × c [1] loadings' plot for mobile phase 70% MeCN in water at pH 6.5.

However, one should not overlook, that the presence of a traditionally basic group (e.g., NH$_2$, NH) will cause an increase in the retention time, but this effect is less important compared to the other N bearing groups that were previously discussed. A more detailed study of the molecules' structure could give the answer to this disagreement.

There are analytes that contain amine groups which will increase their ionization and therefore their delay. Apart from them, there are also compounds which contain amine groups but neighbouring carbonyl groups (e.g., CO, COO). In this case, the basic character of these moieties (NH, NH$_2$) will diminish and in some cases they will act as weak acids. Hence, according to the loadings' plots, the descriptors NH, NH$_2$, are of lower importance in the increase of the elution time while at the same time the presence of groups containing oxygen will favour faster elution. Obviously, the structure of a molecule is of vital importance, since it is the main reason for its ionization ability.

- **The effect of solubility**

Another descriptor that appears to increase the retention of analytes on the HILIC column is the solubility of compounds in water. Explanation of the solubility effect leads to the conclusion that on HILIC column there is another mechanism that works, which is partition. It is known that the solubility of compounds is the minimum amount of a solute that can be dissolved in a given amount of a solvent (in this case water).

At pH value equal to 3 some analytes are encountered in their ionic form. Meanwhile, the solubility of compounds is pH dependent [35]. The solvation of analytes in water is better when a compound is in its ionic form; the solubility of charged species increases. Even in the case of molecules whose ionization is not sufficient (<45%), if their solubility is high the retention is increased. The analytes are dissolved in the water portion of the mobile phase and they are in a way trapped in the water rich layer surrounding the column particle enhancing the partition mechanism and resulting in the delay of the analytes' elution.

Hence, there is higher hydrophilic interaction of analytes with the water layer on the column bed (Figure 3a). The analyte partitions into and out of the adsorbed water layer (partition mechanism) and these forces work along with the ionic interactions and the retention of the analytes will increase.

- **Other descriptors**

Other descriptors that also affect the retention causing a delayed elution are the volume of analytes which is directly related to the carbon atoms on side chains and the number of free rotatable bonds (FRB). Large molecules with long side chains are ponderous but their effect on the retention time is secondary since the ionic interactions prevail.

The w*c loadings' plot in the case of the alcoholic mobile phase shows that the presence of halogen atoms is responsible for faster elution, although this effect is not one of the most important factors. The presence of halogen atoms seems to be a determining factor compared to older studies in lipophilic columns where the effect of halogen atoms was of secondary importance in the retention mechanism and cause an increase in the retention of the analytes [22]. Probably, such molecules are able to form halogen bonds (halogen interactions) with methanol. At the same time the solubility of molecules is working at the same direction and the solvation of analytes in the mobile phase is facilitated not only due to their low solubility in water, but also because of their ability to interact with molecules of methanol. The halogen bond is relatively strong and may favour faster elution.

Another contributing factor to halogen bond strength comes from the short distance between the halogen containing Lewis acid (donor) and the Lewis base (acceptor), where X is a halogen. The attractive forces developed result in the distance between the donor and acceptor to be shorter than the sum of van der Waals radii. The interaction becomes stronger as the distance decreases between the halogen Lewis acid and the Lewis base.

The model compiled in the case that the mobile phase contains acetonitrile but the pH value is adjusted to 6.5 shares several similarities with the previous two models. Among the most important is the effect of the presence of the N atoms in the structure of the analyte which is responsible for the retention of the analytes. The molecules containing nitrogen atoms which are not attached to a carbonyl group seem to enhance the ionization and cause an increase in the retention time.

It is vital to mention at this point that due to the presence of the stagnant water layer on the surface of the column bed, the ionic interactions with the sulfonic group are stronger compared to the interactions of the quaternary ammonium group because of the hindrance caused by the water molecules. Therefore, the sulfonyl group is more active and will be able to retain positively charged molecules longer.

The above is verified due to the fact that in the VIP values the effect of the charge is more determinant as well as the global topological charge (JGT). This increase in the impact of the charge is also based on the fact that the number of analytes that are in their ionic form is bigger; hence the effect in the total mechanism of this descriptor is more vital.

Individual models: Apart from the total models developed for the sum of the observations, individual models were compiled as well (Supplementary Information; Figure D-I). According to the statistical results of Table 1, the most reliable model which corresponds to analytes that have IP% <45% are more reliable in acidic conditions. Of course this result is not associated to the ionization of the analytes but mainly to the ability of MeCN to have a more consistent behaviour. This consistency is determined due to the fact that MeCN is polar but aprotic and does not participate in any type of interactions such as hydrogen bonding between the analytes and the mobile phase or the stationary phase and the mobile phase. On the other hand, MeOH may participate in such interactions and compete with the analyte for the free silanol groups of the stationary phase or create weak bonds with the analyte molecules.

Especially, in the case of MeOH the presence of halogen atoms is a determinant factor causing a decrease in the retention time of analytes. This phenomenon has been explained in a previous paragraph based on the halogen bonds developed.

Concerning the effect of the HOMO-LUMO energies on the retention time, it seems to be a bit complicated. First of all, for ionized species the effect of LUMO is more important causing an increase in the retention time. An analyte with high LUMO value acts as a Lewis base (π-base) which interacts more easily and is retained on the sulfonic group of the stationary phase.

The same effect was observed for the total models but the impact was not of the same importance. As mentioned, a factor affecting the retention of these molecules, despite the fact that their ionization is not high, is their structure and mainly the presence of nitrogen atoms. This can be explained since the presence of such groups promotes the ionization of the analytes. Even when ionization is limited, there are cases that the molecule is able to be retained due to the stabilization of the polar molecules which occurs due to the high energies of HOMO orbitals. At the same time, high HOMO energies will enhance stability

[36,37]. This effect is obvious for compounds whose ionization at this pH is high and it is more probable for compounds that are not ionized but have high polar surface areas due to the presence of moieties where delocalization of electrons can easily occur. The difference in the priority order is logical because the total models contain analytes which are not ionized and they are eluted in short time despite their HOMO - LUMO values.

In the case of group A (Figure 2), high HOMO seems to increase the retention time but the impact of this descriptor is relatively low. The impact of the LUMO energies also seems to be of secondary importance, except in the case where the mobile phase contains methanol. For this model, analytes with increased LUMO energies appear to have faster elution times, more probably because the LUMO of the analyte interacts with the HOMO of MeOH and this electron interaction facilitates faster elution.

When the mobile phase has pH value equal to 6.5, then there are more analytes that are encountered in their ionic form. In this instance, apart from the solubility of the analytes in water which is determinant for their delay, the retention time is also increased when the analyte is a hydrogen bond donor (high HBD value) and in the case that it contains moieties that favor this behavior.

The hydrophilicity of the analytes, due to the presence of hydroxyl groups along with their ability to accept protons (high HBA), facilitates hydrogen bonding between free silanol groups or the water molecules covering the stationary phase and the analyte, or the interaction between the analyte and the charged groups of the stationary phase. The above is directly related to the increased values of the polar surface area of the analytes which is also responsible for the retention time increase.

An important observation is that the lipophilicity of the analytes is not the major factor affecting the retention in the total models. Moreover, in all cases the effect is opposite compared to the reversed phase chromatography. It was obvious that when the molecules could be ionized then the logP has minor effect on the retention (causing a decrease on the retention time). For analytes that are not ionized, logP is also responsible for the decrease on the retention time but this is not the rule because at high pH (6.5) logP is of secondary importance.

Conclusion

The retention behaviour on a ZIC-HILIC column is mainly affected by the ionization of the analytes. Still, a more thorough investigation of several characteristics of the analytes revealed that apart from the structure of a molecule, other parameters affect the retention as well. According to the VIP values and the loadings' plots for both mobile phases at pH 3, the most important factors that increase the retention are the presence of N atoms, the solubility of the compounds in water, as well as the solvent accessible surface area of molecules. Importantly, high logD values of the analytes decreases the elution time since the main mechanism is based on ionic interactions. The stagnant water rich layer that is covering the surface of the stationary phase may limit the interaction of the inner quaternary ammonium group with acidic groups of the analytes. At pH 3 these groups are not charged and the electrostatic interactions are very weak resulting in negligible delay if any.

References

1. Alpert AJ (1990) Hydrophilic-interaction chromatography for the separation of peptides, nucleic acids and other polar compounds. J Chromatogr 499:177-196.

2. Heaton J, Gray N, Cowan DA, Plumb RS, Legido-Quigley C, et al. (2012) Comparison of reversed-phase and hydrophilic interaction liquid chromatography for the separation of ephedrines. J Chromatogr A 1228: 329-337.

3. Buszewski B, Noga S (2012) Hydrophilic interaction liquid chromatography (HILIC)--a powerful separation technique. Anal Bioanal Chem 402: 231-247.

4. Tache F, Udrescu S, Albu F, Micale F, Medvedovici A (2013) Greening pharmaceutical applications of liquid chromatography through using propylene carbonate-ethanol mixtures instead of acetonitrile as organic modifier in the mobile phases. J Pharm Biomed Anal 75: 230-238.

5. Dos Santos Pereira A, David F, Vanhoenacker G, Sandra P (2009) The acetonitrile shortage: is reversed HILIC with water an alternative for the analysis of highly polar ionizable solutes? J Sep Sci 32: 2001-2007.

6. Dai J, Carr PW (2009) Effect of mobile phase anionic additives on selectivity, efficiency, and sample loading capacity of cationic drugs in reversed-phase liquid chromatography. J Chromatogr A 1216: 6695-6705.

7. Bernal J, Ares AM, Pól J, Wiedmer SK (2011) Hydrophilic interaction liquid chromatography in food analysis. J Chromatogr A 1218: 7438-7452.

8. Hemstrom P, Irgum K (2006) Hydrophilic interaction chromatography. J Sep Sci 29: 1784-1821.

9. Jandera P (2011) Stationary and mobile phases in hydrophilic interaction chromatography: a review. Anal Chim Acta 692: 1-25.

10. McCalley DV (2007) Is hydrophilic interaction chromatography with silica columns a viable alternative to reversed-phase liquid chromatography for the analysis of ionisable compounds? J Chromatogr A 1171: 46-55.

11. McCalley DV (2008) Evaluation of the properties of a superficially porous silica stationary phase in hydrophilic interaction chromatography. J Chromatogr A 1193: 85-91.

12. Hao Z, Xiao B, Weng N (2008) Impact of column temperature and mobile phase components on selectivity of hydrophilic interaction chromatography (HILIC). J Sep Sci 31: 1449-1464.

13. Buckenmaier SM, McCalley DV, Euerby MR (2002) Overloading study of bases using polymeric RP-HPLC columns as an aid to rationalization of overloading on silica-ODS phases. Anal Chem 74: 4672-4681.

14. Po HN, Senozan NM (2001) Henderson–Hasselbalch Equation: Its History and Limitations. J Chem Educ 78: 1499-1503.

15. McCalley DV (2010) Study of the selectivity, retention mechanisms and performance of alternative silica-based stationary phases for separation of ionised solutes in hydrophilic interaction chromatography. J Chromatogr A 1217: 3408-3417.

16. Kouskoura MG, Hadjipavlou-Litina D, Markopoulou CK (2014) Elucidation of the retention mechanism on a reverse-phase cyano column by modeling. J Sep Sci 37: 1919-1929.

17. Kouskoura MG, Kachrimanis KG, Markopoulou CK (2014) Modeling the drugs' passive transfer in the body based on their chromatographic behavior. J Pharm Biomed Anal 100: 94-102.

18. Baczek T, Kaliszan R, Novotna K, Jandera P (2005) Comparative characteristics of HPLC columns based on quantitative structure-retention relationships (QSRR) and hydrophobic-subtraction model. J Chromatogr A 1075: 109-115.

19. Buszewski B, Kowalska S, Kowalkowski T, Rozpedowska K, Michel M, et al. (2007) HPLC columns partition by chemometric methods based on peptides retention. J Chromatogr B Analyt Technol Biomed Life Sci 845: 253-260.

20. Visky D, Vander Heyden Y, Ivanyi T, Baten P, De beer J, et al. (2003) Characterisation of reversed-phase liquid chromatographic columns by chromatographic tests. Rational column classification by a minimal number of column test parameters. J Chromatogr A 1012: 11-29.

21. Ivanyi T, Vander Heyden Y, Visky D, Baten P, De beer J, et al. (2002) Minimal number of chromatographic test parameters for the characterisation of reversed-phase liquid chromatographic stationary phases. J Chromatogr A 954: 99-114.

22. Kouskoura MG, Mitan CV, Markopoulou CK (2015) Chemometric of the retention mechanism on butyl column: effect and relation of pH and pKa. Se Pu 33: 1274-1286.

23. Chirita RI, West C, Zubrzycki S, Finaru AL, Elfakir C (2011) Investigations on the chromatographic behaviour of zwitterionic stationary phases used in hydrophilic interaction chromatography. J Chromatogr A 1218: 5939-5962.

24. Euerby MR, Hulse J, Petersson P, Vazhentsev A, et al. (2015) Retention modelling in hydrophilic interaction chromatography. Anal Bioanal Chem 407: 9135-9152.

25. Todeschini R, Consonni V, Pavan M (2012) DRAGON. Talete srl.

26. Sander T (2014) Osiris Property Explorer (2001-2013).

27. Hyper Chem (TM) Professional, 1115 NW 4th Street, Gainesville, Florida 32601, USA.

28. Marvin (2012) ChemAxon.

29. Polak E, Ribiere G (1969) Note sur la convergence de methodes de directions conjuguees. ESAIM 16: 35-43.

30. Kouskoura MG, Mitan CV, Markopoulou CK (2015) Chemometric of the retention mechanism on butyl column: effect and relation of pH and pKa. Se Pu 33: 1274-1286.

31. Haaland DM, Thomas EV (1988) Partial Least Squares Regression for Spectral Analysis 1 Relation to other Quantitative Calibration Methods and the Extraction of Qualitative Information Anal. Chem 60: 1193-1202.

32. Lapins M, Eklund M, Spjuth O, Prusis P, Wikberg JE (2008) Proteochemometric modeling of HIV protease susceptibility. BMC Bioinformatics 9: 181.

33. Umetrics AB (2011) Simca. User guide and tutorial.

34. Balaban AT (1983) Topological indices based on Topological distances in molecular graph. Pure Appl Chem 55: 199-206.

35. Butler JN (1998) Ionic Equilibrium - Solubility and pH calculations. John Wiley and Sons Inc., p: 578.

36. McMurry J (2000) Organic Chemistry. Brooks Cole: Pacific Grove, USA.

37. Scudder PH (2013) Electron flow in Organic Chemistry - A desicion-based guide to organic mechanisms, 2nd edn. Wiley: New Jersey and Canada.

16

Development and Validation of a Reverse Phase Ultra Performance Liquid Chromatographic Method for Simultaneous Estimation of Nebivolol and Valsartan in Pharmaceutical Capsule Formulation

Seetharaman R* and Lakshmi KS

Department of Pharmaceutical Analysis, College of Pharmacy, SRM University, Kattankulathur, Tamilnadu – 603203, India

Abstract

The present study describes the development and validation of sensitive, novel Ultra Performance Liquid Chromatographic technique to simultaneously evaluate Nebivolol and Valsartan in commercial pharmaceutical capsule formulation. The technique was performed utilizing thermo C_{18} (4.6 mm×50 mm, 1.9 μm) column having a mobile phase comprising of 10 mM ammonium dihydrogen phosphate pH adjusted to 3.00 ± 0.02 with dilute orthophosphoric acid as buffer, with a ratio of buffer: acetonitrile 60:40 (v/v) and with a flow rate of 0.4 mL min^{-1}. The eluted components were monitored out at 220 nm utilizing a photo diode array detector. The retention times for NBL and VST were 0.48 and 0.83 min respectively. The developed Ultra Performance Liquid Chromatographic method was validated according to ICH guidelines to confirm specificity, linearity, accuracy and precision. The homoscedasticity of the variances and lack of fit in the evaluation of linearity was established using the Cochran's C test and F-test respectively. The repeatability variances and time different intermediate variances were assessed simultaneously. The time different variances expressed as percentage relative standard deviations (%RSD) and accuracy were well within the limits as prescribed by ICH. In order to designate suitability in the experimental design approach, a robustness test was carried out. To assess robustness 3 aspects were taken into consideration, namely, proportion of flow rate, proportion of acetonitrile in mobile phase and pH; all the three factors have no significant effect on response (assay). Experimental design based robustness with the aid of Full Factorial design (FFD) provided an effective way to simultaneously asses Nebivolol and Valsartan. This method was successfully used to analyze fixed dose capsule samples of Nebivolol as well as Valsartan and can be utilized for regular lab investigation of Nebivolol and Valsartan in capsules.

Keywords: Ultra Performance Liquid Chromatography; Experimental design; Robustness; Nebivolol; Valsartan

Introduction

Nebivolol (NBL) chemically known as α,α-[iminobis(methylene)] bis[6-flouro-3,4-dihydro-2H-1-benzopy-ran-2-methanol] and is shown in Figure 1a [1]. Its empirical formula is $C_{22}H_{25}F_2NO_4$. Nebivolol, an antihypertensive drug is a competitive and cardio selective beta blocker with limited vasodilating properties, probably due to an interaction with the l-arginine/nitric oxide pathway [2].

Valsartan (VST), chemically known as N-(1-oxopentyl)-N-[[2' (1H-tetrazol-5-yl) [1,1'-biphenyl]-4-yl] methyl]-L-valine and is shown in Figure 1b [1]. Its empirical formula is $C_{24}H_{29}N_5O_3$. VST is an active non peptide angiotensin II receptor antagonist VST displaces angiotensin II from the AT_1 receptor and produces its blood pressure lowering effects [2].

The combination therapy of NBL and VST is indicated for the potential treatment of hypertension, to lower blood pressure. The fixed dose combination of nebivolol and valsartan proved to be statistically sound in reducing the diastolic blood pressure against the highest approved doses of both NBL alone (40 mg) and VST alone [3].

The literature reveals few Ultra Performance Liquid Chromatographic (UPLC) techniques were documented to establish VST [4-5] individually or with some other drugs in pharmaceuticals, the review further reveals that no UPLC methods were reported for the estimation of NBL in pharmaceuticals and biological samples. To date, no UPLC technique is reported to concurrently determine NBL and VST in medicinal dose as capsules with short run time. Therefore it is felt necessary to develop a fast liquid chromatographic method with short analysis time.

As per the review made on the literature few HPLC methods [6-12] were reported for the estimation of NBL and VST. The amount of solvent and the time required for the estimation of selected drugs by the reported HPLC methods were more. The technique of UPLC is more

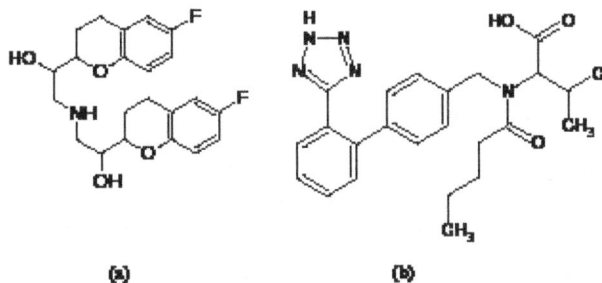

Figure 1: Chemical Structure of (a) NBL (b) VST.

*Corresponding author: R.Seetharaman, Department of Pharmaceutical analysis, SRM College of Pharmacy, SRM University, Kattankulathur, Tamilnadu- 603203, India, E-mail: seerampharm@rediffmail.com

advantageous than HPLC with the amount of solvent and time required for analysis. The capability and quickness of analysis is becoming more significant in the field of pharmaceutical, toxicological and clinical studies [13-15].

A significant reduction in division period and solvent utilization is favoured by UPLC. Research papers reveal that UPLC structure permits approximately nine times reduction in period for investigation in comparison to the conventional HPLC structure utilizing 5 μm unit dimension analytical columns, and approximately 3 times reduction in investigation period when compared to 3 μm unit dimension analytical columns with no concession on the whole division [13-15].

Investigational method was utilized for the substantiation to assess the strength of the method. The objective of this paper is to address the robustness of UPLC assay method and to explore the significant factors from a FFD. It also provides an effective case study on the experimental design application on the assay method of a pharmaceutical dosage form.

Experimental

Instrumentation and apparatus

The UPLC structure utilized for method development and validation happens to be Thermo accela™ equipped with 1050 quaternary pump auto sampler and a photodiode array (PDA) detector. The UPLC PDA detector had 10 mm, 2 μL Light Pipe flow cell. The yield of the detector was documented and developed utilizing Chrome quest software version 5.0, Sonicator (PCI bath sonicator) was utilized for degassing of movable stage as well as sonication of the liquids prepared.

Software

The investigational method and data examination were done by utilizing Unscrambler X edition 10.1; other statistical calculation for the analysis was performed by using Microsoft Excel 2007 software (Microsoft, USA) and Sys Stat 2013 trial version (Sys Stat).

Chemicals, pharmaceutical preparation and reagents

Reference norms of NBL and VST were kindly gifted by Ideal Analytical and Research Institution (Puducherry, India) with stated purity of 99.85% and 99.94%, correspondingly. All the values were used as obtained. Market sample of Nebicard –V capsules claiming 5 mg of NBL and 80 mg of VST were obtained from retail drug store in Chennai. HPLC quality water, acetonitrile, analytical reagent category of orthophosphoric acid was obtained from Rankem, India.

Conditions for chromatographic methods

The chromatographic partition was done on a Thermo C_{18} 50×2.1, 1.9 μm particle size. The mobile phase comprises of mixture of 10 mM ammonium dihydrogen phosphate buffer (pH adjusted to 3.00 with dilute orthophosphoric acid) and acetonitrile to the proportion of 60:40 (v/v). The flow rate and injection volume was 0.4 mL min⁻¹ and 1 μL respectively. The column warmth was ambient and the zeniths were observed at 220 nm.

Preparation of standard solutions

Stock standard mixtures containing NBL and VST (100 μg mL⁻¹ of NBL and 1600 μg mL⁻¹ of VST) were produced by mixing suitable quantities of the compounds in mobile phase. Working mixtures 10 μg mL⁻¹ of NBL, 160 μg mL⁻¹ of VST were produced from the fore said stock mixture in mobile phase for test inference.

Analytical method validation

Solution stability: The key aspect of establishing solution stability is to favor the analytical method to employ auto samplers in the estimation of drugs, as the prepared solutions is allowed to stand overnight and even more than that. The stability of the NBL and VST was assessed by leaving the sample and standard solution in a tightly capped standard flasks at room temperature for 12 hours during which they were assessed for assay at 6 hours intervals. The amount of NBL and VST was calculated for evaluation of solution stability.

System suitability: So as to confirm the system functioning, system appropriateness parameters were measured. With six repeated additions of customary arrangements, system accuracy was decided. Every significant feature together with capability aspect, peak resolution, plus theoretical plate number was calculated.

Specificity: Specificity is the ability to assess unequivocally the analyte in the presence of components which may be expected to be present. Typically these might include impurities, degradants, matrix, [16]. Based on the sample preparation procedure, an investigative placebo solution (including all the inactive substances other than NBL as well as VST) was produced and injected. With the help of this developed method, the interference of these excipients is analyzed for a mixture of inactive ingredients, commercial pharmaceutical preparations including NBL and VST and standard solutions.

Linearity of the calibration line: A standard stock solution claiming 100 μg mL⁻¹ and 1600 μg mL⁻¹ of NBL and VST respectively were prepared. Seven standard solutions were prepared by serial dilution from the stock solution of NBL and VST. An aliquot of 1 μL of each of the calibration solution was injected in the UPLC system. Linearity was performed between 70% and 130% of normal strength utilizing seven calibration intensities (70%, 80%, 90%, 100%, 110%, 120% and 130%) for all the compounds. The linearity was done in triplicate and for three days. Cochrans C test [17] was applied to verify the homogeneity of variances (homoscedasciticity) of the residuals along the line of regression. As the homoscedasticity of the calibration line was satisfactory for the regression line of three analytes the slope and intercept were calculated with their 95% confidence intervals employing Ordinary Least squares. The linearity was assessed visually by observation of calibration line, and statistically by employing F-test [18] for Lack of fit.

Precision: Precision was examined utilizing the proposed method for six genuine commercial samples of Nebicard V. Repeatability and Intermediate precision were evaluated simultaneously [16]. Precision was assessed by under taking six self-determining evaluations NBL and VST at 100% of the target concentration of each compound. The amount of NBL and VST was evaluated against a competent reference benchmark. The assessment was done in duplicate for three days.

Accuracy: With the standard addition method (spiking), revival trials were carried out for verifying the correctness of the proposed technique. 80%, 100% and 120% are the three different standard levels added to pre-analyzed capsule samples. The concentration of each drug injected in UPLC was 8, 10, 12 μg mL⁻¹ of NBL and 128.00 μg mL⁻¹, 160.00 μg mL⁻¹, 192.00 μg mL⁻¹ of VST. The solution for accuracy is prepared in triplicate and assessed. The procedure was done in triplicate and for three ensuing days. The proportion recovery of NBL and VST at every stage was assessed against competent reference benchmark.

Robustness: The robustness of an analytical procedure is a measure of its capacity to remain unaffected by small, but deliberate variations in method parameters and provides an indication of its reliability during normal usage [16]. The robustness was assessed and evaluated by experimental design. The study of robustness with the aid of factorial design [18] involves varying parameters simultaneously rather than one at a time can be more efficient, and allows the effects between parameters to be observed. Comparative, response surface modelling, regression modelling and screening are the four common types of multivariate experimental design based approaches. The focus of the propose work was on screening as it is the most appropriate design for robustness studies. Screening design is an efficient way to identify the critical factors that affect robustness. The most common type of screening experimental design are full factorial, fractional and Plankett Burmann designs.

Full factorial design: Full factorial investigational plan with two or more features wherein all the stages of every feature is connected. It could be further referred to a fully-crossed plan. A full factorial experimental design permits one to understand the impact of every feature on the reaction variables and the impacts of interactions among factors. A common full factorial design is one with all factors set at two levels each, a high and low value The quantity of trials to be carried out is a role of the number of factors and the number of level for every factor. If there are k factors, each at two levels, a full factorial then has 2^k runs. In other words, using four factors, there would be 2^4 or 16 design points or run [18].

Determining of factors: The factors assessed are the flow ratio (A), proportion of acetonitrile (B) as well as pH (C). The selected factors are studied at two levels symmetrically situated around the nominal one. Table 1 illustrate the selected factors and the range investigated

Between two and five center replications are commonly done to ascertain the investigational fault variance and to check the analytical soundness of the method (B). A complete factor factorial plan was used in robustness testing for the selected factors not exceeding three levels ($-1, 0, +1$); the plan employed in robustness tests of NBL as well as VST was a full factorial plan. The investigational domains of the particular variables plus the equivalent reactions are documented in Table 4.

Every one of the trials was carried out in an arbitrary manner to reduce the impacts of unrestrained variables which might bring in a prejudice on the dimensions. Three duplicates of the core features were carried out to assess the investigational fault. The notation for a linear regression method containing three predictor variables with interactions is

$$Y = \beta_0 + \beta_1 X_1 + \beta_2 X_2 + \beta_3 X_3 + \beta_{12} X_1 X_2 + \beta_{13} X_1 X_3 + \beta_{23} X_2 X_3 + \beta_{123} X_1 X_2 X_3 + \varepsilon \quad (1)$$

Wherein Y is the reaction of the model, β is the regression coefficient and X_1, X_2 and X_3 symbolize features A, B and C correspondingly, β_1, β_2 and β_3 are the impact coefficients for the main effects of factors A, B as well as C, correspondingly. β_{12}, β_{13} and β_{23} are the impact coefficients for the AB, AC as well as BC interactions, whereas β_{123} symbolizes the ABC interface.

The equation for the regression method is very suitable, particularly if there is a huge amount of higher order interactions existing.

Factor	Levels		
	(-)	Nominal (0)	(+)
(A) Flow rate (µL min $^{-1}$)	360	400	440
(B) Acetonitrile (%)	36	40	44
(C) pH	2.80	3.00	3.20

Table 1: Selected factors and range investigated during robustness testing.

Preparation of sample solution: The method was applied for estimation of NBL and VST in pharmaceutical capsules. 20 capsules of Nebicard - V were taken; their average nett content was established and powdered to a good homogenous dust. A precisely measured amount of the powder corresponding to one capsule (5 mg of NBL and 80 mg of VST) was kept in a 50 mL volumetric flask. To this flask, approximately 35 mL of mobile phase was included and sonicated for a time of 5 min in a sonicator, later thinned to the mark with mobile phase and blended thoroughly to obtain sample stock solution. The sample stock solution was strained via a whatmann no. 41 filter paper and the remains was saved following the rejecting the initial small number of millilitres. One millilitre of the filtrate was poured into a 10 mL volumetric flask, thinned to capacity with mobile phase and blended thoroughly. The sample solution was prepared in triplicate and the amount of NBL and VST were analysed in accordance with the proposed method.

Results and Discussions

Method development and optimization

The main objective of the proposed method is to provide a fast and reliable analytical procedure for the estimation of NBL and VST. To obtain the best chromatographic condition, different columns like C_8, C_{18} and the mobile phase composed of Ammonium dihydrogen phosphate buffer and organic modifier like methanol and acetonitrile and the pH ranged from 2.6 to 3.8 was used as a starting point. The best chromatographic condition was achieved with a C_{18} Column with a mobile phase comprising 10 mM ammonium dihydrogen phosphate pH adjusted to 3.00 ± 0.02 with dilute orthophosphoric acid as buffer, with a ratio of buffer: acetonitrile 60:40 (v/v) and with a flow rate of 0.4 mL min $^{-1}$. The detection was monitored out at 220 nm. The optimum wavelength for detection was 220 nm at which detector responses were more appropriate for the selected drugs. Under the proposed UPLC chromatographic parameters, NBL and VST were well separated and their analogous peaks were distinctly developed at feasible retention times.

The reputations of organic modifier (concentration) such as acetonitrile and pH were carefully studied. 40 volumes of acetonitrile with 60 volumes of buffer at a flow rate of 0.4 mL min $^{-1}$ gave good separation and reduced retention time of NBL and VST, while the pH of 3.00 ± 0.02 gave good resolution and peak shape.

Solution stability

The variability obtained in the estimation of NBL and VST was within ± 2% during mobile phase and solution stability experiments, which confirmed solution and mobile phase stability up to 12 hours for assay (Table 2).

Results of Validation

Specificity: The chromatogram obtained for the placebo solution did not show any peak (Figure 2) at the retention time of NBL as well as VST, while the chromatogram of the placebo solution spiked with NBL and VST showed well separated peak (Figure 3) of sample which indicates the specificity of the proposed method.

System suitability: The percentage R.S.D. of retention period plus peak region of NBL and VST of six duplicate injections of standard solution was lower than 2.0%. The findings of structure accuracy are illustrated in Table 3. The % R.S.D values were for duplicate injections which that the structure is accurate. Findings of other system appropriateness strictures like capacity factor, resolution as well as hypothetical plates are illustrated in Table 3 and were within the specified limits.

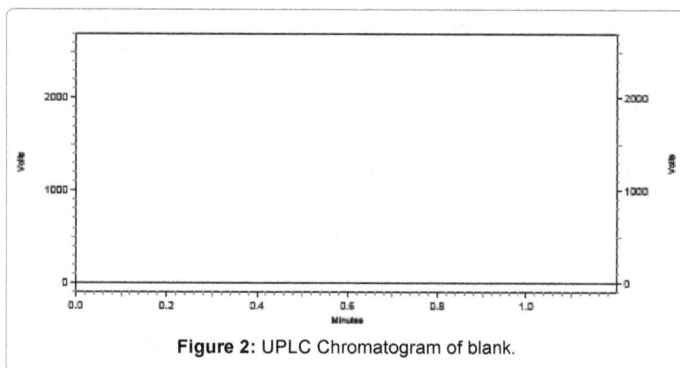

Figure 2: UPLC Chromatogram of blank.

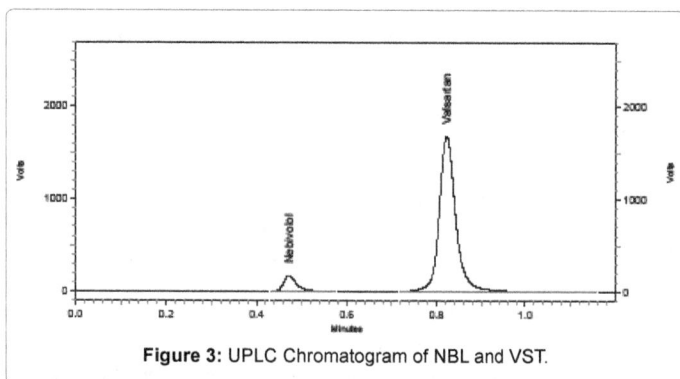

Figure 3: UPLC Chromatogram of NBL and VST.

Analyte	Initial % recovery	6 h		12 h	
		% recovery	D_a	% recovery	D_a
NBL	99.98	98.56	0.01	97.82	0.02
VST	99.95	98.63	0.01	98.01	0.02

^{Da} Percent difference calculated by the difference between two values divided by the average of the two values

Table 2: Results of solution stability.

Parameters	NBL	VST
t_R	0.47	0.82
% RSD of t_R for 6 Injections	0.63	0.66
N	11236	18226
R_s	5.98	
A_s	0.56	0.39

t_R retention time, N Number of theoretical plates, R_s resolution factor, A_s Peak asymmetry

Table 3: Results of system suitability parameters for the estimation of NBL and VST.

Precision: The repeatability variances for NBL and VST at 100% concentration level were 0.28 and 0.27 respectively. The time different intermediate precision variances for the same concentration level were 0.26 and 0.42 for NBL and VST respectively. The corresponding pooled variance and the % RSD were 0.54 and 0.73 for NBL and 0.69 and 0.84 for VST respectively. The % RSD values obtained imply that the precision values obtained were satisfactory.

Linearity: Cochrans C test was employed at first before the regression analysis to verify homoscedascity of the calibration standards and was found to be satisfactory. The C_{calc} values were 0.198 and 0.195 for NBL and VST respectively. The critical value [17] is, $C_{tab(\alpha=0.05;k=7, n=9)}$ = 0.338. The results of the cochrans's C test prove that the variances of the calibration standards were homoscedastic.

The regression lines were evaluated by ordinary least squares. The regression equation of the calibration lines for NBL and VST were Area $_{NBL}$ = 27288C_{NBL} (μg mL^{-1}) + 1050 and Area $_{VST}$ = 23585 C_{VST} (μg mL^{-1}) + 917 respectively and the corresponding values of slopes and intercept at their 95% confidence limits were 27288 ± 2.45 and 1050 ± 77.18 for NBL and 23585 ± 0.24 and 1117 ± 70.36 for VST. The correlation coefficients were 0.9995 and 0.9999 respectively.

The variances of residuals from the calibration data of NBL and VST was evaluated using ANOVA for lack of fit test [17].

The results of the lack of fit test for residuals of the calibration data of NBL and VST were Fcalc = 1.84 and 0.89 respectively. The test values obtained were less that the critical value, $F_{tab\ (\alpha=0.05df1=6,\ df2=56)}$ = 2.265.

The peak area was established linear between 70% and 130% of target concentration. The homoscedasticity of the calibration standards was satisfactory. For every compound the connected coefficient was more than 0.9990 along with cochran's and lack of fit test proves linearity.

Accuracy: The quantity claimed was contained by ± 3% of quantity included that showed that the technique is precise and in addition exclude the intrusion owing to excipients existing in capsules. The mean percentage recovery for NBL at 80%, 100% and 120% concentration levels were 99.6% ± 0.49, 100.03% ± 0.64 and 99.97% ± 0.39. The % recovery range was 99.37 -100.49% and the mean recovery for all the concentration levels at 95% confidence limits were 99.89% ± 0.38.

The % recoveries for VST at the same concentration levels were 99.91% ± 0.36, 100.03% ± 0.64 and 99.69% ± 0.29. The percentage recovery range was 98.55-100.31% and the mean recovery for all the concentration levels at 95% confidence limits were 99.56% ± 0.37.

Robustness: The method was authenticated by the examination of variance (ANOVA) using Unscrambler X. The numerical examination illustrated (Table 5) that the method symbolizes the occurrence excellently and the difference of the reaction was accurately connected to the difference of the features.

The ANOVA chart obtained is a synopsis of the importance of the worldwide method. If p-value for the worldwide method is lesser than 0.05, it discloses the method is noteworthy at 5% level. That is a lower P-value the additionally important is the method. As the p values obtained are more than 0.05 null hypotheses Ho is accepted [17]. The effect summary is reported in Table 5 which offers an outline of the importance of every impact for all reactions.

Run	Design			Response	
	Coded factors^a			Assay (%)	
	Factor – A Flow rate (μL min^{-1})	Factor – B Acenonitrile (%)	Factor – C pH	NBL	VST
1	-1	-1	1	98.23	98.95
2	-1	1	-1	99.56	99.61
3	-1	-1	-1	99.52	99.54
4	0	0	0	99.89	98.73
5	1	-1	-1	100.56	99.52
6	1	1	-1	100.16	98.56
7	1	1	1	99.61	98.93
8	0	0	0	99.28	99.86
9	-1	1	1	99.47	99.85
10	0	0	0	98.66	98.23
11	1	-1	1	98.36	99.56

`Randomized.

^{a}The low, middle (central) and high level of the factors were designated as (-1), (0) and (1), respectively.

Table 4: 2^3 Full factorial experimental plans for robustness testing and obtained responses`

Analyte	Variables	Sum of squares	Degree of freedom	Mean square	F value	P value	Effect value	Significance
NBL	A	1.103	2	0.551	5.010	0.11	0.74	
	B	1.103	2	0.551	5.010	0.11	0.74	
	C	0.637	2	0.318	2.800	0.18	-0.56	
	AB	0.002	4	0.001	0.000	0.92	0.03	NS*
	AC	0.007	4	0.002	0.030	0.87	0.05	
	BC	0.300	4	0.075	1.360	0.32	0.38	
	ABC	0.160	8	0.020	0.720	0.45	-0.28	
	Pure error	0.659	27					
	corrected total		53					
VST	A	0.016	2	0.008	0.126	0.74	-0.09	
	B	0.027	2	0.013	0.207	0.67	-0.11	
	C	0.151	2	0.076	1.185	0.35	-0.27	
	AB	0.039	4	0.010	0.307	0.61	-0.14	NS*
	AC	0.259	4	0.065	2.031	0.24	0.36	
	BC	0.594	4	0.149	4.654	0.11	-0.54	
	ABC	0.500	8	0.063	3.917	0.14	-0.50	
	Pure error	0.383	27					
	corrected total		53					

*Not significant

Table 5: Results of ANOVA and effect summary.

Product name (Composition)	Manufacturer	Percentage found	
		NBL	VST
Nebicard - V (Nebivolol 5 mg and Valsartan 80 mg)	Torrent pharmaceuticals	99.63*	99.86*

*average of six readings

Table 6: Results of Analysis of Marketed formulations.

The regression equation model for NBL, VST is in equation 2 and 3

$$Y_{NBL} = 99.39 + 0.37X_1 + 0.37X_2 - 0.28X_3 + 0.01\ X_1X_2 + 0.02\ X_1X_3 + 0.19\ X_2X_3 - 0.14\ X_1X_2X_3 \quad (2)$$

$$Y_{VAL} = 99.28 - 0.04X_1 - 0.05X_2 - 0.13X_3 - 0.07\ X_1X_2 + 0.18\ X_1X_3 - 0.27\ X_2X_3 - 0.25\ X_1X_2X_3 \quad (3)$$

In conclusion, by examining the ANOVA results confirms that Y_{NBL} and Y_{VST} are robust for all the three factors.

Results of analysis in capsule formulation: The results obtained with the analysis of marketed capsule formulation was carried out and the recovered amount each drug component were expressed as percentage amount of label claim. The results are presented in Table 6, which shows that in all the selected formulations for the study NBL and VST ranged between 99.45 to 99.69%, and 99.63 to 99.86% respectively. These values comply with the assay specifications for active drugs in the USP pharmacopeia (90.0–110.0%) [19], which are required to be met by most drug formulations.

Conclusions

The selected analytes such as NBL and VST have been simultaneously analyzed in pharmaceutical formulation (capsules) with UPLC. The entire run time happened to be 1 min, wherein the two peaks NBL and VST were well separated. The proposed rapid UPLC method had been assessed on the linearity, accuracy, precision, specificity and robustness and established to be suitable and effectual in the quality assessment of NBL as well as VST in Pharmaceutical capsule hence can be used in QC laboratories for the estimation of NBL and VST. All the validation results were comparable with the already reported HPLC method. The developed UPLC method is more advantageous when compared to the reported HPLC method [6-12] in terms of analysis time, cost

and consumption of solvents, sensitivity etc. The finding of the attempt reveal the advantage of utilizing experimental design based robustness study in method validation.

Acknowledgement

The authors are highly grateful to the management of SRM College of Pharmacy, SRM University, Chennai and Ideal Analytical and Research Institution, Villianur, Puducherry, for offering conveniences to conduct the study.

References

1. Maryadele J. O' Neil (2006) The Merck Index: Encyclopedia of Chemicals, Drugs, and Biologicals. (14thedn), Merck research laboratories, New Jersey, USA.

2. Craiag CR, Stit RE (2004) Modern pharmacology with Clinical Applications. (6thedn), Lippincott Williams Wilkins, USA.

3. http://www.businesswire.com/news/home/20140530005856/en

4. Krishnaiah Ch, Ragupathy R, Ramesh K, Mukkanti K (2010) Stability-indicating UPLC method for determination of valsartan and their degradation products in active pharmaceutical ingredient and pharmaceutical dosage forms. J Pharm Biomed Anal 53: 483-489.

5. Antil P, Kaushik D, Jain G, Srinivas KS, Thakur I (2013) UPLC Method for Simultaneous Determination of valsartan and hydrochlorothiazide in drug Products. J Chromatogr Separation Techniq 4: 182.

6. Selvan PS Gowda KV, Mandal U, Solomon WD, Pal TK (2007) Simultaneous determination of fixed dose combination of nebivolol and valsartan in human plasma by liquid chromatographic-tandem mass spectrometry and its application to pharmacokinetic study. J Chromatogr B Analyt Technol Biomed Life Sci 858: 143-150.

7. Singh SM, Amol B , Rahul D, Mrinalini C (2008) Simultaneous estimation of nebivolol hydrochloride and valsartan in capsules by reverse phase high performance liquid chromatography. Indian drugs 46:257-260.

8. Doshi AS Bhagwan SS, Mehta TN, Gupta VK, Subaaiah G (2008) Determination of nebivolol and valsartan in a fixed-dose combination by liquid chromatography. J AOAC Int 91: 292-298.

9. Kokil SU Bhatia MS (2009) Simultaneous Estimation of Nebivolol Hydrochloride and Valsartan using RP HPLC. Indian J Pharm Sci 71: 111-114.

10. Arunadevi S, Meyyanathan SN, Suresh B (2011) Simultaneous determination of nebivolol HCl and valsartan in solid dosage form by spectrophotometric and RP-HPLC method. IJPSR 2: 424-431.

11. Madhavi C, Siddartha B, Parthiban C (2014) Simultaneous estimation and validation of nebivolol and valsartan in tablet dosage form by RP-HPLC. Int J Pharm Pharm Sci 6:1–5.

12. Siddartha B, Sudheer B, Ravichandra G, Parthiban C (2014) Analytical method development and validation for simultaneous estimation of nebivolol and valsartan in bulk and pharmaceutical dosage form by RP-HPLC method. Int J Pharm 4: 340-346.

13. Nováková L Matysová L, Solich P (2006) Advantages of application of UPLC in pharmaceutical analysis. Talanta 68: 908-918.

14. Cielecka-Piontek J Zalewski P, JeliÅ„ska A, Garbacki P (2013) UHPLC: The Greening Face of Liquid Chromatography. Chromatographia 76: 1429-1437.

15. Gumustas M, Kurbanoglu S, Uslu B, Ozkan S (2013) UPLC versus HPLC on Drug Analysis: Advantageous, Applications and Their Validation Parameters. Chromatographia 76: 1365–1427.

16. Revathi R Ethiraj T, Thenmozhi P, Saravanan VS, Ganesan V (2011) High performance liquid chromatographic method development for simultaneous analysis of doxofylline and montelukast sodium in a combined form. Pharm Methods 2: 223-228.

17. Swart Ellison SLR, Barwick VJ, Farrant TJD (2009) Practical statistics for the Analytical Scientist. (1stedn), RSC publisher, United Kingdom.

18. Swart ME, Krull IS (2012) Hand book of Analytical validation. (1stedn), CRC Press, Finland.

19. The United States Pharmacopoeia 29 (2006) National Formulary 24, United States. Pharmacopoeia Convention, Rockville, MD, USA.

17

Preparation and Characterization TiO$_2$ Microspheres for the Liquid Chromatography Stationary Phase

Heyong Huang[1], Yan Zhou[2], Jiahong Zhou[1] and Yuying Feng[1]*

[1]*Analysis and Testing Center, School of Geography Science, Nanjing Normal University, Nanjing 210023, China*
[2]*Department of Chemistry and Environmental Science, Nanjing Normal University, Nanjing 210023, China*

Abstract

TiO$_2$ microspheres with a uniform particle size distribution were synthesized by hydrothermal method. After being sintered, the titania microspheres that are obtained have an average diameter of 5 μm, a surface area of 383.5 m^2/g, an average pore volume of 0.25 cm^3/g, and an average pore diameter of 35.9 nm. Normal phase chromatography was separated the mixture of benzene, nitrobenzene and nitro-anisole, three substances were separated well on the titania column. The microspheres possess enough rigidity to withstand high packing pressure and are very useful as a new kind of chromatographic packing material for high performance liquid chromatography (HPLC).

Keywords: Hydrothermal treatment; TiO$_2$ microspheres; Liquid chromatography stationary phase

Introduction

TiO$_2$ has been extensively used in photocatalysis [1], lithium battery [2], sensor materials [3], dye degradation [4], cosmetics [5] and other technical fields [6]. In recent years titania materials for high performance liquid chromatography (HPLC) has attracted considerable attentionsattributed to the good chemical stability, mechanical strength, enough rigidity and amphoteric ion-exchange properties [7-9]. By far silica is the most popular material for manufacturing chromatographic columns. While the chemical and thermal stability of silica has limited application in commercial use. The silica only can be used in a narrow pH range as the strong hydrolysis and the instable Si-O band when pH>8 or pH<2 that leads to the changes of peaks sharp and retention characteristics of the separations [10]. In addition, bare Si-OH of the silica surface affected the mesidino, chelate and hydrogen bond will generate irreversible effect that cause serious detail and reduce the efficiency of separation. Both these conditions reduce the lifetime of the HPLC column.

Titania has greater mechanical and pH stability (pH 1-14) that may compensate for the narrow pH range of silica and it is suitable for the separation of the alkaline substances especially the biological macromolecules [11]. Nowadays, the research about the TiO$_2$ used as the chromatographic packings instead of the silica column is less reported, not to mention the practical application. The main difficult is due to how to synthesize the uniform and controllable size TiO$_2$ microspheres. So the preparation of the TiO$_2$microspheres that meet the conditions is the key.

Recently the methods about how to prepare the TiO$_2$ microspheres are broadly reported, including the sol-gel [12,13], hydro-thermal method [14], Polymerization-Induced Colloid Aggregation (PICA) [15], non-aqueous emulsions [16], co-precipitation [17] and so on. While the latter three methods cannot be applied in the extensive production as the ways are so verbose and thereagent are expensive. Thus in our paper, the TiO$_2$ microspheres are prepared by the hydro-thermal method. Through a large number of experiments, we have discovered the optimal experimental conditions of synthesizing the TiO$_2$and the morphology and particle size of the microspheres are controlled. So TiO$_2$can be made into the normal phase packing. In order to test the chromatographic performance, we plan to separate some basic materials.

Experimental Section

Apparatus

A scanning electron microscope, model JSM-5610LV (Electron Optics Corporation, Japan) was used to record the electron micrograph. An X-ray diffractometer model D/max-2500 VL/PC (Science Instruments, Japan) was used to determine the crystal shape of the titania. An oven, model DGF-30 (Experiment Instruments Corporation, Nanjing, China) was used to control the reaction temperature. An autoclave of a 50 mL polyfluortetraethylene cup was to be the reaction vessel. A Micromeric model ASAP-2010 surface analysis instrument (Micromeric Corporation, U.S.A) was used to collect nitrogen adsorption/desorption isotherms. A laser particle size analyzer, model Master Sizer2000 (Malvern Company, U.K) was used to determine pore size distribution (PSD). A HPLC/MS model 1290-6460 (Agilent Corporation, American) was used to separate the special substances.

Reagents

Titanium tetrachloride, ammonium sulphate, urea and other reagents were all analytical grade. n-hexane,ethanol, benzene, nitrobenzene, o-nitroanisole are HPLC grade. Water (H$_2$O) was distilled and deionized using a Millipore Milli-Q system (Bedfold, MA, USA).

Synthesis of Porous Titania Microspheres

The synthesis of TiO$_2$ microspheres are prepared by the following steps: 0.72 g of ammonium sulphate, 8.5 g of urea, 8mLof water and 8mL of ethanol were added into the conical beaker. Under continuous stirring, 0.6 mL of titanium tetrachloride was added drop by drop. After stirring for 5h, the transparent solution was obtained and the pH was

***Corresponding author:** Yuying Feng, Analysis and Testing Center, School of Geography Science, Nanjing Normal University, Nanjing 210023, China
E-mail: huangheyong@njnu.edu.cn

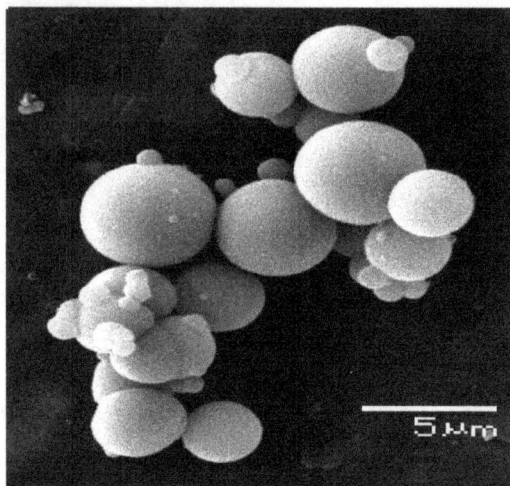

Figure 1: SEM image of TiO$_2$ microspheres.

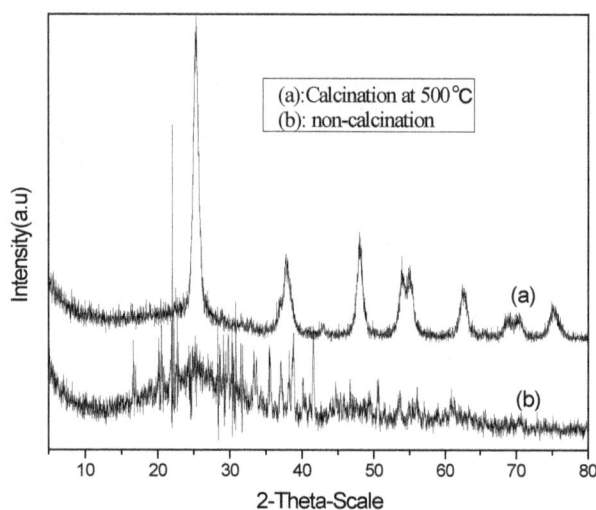

Figure 2: XRD patterns of TiO$_2$ microspheres.

about 2. Then the solution was transferred into an autoclave of a 50 mL polyfluortetraethylenecup. The autoclave was heated and maintained at 393 K for approximate 5h and then the autoclave was cooled to the room temperature. After the suspension was filtered, the precipitate was washed twice with deionized water and washed third with ethanol. The microspheres was dried at 353K and carbonized at 773K in a muffle furnace after grinded. Then, the TiO$_2$ microspheres sample can be obtained.

Column packing

The chromatographic column (250×4.6 mm i.d.) was filled with titanium oxide and a pressure of 450psi was applied by the use of a high-pressure pump, model DSTV-150 (HASKEL Technical Company, American). Isopropanol and methanol was used as propulsion solvent.

Results and Discussion

SEM images and X-ray diffraction of porous titania microspheres

TiO$_2$ synthesized by hydrothermal treatment are spherical and free

from clustering. The SEM micrograph is shown in Figure 1. It can be seen the TiO$_2$ microspheres are 6-10 μm in diameter.

Figure 2 shows the X-ray diffraction of the TiO$_2$. It is clearly seen from Figure 2, the X-ray diffraction intensity of the TiO$_2$ without the heat treatment is very lower and the crystal structure is amorphous. After being sintered, the diffraction intensity sharply increase and the diffraction peaks in 25.32°C, 37.88°C, 48.11°C, 53.94°C, 62.51° are the characteristic peaks of anatase TiO$_2$. It is proved that the crystal shape of titania microspheres is anatase-type.

The specific surface area of TiO$_2$ microspheres

The specific surface area of the stationary phase is one of the most important parameters in chromatogram [7] Chromatographic separation requires the packing be a larger surface area in order to facilitate the carrying load lots of more samples and provide sufficient separation efficiency. The specific surface area of titania that being sintered is 383.5 m^2/g, whereas the microspheres without heat treatment is 276.0 m^2/g, which was calculated using the standard Brunauer Emmett Teller (BET) method. This is because the water adsorbed in the TiO$_2$ microspheres will dissociate to Ti-OH bond formation owing to the heat treatment. With the temperature rising, the adjacent Ti-OH bond will concentrate to Ti-O-Ti bond. Though this agglomeration leads to decrease the specific surface areas, the TiO$_2$ microspheres synthesized byhydrothermal treatment still have large surface areas.

Pore Size Distribution (PSD) of porous TiO$_2$ microsphere

According to the theoretical analysis and practical application, Carr [12] has considered that the ideal packings should be in accordance with the suitable physical and chemical parameters. The diameter of the microspheres must be in the range of 2 to 50 nm and the pore diameter should show a narrow pore size distribution. Meanwhile it is necessary the appropriate pore size is supposed from 0.2 to 0.7 cm^3/g. Figure 3 shows nitrogen adsorption /desorption isotherms of porous TiO$_2$ microspheres after heat treatment. As is shown, it is intuitively seen that the adsorption branch is consistent with the IV isothermal map of the ideal chromatography. In accordance withthe IUPAC classification, the most favorable stationary phaseshould have a type IV adsorption isotherm and a H type hysteresis loop [18]. According to the desorption isotherm the average pore volume of TiO$_2$ is 0.25 cm^3/g and

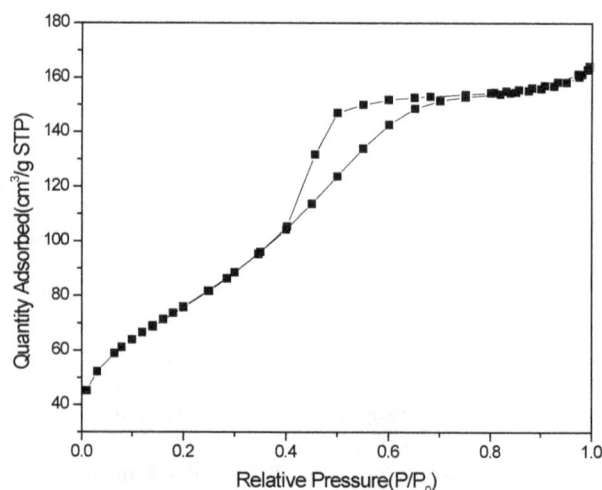

Figure 3: Nitrogen adsorption / desorption isotherms of porous titania microspheres.

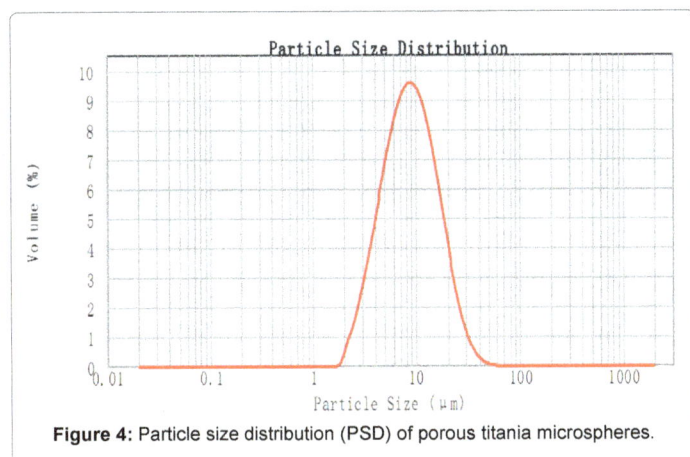

Figure 4: Particle size distribution (PSD) of porous titania microspheres.

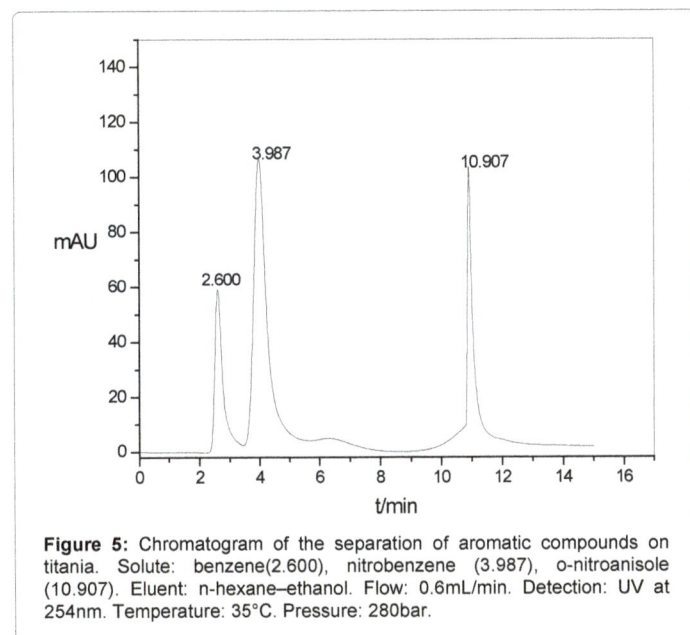

Figure 5: Chromatogram of the separation of aromatic compounds on titania. Solute: benzene(2.600), nitrobenzene (3.987), o-nitroanisole (10.907). Eluent: n-hexane–ethanol. Flow: 0.6mL/min. Detection: UV at 254nm. Temperature: 35°C. Pressure: 280bar.

the pore diameter is 35.9 nm based on the BJH models. The pore size distribution of TiO_2 microspheres is shown in Figure 4. It is seen that the PSD of TiO_2 is narrow and relatively uniformity. The average size of TiO_2 is 7 μm which is consistent with the results of scanning electron microscope.

Chromatographic condition

Normal-phase separation of aromatic compounds was performed on the titania column using n-hexane and ethanol as the mobile phase. The mixtures contained benzene (10 μg/mL), nitrobenzene (0.4 μg/mL) and o-nitroanisole (0.2 μg/mL). Injections of 5 μL of appropriate concentrations of these mixtures produced satisfactory chromatographic peaks with detection at 254 nm and the mobile phase was controlled at 0.6 mL/min. The mobile phase was gradient elution :0~7 min,100% n-hexane, 7~8 min,100%~90%n-hexane, 8~20 min, 90% n-hexane. The column temperature was controlled at 35% and the initial pressure of the column was about 280 bar. The separation results are shown in Figure 5. As can be seen the three substances were separated well on the titania column. Benzene and nitrobenzene were retained on the stationary phase as Lewis base of the π electron of the aromatic compounds was reacted with the Lewis acid of the TiO_2. o-nitroanisole was retained longer in the chromatographic column

due to the strong polarity. As can be seen, titania that is synthesized by hydro-thermal method is useful as a packing material for HPLC.

Conclusions

The traditional silica column will enhance solubility in basic mobile phases and at higher temperatures. Meanwhile the tailed peaks are appearedwhen the alkaline substances are separated on the column. Thus our work hammers at exploring a new liquid chromatography stationary phase. The TiO_2 microspheres synthesized byhydrothermal treatment was about 5 μm. The PSD of $TiO_2$was relative narrow and uniform. The pore size of TiO_2 was 35.9 nm which is in the range of mesopore. Simultaneously specific surface area was 383.5 m^2/g and pore volume was 0.25 cm^3/g. Benzene, nitrobenzene and o-nitroanisole were separated well in the titania column. All these parameters are showed that $TiO_2$is the excellent normal phase for the liquid chromatography.

Acknowledgment

This work was supported by the Natural Science Foundation of China (Grant No.21201102) and Science Foundation of Jiangsu (Grant No. BM2007132), China. We gratefully acknowledge the anonymous reviewers whose comments helped to improve the manuscript.

References

1. Jiang ZT, Zuo YM (2001) Synthesis of porous titania microspheres for HPLC packings by polymerization-induced colloid aggregation. Anal Chem 73: 686-688.

2. Li YT, Li J, Wei YJ, Wei Y, Hu YZ (2007) Preparation and properties of Titania based supports for HPLC. Chemical Research and Application 19: 882-885.

3. Ellwanger A, Matyska MT, Albert K, Pesek JJ (1999) Comparison of octadecyl bonded titania phases, Chromatographia 49: 424-430.

4. Li ZX, Kawashita M (2010) Sol–gel synthesis and characterization of magnetic TiO2 microspheres. J Ceram Soc Jpn 118: 467-473.

5. Guo CW, Cao Y, Xie SH, Dai WL, Fan KN (2003) Fabrication of mesoporous core-shell structured titania microspheres, Chem Comm 21: 700-701.

6. Wu TX, Liu GM, Zhao JC (1998) Rhodamine B under Visible Light Irradiation in Aqueous TiO2 Dispersions. J Phys Chem B 102: 5845-588-51.

7. Collins AM, Mann S (2003) Synthesis of titania hollow microspheres using non-aqueous emulsions.J Mater Chem 13: 1112-1114.

8. Huang HY, Zhou JH, Liu HL, Zhou YH, Feng YY (2010) Selective photoreduction of nitrobenzene to aniline on TiO2 nanoparticles. J Hazard Mater 178:994-998.

9. Deng D, Kim MG, Lee JY, Cho J (2009) Green energy storage materials Nanostructured TiO2 and Sn-based anodes for lithium-ion batteries. Energy Environ Sci 2: 818-837.

10. Kalanur SS, Prashanth SN (2010)Voltammetric sensor for buzepidemethiodide determination based on TiO2 nanaparticle-modified carbon paste electrode. Colloids and Surfaces B:Biointerfaces 78: 217-221.

11. Kim JB, Seol DH, Shon HK, Kim GJ, Kim JH (2010) Preparation and Characterization of Titania Nanoparticles from Titanium Tetrachloride and Titanium Sulfate Flocculation of Dye Wastewater. Journal of Japan Petroleum Institute 53: 167-172.

12. Auffan M, Pedeutour M, Bottero JY (2010) Structural Degradation at the surface of a TiO2-Based Nanomaterial Used in Cosmetica. Environ Sci Technol 44: 2689-2694

13. Winkler J, Marme S (2000)Titania as a sorbent in normal-phase liquid chromatography. J Chromatogr A 888: 51-62.

14. Maier NM, Franco P, Lindner W (2001) Separation of enantiomers: needs, challenges, perspectives.J Chromatogr A 906: 3-33.

15. Zizkocsky V, Kucera R, Klimers J, Dohnal J (2008) Titania-based stationary phase in separation of ondansetron and its related compounds.J Chromatogr A 1189: 83-91.

16. Xie YB, Yuan CW, Li XZ (2005) Photocatalytic degradation of X-3B dye by visible light using lanthanide ion modified titanium dioxide hydrosol system. Colloids and Surface A: Physico chem Eng Aspects 252: 87-94.

17. Tani K, Kubojima H (1998) Separation of Inorganic Anions and Cations on Titania by Use of Acetic Acid-Sodium Acetate and Bicine-Sodium Hydroxide Buffers. Chromatographia 47: 655-658.

18. Tani K, Suzuki Y (1997) Investigation of the Ion-Exchange Behaviour of Titania: Application as a Packing Material for Ion Chromatography. Chromatographia 46: 623-627.

18

Use of LC-MS and GC-MS Methods to Measure Emerging Contaminants Pharmaceutical and Personal Care Products (PPCPs) in Fish

M Abdul Mottaleb[1,2*], Michael K Bellamy[2], Musavvir Arafat Mottaleb[2] and M Rafiq Islam[2]

[1]*Center for Innovation and Entrepreneurship, 1402 North College Drive, Northwest Missouri State University, Maryville, Missouri 64468, USA*
[2]*Department of Natural Sciences, 800 University Drive, Northwest Missouri State University, Maryville, Missouri 64468, USA*

Abstract

Aquatic ecosystems are continuously contaminated by manufactured pharmaceutical and personal care products (PPCPs). Non-regulated, multi-purpose PPCP contaminants enter aquatic systems through sewage/wastewater treatment plants after consumption and use by humans and animals. These micro-pollutants receive increased attention worldwidesince significant levels of contamination have been found in various environmental compartments and organisms. Highly sophisticated equipment such as liquid chromatography-mass spectrometry (LC-MS) and gas chromatography-mass spectrometry (GC-MS) are reliable ways to determine PPCPs at sub-ppb levels from air, waters, sediments, effluents, aquatic organisms and human body fluids. Although the consequences of these pollutants are gradually becoming visible, their potential impacts on aquatic ecosystems and organisms are poorly known. Some studies have suggested that PPCPs are persistent and have bioaccumulation potential leading to ecological effects and abnormalities in fish. Other findings illustrate that PPCPs can impair swimming behavior in fathead minnow (*Pimephales promelas*), interfere with thyroid axis in the zebra fish (*Danio rerio*), or form adducts with hemoglobin and/or suitable protein breakdown PPCPs products. Thus, this review focuses on PPCPs emerging contaminants concern with regards to sources, occurrences, analytical methods, fate and biological transformation.

Keywords: PPCPs; Micro-pollutants; Emerging contaminants; Biomarkers; Aquatic organisms; LC-MS and GC-MS

Introduction

In modern life, one of the most important issues in the world is the exposure to man-made chemicals that cause interference of regular activities such as reproduction and development of different organisms in the environment [1,2]. Some of them are hazardous and present potential or actual threat to human health, wildlife, aquatic organisms and/or soundings [2]. Newer analytical techniques have made it possible to identify these compounds at extremely low level of the order of sub-ng/g. These are frequently detected in different environmental compartments including surface waters, wastewaters, air, wildlife and fish, and had not been recognized previously at such low levels. These compounds are often referred to as "emerging contaminants" (ECs) because adequate information associated with their presence, occurrence, fate, transport and mechanisms are not available to assess their risk to human health and the ecosystem [3]. ECs are used daily in homes, farms, businesses and industry as detergents, fragrances, prescription and non-prescription drugs, disinfectants, and pesticides etc. Some ECs have been commonly found in water resources around the world and across the USA [4-6].

PPCPs as Endocrine Disruptors

Emerging evidence from wildlife and laboratory studies indicates that some chemicals may interfere with the endocrine system. Compounds identified include pesticides, polychlorinated biphenyls, dioxins, furans, alkyl phenols, and steroid hormones. These chemicals are routed to ecosystems through wastewater treatment plants. Several studies reported that many ECs present in municipal wastewater effluent can act as endocrine disruptors at concentrations capable of inducing fish feminization [7,8]. The feminization has been linked to exposure to compounds that mimic estrogen activity. However, it has also been determined that thousands of the compounds have the potential to interact with components of the endocrine system, altering the natural action of hormones [9,10] in both freshwater and marine fish species [4,8,11-13]. The occurrence of some ECs correlates with ecological effects and sexual abnormalities in fish [14-16]. In other studies, complex mixtures of ECs at environmentally relevant concentrations were reported to inhibit the growth of human embryonic cells [17,18]. Other evidence suggests that some ECs are persistent in the environment and survive through conventional water treatment plant and ultimately reaching the aquatic organisms [18]. Overall, an important concern, posed by ECs, is the interference of reproduction and development of aquatic organisms and wildlife [1,2,19].

Release of PPCPs into the Environment

Pharmaceuticals and personal care products (PPCPs) are a major class of ECs commonly used in human and animal applications. PPCPs are many chemical compounds with a variety of chemical structures, conformations, functional groups, polarities and characteristics. PPCPs include prescription and non-prescription drugs together with fragrances, cosmetic ingredients, diagnostic agents, biopharmaceuticals, and growth enhancing constituents used in livestock operations. Tons of these chemicals are produced annually worldwide [20]. After consumption, PPCPs are released into ecosystems via urine, feces or residues as either parent compounds or their metabolites. PPCPs enter the environmental system through effluent discharge from wastewater/sewage treatment plants, inappropriate disposal of expired or unused

*Corresponding author: M Abdul Mottaleb, Center for Innovation and Entrepreneurship, 1402 North College Drive, Northwest Missouri State University, Maryville, Missouri 64468, USA
E-mail: mmottaleb@yahoo. com and mmottal@nwmissouri. edu

drugs, shower drain, residues from drug manufacturing companies, nursing homes and hospital facilities.

Wastewater treatment plants are not generally designed to eliminate the PPCPs because they are non-regulated water pollutants [21]. Based on the process design of the treatment plants, the elimination rates of drugs range from <10% (e. g. atenolol and carbamazepine) to almost complete removal (e. g. propranolol) [22]. As these compounds are continuously released into aquatic systems, the effluents from the wastewater treatment plants are considered as the main routes of human pharmaceuticals into the environment [23], reaching concentrations of ng/L to μg/L [24]. Aquatic organisms are consistently exposed to these non-regulated PPCPs, as environmental contaminants.

Most PPCPs/ECs ultimately end up in the aquatic environment. Consequently, these compounds frequently appear in ecosystems and are frequently detected in different environmental compartments and organisms at different concentration levels [25]. PPCPs have been frequently detected in different environmental matrices such as air, waters, sediments, sewage sludge, humans and fish [26-45]. The high rate of occurrence of PPCPs is due to the fact that they are easily dissolved in aquatic environment and do not evaporate at normal temperature and pressures. Moreover some PPCPs appear to show low biodegradation rate and high lipophilicity, and have high bioaccumulation potential as environmental contaminants [21]. Several studies have been published that further discuss the sources of PPCPs and their transportation from personal usages to waters and aquatic organisms [3,18, 21,22-27].

Until recently, many around the world were unaware that a new environmental health concern had emerged. Now, regulatory authorities, health agencies, and professional organizations, all over the globe are informed of the growing PPCP problem which drives research on the presence, occurrence, fate of the PPCPs and metabolites [26,27]. Today, the U. S. Environmental Protection Agency (EPA) and other organizations are working together to improve its understanding of a number of ECs, particularly PPCPs.

Occurrences and Effects of PPCPs on Organisms

PPCPs and their metabolites have been detected in aquatic and terrestrial organisms [28,29], surface water [30], air [31], sewage effluent [32], lake Michigan water and sediments [33], industrial sewage sludge [34], municipal effluents [35], marine sediments [36], marine mammals [37], effluent-dominated river water fish [38,39], Pecan Creek fish [40] and German fish specimen bank [41], fish-eating birds and fish [42], receiving marine waters and marine bivalves [43]. PPCPs and metabolites have also been identified in human milk [44], and human blood [45]. More importantly, multiple studies have indicated that PPCPs are not only accumulated but also subsequently metabolized to reactive intermediates that form covalently-bound protein adducts in human [46] and aquatic organisms such as fish [47,48].

Studies indicate that many PPCPs are environmentally persistent, bioactive, and have bioaccumulation potential [49-52]. For example, the PPCPs known astriclosan (TCS), an antimicrobial agent, has been widely used in dental care products, disinfectants, hand soaps, footwear, skin care creams and textiles. TCS and its methyl metabolites were detected in surface waters [53], biosolids [54], fish [55], and algae [56]. Still, the fate and chemistry of TCS are not fully understood. TCS is quite stable to hydrolysis; however its photolysis was identified as one of the major pathways of degradation in surface waters [57]. Other research groups have shown that TCS in surface water may be toxic to certain algae species. Specifically, Orvos et al. [58] found

no observer-effect concentration (72-h growth) at 500 ng/L for algae *Scenedesmus subspicatus* while Wilson [59] reported that TCS may cause significant increase in Synedra algae and a substantial reduction of the rare genus Chlamydomonas algae at 15 ng/L and 150 ng/L. Levy et al. [60] demonstrated that TCS can block bacterial lipid biosynthesis inhibiting the enzyme enoyl-acyl carrier protein reductase, which leads to a possible development of bacterial resistance to TCS. Recent studies have shown that TCS impaired swimming behavior and altered expression of excitation-contraction coupling proteins in fathead minnow (*Pimephales promelas*) [61] and interfered with thyroid axis in the zebra fish (*Danio rerio*) [62].

Analytical Methods

Modern equipment has made it made possible to detect PPCPs from different matrices at sub-ng/g levels. The main advances in PPCPs analysis have been made using liquid chromatography-mass spectrometry (LC-MS) and gas chromatography-mass spectrometry (GC-MS) techniques. Pharmaceuticals comprised of polar compounds are easily dissolved in water or polar solvents, which is special advantage for LC-MS analysis. Employing isocratic or gradient elution in LC method, complex composites/mixtures of sample can be separated using different polarities (polar or medium polar or mixed polar) mobile phases with an analytical column such as C18. The separated compounds are characterized with MS detection. LC - tandem mass spectrometry (LC-MS/MS) with positive- and or negative modes of operations using electrospray ionization (ESI) and atmospheric pressure chemical ionization (APCI) are able to detect PPCPs up to sub-ppb level. The presence of unknown compounds are confirmed and identified by comparing the mass signals and retention times of unknown samples to known standards.

On the other hand, personal care products (PCPs) are relatively non-polar and are more easily dissolved and extracted in relatively non-polar organic solvents. After cleaned up, the samples are analyzed by GC- MS/MS or - selected ion monitoring (SIM) modes with election ionization (EI) or negative ion chemical ionization (NICI) based on sample nature. The GC-MS/MS or GC-SIM-MS methods are capable of detecting PPCPs down to sub-ng/g levels. The presence of unknown compounds are characterized establishing over 80% to 99% agreement with standard compounds of the respective samples [40,63]. Examples of some analyses performed by LC-MS and GC-MS are illustrated below.

Analysis of PPCPs by LC-MS and GC-MS

Many researchers have reported the identification and analysis of emerging PPCPs contaminants [39,42,64]. Our research group [39,64] developed LC-MS/MS methods for determination of pharmaceuticals and metabolites from environmental fish. Specifically, non-linear gradient elution of water and methanol solvents consisting of 0. 1% (v/v) formic acid in water and 100% methanol are passed through a C18 analytical column at a flow rate 350 μL/min to achieve the separation of the complex mixture of PPCPs. An auto-sampler is used to inject 10 μL sample solution. Column effluents are monitored by MS/MS equipped with an electrospray interface (ESI). Figure 1 displays chromatographic separation of 25 target pharmaceutical drugs and metabolites, 5 surrogates and 2 internal standards that were spiked to the clean fish tissues by LC-MS/MS using electrospray ionization (ESI) positive (+) and negative (-) modes.

Employing the extraction protocol and LC-MS/MS method [64] all target compounds were analyzed from environmental fish samples that were obtained from the Pecan Creek, Denton, Texas, downstream from the effluent discharge. The presence and characterization of the target analytes were confirmed based on comparison of retention time and relative intensities of fragment ions observed from spiked and environmental fish specimens. Four pharmaceutical compounds were detected over method detection limits (MDLs). Figure 2 displays a typical LC-MS/MS ion chromatogram for identification and determined the concentration of diphenhydramine (0. 66-1. 32 ng/g) ,diltiazem (0. 11-0. 27 ng/g), carbamazepine (0. 83-1. 44 ng/g), norfluoxetine (3. 49-5. 114 ng/g) drugs [64]. Method performance associated with method detection limits (MDLs), limit of detection (LOD) and limit of quantitation (LOQ) of 25 target compounds were compared and illustrated in Table 1. It was estimated that MDLs were approximately 3 to 10 times higher than LOD for a majority of the target analytes. In the LC-ESI-MS/MS analyses, matrix influences played a critical role that was essential to consider. These matrix effects are caused by co-extracted constituents that affect analyte ionization using either ESI positive or negative modes [42]. An approach, developed by our group [64] was used to measure the matrix influence for extraction solvents that promoted recoveries. The matrix effect in analyzing the samples that influence mass signal generation, matrix-match calibration curve was proven to be excellent method that minimized the matrix effect in quantitation of analytes from the environmental samples such as fish [64].

GC-MS is a highly efficient tool widely used to analyze semi-volatile and volatile organic personal care products at extremely low levels from environmental samples. Sample nature and complexity are critical considerations in choosing the GC-MS techniques. Extraction/pre-concentration and clean-up steps are required in preparation of samples for GC-MS examination. As fish samples are complex matrices containing lipids, fat etc. , a wide range of extraction and clean-up techniques are needed to handle the samples prior to analysis [40,63,65-68]. Extraction methodologies include Soxhlet extraction, microwave assisted extraction, ultrasound-solid liquid extraction, and pressurized liquid extraction (PLE) and clean-up approaches are silica gel, florisil, and/or gel permeation chromatography (GPC) [40,63,65-68]. Figure 3 shows a schematic diagram of fish sample extraction, pre-concentration and clean-up protocols for GC-MS analysis [63].

Our research group developed GC-SIM-MS and GC-MS/MS

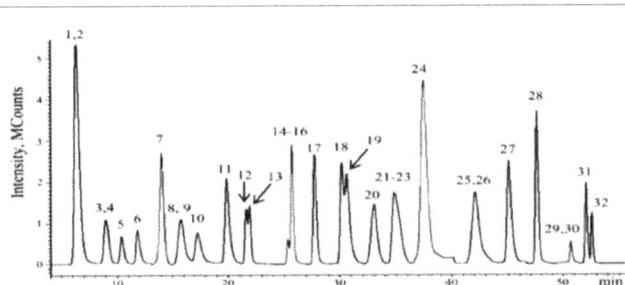

Figure 2: LC-MS/MS reconstituted ion chromatograms displaying analyte-specific quantitation and qualifier ions monitored for (A) a tissue extract from a fish (Lepomis sp.) collected in Pecan Creek and (B) an extract from clean' tissue spiked with known amounts of diphenhydramine (1.6 ng/g), diltiazem (2.4 ng/g), carbamazepine (16 ng/g), and norfluoxetine (80 ng/g). The higher m/z fragment is more intense in all cases. (Reproduced with permission from (64), © American Chemical Society).

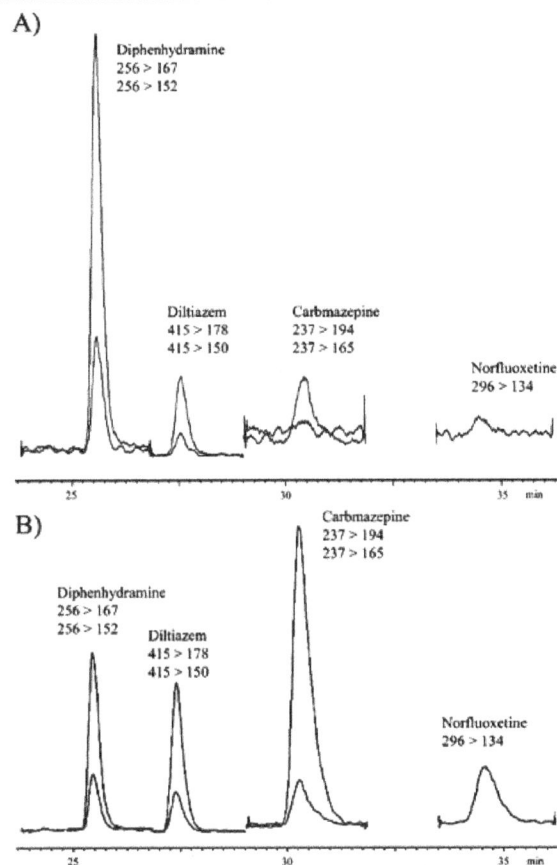

Figure 1: LC-MS/MS total ion chromatogram resulting from analysis of clean tissue spiked with a mixture of pharmaceutical standards. Peak identifications are as follows: (1) acetaminophen-d4, (2) acetaminophen, (3) atenolol, (4) cimetidine, (5) codeine, (6) 1,7-dimethylxanthine, (7) lincomycin, (8) trimethoprim, (9) thiabendazole, (10) caffeine, (11) sulfamethoxazole, (12) 7-aminoflunitrazepam-d7 (+IS), (13) metoprolol, (14) propranolol, (15) diphenhydramine-d3, (16) diphenhydramine, (17) diltiazem, (18) carbamazepine-d10, (19) carbamazepine, (20) tylosin, (21) fluoxetine-d6, (22) fluoxetine, (23) norfluoxetine, (24) sertraline, (25) erythromycin, (26) clofibric acid, (27) warfarin, (28) miconazole, (29) ibuprofen-13C3, (30) ibuprofen, (31) meclofenamic acid (-IS), and (32) gemfibrozil. (Reproduced with permission from (64), © American Chemical Society).

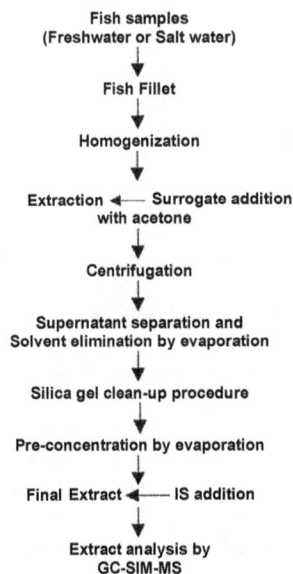

Figure3: A simplified schematic diagram of the experimental procedure used for extraction and analysis of nitromusks, antimicrobial agent and antihistamine from edible fish fillets. (*Reproduced with permission from (63), © Elsevier*).

Analyte	Linear range,[a] (ng/g)	LOD,[b](ng/g)	LOQ[c] (ng/g)	MDL[d] (ng/g)
Acetaminophen	3.12 - 400	0.30	0.99	4.40
Atenolol	1.25 - 160	0.48	1.62	1.48
Cimetidine	0.625 - 80	0.24	0.81	1.04
Codeine	4.60 - 600	1.07	3.55	6.11
1,7-dimethylxanthine	0.625 - 80	0.17	0.58	1.02
Lincomycin	3.12 - 400	0.63	2.09	5.53
Trimethoprim	1.25 - 160	0.79	2.63	2.15
Thiabendazole	1.25 - 160	0.14	0.47	2.63
Caffeine	3.12 - 400	0.34	1.15	3.93
Sulfamethoxazole	1.25 - 160	0.23	0.76	2.29
Metoprolol	1.25 - 160	0.25	0.85	2.50
Propranolol	0.625 - 80	0.01	0.03	1.07
Diphenhydramine	0.0625 - 8	0.01	0.03	0.05
Diltiazem	0.09 - 12	0.04	0.13	0.12
Carbamazepine	0.625 - 80	0.03	0.12	0.54
Tylosin	3.12 - 400	1.18	3.93	5.02
Fluoxetine	4.69 - 600	0.76	2.54	6.73
Norfluoxetine	3.12 - 400	0.32	1.08	2.90
Sertraline	3.12 - 400	0.21	0.71	3.57
Erythromycin	3.12 - 400	0.85	2.84	6.42
Clofabric acid	1.25 - 160	0.10	0.32	2.69
Warfarin	0.625 - 80	0.09	0.29	0.86
Miconazole	3.12 - 400	0.39	1.32	10.8
Ibuprofen	25 - 3200	3.14	10.4	45.9
Gemfibrozil	3.12 - 400	0.25	0.85	6.68

Table 1: Investigated linear range, LOD, LOQ, and MDL for target analytes in fish muscle tissue[a].
[a]clean tissues employed in the determination of these parameters were extracted using a 1:1 mixture of 0.1 M acetic acid (pH 4) and methanol. [b]LOD, calculated as3 times the standard deviation in the background signal observed for the replicate analysis of a tissue blank. [c]LOQ, calculated as 10 times the standard deviation in the background signal observed for the replicate analysis of a tissue blank, d MDL, determined by multiplying the one-sided Student's t-statistics at 99% confidence limit times the standard deviation observed for eight replicate analyses of a matrix spike (spiking level ≤ 10 x MDL). (*SOURCE Reproduced with permission from reference 64, Copyright© American Chemical Society*).

methods [40,41] for PPCPs analysis from fish and other environmental samples following U. S. EPA protocols. For example, Mottaleb et al. [40] detected the presences of PCPs in environmental fish collected from Pecan Creek and Clear Creek streams, Denton, TX, USA. Figure 4 represents a GC-SIM-MS total ion chromatogram for standard solution displaying the separation of compounds that were targeted in the environmental fish collected from the Pecan Creek, Denton, Texas.

Employing the extraction protocols illustrated (Figure 3) [63],we recently reported the concentration of four frequently observed PPCPs in edible fish fillets obtained from grocery stores by GC-SIM-MS [63]. In this investigation, the presence of the target compounds in fish extracts was confirmed based on similar mass spectral features and retention times compared to pure standards. Mass spectral features and retention times of the target compounds obtained from the fish extracts were used for characterization by comparing with the authentic standards. Figure 5 is a GC-SIM-MS ion chromatogram for (A) a standard solution containing, 100 pg/μL of triclosan, an antimicrobial agent and (B) a Whiting (genus *Merlangius*) fillet extract. Individual identities of the compounds extracted from the grocery store fish fillet were characterized based on comparison of the relative ion abundance ratios between the quantification and the qualifier ions mass signals. The presence of target compounds in the fish fillet extracts was confirmed when the difference of the relative abundance ratio was less than or equal to approximately ± 20%, or an agreement of the relative

abundance ratio of 80% or over. Figure 5 compares typical mass spectra derived from the ion chromatogram. These spectra show an excellent agreement of the mass spectral features/mass signals with a variation of about ± 10%. When similar agreement of ion relative abundance ratio and retention features were observed, then the presence, characterization and quantification of other compounds in the different fish samples were established. Table 2 shows the concentrations of the compounds that were characterized and quantified in grocery stores fish species. The values of detected compounds from grocery store samples are approximately 1 to 3 orders of magnitude lower than the fish that were collected from the environmental sites [39-42,64].

Biological Transformation and Effects of PPCPs

Biotransformation of any chemicals is a critical consideration because the compounds get metabolized in biological system forming different species or breakdown products that may induce numerous issues over the period of time. The effects of PPCPs differ from those of conventional pollutants because drugs are intentionally designed to interact with cellular receptors at low concentrations and to cause specific biological effects. Unintended adverse effects can also occur from interaction with non-target receptors. Environmental toxicology focuses on acute effects of exposure rather than chronic effects. Effects on aquatic life are a major concern because aquatic organisms receive more exposure risks than do human, and are exposed with continual and multi-generational basis with higher concentrations of PPCPs in untreated water possible low dose effects. The risks posed to aquatic organisms by trace level concentration of the PPCPs are largely unknown. Some of the known potential impacts on organisms include delayed development in fish, delayed metamorphosis in frogs, and a variety of reactions including altered behavior and reproduction [1,2]. Overall, the behavior and fate of pharmaceuticals and their metabolites in the aquatic environment and organism are not well established. The low volatility and increase polarity of pharmaceuticals indicates that distribution in the environment occurs primarily through aqueous transport, but also via food chain dispersal. Recent studies have indicated that many pharmaceuticals and metabolites are environmentally persistent, bioactive, and have potential for bioaccumulation [49,50]. Acute aquatic toxicities of drugs and metabolites were examined on marine bacterium (*Vibrio fischeri*), a freshwater invertebrate (*Daphnia magna*), and the Japanese medaka fish (*Oryzia slatipes*) by Kim et al. [69]. They demonstrated that *Daphnia* was the most susceptible among the tested organisms. Correa and Hoffmann [70] studied the variation of magnitude of response effect of drugs d-amphetamine, sodium pentobarbital, diazepam, β-carboline, and saline before and after inducing into of knife-fish (*Gymnotus carapo*). They concluded a reduction of the degree of alertness by the barbiturate and a decrease in emotionality and/or stress by the benzodiazepine with the novelty response. Brandao et al. [71] evaluated the biochemical and behavioral effects of neuro-active anticonvulsant drugs (diazepam, carbamazepine, and phenytoin) on pumpkin-seed sunfish (*Lepomis gibbosus*) and showed behavioral changes of sunfish through oxidative stress parameters such as glutathione reductase, glutathione S-transferases, catalase and lipid peroxidation.

Mottaleb et al. [47,72,73] investigated the biotransformation and toxicokinetics of PPCPs known as nitro musks fragrance ingredients, musk xylene (MX), and musk ketone (MK) using trout fish as model. The fish were exposed to nitro musks compounds. Details regarding fish exposure, extraction, and analysis of breakdown product or metabolites have been reported [47,72,73]. Previously we demonstrated the formation of amine (cysteine - hemoglobin) adducts through enzymatic biotransformation reaction between MX/MK and

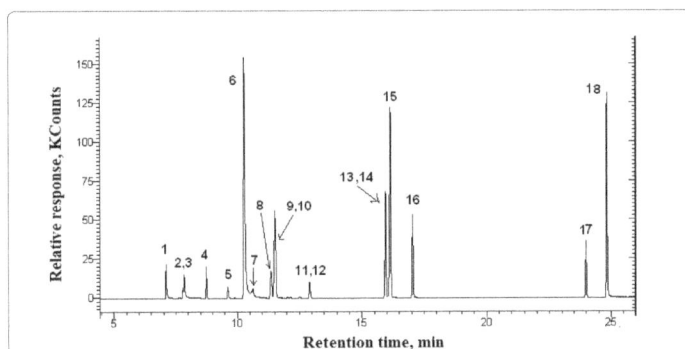

Figure 4: GC-SIM-MS representative total ion chromatogram for a calibration standard. Peak identifications are as follows: (1) *m*-toluamide, (2) benzophenone, (3) benzophenone-d$_{10}$, (4) celestolide, (5) pentachloronitro benzene, (6) phenanthrene-d$_{10}$, (7) *p-n*-octylphenol, (8) galaxolide, (9) tonalide, (10) musk xylene, (11) *p-n*-nonylphenol, (12) [^{13}C$_6$]*p-n*-nonylphenol, (13) 4-methylbenxylidine camphor, (14) 2,2'- dinitrobiphenyl, (15) musk ketone, (16) triclosan, (17) mirex, and (18) octocrylene. (*Reproduced with permission from (40), 2009, © Elsevier*).

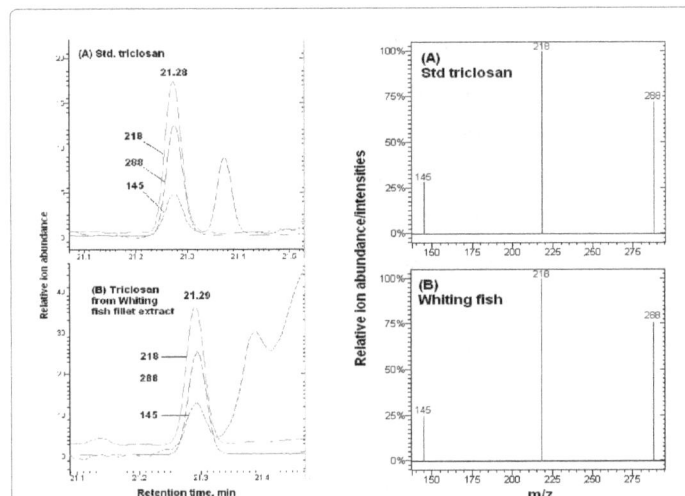

Figure 5: (i) Representative GC-SIM-MS ion chromatograms for (A) a standard solution containing, 100 pg/μl of triclosan and (B) a Whiting (genus *Merlangius*) fillet extract (left). (5 ii) Typical GC-SIM-MS mass spectra for (A) standard triclosanand (B) Whiting fillet extract solution derived by selecting the data file from Fig. 5 (i) (A), retention time 21.28 min and Fig. 5 (i) (B), retention time 21.29 min, respectively (right). (*Reproduced with permission from (63), © Elsevier*).

Name of fish	Concentration of analytes in edible fish fillets (ng g^{-1})				
	HHCB	AHTN	DPH	MK	TCS
Tilapia	0.876	0.813	0.679	nd	4.122
Catfish	0.276	0.429	0.939	nd	2.086
Swai	0.336	0.190	0.189	nd	1.782
Flounder	0.892	0.904	1.182	nd	7.472
Salmon	0.250	0.068	1.037	nd	0.942
Whiting	0.263	0.431	0.503	nd	3.699
Pollock	0.163	0.304	0.692	nd	1.011
Yellow fin Tuna	0.343	0.269	0.811	nd	2.292

Table 2: Concentration of target analytes in fish fillets received from local grocery stores, Maryville, Missouri, USAand their comparison with environmental fish samples.
nd – not detected. (*Source:Reproduced with permission from (63), © Elsevier*).

trout [47,74]. The reduction of a nitro-group in MX and MK led to the formation of amine adducts of hemoglobin that could be suitable as a biochemical endpoint useful for exposure monitoring and assessment of potential hazards resulting from MX and MK compounds [47,72-74].

Recently our group [48] also reported that the bound metabolites obtained from liver proteins may be used as indicators of internal exposure to chemical carcinogens. Table 3 illustrates the relationship between exposure time and the uptake of different dosages of MX and MK compounds over a period of 7 days. The metabolites of nitro musks or other related nitroarenes, bound to the cysteine sulfhydryl group (– SH) of proteins in liver as biomarkers of exposure, could potentially be used to assess continuous exposure over a longer time range, and thus, may be better suited for risk assessment than quantitation of urinary metabolites [75]. The biological transformation processes of MX and MK to their corresponding amine metabolites, with cysteine containing proteins in the liver results in adducts formation are shown in Figure 6. Nitroarenes are enzymatically reduced to nitroso reactive intermediates, nitrosoarenes, capable of covalently binding with the – SH group of cysteine amino acids in proteins to form an acid/base labile sulfonamide adducts that hydrolyzes to aromatic amines in the presence of aqueous base [76]. The aromatic amines were considered to be good dosimeters for the target tissue [77].

Spike recovery studies and limit of detection (LOD) measurements of 2-amino-MX, 4-amino-MX and 2-amino MK metabolites were accomplished as 95-114% with relative standard deviation <10% and 0. 91-3. 8 ng/g, respectively. Table 4 illustrated the concentration of metabolites observed in the trout liver [48]. The half-lives of 2-AMX, 2-AMK and 4-AMX metabolites were estimated to be 2 - 9 days in the trout liver based on the assumption of first-order kinetics. The individual values of the elimination rate constants and the half-lives of the 2-AMX, 2-AMK and 4-AMX metabolites in the fish liver suggested that the toxico kinetics are more complex than a simple first-order reaction because additional internal biological processes transpire in the living organisms. Nitromusks (MX and or MK) were identified as inducers of hepatic cytochrome P450 2B enzymes and P450 1A1 and 1A2 isoenzymes [78,79] and are non-genotoxic [80,81]. Although the non-genotoxic carcinogenesis is not fully understood, it is believed that a non-genotoxic mechanism, such as increased cell proliferation, might be responsible for the increase in the liver tumors [82]. In our investigation [48], the MX and or MK-cysteine-protein adducts in fish liver were used to monitor nitromusks hazards as biomarkers of exposure.

Conclusions

PPCPs are increasingly being used in human and animal applications for numerous purposes. The ultimate fates of these chemicals are in aquatic systems, where organisms get exposed over an extended period of time through wastewater treatment plants and other sources. Continuous loading of the parent compounds and metabolites of PPCPs will reach harmful concentrations that adversely affect the freedom of aquatic creatures. Thus, periodic measurements of exposure level of those compounds are very important. Modern analytical techniques such as LC-MS and GC-MS have made it possible to detect extremely low levels of those chemicals.

To address the challenge of PPCPs as emerging contaminants, regulatory authorities or agencies or health care professionals need to work collectively. An important and effective way aspect to reduce the load of PPCPs and their metabolites in wastewaters and surface

Exposure time, Day	MX exposure					MK exposure				
	MX conc. (mg/mL)	Fish wet weight, (g)	Liver wet weigh, (g)	MX dose / trout, (mg)	Average dosing level, (mg/g)	MK conc. (mg/mL)	Fish wet weigh, (g)	Liver wet weigh, (g)	MK dose / trout, (mg)	Average dosing level, (mg/g)
1- Day	10	202	0.75	2.0	0.01	10	257	0.88	2.6	0.01
		256	0.87	2.5			237	0.77	2.4	
		165	0.50	1.6			222†	NC	2.2	
	30	180	0.75	5.4	0.03	30	199	0.70	6.0	0.03
		256	0.92	7.5			230	0.76	6.9	
		280‡	1.41	8.4			212	0.69	6.3	
	100	236	0.76	24.0	0.10	100	272	1.26	27.0	0.10
		264	0.92	26.0			271	1.26	27.0	
		204	0.74	20.0			197	0.78	20.0	
	300	250	0.88	75.0	0.30	300	190	0.70	57.0	0.30
		310	1.52	90.0			270	1.18	81.0	
		227	0.71	69.0			250	1.00	75.0	
	Control	206	0.76	0.20 mL, exposed with salmon oil only						
		304	1.52	0.30 mL, exposed with salmon oil only						
		184	0.70	0.18 mL, exposed with salmon oil only						
3-Days	30	208	0.79	6.3	0.03	30	278	1.22	8.4	0.03
		244	0.81	7.2			156	0.46	4.5	
		193	0.71	6.0			196	0.68	6.0	
	Control	253	0.89	0.25 mL, exposed with salmon oil only						
		272	1.23	0.27 mL, exposed with salmon oil only						
		233	0.75	0.23 mL, exposed with salmon oil only						
7-Days	30	212‡	0.71	6.3	0.03	30	121	0.30	3.6	0.03
		230	0.76	6.9			241	0.88	7.2	
		204	0.74	6.0			167	0.53	5.1	
	Control	273	1.12	0.27 mL, exposed with salmon oil only						
		305	1.45	0.30 mL, exposed with salmon oil only						
		250	0.89	0.25 mL, exposed with salmon oil only						

†Trout was found dead and the liver sample was not collected (NC). ‡Trout was found sick (not equilibrium condition) and the collected liver specimen was not used in this study for composite preparation. All control liver samples were mixed together to make one control composite specimen. The collected MX and or MK exposed liver samples were composited mixing three liver for each dosing level with exception of sick or dead fish liver. (*Source: Reproduced with permission from (48), 2012, © Elsevier*).

Table 3: *In vivo* trout exposure dosing schedule with nitro musk compounds and salmon oil vehicle.

Exposure period (Day)	Exposure level MX or MK (mg/g)	Nitro musk metabolites (ng/g)		
		2-AMX	4-AMX	2-AMK
1-Day	0.01	94.0	2404.4	115.4
	0.03	492.0	12588.5	505.5
	0.10	444.1	10325.9	426.6
	0.30	259.1	5147.3	396.1
3-Days	0.03	213.6	6097.6	357.8
7-Days	0.03	113.5	2988.3	298.0
Controls	None	Not detected		

(*Source: Reproduced with permission from (48), © Elsevier*).

Table 4: Concentration of nitro musk metabolites in trout liver samples using hydrolyzed extraction.

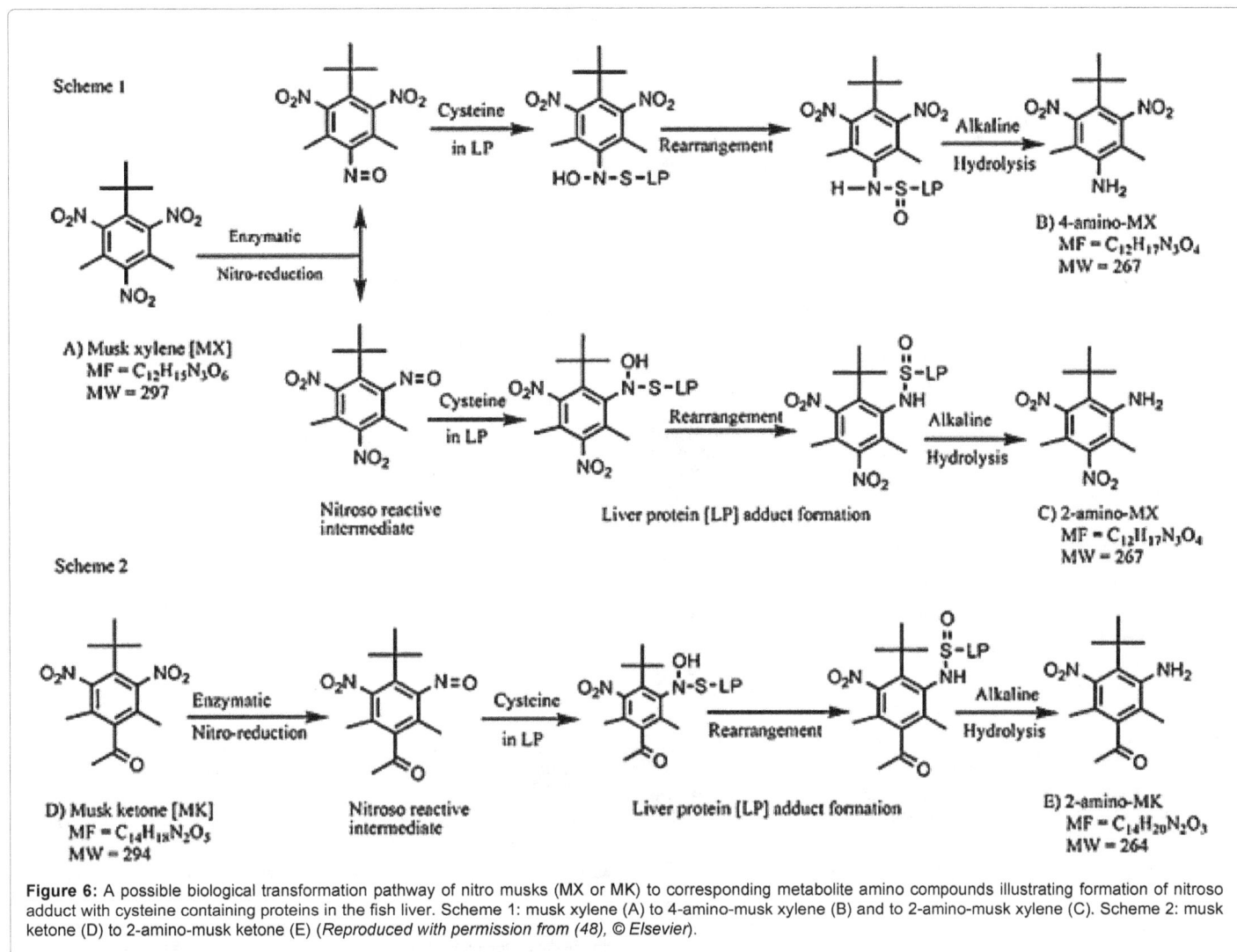

Figure 6: A possible biological transformation pathway of nitro musks (MX or MK) to corresponding metabolite amino compounds illustrating formation of nitroso adduct with cysteine containing proteins in the fish liver. Scheme 1: musk xylene (A) to 4-amino-musk xylene (B) and to 2-amino-musk xylene (C). Scheme 2: musk ketone (D) to 2-amino-musk ketone (E) (*Reproduced with permission from (48), © Elsevier*).

waters is to develop new sewage treatment processes. This requires understanding the fate of PPCPs during sewage treatment plants for implementation of better removal techniques. Consumers need to be aware of the consequences of PPCPs to aquatic organisms and ecosystems, and should follow the regulatory agencies disposal guidelines to make our environment friendly for all living organisms. At the same time, scientists and toxicologists should continue to investigate the transport, fate, toxicity and their potential physiological and psychological effects on humans and wildlife as well as relationship between bioaccumulation and diseases.

References

1. Stone R (1994) Environmental estrogens stir debate. Science 265: 308-310.

2. Ashby J, Houthoff E, Kennedy SJ, Stevens J, Bars R, et al. (1997) The challenge posed by endocrine-disrupting chemicals. Environ Health Perspect 105: 164-169.

3. Daughton CG, Ternes TA (1999) Pharmaceuticals and personal care products in the environment: agents of subtle change? Environ Health Perspect 107 Suppl 6: 907-938.

4. Hinck JE, Blazer VS, Schmitt CJ, Papoulias DM, Tillitt DE (2009) Widespread occurrence of intersex in black basses (Micropterus spp.) from U.S. rivers, 1995-2004. Aquat Toxicol 95: 60-70.

5. Loraine GA, Pettigrove ME (2006) Seasonal variations in concentrations of pharmaceuticals and personal care products in drinking water and reclaimed wastewater in southern California. Environ Sci Technol 40: 687-695.

6. Barnes KK, Kolpin DW, Furlong ET, Zaugg SD, Meyer MT, et al. (2008) A national reconnaissance of pharmaceuticals and other organic wastewater contaminants in the United States - I) Groundwater. Sci Total Environ 402: 192-200.

7. Hahlbeck E, Griffiths R, Bengtsson BE (2004) The juvenile three-spined stickleback (Gasterosteus aculeatus L.) as a model organism for endocrine disruption. I. Sexual differentiation. Aquat Toxicol 70: 287-310.

8. Harris CA, Hamilton PB, Runnalls TJ, Vinciotti V, Henshaw A, et al. (2011) The consequences of feminization in breeding groups of wild fish. Environ Health Perspect 119: 306-311.

9. Xu P, Drewes JE, Kim T, Bellona C, Amy G (2006) Effect of membrane fouling on transport of emerging organic contaminants in NF/RO membrane applications. J Membrane Sci 279: 165-175.

10. Drewes JE, Hoppe C, Jennings T (2006) Fate and transport of N-nitrosamines under conditions simulating full-scale groundwater recharge operations. Water Environ Res 78: 2466-2473.

11. Jakimska A, SliwkaKaszynska M, Nagorski P, KotWasik A, Namiesnik J (2014) Environmental fate of two psychiatric drugs, diazepam and sertraline: Phototransformation and investigation of their photoproducts in natural waters. J Chromatogr Sep Tech 5: 253.

12. Blazer VS, Iwanowicz LR, Iwanowicz DD, Smith DR, Young JA, et al. (2007) Intersex (testicular oocytes) in smallmouth bass from the Potomac River and selected nearby drainages. J Aquat Anim Health 19: 242-253.

13. De Metrio G, Corriero A, Desantis S, Zubani D, Cirillo F, et al. (2003) Evidence of a high percentage of intersex in the Mediterranean swordfish (Xiphias gladius L.). Mar Pollut Bull 46: 358-361.

14. Jobling S, Nolan M, Tyler CR, Brighty G, Sumpter JP (1998) Widespread sexual disruption in wild fish. Environ SciTechnol 32: 2498-2506.

15. Vajda AM, Barber LB, Gray JL, Lopez EM, Woodling JD, et al. (2008) Reproductive disruption in fish downstream from an estrogenic wastewater effluent. Environ Sci Technol 42: 3407-3414.

16. Vajda AM, Barber LB, Gray JL, Lopez EM, Bolden AM, et al. (2011) Demasculinization of male fish by wastewater treatment plant effluent. Aquat Toxicol 103: 213-221.

17. Pomati F, Orlandi C, Clerici M, Luciani F, Zuccato E (2008) Effects and interactions in an environmentally relevant mixture of pharmaceuticals. Toxicol Sci 102: 129-137.

18. Westerhoff P, Yoon Y, Snyder S, Wert E (2005) Fate of endocrine-disruptor, pharmaceutical, and personal care product chemicals during simulated drinking water treatment processes. Environ SciTechnol 39: 6649-6663.

19. Lange A, Katsu Y, Ichikawa R, Paull GC, Chidgey LL, Coe TS (2008) Altered sexual development in roach (Rutilusrutilus) exposed to environmental concentrations of the pharma¬ceutical 17a-ethinylestradiol and associated expression dynamics of aromatases and estrogen receptors. Toxicol Sci 106: 113–123.

20. Christen V, Hickmann S, Rechenberg B, Fent K (2010) Highly active human pharmaceuticals in aquatic systems: A concept for their identification based on their mode of action. Aquat Toxicol 96: 167-181.

21. Daughton CG (2004) Non-regulated water contaminants: emerging research. Environ Impact Assess Rev 24: 711-732.

22. Miege C, Choubert JM, Ribeiro L, Eusebe M, Coquery M (2009) Fate of pharmaceuticals and personal care products in wastewater treatment plants--conception of a database and first results. Environ Pollut 157: 1721-1726.

23. Fent K, Weston AA, Caminada D (2006) Ecotoxicology of human pharmaceuticals. Aquat Toxicol 76: 122-159.

24. Mompelat S, Le Bot B, Thomas O (2009) Occurrence and fate of pharmaceutical products and by-products, from resource to drinking water. Environ Int 35: 803-814.

25. Kolpin DW, Furlong ET, Meyer MT, Thurman EM, Zaugg SD, et al. (2002) Pharmaceuticals, hormones, and other organic wastewater contaminants in U.S. streams, 1999-2000: a national reconnaissance. Environ Sci Technol 36: 1202-1211.

26. Brodin T, Fick J, Jonsson M, Klaminder J (2013) Dilute concentrations of a psychiatric drug alter behavior of fish from natural populations. Science 339: 814-815.

27. Kelly BC, Ikonomou MG, Blair JD, Morin AE, Gobas FA (2007) Food web-specific biomagnification of persistent organic pollutants. Science 317: 236-239.

28. Snyder SA, Westerhoff P, Yoon Y, Sedlak DL (2003) Pharmaceuticals, personal care products and endocrine disruptors in water: implications for the water industry. Environ Eng Sci 20: 449–469.

29. van der Oost R, Beyer J, Vermeulen NP (2003) Fish bioaccumulation and biomarkers in environmental risk assessment: a review. Environ Toxicol Pharmacol 13: 57-149.

30. Wu C, Huang X, Witter JD, Spongberg AL, Wang K, et al. (2014) Occurrence of pharmaceuticals and personal care products and associated environmental risks in the central and lower Yangtze river, China. Ecotoxicol Environ Saf 106: 19-26.

31. Kallenborn R, Gatermann R, Planting S, Rimkus GG, Lund M, et al. (1999) Gas chromatographic determination of synthetic musk com-pounds in Norwegian air samples. J Chromatogr A 846: 295-306.

32. Osemwengie LI, Steinberg S (2001) On-site solid-phase extraction and laboratory analysis of ultra-trace synthetic musks in municipal sewage effluent using gas chromatography-mass spectrometry in the full-scan mode. J Chromatogr A 932: 107-118.

33. Blair BD, Crago JP, Hedman CJ, Klaper RD (2013) Pharmaceuticals and personal care products found in the Great Lakes above concentrations of environmental concern. Chemosphere 93: 2116-2123.

34. Berset JD, Bigler P, Herren D (2000) Analysis of nitro musk compounds and their amino metabolites in liquid sewage sludges using NMR and mass spectrometry. Anal Chem 72: 2124-2131.

35. Gagne F, Blaise C, Andre C (2006) Occurrence of pharmaceutical products in a municipal effluent and toxicity to rainbow trout (Oncorhynchusmykiss) hepatocytes. Ecotoxicol Environ Safety 64: 329-336.

36. Beretta M, Britto V, Tavares TM, Silva SMT, Pletsch AL (2014) Occurrence of Pharmaceutical and personal care products (PPCPs) in marine sediments in the Todosos Santos Bay and the north coast of Salvador, Bahia, Brazil. J Soils Sediments 14: 1278-1286.

37. Nakata H (2005) Occurrence of synthetic musk fragrances in marine mammals and sharks from Japanese coastal waters. Environ Sci Technol 39: 3430-3434.

38. Mottaleb MA, Brumley WC, Pyle SM, Sovocool GW (2004) Determination of a bound musk xylene metabolite in carp hemoglobin as a biomarker of exposure by gas chromatography-mass spectrometry using selected ion monitoring. J Anal Toxicol 28: 581-586.

39. Ramirez AJ, Brain RA, Usenko S, Mottaleb MA, O'Donnell JG, et al. (2009) Occurrence of pharmaceuticals and personal care products in fish: results of a national pilot study in the United States. Environ Toxicol Chem 28: 2587-2597.

40. Mottaleb MA, Usenko S, O'Donnell JG, Ramirez AJ, Brooks BW, et al. (2009) Gas chromatography-mass spectrometry screening methods for select UV filters, synthetic musks, alkylphenols, an antimicrobial agent, and an insect repellent in fish. J Chromatogr A 1216: 815-823.

41. Subedi B, Mottaleb MA, Chambliss CK, Usenko S (2011) Simultaneous analysis of select pharmaceuticals and personal care products in fish tissue using pressurized liquid extraction combined with silica gel cleanup. J Chromatogr A 1218: 6278-6284.

42. Tanoue R, Nomiyama K2, Nakamura H3, Hayashi T4, Kim JW5, et al. (2014) Simultaneous determination of polar pharmaceuticals and personal care products in biological organs and tissues. J Chromatogr A 1355: 193-205.

43. McEneff G, Barron L, Kelleher B, Paull B, Quinn B (2014) A year-long study of the spatial occurrence and relative distribution of pharmaceutical residues in sewage effluent, receiving marine waters and marine bivalves. Sci Tot Environ 476-477: 317-326.

44. Liebl B, Ehrenstofer S (1993) Nitro musk in human milk. Chemosphere27: 2253-2260.

45. Hu Z, Shi Y, Niu H, Cai Y, Jiang G, et al. (2010) Occurrence of synthetic musk fragrances in human blood from 11 cities in China. Environ Toxicol Chem 29: 1877-1882.

46. Riedel J, Dekant W (1999) Biotransformation and toxicokinetics of musk xylene in humans. Toxicol Appl Pharmacol 157: 145-155.

47. Mottaleb MA, Brumley WC, Curtis LR, Sovocool GW (2005) Nitro musk adducts of rainbow trout hemoglobin: dose-response and toxicokinetics determination by GC-NICI-MS for sentinel species. Am Biotech Lab 23: 26-29.

48. Mottaleb MA, Osemwengie LI, Islam MR, Sovocool GW (2012) Identification of bound intro musk-protein adducts in fish liver by gas chromatography-mass spectrometry: Biotransformation, dose-response and toxicokinetics of nitro musk metabolites protein adducts in trout liver as biomarkers of exposure. Aquat Toxicol 106-107: 164-172.

49. Valdes ME, Ames MV, Bistoni Mde L, Wunderlin DA (2014) Occurrence and bioaccumulation of pharmaceuticals in a fish species inhabiting the SuquÃa River basin (CÃ³rdoba, Argentina). Sci Total Environ 472: 389-396.

50. Gomez E, Bachelot M, Boillot C, Munaron D, Chiron S, et al. (2012) Bioconcentration of two pharmaceuticals (benzodiazepines) and two personal care products (UV filters) in marine mussels (Mytiluisgalloprovincialis) under controlled laboratory conditions. Environ Sci Pollut Res19: 2561-2569.

51. Peck AM (2006) Analytical methods for the determination of persistent ingredients of personal care products in environmental matrices. Anal Bioanal Chem 386: 907-939.

52. Mackay D, Barnthouse L (2010) Integrated risk assessment of household chemicals and consumer products: addressing concerns about triclosan. Integr Environ Assess Manag 6: 390-392.

53. Wu JL, Lam NP, Martens D, Kettrup A, Cai Z (2007) Triclosan determination in water related to wastewater treatment. Talanta 72: 1650-1654.

54. Behera SK, Oh SY, Park HS (2010) Sorption of triclosan onto activated carbon,

kaolinite and montmorillonite: effects of pH, ionic strength, and humic acid. J Hazard Mater 179: 684-691.

55. Rudel H, Bohmer W, Muller M, Fliedner A, Ricking M, et al. (2013) Retrospective study of triclosan and methyl-triclosan residues in fish and suspended particulate matter: results from the German Environmental Specimen Bank. Chemosphere 91: 1517-1524.

56. Coogan MA, Edziyie RE, La Point TW, Venables BJ (2007) Algal bioaccumulation of triclocarban, triclosan, and methyl-triclosan in North Texas wastewater treatment plant receiving stream. Chemosphere 67: 1911-1918.

57. Aranami K, Readman JW (2007) Photolytic degradation of triclosan in freshwater and seawater. Chemosphere 66: 1052-1056.

58. Orvos DR, Versteeg DJ, Inauen J, Capdevielle M, Rothenstein A, et al. (2002) Aquatic toxicity of triclosan. Environ Toxicol Chem 21: 1338-1349.

59. Wilson BA, Smith VH, deNoyelles F Jr, Larive CK (2003) Effects of three pharmaceutical and personal care products on natural freshwater algal assemblages. Environ Sci Technol 37: 1713-1719.

60. Levy CW, Roujeinikova A, Sedelnikova S, Baker PJ, Stuitje AR, et al. (1999) Molecular basis of triclosan activity. Nature 398: 383-384.

61. Fritsch EB, Connon RE, Werner J, Davies RE, Beggel S, Feng W, Pessah IN (2013) Triclosan impairs swimming behavior and alters expression of excitation-contraction soupling proteins in Fathead Minnow (Pimephalespromelas). Environ SciTechnol 47: 2008-2017.

62. Pinto PIS, Guerreiro EM, Power DM (2013) Triclosan interferes with the thyroid axis in the zebrafish (Daniorerio). Toxicol Res 2: 60-69.

63. Foltz J, Mottaleb MA, Meziani MJ, Rafiq Islam M (2014) Simultaneous detection and quantification of select nitromusks, antimicrobial agent, and antihistamine in fish of grocery stores by gas chromatography-mass spectrometry. Chemosphere 107: 187-193.

64. Ramirez AJ, Mottaleb MA, Brooks BW, Chambliss CK (2007) Analysis of pharmaceuticals in fish using liquid chromatography-tandem mass spectrometry. Anal Chem 79: 3155-3163.

65. Balmer ME, Buser HR, Muller MD, Poiger T (2005) Occurrence of some organic UV filters in wastewater, in surface waters, and in fish from Swiss Lakes. Environ Sci Technol 39: 953-962.

66. Rudel H, Bohmer W, Schroter-Kermani C (2006) Retrospective monitoring of synthetic musk compounds in aquatic biota from German rivers and coastal areas. J Environ Monit 8: 812-823.

67. Bohmer W, Muller J, Rudel H, Schroter-Kermani C (2004) Retrospective monitoring of alkylphenols and alkylphenol monoethoxylates in aquatic biota from 1985 to 2001: results from the German Environmental Specimen Bank. Environ Sci Technol 38: 1654-1661.

68. Navarro P, Bustamante J, Vallejo A, Prieto A, Usobiaga A, et al. (2010) Determination of alkylphenols and 17beta-estradiol in fish homogenate.

Extraction and clean-up strategies. J Chromatogr A 1217: 5890-5895.

69. Kim Y, Choi K, Jung J, Park S, Kim PG, Park J (2007) Aquatic toxicity of acetaminophen, carbamazepine, cimetidine, diltiazem and six major sulfonamides, and their potential ecological risks in Korea. Environ Int 33: 370-375.

70. Correa L (1999) Effect of drugs that alter alertness and emotionality on the novelty response of a weak electric fish, Gymnotus carapo. Physiol Behav 65: 863-869.

71. Brandao FP, Rodrigues S, Castro BB, Goncalves F, Antunes SC, et al. (2013) Short-term effects of neuroactive pharmaceutical drugs on a fish species: biochemical and behavioural effects. Aquat Toxicol 144-145: 218-29.

72. Mottaleb MA, Zimmerman JD, Moy TW (2008) Biological transformation, kinetics and dose-response assessments of bound musk ketone hemoglobin adducts in rainbow trout as biomarkers of environmental exposure. J Environ Sci 20: 878-884.

73. Mottaleb MA, Moy TW, Zimmerman JH (2006) Biotransformation of musk xylene in trout haemoglobin: does-response and toxicokinetics of musk xylene metabolites haemoglobin adducts by gas chromatography-mass spectrometry. Int J Enviorn Anal Chem 86: 743-756.

74. Mottaleb MA, Zhao X, Curtis LR, Sovocool GW (2004) Formation of nitro musk adducts of rainbow trout hemoglobin for potential use as biomarkers of exposure. Aquat Toxicol 67: 315-324.

75. Farmer PB, Neumann HG, Henschler D (1987) Estimation of exposure of man to substances reacting covalently with macromolecules. Arch Toxicol 60: 251-260.

76. Sabbioni G (1994) Hemoglobin binding of nitroarenes and quantitative structure-activity relationships. Chem Res Toxicol 7: 267-274.

77. Skipper PL, Tannenbaum SR (1994) Molecular dosimetry of aromatic amines in human populations. Environ Health Perspect 102 Suppl 6: 17-21.

78. Minegishi K, Nambaru S, Fukuoka M, Tanaka A, Nishimaki-Mogami T (1991) Distribution, metabolism, and excretion of musk xylene in rats. Arch Toxicol 65: 273-282.

79. Mersch-Sundermann V, Reinhardt A, Emig M (1996) [Mutagenicity, genotoxicity and cogenotoxicity of environmentally relevant nitro musk compounds]. Zentralbl Hyg Umweltmed 198: 429-442.

80. Emig M, Reinhardt A, Mersch-Sundermann V (1996) A comparative study of five nitro musk compounds for genotoxicity in the SOS chromotest and Salmonella mutagenicity. Toxicol Lett 85: 151-156.

81. Api AM, Ford RA, San RH (1995) An evaluation of musk xylene in a battery of genotoxicity tests. Food Chem Toxicol 33: 1039-1045.

82. Lehman-McKeeman LD, Johnson DR, Caudill D, Stuard SB (1997) Mechanism-based inactivation of mouse hepatic cytochrome P4502B enzymes by amine metabolites of musk xylene. Drug Metab Dispos 25: 384-389.

19

Simultaneous Detection Method for Mycotoxins and their Metabolites in Animal Urine by using Impurity Adsorption Purification followed by Liquid Chromatography-Tandem Mass Detection

Wang Rui Guo[1], Su Xiao Ou[1], Wang Pei Long[1]*, Zhang Wei[1], Xue Yan[2] and Li Yu[2]

[1]*Institute of Quality Standards and Testing Technology for Agricultural Products, Chinese Academy of Agricultural Science, Key Laboratory of Agrifood Safety and Quality, Ministry of Agriculture, Beijing, PR China*
[2]*Alltech Biological Products Co. Ltd., Beijing 100600, China*

Abstract

A novel method for simultaneous detection of mycotoxins (e.g., aflatoxin B$_1$) or their metabolic residues in animal urine with impurity adsorption purification followed ultra-performance liquid chromatography coupled tandem mass spectrometry (UPLC-MS/MS) detection has been developed. Extraction of mycotoxins or their metabolites in animal urine sample was performed with 0.1% formic acid-acetonitrile solution after addition of sodium chloride. The extract was then dehydrated and purified with hydrous magnesium sulfate, C$_{18}$, primary secondary amine, and alumina-A. 3 mL of the supernatant was evaporated and re-dissolved by 0.5 mL of 0.1% formic acid aqueous solution/acetonitrile (70:30, *V/V*) for UPLC-MS/MS detection. A C$_{18}$ reversed-phase chromatographic column was employed for separation of target analytes. In the chromatographic separation of target analytes, 0.1% formic acid aqueous solution and 0.1% formic acid-methanol solution were used as the mobile phases with the optimum gradient elution procedures. Multiple-reaction monitoring (MRM) mode was applied for qualitative and quantitative analysis, and matrix calibration curves obtained with the external-standard method was used for quantitation of target analytes. Under optimized conditions, the linearity range was 0.05-100 ng/mL, and the limit of quantification of the developed method was 0.05-0.25 ng/mL. The recoveries of mycotoxins and their metabolites spiked in urine samples were from 80.8% to 114.3%, and the relative standard deviation was <15%.

Keywords: Impurity adsorption purification; UPLC-MS/MS; Mycotoxins; Animal urine

Introduction

Mycotoxins are the secondary toxic metabolites generated during the growth process of mold [1]. So far, the number of mycotoxin species with an identified chemical structure have reached over 400 [2], and they are mainly the metabolites of *Aspergillus*, *Fusarium*, *Penicillium*, *Alternaria*, and *Claviceps* [3]. Cereals, oil seeds, and animal feeds made thereof are subject to contamination by various mycotoxins, thereby acutely or chronically poisoning humans and animals [4]. Research indicates that about 25% of agricultural products across the world may be contaminated by mycotoxins, so mycotoxins in the food chain are necessary to monitor and control [5]. World health organization (WHO) has regarded mycotoxins as items that should be monitored under the food-safety system [6], and China has established the maximum limit for mycotoxins in food and animal feeds [7-11]. Generally, mycotoxin exposure is measured by analyzing mycotoxin content in food or animal feeds [12]. In recent years, some researchers have revealed mycotoxin exposure and the *in vivo* absorption and metabolic process of mycotoxins by using mycotoxins in human or animal urine or their metabolites as biomarkers [13-15]. However, most available methods can be used to detect merely a single type of mycotoxin or mycotoxins with similar structure, and simultaneous contamination by different types of mycotoxins is inevitable because animal feeds are made of various raw materials [16]. Thus, a method that can simultaneously detect the various mycotoxins in animal urine is urgent and significant to develop for the purpose of food safety and mycotoxins exposure study.

In this study, a novel and simple clean-up method based on impurity adsorption mechanism has been developed to purify 25 mycotoxins and their metabolites in animal urine. The developed clean-up method with UPLC-MS/MS detection was easy and quick to operate, low cost, can achieve high quantitative accuracy. The proposed simultaneous detection method can also be used to evaluate the coexposure of mycotoxins on animals and to perform toxic kinetic studies of mycotoxins.

Experiment

Reagents and chemicals

The stock standard solutions of mycotoxins and their metabolites were prepared with acetonitrile as solvent and the detailed information of preparation has been listed in Table 1. The stock solution was stored at -20°C. When it was used, an appropriate amount of solution was dried with the vacuum centrifugal concentrator and then dissolved in 0.1% formic acid aqueous solution-acetonitrile (70:30, *v/v*) to prepare standard working solutions of different concentrations. Acetonitrile, methanol and formic acid are chromatographic pure (Fisher, USA), and Milli-Q ultrapure water were used in the experiment. Sodium chloride and anhydrous magnesium sulfate (analytical grade) were purchased from Sinopharm Chemical Reagent Co., Ltd. Octadecyl bonded silica (C$_{18}$), primary secondary amine (PSA), alumina-A (A-AL) and graphitized carbon black (GCB) were purchased from Bonna-Agela Technologies Company, China. Blank pig-urine samples were donated by the National Feed Quality Control Center (Beijing).

*Corresponding author: Wang Pei Long, Institute of Quality Standards and Testing Technology for Agricultural Products, Chinese Academy of Agricultural Science, Key Laboratory of Agrifood Safety and Quality, Ministry of Agriculture, Beijing 10081, PR China, E-mail: wplcon99@163.com

No	Analyte	Formula	CAS number	Purity (%)	Concentration in stock solution (µg/mL)
1	Aflatoxin B₁ (AFB₁)	$C_{17}H_{12}O_6$	1162-65-8	99	1
2	Aflatoxin B₂ (AFB₂)	$C_{17}H_{14}O_6$	7220-81-7	99	1
3	Aflatoxin G₁ (AFG₁)	$C_{17}H_{12}O_7$	1165-39-5	99	1
4	Aflatoxin G₂ (AFG₂)	$C_{17}H_{14}O_7$	7241-98-7	99	1
5	Aflatoxin M₁ (AFM₁)	$C_{17}H_{12}O_7$	6795-23-9	99	1
6	Aflatoxin M₂ (AFM₂)	$C_{17}H_{14}O_7$	6885-57-0	99	1
7	Sterigmatocysin (STE)	$C_{18}H_{12}O_6$	10048-13-2	99	1
8	T-2 toxin (T-2)	$C_{24}H_{34}O_9$	21259-20-1	98	1
9	Lysergol (LYS)	$C_{16}H_{18}N_2O$	602-85-7	99	1
10	Methylergonovine (MET)	$C_{20}H_{25}N_3O_2$	113-42-8	99	1
11	Roquefortine C (RC)	$C_{22}H_{23}N_5O_2$	58735-64-1	99	1
12	Diacetoxyscirpenol (DIA)	$C_{19}H_{26}O_7$	2270-40-8	99	1
13	Deoxynivalenol (DON)	$C_{15}H_{20}O_6$	51481-10-8	99	5
14	3-AcetylDeoxynivalenol (3-AcDON)	$C_{17}H_{22}O_7$	50722-38-8	99	5
15	15-AcetylDeoxynivalenol (15-AcDON)	$C_{17}H_{22}O_7$	88337-96-6	99	5
16	Neosolaniol (NEO)	$C_{19}H_{26}O_8$	36519-25-2	99	5
17	Wortmannin (WOR)	C23H24O8	19545-26-7	99	5
18	Verruculogen (VER)	$C_{27}H_{33}N_3O_7$	12771-72-1	98	5
19	HT-2 toxin (HT-2)	$C_{22}H_{32}O_8$	26934-87-2	98	5
20	Zearalenone (ZEN)	$C_{18}H_{22}O_5$	17924-92-4	99	5
21	α-Zearalenol (α-ZEL)	$C_{18}H_{24}O_5$	6455-72-8	98	5
22	β-Zearalenol (β-ZEL)	$C_{18}H_{24}O_5$	71030-11-0	98	5
23	Zearalanone (ZAN)	$C_{18}H_{24}O_5$	5975-78-0	99	5
24	α-Zearalanol (α-ZAL)	$C_{18}H_{26}O_5$	26538-44-3	97	5
25	β-Zearalanol (β-ZAL)	$C_{18}H_{26}O_5$	42422-68-4	97	5

Table 1: Information of standards of 25 mycotoxins.

Instrumentation

Ultra performance liquid chromatography (Acquity UPLC) coupled tandem mass spectrometer (XEVO TQ-S, Waters, USA) was employed to detect mycotoxins and their metabolites with MRM mode.

For the chromatographic conditions, the BEH RP18 chromatographic column (Acquity UPLC, 100 mm × 2.1 mm; 1.7 µm; Waters, USA) was used as analytical column to separate target analytes. The column temperature was 40°C and flow rate was 0.3 mL/min. 0.1% formic acid aqueous solution (mobile phase A) and 0.1% formic acid-methanol (mobile phase B) were used as mobile phase with gradient elution. The gradient elution procedures were as followed: 0-2 min, 95% A; 2-4 min, 95%-80% A; 4 min-12 min, 80%-5% A; 12-12.1 min, 5%-1% A; 12.1-14 min, 1% A; 14-14.1 min, 1%-95% A; and 14.1-16 min, 95% A. The injection volume was 5 µL.

The detection of target analytes was carried out on Xevo TQ tandem MS platform fitted with ESI probe operated in the positive ion and negative ion mode and the target analytes was monitored under MRM mode. The optimal parameters as followed, capillary voltage, 2800 V; ion source temperature, 150°C; desolvation gas temperature, 400°C; desolvation gas flow rate, 800 L/h. Argon was used as the collision gas, and the collision cell pressure was 4 m Bar. Other parameters were shown in Table 2. Instrument control, data acquisition and data processing were carried out with Masslynx V 4.1 software (Waters, MA, USA). Positive-ion monitoring mode was used to monitor mycotoxin nos. 1-19, and negative-ion monitoring mode was used to monitor mycotoxin nos. 20-25. Multiple-reaction monitoring (MRM) mode was used to detect and monitor ions, and the parameters including collision energy and cone voltage are shown in Table 2.

The urine sample preparation was performed with RVC 20-18 desktop vacuum centrifugal concentrator (CHRIST, Germany), 3K15 high-speed refrigerated centrifuge (Sigma, USA), D37520 high-speed centrifuge (Kendro, USA), and VX-III multitube vortexer (Targin Technology, China).

Sample preparation

Exactly 2.0 mL of pig-urine sample was transferred into a 50 mL plastic centrifuge tube with a 5 mL pipette. 6 mL of 0.1% formic acid-acetonitrile solution was added after addition of 0.8 g of sodium chloride. The mixture was shook with a high-speed vortexer for 1 min and then centrifuged at 8000 r/min for 5 min. 4 mL of upper supernatant was transferred into 10 mL plastic centrifuge tube. 500 mg of anhydrous magnesium sulfate, 50 mg of C_{18}, 50 mg of PSA, and 50 mg of aluminia A were added, successively. The mixture was shook with a high-speed vortexer about 1 min and then centrifuged at 10000 r/min for another 5 min. The supernatant was passed through a 0.22 µm nylon membrane filter. Exactly 3 mL of the filtrate was placed in a 10 mL plastic centrifugal tube, which was then placed in the vacuum centrifugal concentrator to drain the liquid at 60°C and 1500 r/min. Residues were re-dissolved with 0.5 mL of 0.1% formic acid aqueous solution-acetonitrile (70:30, v/v) and then centrifuged at 13000 r/min for 5 min. The supernatant was collected in a vial for further detection.

Results and Discussion

Optimization of instrumental parameters

The UPLC-MS/MS is a powerful separation and detection platform in multi residues analysis. The chromatographic separation condition and MS parameters were important factors that needed to optimize in different instrument. In this study, Xevo TQ UPLC-MS/MS was used to determine mycotoxins and their metabolites in throughout study. UPLC separation and MS/MS detection conditions for target analytes were examined.

Methanol-water (50:50, v/v) was used as the mobile phase. Combined sample injection was adopted to optimize mass-spectrometry conditions for the 25 mycotoxins and their metabolites. The appropriate molecular ion peaks and ionization mode were selected after a full scan under the positive- and negative-ion modes. [M+H]⁺, [M+NH₄]⁺, or [M+Na]⁺ were obtained under positive-ion mode for parts of mycotoxins (1-19), and [M-H]⁻ was obtained under negative-ion mode for another parts of mycotoxins (20-25). Based on the scan of the ions in the blank matrix solution and matrix standard solution, the parameters were further optimized, and the characteristic ion pairs collected under MRM mode for each mycotoxin type were identified (Table 2). For isomers featuring identical retention times, i.e., 3-AcDON and 15-AcDON, their respective unique ion fragments were used as product ions for monitoring.

Based on early findings of the experiment [17], 0.1% formic acid aqueous solution and 0.1% formic acid-methanol were used as mobile phases. Gradient elution was carried out, and peak signals of analytes were collected by stages, so the method was applicable to the simultaneous detection of various mycotoxins. Furthermore, isomers with identical ion fragments, i.e., α-ZEL, β-ZEL, and ZAN, as well as α-ZAL and β-ZAL, could be well separated according to their retention times. Figure 1 shows the chromatograms of quantification transitions of the blank pig urine matrix-matched standard.

Optimization of clean-up procedures

Usually, sample preparation is the most crucial step in a simultaneous method due to the different properties of the target analytes that have

No	Analyte	Retention time (min)	Precursor ion (*m/z*)	Product ions (*m/z*)	Cone voltage (V)	Collision energy (eV)	Dwell time (ms)
1	AFB_1	8.27	$313.14[M+H]^+$	285.03*/241.15	60/60	36/22	25
2	AFB_2	7.95	$315.14[M+H]^+$	287.15*/259.13	52/52	28/26	25
3	AFG_1	7.75	$329.14[M+H]^+$	199.94*/214.90	40/40	40/30	25
4	AFG_2	7.41	$331.14[M+H]^+$	189.14*/245.18	64/64	38/28	25
5	AFM_1	7.48	$329.10[M+H]^+$	273.15*/311.12	2/2	22/18	25
6	AFM_2	7.05	$331.10[M+H]^+$	273.14*/285.16	2/2	22/22	25
7	STE	10.85	$325.14[M+H]^+$	253.20*/115.11	66/66	60/42	25
8	T-2	9.65	$484.28[M+NH_4]^+$	305.23*/185.17	6/6	18/12	25
9	LYS	4.34	$255.23[M+H]^+$	240.19*/197.15	48/48	22/20	25
10	MET	5.10	$340.21[M+H]^+$	223.22*/208.07	56/56	28/24	25
11	RC	7.82	$390.21[M+H]^+$	193.12*/322.24	52/52	26/20	25
12	DIA	8.10	$384.24[M+NH_4]^+$	247.22*/105.09	6/6	30/14	25
13	DON	4.46	$297.20[M+H]^+$	249.19*/91.05	28/28	42/10	25
14	3-AcDON	6.78	$339.19[M+H]^+$	231.16*/203.21	2/2	14/10	25
15	15-AcDON	6.78	$339.24[M+H]^+$	321.27*/261.19	8/8	12/6	25
16	NEO	6.11	$400.24[M+NH_4]^+$	185.17*/305.24	2/2	20/12	25
17	WOR	9.05	$447.33[M+NH_4]^+$	345.23*/285.23	4/4	20/16	25
18	VER	10.73	$534.25[M+Na]^+$	392.23*/191.10	44/44	20/12	25
19	HT-2	9.05	$442.34[M+NH_4]^+$	263.20*/215.22	10/10	10/16	25
20	ZEN	10.90	$317.18[M-H]^-$	175.04*/131.03	68/68	30/26	25
21	α-ZEL	10.75	$319.22[M-H]^-$	160.09*/130.12	68/68	36/30	25
22	β-ZEL	10.98	$319.22[M-H]^-$	160.09*/275.24	68/68	30/20	25
23	ZAN	10.36	$319.22[M-H]^-$	275.24*/205.15	68/68	22/20	25
24	α-ZAL	10.72	$321.24[M-H]^-$	277.26*/303.26	68/68	22/24	25
25	β-ZAL	10.13	$321.24[M-H]^-$	277.26*/303.26	68/68	28/24	25

Table 2: Optimized MS/MS parameters of 25 mycotoxins.

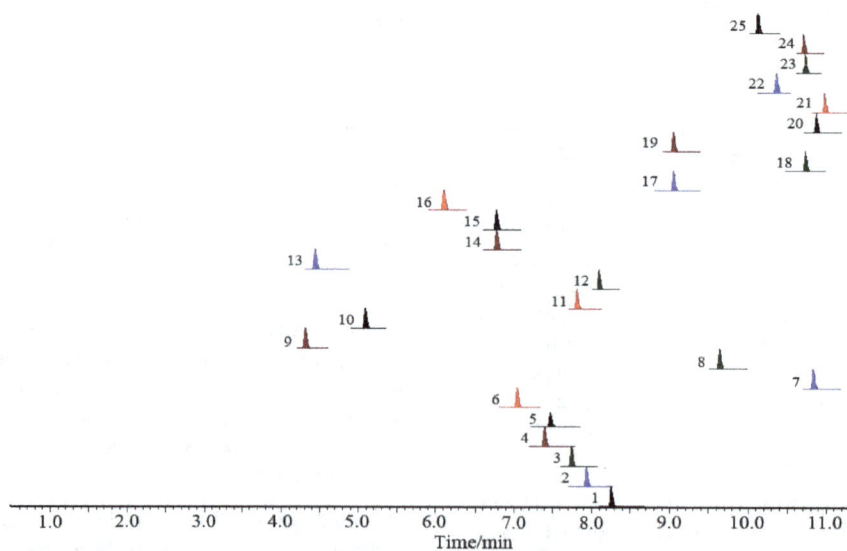

Figure 1: Chromatograms of quantification transitions of the blank pig urine matrix-matched standard (1 ng/g for analyte nos. 1-12; 5 ng/g for 13-25).

to be extracted and cleaned simultaneously. Traditionally, the solid phase extraction and immuno affinity were usually clean-up method for mycotoxins and their metabolite in various samples. In this study, we developed a novel impurity adsorption clean-up method for 25 mycotoxins and their metabolites in urine sample. Firstly, the clean-up effect of C_{18}, PSA, A-AL, and GCB on the target analytes in pig-urine samples was initially evaluated. Results indicated that GCB resulted in poor recovery for most of target analytes due to strong absorbing for mycotoxins and their metabolites. Instead, C_{18}, PSA, and A-AL had no

significant absorption effect, with recovery of 80%-120% for all target mycotoxins and their metabolites. C_{18} could significantly reduce the odor of pig-urine samples and can absorb the pigment substance in the samples. PSA could effectively remove the pigment substance from the samples and exerted an enhancing effect on the mass spectrum peak signals of most analytes. A-AL could significantly decrease matrix interference to the analytes including AFB_1, T-2, DON, and ZEA and also improve detection sensibility, although it cannot effectively remove the odor or pigment substance from the samples.

Then the amount of C_{18}, PSA, and A-AL was optimized by examining the recovery of target analytes and baseline of extraction solution. The results indicated that the optimum amount of C_{18}, PSA and A-AL was 25 mg per millitre of urine sample. The developed sample preparation procedures could obtain clean extracts and consistent chromatographic responses, and in addition, the sample treatment time was reduced.

Evaluation of matrix effects

When ESI is used as the ionization technique in mass spectrometry, one of the main problems is the signal suppression or enhancement of the analytes due to the other components present in the matrix (matrix effect). To evaluate matrix effects, 0.1% formic acid aqueous solution-acetonitrile (70:30, v/v) was used to prepare mixed solvent standard solutions with different concentrations of target analytes (0.05, 0.1, 0.5, 1.0, 5.0, and 20.0 ng/mL for analyte nos. 1-12; 0.25, 0.5, 2.5, 5.0, 25.0 and 100.0 ng/mL for analyte nos. 13-25). Meanwhile, the pig-urine sample was extracted, purified, and analyzed according to the developed impurity adsorption clean-up procedures. After confirming that no trace analyte remained, the blank sample was used to dilute the matrix-matched standard sample that had the same concentration as the matrix-matched standard sample. The slope and correlation coefficient (R^2) of solvent and matrix-matched calibration curves are shown in Table 3. The slope ratios between pig urine matrix-matched and solvent calibration of the target mycotoxins (Figure 2) reflected the intensity of matrix effect, and the range 80%-120% was regarded as acceptable. A range that exceeded 80%-120% indicated a strong matrix suppression effect or a strong matrix-induced response-enhancement effect. The slopes of target compounds were compared (Table 3), and it can be observed that a matrix effect was noticed for some compounds such as AFM_1, DIA, DON, 15-AcDON and 3-AcDON etc. So it was necessary to avoid the matrix effect, matrix-matched calibration standard curves were selected to quantify target compounds in urine. Therefore, blank pig-urine was selected as representative sample matrix to analysis spiked blank samples of urine at concentration level of 5.0 ng/mL. The recoveries of different class compounds were above 80%. It was demonstrated that the developed extraction and clean-up protocol based on impurity adsorption is suitable for the tested mycotoxins and their metabolites.

Analytical performance of the developed method

The blank sample extract was used to prepare the mixed standard solutions with concentrations ranging from 0.05 ng/mL to 20 ng/mL for analytes nos. 1-12 and from 0.25 ng/mL to 100 ng/mL for analytes nos. 13-25. The limit of quantitation (LOQ) of the method was determined according to 10 times of the ratios of signal to noise (S/N). The R^2 that was obtained exceeded 0.99. Results are shown in Table 4.

Recovery studies were performed at three spiked concentration levels (0.5, 5.0 and 50 ng/mL), fortifying six blank pig-urine samples with target analytes at each fortification level. The obtained results are shown in Table 4 and the chromatogram of spiked sample (0.5 ng/mL) was shown in Figure 3. It can be observed that the recoveries of target analytes in pig-urine samples varied from 80.8% to 114.3%. Repeatability (intra-day precision) was evaluated at the three concentration levels of the recovery study, performing six replicates for each level (Table 4), it can be observed that repeatability, expressed as RSD, were lower than 10.3% for all the cases. Whereas reproducibility (inter-day precision) was studied analyzing ten spiked samples at 50 µg/kg during five consecutive days, the RSD were below 12.7%.

Finally the selectivity of the method was evaluated by analyzing control blank samples. The absence of any signal at the same retention time as the selected compounds indicated that there were no matrix interferences that may give a false positive signal (Figure 4).

Application in real urine samples

The developed method was successfully applied to detect target analytes in real samples that were obtained from growing-finishing pigs that were administered DON in different dose by feed (low concentration was 200 µg/kg and high concentration was 2500 µg/kg in feed respectively). The analytical results showed that the low level (133 ng/ml) and high level (1207 ng/ml) of DON were found in low and high concentration diet groups respectively. It was demonstrated that the concentration of DON in pig urine and diet had positive correlation. Meantime, some other mycotoxins (e.g., LYS, ZEA) were found in both groups samples with very low concentration. So, the results indicated that this method can be used for the routine analysis of exposure assessment of the mycotoxins in animal urine samples.

No.	Analyte	Solvent calibration curves		Matrix-matched calibration curves	
		Slope	R^2	Slope	R^2
1	AFB_1	12341.8	0.9993	11264.0	0.9996
2	AFB_2	8711.6	0.9990	8174.8	0.9993
3	AFG_1	11440.5	0.9994	10750.2	0.9996
4	AFG_2	7276.0	0.9992	6264.2	0.9998
5	AFM_1	6212.1	0.9991	4468.1	0.9993
6	AFM_2	20886.7	0.9996	17087.3	0.9999
7	STE	2155.2	0.9998	2333.7	0.9997
8	T-2	1928.4	0.9993	1662.5	0.9996
9	LYS	10975.8	0.9989	11430.2	0.9997
10	MET	9728.3	0.9996	9353.8	0.9998
11	RC	16818.7	0.9997	12530.1	0.9993
12	DIA	2197.7	0.9987	1432.4	0.9996
13	DON	370.2	0.9998	227.8	0.9995
14	3-AcDON	522.7	0.9995	312.3	0.9998
15	15-AcDON	626.6	0.9975	281.8	0.9992
16	NEO	3418.7	0.9988	2238.3	0.9987
17	WOR	283.2	0.9993	203.2	0.9997
18	VER	103.7	0.9990	95.5	0.9993
19	HT-2	262.1	0.9996	172.7	0.9997
20	ZEN	102.5	0.9990	45.7	0.9991
21	α-ZEL	21.7	0.9989	9.6	0.9981
22	β-ZEL	77.4	0.9993	30.3	0.9997
23	ZAN	102.1	0.9991	51.0	0.9995
24	α-ZAL	235.4	0.9993	116.8	0.9997
25	β-ZAL	195.7	0.9989	99.7	0.9996

Table 3: Slope and R^2 of solvent and matrix-matched calibration curves.

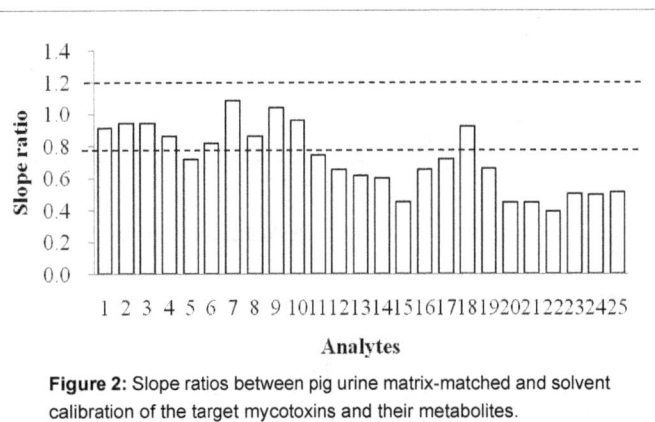

Figure 2: Slope ratios between pig urine matrix-matched and solvent calibration of the target mycotoxins and their metabolites.

No.	Analyte	LOQ (ng/mL)	Concentration range (ng/mL)	Spiked levels (ng/mL)	Recovery (%)	RSD(%, n=6)
1	AFB$_1$	0.05	0.05-20	0.1, 1.0, and 10	96.2-102.4	1.4-3.6
2	AFB$_2$	0.05	0.05-20	0.1, 1.0, and 10	94.0-100.8	1.9-5.5
3	AFG$_1$	0.05	0.05-20	0.1, 1.0, and 10	98.3-107.1	2.4~4.6
4	AFG$_2$	0.05	0.05-20	0.1, 1.0, and 10	91.8-104.2	1.8-3.2
5	AFM$_1$	0.05	0.05-20	0.1, 1.0, and 10	98.3-110.1	0.8-3.3
6	AFM$_2$	0.05	0.05-20	0.1, 1.0, and 10	97.7-106.2	1.5-4.5
7	STE	0.05	0.05-20	0.1, 1.0, and 10	91.2-99.4	2.0-6.4
8	T-2	0.05	0.05-20	0.1, 1.0, and 10	88.1-98.6	2.8-5.9
9	LYS	0.05	0.05-20	0.1, 1.0, and 10	86.1-95.7	1.8-4.5
10	MET	0.05	0.05-20	0.1, 1.0, and 10	90.1-101.3	0.8-2.8
11	RC	0.05	0.05-20	0.1, 1.0, and 10	91.2-97.7	1.1-3.7
12	DIA	0.05	0.05-20	0.1, 1.0, and 10	84.3-95.0	1.8-3.5
13	DON	0.25	0.25-100	0.5, 5.0, and 50	89.1-102.4	2.1-5.7
14	3-AcDON	0.5	0.5-100	0.5, 5.0, and 50	90.4-96.5	3.6-8.9
15	15-AcDON	0.5	0.5-100	0.5, 5.0, and 50	89.1-98.5	2.3-7.8
16	NEO	0.25	0.25-100	0.5, 5.0, and 50	88.0-95.5	0.6-2.7
17	WOR	0.25	0.25-100	0.5, 5.0, and 50	84.6-92.4	2.0-5.7
18	VER	0.25	0.25-100	0.5, 5.0, and 50	88.7-96.8	1.8-5.5
19	HT-2	0.25	0.25-100	0.5, 5.0, and 50	92.5-110.2	2.8-4.4
20	ZEN	0.25	0.25-100	0.5, 5.0, and 50	84.2-95.6	2.4-7.4
21	α-ZEL	0.5	0.5-100	0.5, 5.0, and 50	80.8-94.0	3.8-10.3
22	β-ZEL	0.5	0.5-100	0.5, 5.0, and 50	82.6-96.3	2.5-7.7
23	ZAN	0.5	0.5-100	0.5, 5.0, and 50	88.4-96.5	3.2-9.2
24	α-ZAL	0.5	0.5-100	0.5, 5.0, and 50	86.1-105.2	4.2-8.6
25	β-ZAL	0.5	0.5-100	0.5, 5.0, and 50	92.1-114.3	3.6-7.1

Table 4: Performance parameters of the method, including LOQ, concentration range, average recoveries, and relative standard deviation (RSD), of 25 target mycotoxins in pig-urine sample.

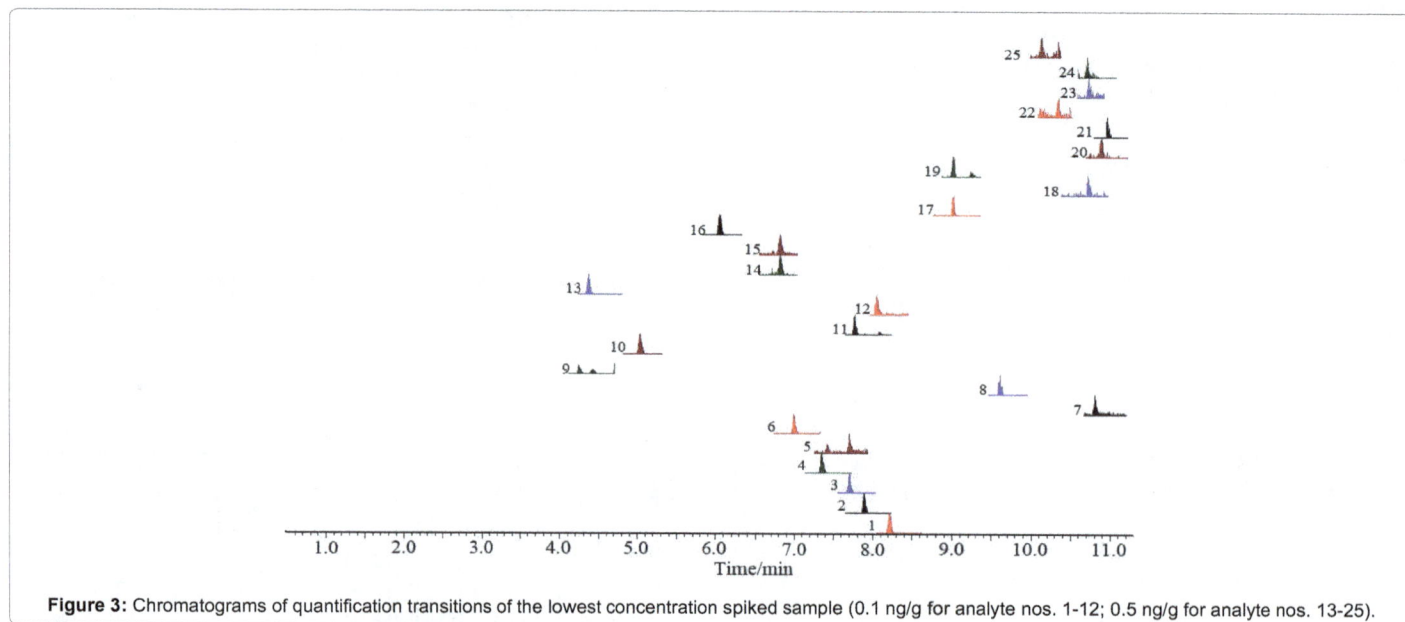

Figure 3: Chromatograms of quantification transitions of the lowest concentration spiked sample (0.1 ng/g for analyte nos. 1-12; 0.5 ng/g for analyte nos. 13-25).

Conclusions

In conclusion, a novel simultaneous detection method for mycotoxins and their metabolites in animal urine by using a simple preparation method based on impurity adsorption mechanism followed UPLC-MS/MS detection was presented. The developed sample preparation procedures could obtain clean extracts and consistent chromatographic responses, and in addition, the sample treatment time was reduced. The compounds examined in this study represent a wide range of physico-chemical properties indicating the potential of impurity adsorption for the extraction of target analytes in animal urine samples.

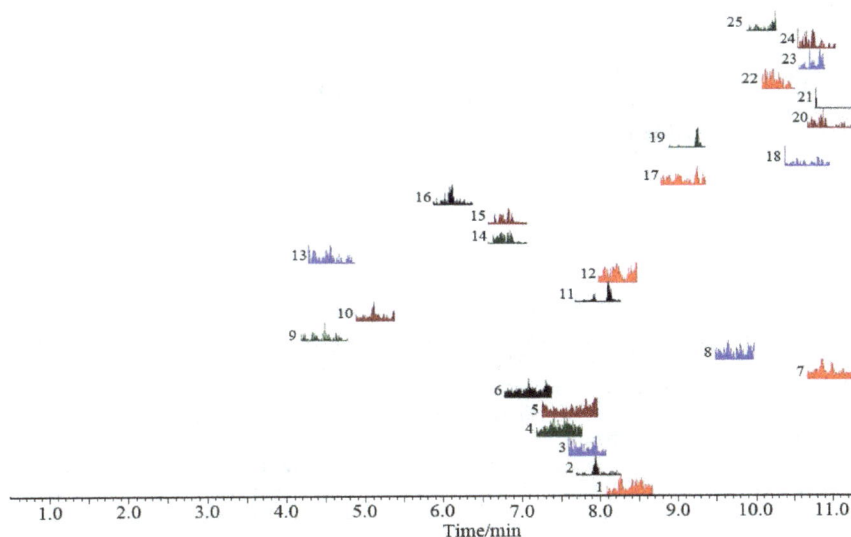

Figure 4: Chromatograms of quantification transitions of the blank pig-urine sample.

Acknowledgements

The authors would like to acknowledge the China National S&T Project (2011BAD26B0405), Special Fund for Agro-scientific Research in the Public Interest (No. 201203088) and S&T innovation project of Chinese Academy of Agricultural Sciences for financially supporting this research.

References

1. Cao X, Wu S, Yue Y, Wang S, Wang Y, et al. (2013) A high-throughput method for the simultaneous determination of multiple mycotoxins in human and laboratory animal biological fluids and tissues by PLE and HPLC–MS/MS. J Chromatogr B Analyt Technol Biomed Life Sci 942-943: 113-125.

2. Zheng CM, Zhang Y, Wang XS, Xie G, Zhang GM, et al. (2012) Study on application of a liquid chromatography-mass spectrometric technology in simultaneous determination of multi-mycotoxin in grain. Grain Science and Technology and Economy China 37: 45-49.

3. Marin S, Ramos AJ, Cano-Sancho G, Sanchis V (2013) Mycotoxins: occurrence, toxicology, and exposure assessment. Food Chem Toxicol 60: 218-237.

4. Binder EM, Tan LM, Chin LJ, Handl J, Richard J (2007) Worldwide occurrence of mycotoxins in commodities, feeds and feed ingredients. Animal feed science and technology 137: 265-282.

5. Bennett J W, Klich M (2013) Mycotoxins. Clin Microbiol Rev 16: 497-516.

6. World Health Organization (2003) WHO global strategy for food safety: safer food for better health. 20 Avenue Appia, 1211 Geneva 27, Switzerland.

7. GB 2761-2011 (2011) Maximum levels of mycotoxins in foods, China.

8. GB 13078-2001 (2001) Hygienical standard for feeds, China.

9. GB 13078.2-2006 (2006) Hygienical standard for feeds-Toleration of ochratoxin A and zearalenone in feeds, China.

10. GB 13078.3-2007 (2007) Tolerance limits for deoxynivalenol in formula feeds, China.

11. GB 21693-2008 (2008) Tolerance limits for T-2 toxin in formula feeds, China.

12. Zachariasova M, Dzuman Z, Veprikova Z, Hajkova K, Jiru M, et al. (2014) Occurrence of multiple mycotoxins in European feedingstuffs, assessment of dietary intake by farm animals. Animal Feed Science and Technology 193: 124-140.

13. Turner PC, Hopton RP, White KLM, Fisher J, Cade JE, et al. (2011) Assessment of deoxynivalenol metabolite profiles in UK adults. Food Chem Toxicol 49: 132-135.

14. Cunha SC, Fernandes JO (2012) Development and validation of a gas chromatography–mass spectrometry method for determination of deoxynivalenol and its metabolites in human urine. Food Chem Toxicol 50: 1019-1026.

15. Meky FA, Turner PC, Ashcroft AE, Miller JD, Qiao YL, et al. (2003) Development of a urinary biomarker of human exposure to deoxynivalenol. Food Chem Toxicol 41: 265-273.

16. Monbaliu S, Van Poucke C, Detavernier C, Dumoulin F, Van De Velde M, et al. (2010) Occurrence of mycotoxins in feed as analyzed by a multi-mycotoxin LC-MS MS method. J Agric Food Chem 58: 66-71.

17. Wang RG, Su XO, Cheng FF, Wang PL, Fan X, et al. (2015) Determination of 26 Mycotoxins in Feedstuffs by Multifunctional Clean-up Column and Liquid Chromatography-Tandem Mass Spectrometry. Chinese Journal of Analytical Chemistry 43: 264-270.

20

Stability Indicating HPLC Method for Simultaneous Estimation of Entacapone, Levodopa and Carbidopa in Pharmaceutical Formulation

Bhatnagar P[1,2], Vyas D[1], Sinha SK[2] and Chakrabarti T[1]

[1]Sir Padampat Singhania University, Udaipur, India
[2]Macleods Pharmaceuticals Limited, Mumbai, India

Abstract

In pharmaceutical industry, researchers aim at catering to the need of robust analytical methods for analysis of generic drug products. The paper deals with the pharmaceutical formulation - Entacapone, Levodopa and Carbidopa tablets for the treatment of Parkinson's disease. The paper presents a simple and efficient stability indicating HPLC method that has been developed in a multi component drug formulation for simultaneous estimation of Entacapone, Levodopa and Carbidopa in presence of their related impurities. This HPLC method uses 'Cosmosil PE 150 × 4.6 mm, 5 μ' HPLC column, phosphate buffer pH 2.5 and Methanol as mobile phase in gradient mode with UV detection at 280 nm. The method was validated and found to be precise, robust, accurate, linear (in range 0.05 to 0.15 mg/ml, 0.012 to 0.15 mg/ml and 0.003 to 0.038 mg/ml of Entacapone, Levodopa and Carbidopa respectively), and specific for 15 known impurities ensuring suitability of the method for quantitative determination of Entacapone, Levodopa and Carbidopa.

Keywords: Pharmaceutical formulation; HPLC method; Simultaneous estimation; Assay test; Multi component drug formulation; Parkinson

Introduction

Parkinson's disease is a progressive, neurodegenerative disorder of the extrapyramidal nervous system affecting the mobility and control of the skeletal muscular system. Symptoms of Parkinson's disease are related to depletion of dopamine. But administration of dopamine is ineffective in the treatment of Parkinson's disease. This is because it does not cross the blood-brain barrier. However, levodopa, the metabolic precursor of dopamine, does cross the blood-brain barrier, and presumably is converted to dopamine in the brain. Carbidopa inhibits the decarboxylation of peripheral levodopa, making more levodopa available for transport to the brain. Entacapone is a selective and reversible inhibitor of catechol-O-methyltransferase (COMT). When entacapone is given in conjunction with levodopa and carbidopa, plasma levels of levodopa are greater and more sustained than after administration of levodopa and carbidopa alone.

It is very difficult to develop a stability indicating method for such triple combination drug products that is capable of analyzing each active ingredient in presence of their related impurities.

Literature survey revealed few methods for individual or combination product analysis, Spectroscopic methods for simultaneous estimation of Levodopa and Carbidopa [1], and LC estimation of Entacapone in tablets [2]. A method for in-vitro release of drugs is also found [3-5]. Publications were found on LC method for estimation using electrochemical detector [6] and LC method for estimation of levodopa and carbidopa using fluorescence detector [7-18]. Few official pharmacopial monographs for single and dual drug combinations [19-24] were also found. In the present study, we propose a rapid and stability indicating HPLC method for simultaneous estimation of Levodopa [(2S)-2-amino-3-(3,4-dihydroxyphenyl) propanoic acid], Carbidopa [(2S)-3-(3,4-dihydroxyphenyl)-2-hydrazino-2-methylpropanoic acid] and Entacapone [(2E)-2-cyano-3-(3,4-dihydroxy-5-nitrophenyl)-N,N-diethyl-2-propenamide] in presence of their 15 process related or degradation impurities [25-29].

Materials and Methods

Reagents and materials

All analytical reagent grade (AR Grade) reagents were used for method development purpose. Acetonitrile (Merck) and Tetrahydrofuran (Merck) were used for standard and sample solution preparation. Orthophosphoric acid (Rankem) and Potassium dihydrogen orthophosphate (Merck) were used for mobile preparation. Milli-Q water (HPLC grade) was used for all solution preparations. Impurities and working standards of Entacapone, Levodopa and Carbidopa were obtained from Macleods Pharmaceuticals Limited, Mumbai, India.

Chromatographic system and conditions

Development study was performed on Shimadzu HPLC, consisting of UV-Visible, photodiode array detector and a quaternary gradient pump. Sample loop in the system was of 100 μl capacity. Cosmosil 5PE 150 × 4.6 mm, 5 μ (Nacalai Tesque, USA) HPLC column was used for chromatographic separation. Mobile phase consisted of phosphate buffer and methanol in gradient mode. Buffer consisted of 10mM potassium dihydrogen orthophosphate solution with pH adjusted to 2.5 using orthophosphoric acid. Flow rate was 1.0 mL/min and detection was carried out at 280 nm based on there wavelength maxima as per UV spectrum. Labsolutions software was used for data collection. For intermediate precision study, Agilent HPLC system

*Corresponding author: Prasoon Bhatnagar, Sir Padampat Singhania University, Udaipur, India, E-mail: prasoon.bhatnagar@spsu.ac.in

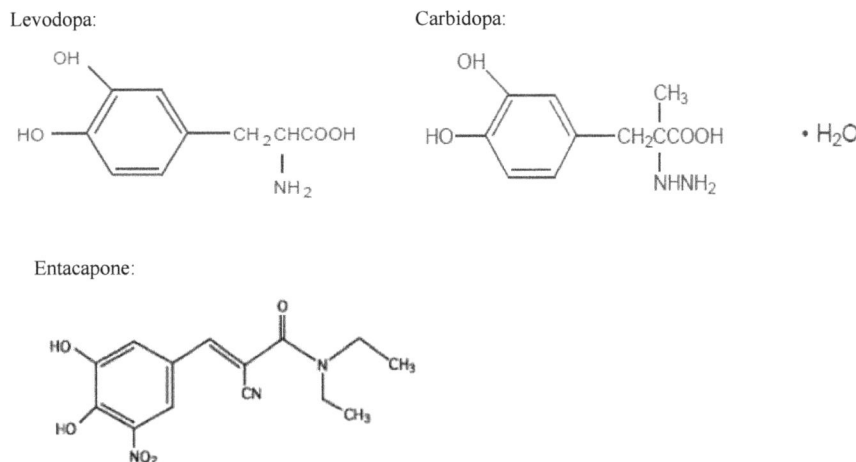

Figure 1: Chemical structures of analyzed substances.

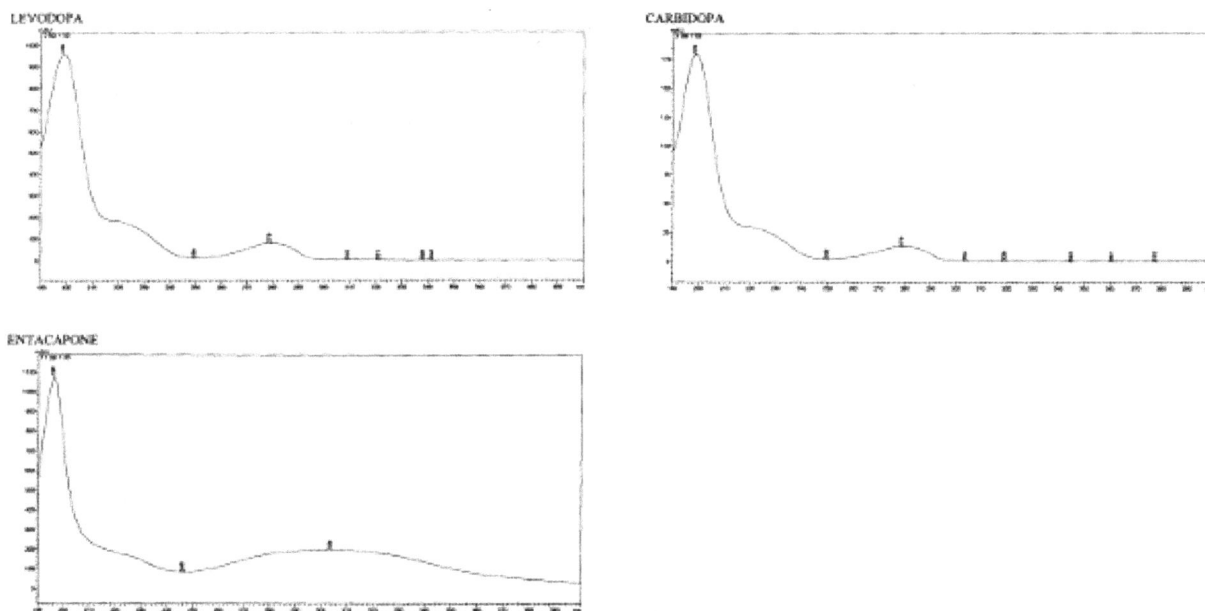

Figure 2: UV Spectrum.

with gradient pump, UV-visible detector and Chemstation software was used (Figure 1).

Results and Discussion

Preliminary studies

There is no pharmacopoeial or literature reference of a suitable stability indicating assay test for the proposed triple combination formulation. The development of Assay method was initiated from USP method for Levodopa tablets which use a simple HPLC method including octadecyl stationary phase and phosphate buffer pH 2.0/ acetonitrile for mobile phase.

Selection of wavelength: Wavelength was selected based on absorbance maxima of three drugs as per UV spectrum. 280 nm was optimum for all the active ingredients (Figure 2).

Selection of mobile phase: Due to difference in acidity of levodopa/ carbidopa and entacapone, low pH was selected to achieve optimum separation of all the peaks. Looking at the pH range of HPLC column, pH 2.5 was evaluated and found to be optimum.

Selection of HPLC column: Entacapone elute late on a C_{18} column even with 60% solvent in mobile phase (Figure 3). Levodopa and Carbidopa are polar in nature which makes them elute early on a non polar octadecyl (Inertsil ODS 250 mm × 4.6 mm, 5 μ) phase. In order to elute Entacapone early, a more polar phase was evaluated and selected for method development. Cosmosil PE, 150 × 4.6 mm, 5 μ was the column of choice. Phenyl phase is polar in nature but do not last long at low pH due to its weak bonding. Cosmosil PE column has an ethyl group attached to phenyl group which makes this column a rugged stationary phase with better column life than phenyl column. A 150 mm column was chosen to achieve a shorter run time.

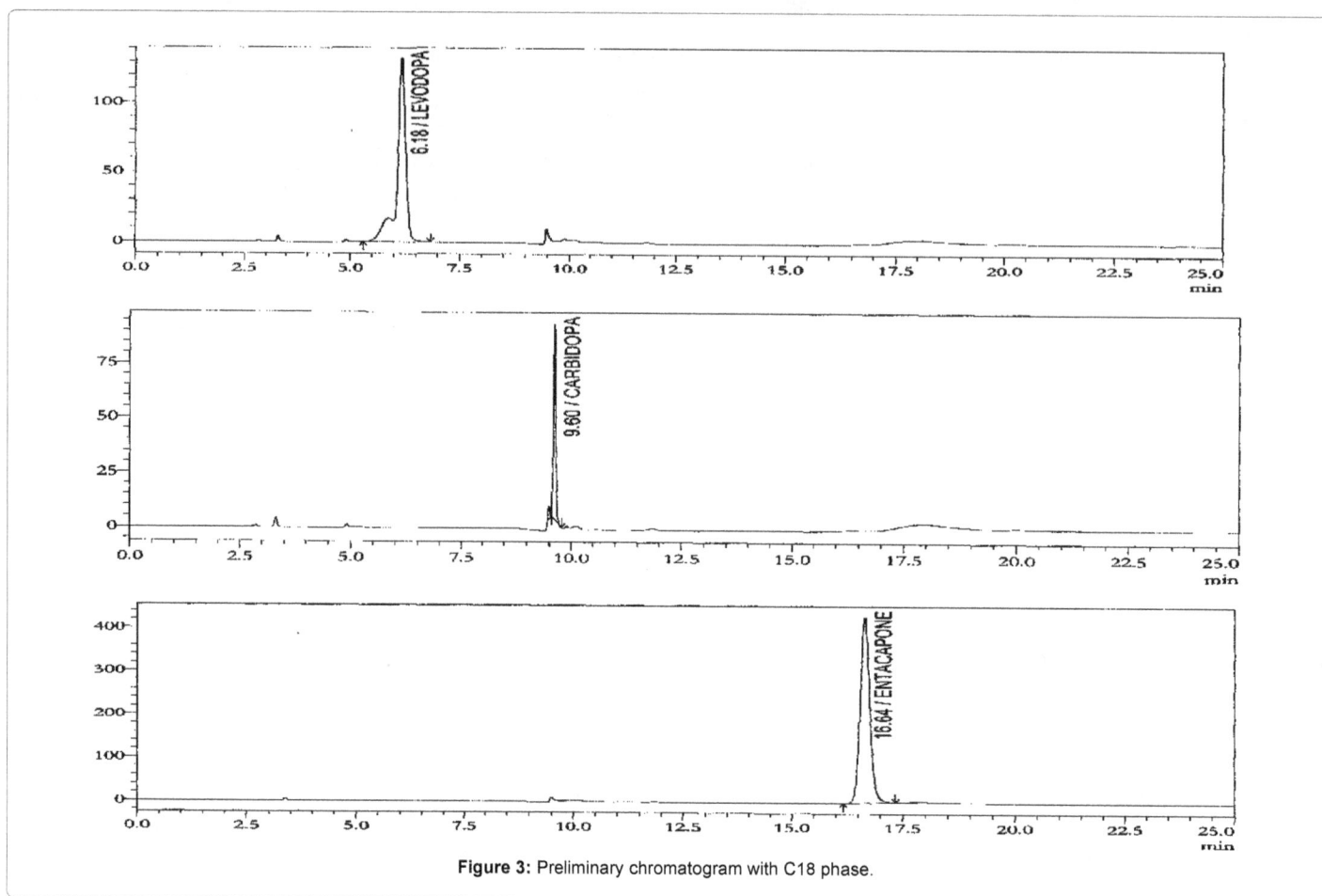

Figure 3: Preliminary chromatogram with C18 phase.

Time (min)	Phosphate Buffer (% v/v)	Methanol (% v/v)
0 → 5	98	2
5 → 6	98→40	2→60
6 → 13	40	60
13 → 14	40→98	60→2
14 → 18	98	2

Table 1: Gradient time program.

Selection of HPLC pump mode: Entacapone do not elute early with a low solvent mobile phase. Hence, gradient mode was chosen and optimized for separation of active ingredients with a flow rate of 1 ml/min and run time of 17 minutes (Table 1).

Selection of diluent: The difference in solubility of the active ingredients makes it difficult to finalize an optimum diluent. Levodopa and Carbidopa dissolves in acidic and aqueous condition whereas entacapone dissolves in less polar solvent like acetonitrile. Looking at the difference in solubility, a combination of 1% orthophosphoric acid and acetonitrile in the ratio 60:40 (diluent 1) was suggested for stock solution preparation. But entacapone has a tendency to precipitate on standing with acetonitrile and methanol in diluent. For better solubility of entacapone and stability of solutions with improved peak shapes, second dilution was performed in 1% orthophosphoric acid and tetrahydrofuran in the ratio 80:20 (diluent 2). Higher percent of tetrahydrofuran is not recommended due to its corrosive nature.

Solution preparation

Standard preparation: About 50 mg of Entacapone and about 31.25 mg of Levodopa was accurately weighed and dissolved in 50 ml of Diluent 1 (Solution A). About 31.25 mg of Carbidopa was dissolved in 100 ml of diluent 2 (Solution B). Further, 10 ml of (Solution A) and 5 ml of (Solution B) was diluted to 100 ml with diluent 2.

Sample Preparation: To prepare the sample, 5 intact tablets were transferred to a volumetric flask of 500 mL; 250 ml of diluent (1) was added to it and was sonicated for 30 minutes to dissolve. It was then cooled to room temperature and made up to mark with the same diluent. Filtered through 0.45 µ nylon filter, and further diluted 5 ml of the above solution to 100 mL with diluent (2), and mixed.

Method validation

Once optimum separation conditions are achieved, method is validated to ensure its suitability and reliability for routine use in estimation of % content of active ingredients in a pharmaceutical formulation. Validation parameters adopted are as follows:

Specificity: Specificity for blank, placebo, and known impurities was established by injecting known concentration of impurity solutions

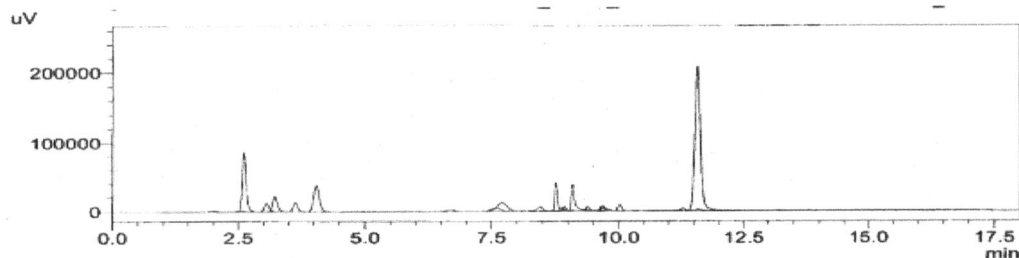

Figure 4: Overlaid Chromatogram of impurity solution for specificity.

Sr. No.	Sample Details	Retention Time (min)
1	Blank	No peak detected
2	Placebo Solution	No peak detected
3	Levodopa Related compound A	2.4
4	L-Tyrosine	3.1
5	Methyldopa	3.3
6	3-methoxy tyrosine	4.1
7	Methyldopa Methyl ester	7.6
8	Carbidopa Related compound A	7.9
9	1-veratryglycine	8.9
10	3-Acetyl-L-Tyrosine	9.0
11	3,4-Dihydoxyphenylacetone	9.2
12	Entacapone Impurity C	9.7
13	Entacapone Impurity F	9.7
14	Cyclohexylidine Carbidopa -Methyl Ester	10.0
15	Entacapone De-nitro	10.4
16	Entacapone Impurity A	11.3
17	Entacapone Methoxy Impurity	13.2
18	Levodopa	2.7
19	Carbidopa	3.9
20	Entacapone	11.8

Table 2: Values of retention time obtained.

(Figure 4). Specificity for unknown impurities was established by performing forced degradation study on tablet formulation as shown in Table 2. Peaks of interest were subjected to peak purity assessment using photodiode detector. All the peaks were found to be spectrally pure and no co-elution of any impurity was observed.

As shown in Table 3, No interference from blank, placebo and known impurities was observed at retention times of Entacapone, Levodopa and Carbidopa peaks.

Solution stability: Solution stability was evaluated by storing solutions at 25°C and 10°C. Carbidopa degraded by 2% at 25°C, whereas, solutions were found to be stable till 24 h when stored at 10°C (Tables 4 and 5).

Accuracy: Since Entacapone, Levodopa, and Carbidopa tablets have 7 strengths [(200+200+50), (200+175+43.75), (200+150+37.5), (200+125+31.25), (200+100+25), (200+75+18.75), and (200+50+12.5)], accuracy study was performed at 50% of the lowest concentration and 150% of the highest concentration of individual active ingredient. Recovery solutions were prepared by spiking Entacapone; Levodopa and Carbidopa API to placebo powder to obtain solutions of desired concentration (Table 6).

Linearity: A series of solutions were prepared by quantitative dilutions of the stock solution of standard to obtain solutions as mentioned in the following table. Each solution was injected and the peak area was recorded. Slope, Y-intercept and Correlation coefficient of the regression line were calculated (Table 7). In above, 200+200+50 mg strength was taken into consideration. By establishing linearity in entire working range, samples of all the 7 strengths can be analyzed against a single standard corresponding to any strength.

Precision

Repeatability: Six sample preparations were prepared and injected. The mean and relative standard deviation of the results was calculated. The results obtained for assay are tabulated in Table 8.

Intermediate precision: For intermediate precision analysis was carried out different day, using a different HPLC and different column. The absolute difference between the mean assay results obtained in repeatability and intermediate precision was calculated (Figure 5).

The obtained results for % assay and overall comparative data presented in the following Table 9.

The absolute difference between the mean assay results obtained in repeatability and intermediate precision is within the acceptance criteria of not more than 2.0. Hence, the method is precise.

Robustness: The Assay method was carried out as described

Forced Degradation Condition	% Degradation	% Impurity
Acid Hydrolysis: Exposure to acidic condition with 5M hydrochloric acid	About 5.7% Degradation of carbidopa observed	Methyldopa: About 1.16% DHP: About 1.16% Total Impurity: 5.34%
Base Hydrolysis: Exposure to basic condition with 5M sodium hydroxide	About 5.5% Degradation of carbidopa observed	Methyldopa: About 1.06% DHP: About 3.18% Total Impurity: 4.24%
Oxidative Degradation: Exposure to Oxidative condition with 3% hydrogen peroxide	About 9.5% Degradation of carbidopa observed	Methyldopa: About 1.20% DHP: About 2.24% Unknown Impurity about 6.5% Total Impurity: 9.94%
Thermal Degradation: Exposure to 80°C for 24 hrs	About 2.02% Degradation of carbidopa observed	Methyldopa: About 2.83% Total Impurity: 2.83%
Photostability: Exposure to UV Radiation NLT 1.2 million lux hours	About 1.6% Degradation of carbidopa observed	Methyldopa: About 2.13% Total Impurity: 2.13%
Humidity Degradation: Exposure to 40°C temperature and 75% Relative humidity	About 5.6% Degradation of carbidopa observed	Methyldopa: About 2.7% DHP: About 3.8% Total Impurity: 6.5%

Table 3: Observations of forced degradation study.

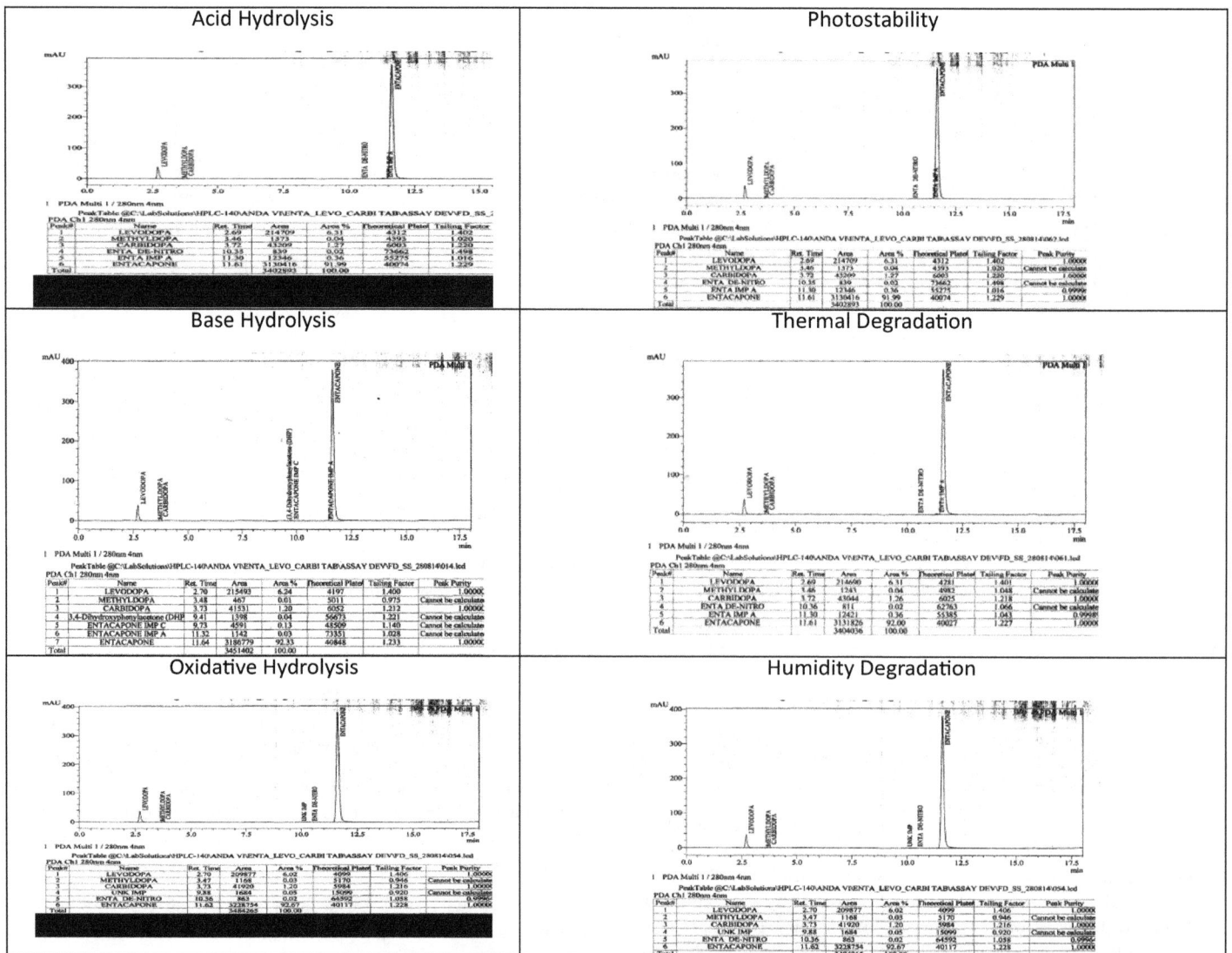

Table 4: Chromatograms of Forced Degradation.

Time (hours)	Entacapone				
	AT 25°C		AT 10°C		
				Area	% Difference w.r.t. Initial
Initial	3163365	-		3208318	-
24 hours	3161080	0.07		3199728	0.26
Time (hours)	Levodopa				
	AT 25°C		AT 10°C		
				Area	% Difference w.r.t. Initial
Initial	213017	-		218864	-
24 hours	214085	-		218927	-
Time (hours)	Carbidopa				
	AT 25°C		AT 10°C		
				Area	% Difference w.r.t. Initial
Initial	44007	-		43818	-
24 hours	42816	2.71		43589	0.52

Table 5: Observation of solution stability.

| | % Recovery | | |
Level	Entacapone	Levodopa	Carbidopa
50%	99.2	100.0	98.7
100%	99.4	100.2	99.0
150%	99.6	98.5	99.2

Table 6: Accuracy results of the proposed method.

| Entacapone | | | Levodopa | | | Carbidopa | | |
% Level	Concen-tration (ppm)	Area	% Level	Concen-tration (ppm)	Area	% Level	Concen-tration (ppm)	Area
50	50.33	1644591	12	12.08	105951	12	3.05	19826
60	60.39	1974936	25	24.17	211972	24	6.09	40184
66	66.43	2161375	50	50.35	436302	50	12.69	83963
75	74.48	2445512	75	74.52	631283	75	18.78	124561
100	100.65	3276792	100	100.70	860553	100	25.38	167461
120	120.78	3943031	120	120.84	1020742	120	30.46	201900
150	150.98	4926510	150	151.05	1295253	150	38.07	253235
Slope	2590.1783		Slope	8494.4494		Slope	6650.2203	
Y-Intercept	5024.5917		Y-Intercept	4069.0608		Y-Intercept	493.9482	
Correlation coefficient		1.000	Correlation coefficient		0.9998	Correlation coefficient		1.0000

Table 7: Linearity results of the proposed method.

| | % Assay | | |
	Entacapone	Levodopa	Carbidopa
Sample-1	98.8	100.9	95.6
Sample-2	99.4	101.0	96.3
Sample-3	99.3	101.2	95.5
Sample-4	98.8	101.1	95.7
Sample-5	99.0	101.4	96.4
Sample-6	98.5	101.3	96.1
Mean	98.96	101.15	95.9
% RSD	0.06	0.03	0.03

Table 8: Precision results of the proposed method.

in the methodology and by making the following alterations in the chromatographic conditions.

> Changing the flow rate of mobile phase (0.8 mL/min, 1.2 mL/min)

> Changing the pH of buffer of mobile phase (pH=2.3, pH=2.7)

The results for system suitability are presented in Tables 10-12.

The system suitability parameters like % relative standard deviation for five replicate injections of standard solution, tailing factor and theoretical plates were not significantly changed with altered conditions. Hence, it is concluded that, the method is robust to deliberate changes made in analytical method.

Conclusion

A simple and efficient stability indicating HPLC method for simultaneous estimation of Entacapone, Levodopa and Carbidopa in

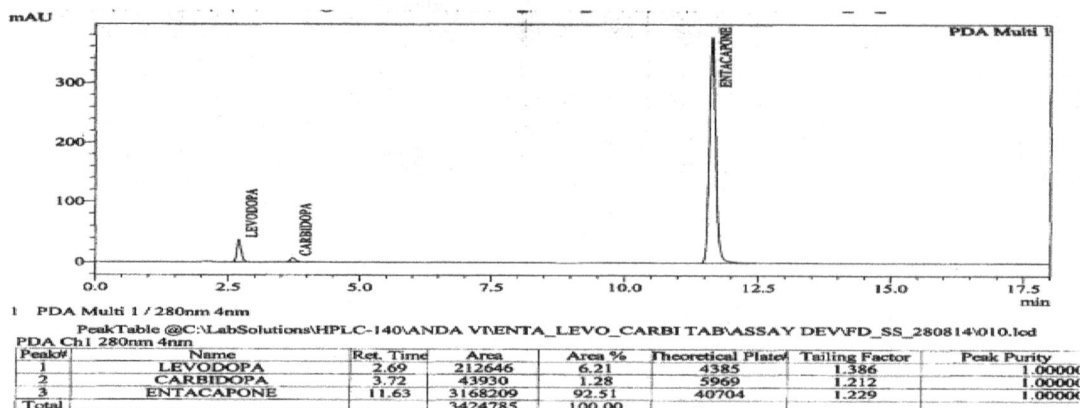

Figure 5: Representative chromatogram of sample.

	% Assay		
	Mean Assay in Repeatability (Precision)	Mean Assay in Intermediate Precision	Absolute difference
Entacapone	98.96	100.15	1.19
Levodopa	101.15	101.18	0.03
Carbidopa	95.9	96.6	0.6

Table 9: Comparison of precision and intermediate precision results of the proposed method.

Altered condition	Retention time (min)	Tailing Factor	Theoretical plates	% RSD	%Assay
Normal (unaltered) (Repeatability)	11.6	1.19	44742	0.08	98.9
Flow rate of mobile phase (0.8 mL/min)	12.94	1.20	43095	0.08	100.5
Flow rate of mobile phase (1.2 mL/min)	10.61	1.18	47443	0.09	99.4
pH of buffer of mobile phase (pH=2.3)	10.30	0.98	71000	0.07	100.0
pH of buffer of mobile phase (pH=2.7)	10.30	1.01	69703	0.03	100.8

Table 10: Robustness results of the proposed method (For Entacapone).

Altered condition	Retention time (min)	Tailing Factor	Theoretical plates	% RSD	%Assay
Normal (unaltered) (Repeatability)	2.60	1.37	5402	0.15	101.15
Flow rate of mobile phase (0.8 mL/min)	3.22	1.36	5928	0.22	101.9
Flow rate of mobile phase (1.2 mL/min)	2.18	1.32	4493	0.12	102.3
pH of buffer of mobile phase (pH=2.3)	2.37	1.00	7508	0.11	100.4
pH of buffer of mobile phase (pH=2.7)	2.33	1.04	7097	0.08	99.4

Table 11: Robustness results of the proposed method (For Levodopa).

Altered condition	Retention time (min)	Tailing Factor	Theoretical plates	% RSD	%Assay
Normal (unaltered) (Repeatability)	3.62	1.19	6330	0.10	95.9
Flow rate of mobile phase (0.8 mL/min)	4.51	1.18	6907	0.12	96.9
Flow rate of mobile phase (1.2 mL/min)	3.04	1.17	5204	0.12	96.6
pH of buffer of mobile phase (pH=2.3)	3.27	0.94	6436	0.94	96.0
pH of buffer of mobile phase (pH=2.7)	3.05	1.01	6201	0.09	96.5

Table 12: Robustness results of the proposed method (For Carbidopa).

presence of 15 impurities has been developed. Method was validated for specificity, accuracy, linearity, precision and robustness ensuring suitability of the method for quantitative analysis. The results indicated that this method is suitable for simultaneous estimation of Entacapone, Levodopa and Carbidopa in a pharmaceutical formulation.

Acknowledgements

The authors express sincere gratitude to the Research & Development Center of Macleods Pharmaceuticals Limited, Mumbai, India, for granting permission to use the Analytical Research and Development section for this research work. The first author is also thankful to his colleagues for their consistent support to carry out this research.

References

1. Madrakian T, Afkhami A, Borazjani M, Bahram M (2004) Simultaneous derivative spectrophotometric determination of levodopa and Carbidopa in pharmaceutical preparations. Bull Korean Chem Soc 25: 1764-1768.

2. Paim CS, Gonc HM, Miron D, Sippel J, Steppe M (2007) Stability-indicating LC determination of Entacapone in tablets. Chromatographia 65: 595-599.

3. Doshi AS, Upadhyay KJ, Mehta TN, Nanda N (2009) Development and application of a high- performance liquid chromatographic method for the determination of in vitro drug release of levodopa, Carbidopa, and Entacapone from a tablet formulation. JAOAC Int 92: 394- 403.

4. Gadallah MN, Mohamed MS (2010) Determination of antiparkinsonism drug Entacapone. J Chil Chem Soc 55: 85-89.

5. Paim CS, Martins MT, Malesuik MD, Steppe M (2010) LC determination of entacapone in tablets: in vitro dissolution studies. J Chromatogr Sci 48: 755-759.

6. Bugamelli F, Marcheselli C, Barba E, Raggi MA (2011) Determination of L-dopa, carbidopa, 3-O-methyldopa and entacapone in human plasma by HPLC-ED. J Pharm Biomed Anal 54: 562-567.

7. Prafullachandra T, Vinayak MS, Pai NR, Chandrabhanu M, Tekale S (2011) Estimation of Entacapone tablets by reverse phase high performance liquid chromatographic method. Bioscience Discovery 2: 294-298.

8. Issa YM, Hassoun ME, Zayed AG (2011) Application of high performance liquid chromatographic method for the determination of levodopa, Carbidopa, and Entacapone in tablet dosage forms. Journal of Liquid Chromatography & Related Technologies 34: 2433-2447.

9. Prafullachandra T, Smruti T, Vinayak SM (2011) Determination of impurities in formulated form of entacapone by using rp - hplc method. Der Pharma Chemica 3: 63-68.

10. Chalikwar SS, Shirkhedkar AA, Bagul MA, Jain PS, Surana SJ (2012) Development and validation of zero and first-order derivative area under curve spectrophotometric methods for the determination of entacapone in bulk material and in tablets. Pharm Methods 3: 14-17.

11. Bujji BN, Srinivasa Rao P, Ramesh RR (2012) Development of new robust rp-hplc method for analysis of levodopa in formulations. IJSIT 1: 130-143.

12. DR Kumar, Ravi R, Subburaju T, Revathi H, Arul C, et al. (2013) Development and validation of stability indicating assay method for levodopa and carbidopa in levodopa, carbidopa and entacapone ER tablets. International Journal of Pharmaceutical Research and Development 4: 98-105.

13. Kumar DR, Ravi R, Subburaju T, Revathi H, Arul C, et al. (2013) Development and validation of stability-indicating rp-hplc assay method for Entacapone in entacapone tablets. International Journal of Pharmaceutical Sciences and Research 4: 1227-1232.

14. Sravanthi D, Anusha M, Madhavi S, Shaik F, Buchi N (2013) Simultaneous estimation of levodopa and Carbidopa in bulk, pharmaceutical dosage forms and dissolution sample analysis by RP-HPLC-PDA method. Journal of Chemical and Pharmaceutical Research 5: 422-428.

15. Raja AM, Shyamsunde CH, Banji D, Rao KNV, Sevalkumar (2013) Analytical method development and validation for simultaneous estimation of Carbidopa, levodopa and Entacapone in its bulk drug and tablet dosage form by UPLC. International Research journal of Pharmacy 4: 53-56.

16. Jain R, Jain N, Jain DK, Jain SK (2013) A Novel Approach using Hydrotropic Solubalization Technique for Quantitative Estimation of Entacapone in Bulk Drug and Dosage Form. Adv Pharm Bull 3: 409-413.

17. Vemić A, Jančić Stojanović B, Stamenković I, Malenović A (2013) Chaotropic agents in liquid chromatographic method development for the simultaneous analysis of levodopa, carbidopa, entacapone and their impurities. J Pharm Biomed Anal 77: 9-15.

18. Raut PP, Charde SY (2014) Simultaneous estimation of levodopa and carbidopa by RP-HPLC using a fluorescence detector: its application to a pharmaceutical dosage form. Luminescence 29: 762-771.

19. Monograph for levodopa and carbidopa orally disintegrating tablets (2014) USP 37 NF 32: 2134-2136.

20. Monograph for levodopa and carbidopa extended release tablets (2014) USP 37 NF 32: 2129-2134.

21. Monograph for levodopa and Carbidopa tablets (2014) British pharmacopoeia 3: 357-358.

22. Monograph for levodopa and Carbidopa tablets (2014) Indian pharmacopoeia 2081-2084.

23. Monograph for Entacapone tablets (2014) USP 37 NF 32: 2801-2802.

24. Monograph for levodopa tablets (2014) USP 37 NF 32: 3535-3536.

25. ICH guidance (2006) Impurities in new drug products, Q3B (R2), June 2006.

26. ICH guidance (2005) Validation of analytical procedures, text and methodology Q2 (R1).

27. ICH guidance (2006) Impurities in New Drug Substances, Q3A (R2).

28. Lloyd RS, Josheph JK, Josheph LG, Handbook of "Practical HPLC method development". 2nd edn.

29. Stalevo (carbidopa, levodopa and entacapone) tablets, prescribing information, Novartis pharmaceuticals limited.

21

Stability Indicating LC Method for the Estimation of Benazepril HCl and Hydrochlorthiazide in Pharmaceutical Dosage Form

Chhalotiya UK*, Varsha LP, Dimal AS, Kashyap KB and Sunil LB

Department of Pharmaceutical Analysis, Indukaka Ipcowala College of Pharmacy, Gujarat, India

Abstract

A rapid, specific and sensitive reverse phase high performance liquid chromatographic method has been developed and validated for analysis of benazepril hydrochloride and hydrochlorothiazide in both bulk and pharmaceutical dosage form. A sunfire C-18, 250×4.6 mm i.d. and 5 μm particle size column with mobile phase containing water: methanol (55:45, v/v, pH 7). The flow rate was 1.0 mL min^{-1} and effluents were monitored at 233 nm. The retention time of benazepril hydrochloride and Hydrochlorthiazide was 9.19 min and 3.10 min respectively. Benazepril hydrochloride and hydrochlorthiazide was subjected to acid and alkali hydrolysis, chemical oxidation, wet hydrolysis, dry heat degradation and sun light degradation. The degraded product peaks were well resolved from the pure drug peak with significant difference in their retention time values. Stressed samples were assayed using developed LC method. The proposed method was validated with respect to linearity, accuracy, precision and robustness. The method was successfully applied to the estimation of benazepril hydrochloride and hydrochlorthiazide in tablet dosage forms.

Keywords: Benazepril hydrochloride; Hydrochlorthiazide; Liquid chromatography; Forced degradation; Validation

Introduction

Benazepril hyrochloride (BEN) is chemically 3-[[1-(ethoxycarbonyl)-3-phenyl-(1S)-propyl] amino]-2,3,4,5-tetrahydro-2-oxo-1H-1-(3S)-benza zepine-1-acetic acid monohydrochloride (Figure 1A). The empirical formula of BEN is $C_{24}H_{28}N_2O_5 \bullet HCl$ with a molecular weight 460.96 g/mole. It is a angiotensin converting enzyme. It is used as antihypertensive agent. Hydrochlorthiazide (HCT) is chemically 6-chloro-3, 4-dihydro-2H-1,2,4-benzothiadiazine-7-sulfonamide 1,1-dioxide (Figure 1B). The empirical formula is $C_7H_8ClN_3O_4S_2$ with a molecular weight 297.73 g/mole. It is a diuretic agent [1-3].

In the proposed study, attempt has been made to develop sensitive stability indicating RP-LC method for the estimation of BEN and HCT in bulk and pharmaceutical dosage form.

Comprehensive literature survey reveals that several analytical methods have been reported for the estimation of BEN which includes high performance liquid chromatography (HPLC) (1), and HCT which includes potentiometry (1), HPTLC (High performance thin layer chromatography) (2), UV–visible simultaneous estimation method (3), RP-HPLC has been reported for the estimation of BEN with another drug combination instead of HCT in pharmaceutical dosage form [4,5]. RP–HPLC, HPTLC, and UV–Visible spectrophotometric methods has been reported for estimation of HCT with another drug combination instead of BEN in bulk and pharmaceutical dosage form [6-15]. LC and HPTLC method has been reported for simultaneous estimation of BEN and HCT in bulk and pharmaceutical dosage form [16,17]. In the proposed study, attempt has been made to develop stability indicating liquid chromatographic method for the estimation of BEN and HCT in pharmaceutical dosage form as per ICH guide lines [18,19].

Experimental

Instrumentation

High performance liquid chromatography: The liquid chromatographic system of Perkin Elmer, containing HPLC isocratic pump (515), UV detector and rheodyne injector with 20 μl fixed loop

was used. A Sunfire C18 column with 250×4.6 mm i.d. and 5 μm particle size was used as stationary phase.

Reagents and materials: Analytically pure benazepril hydrochloride (BEN) and hydrochlortiazide (HCT) was procured from Dishman Pharmaceutical Pvt. Ltd and Cadila pharmaceutical Ltd, (Ahmedabad, India). Methanol, water (E. Merck, Mumbai, India) used for the preparation of mobile phase was of LC grade. Triethylamine (Sissco research laboratories, Mumbai, India) was of analytical reagent grade. Tablet formulation A (Lotencin HCT–(10 mg Benazepril hydrochloride and 12.5 mg Hydrochlortiazide), Novartis Pharmaceutical was purchased from local market.

Preparation of mobile phase and stock solution: Mobile phase was prepared by mixing 550 ml of Water with 450 ml of methanol in 1000 ml volumetric flask. The pH was adjusted to 7.0 using triethylamine (1%) and the solution was filtered through Whatman filter paper No.42 (0.45 μm) and it was sonicated for 15 min prior to use for degassing. This solution was used as a mobile phase.

BEN and HCT were weighed accurately (25 mg each) and transferred to separate 25 ml volumetric flasks containing few ml of methanol. Volumes were adjusted up to the mark with methanol to yield a solution containing 1000 μg/ml of BEN and HCT respectively. Aliquot (1.0 ml) from the above solutions of BEN and HCT were appropriately diluted with methanol to obtain working standards stock solution of 100 μg/ml of BEN and HCT respectively.

---

*Corresponding author: Chhalotiya UK, Indukaka Ipcowala College of Pharmacy, Beyond GIDC Phase IV, Vithal Udyognagar, New Vallabh Vidyanagar -388121, Anand, Gujarat, India, E-mail: usmangani84@gmail.com

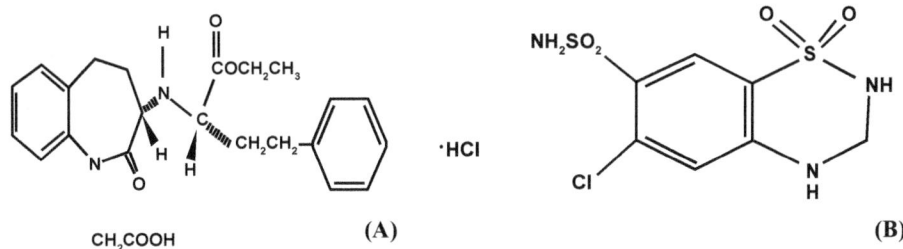

Figure 1: Chemical Structure of (A) Benazepril hydrochloride (B) Hydrochlorthiazide.

Chromatographic conditions: A reversed phase C18 column (Sunfire) equilibrated with mobile phase comprising of water:methanol (55:45 v/v; pH 7) was used. Mobile phase flow rate was maintained at 1 ml/ min and effluents were monitored at 233 nm. A 20 µL of sample was injected using a fixed loop, and the total run time was 10 min. All the chromatographic separations were carried out at controlled room temperature (25 ± 2°C).

Calibration curves for BEN and HCT: Appropriate aliquots of BEN and HCT working standard solutions were taken in different 10 ml volumetric flasks. The volume was made up to the mark with mobile phase to obtain final concentrations 0.1, 0.5, 1, 5, 10, 20 µg/ml of BEN and 0.5, 1, 5, 10, 20, 30 µg/ml of HCT respectively. The solutions were injected using a 20 µL fixed loop system and chromatograms were recorded. Calibration curves were constructed by plotting peak area versus concentrations of the drug. The non-weighted linear regression equation was computed for BEN and HCT.

Analysis of Marketed Formulations: Twenty tablets were weighed accurately and finely powdered. Tablet powder equivalent to 10 mg BEN and 12.5 mg of HCT was taken in 100 ml volumetric flask. A few ml of methanol was added to the above flask and the flask was sonicated for 10 min. The solution was filtered using Whatman filter paper No.42 (0.45 µm) and volume was made up to the mark with the methanol. Aliquot (0.4) was transferred to a 10 ml volumetric flask and the volume was made up to the mark with the mobile phase to obtain a solution containing 4 µg/ml of BEN and 5 µg/ml of HCT. This solution was used for the estimation of BEN and HCT. Both the solutions were sonicated for 10 min. Solutions were injected as per the above chromatographic conditions and peak areas were recorded. The quantifications were carried out by keeping these values to the straight line equation of calibration curve.

Validation of method : The method was validated as per ICH guideline for accuracy, precision, specificity, detection limit, quantitation limit and robustness.

Accuracy: Known amount of BEN (0, 0.5, 1, 1.5 µg/ml) and HCT (0, 1, 2, 3 µg/ml) was added to a pre quantified sample solutions. The amount of BEN and HCT was estimated using linear regression equation.

Precision: The instrument precision was evaluated by injecting the solution containing BEN (0.1, 1, 10 µg/ml) and HCT (0.5, 5, 20 µg/ml) six times repeatedly and peak area was measured. The results are reported in terms of % relative standard deviation. The intra-day and inter-day precision study of BEN and HCT was carried out by estimating the corresponding responses 3 times on the same day and on 3 different days (first, second and third day) for 3 different concentrations of BEN (0.1, 1, 10 µg/ml) and HCT (0.5, 10, 20) within the calibration range and the results are reported in terms of % relative standard deviation (%RSD).

Specificity: The specificity was estimated by spiking commonly used excipients (starch, talc and magnesium stearate) into a pre weighed quantity of drug. The chromatogram was taken by appropriate dilutions and the quantities of drugs were determined.

Limit of detection and quantification: The detection limit is defined as the lowest concentration of an analyte that can reliably be differentiated from background levels. Limit of quantification of an individual analytical procedure is the lowest amount of analyte that can be quantitatively determined with precision and accuracy. LOD and LOQ were calculated using following equation as per ICH guidelines. LOD=$3.3 \times \sigma/S$ and LOQ=$10 \times \sigma/S$, where σ is the standard deviation of y-intercepts of regression lines and S is the slope of the calibration curve.

Robustness: Robustness of the method was studied by deliberately changing the experimental conditions like flow rate, percentage of organic phase, and also by observing the stability of the sample solution at 25 ± 2° for 24 h. The sample solution was assayed at every 2 h interval up to 24 h.

Forced degradation study: Stress degradation study using acid and alkali hydrolysis, chemical oxidation, wet hydrolysis exposure to sun light and dry heat degradation was carried out and interference of the degradation products was investigated. BEN and HCT was weighed (10 mg) and transferred to 10 ml volumetric flasks and expose to different stress conditions.

Heat induced alkali hydrolysis: To the 10 ml volumetric flask, 10 mg of BEN and HCT was taken and 2 ml of 0.1 N NaOH was added to perform heat induced base hydrolysis. The flask was heated at 80°C for 4 hrs and allowed to cool to room temperature. Solution was neutralized with 0.1 N HCl and volume was made up to the mark with methanol. 0.1 ml of aliquots was taken from the above solution and diluted with mobile phase to obtain final concentration of 10 µg mL^{-1} of BEN and HCT.

Heat induced acid hydrolysis: To the 10 ml volumetric flask, 10 mg of BEN and HCT was taken and 2 ml of 0.1 N HCl was added to perform heat induced acid hydrolysis. The flask was heated at 80°C for 4 hrs and allowed to cool to room temperature. Solution was neutralized with 0.1 N NaOH and volume was made up to the mark with methanol. 0.1 ml of aliquot was taken from the above solution and diluted with mobile phase to obtain final concentration of 10 µg mL^{-1} of BEN and HCT.

Heat induced wet hydrolysis: To the 10 ml volumetric flask, 10 mg of BEN and HCT was taken and 2 ml of HPLC grade water was added to perform heat induced wet hydrolysis. The flask was heated at 80°C for 4 hrs and allowed to cool to room temperature and volume was made up to the mark with methanol. 0.1 ml of aliquot was taken from the above solution and diluted with mobile phase to obtain final concentration of 10 µg mL^{-1} of BEN and HCT.

Heat induced oxidative stress degradation: To heat induced perform oxidative stress degradation, 10 mg of BEN and HCT was taken in 10 ml volumetric flask and 2 ml of 6% hydrogen peroxide was added. The mixture was heated in a water bath at 80°C for 4 hrs and allowed to cool to room temperature and volume was made up to the mark with methanol. 0.1 ml of aliquot was taken from above solution and diluted with mobile phase to obtain final concentration of 10 µg mL^{-1} of BEN and HCT.

Photolytic degradation: Analytically pure 10 mg of drugs were exposed to sunlight for 72 hrs. The solid was allowed to cool and transferred to volumetric flask (10 ml) and dissolve in few ml of methanol. Volume was made up to the mark with the methanol. Solution was further diluted with the mobile phase to obtain final concentration of 10 µg mL^{-1} of BEN and HCT. All the solutions were injected in the liquid chromatographic system and chromatograms were recorded.

Result

Validation of the proposed methods

Linearity: Linearity of an analytical method is its ability, within a given range, to obtain test results that are directly, or through a mathematical transformation, proportional to the concentration of the analyte. The calibration curve for BEN was found to be linear in the range of 0.1-20 µg/ml with a correlation coefficient of 0.9923. The calibration curve for HCT was found to be linear in the range of 0.5-30 µg/ml with a correlation coefficient of 0.9977. The regression data shown in table confirms the linearity of the method over the concentration range studied (Table 1). Summary of validation parameters shown in (Tables 2 and 3).

Precision: Repeatability was determined by performing injection repeatability test and the % RSD values for BEN and HCT were found to be 0.30-1.68 and 0.20-1.43 respectively.

The intraday and interday precision studies were carried out on the same day and three different days. The results are reported in terms of %RSD. The low % RSD values indicate that the method is precise.

Parameters	BEN	HCT
Linearity range (µg/ml)	0.1-20 µg/ml	0.5-30 µg/ml
Correlation coefficient (r)	0.9923	0.9977
Slope	24072	30111
Standard deviation of slope	491.8421	3.34664
Intercept of regression	20734	6899.1
Standard deviation of intercept	485.9614	132.5796

Table 1: Statistical analysis data of calibration curve.

Parameters	BEN	HCT
Limit of Detection	0.065	0.0145
Limit of Quantitation	0.1	0.5
Accuracy (%)	98.66-99.32	98.93-99.73
Repeatability (%RSD, n=6)	0.30-1.68	0.20-1.43
Precision Intraday (n=3)	0.11-0.19	0.11-0.48
Interday (n=3)	1.11-1.72	1.22-1.70
Specificity	Specific	Specific
Robustness	Robust	Robust
Solvent suitability	Suitable for 24 hrs	Suitable for 24 hrs

Table 2: Summary of validation Parameters of RP-HPLC.

%Level	Amount of sample taken (µg/ml)	Amount of standard drug added (µg/ml)	Amount of drug recovered (µg/ml)	% Recovery
0%	1	0	98.66	98.66 ± 0.57
50%	1	0.5	1.49	99.2 ± 0.91
100%	1	1	1.97	98.83 ± 1.55
150%	1	1.5	2.49	99.32 ± 1.06

Table 3: Accuracy study of BEN by proposed RP-HPLC method.

%Level	Amount of sample taken (µg/ml)	Amount of standard drug added (µg/ml)	Amount of drug recovered (µg/ml)	% Recovery
0%	1.25	0	1.24	99.2 ± 0.5
50%	1.25	0.625	1.85	98.93 ± 1.32
100%	1.25	1.25	2.49	99.73 ± 0.38
150%	1.25	1.875	3.10	99.2 ± 0.76

Table 4: Accuracy study of HCT by proposed RP-HPLC method.

Method parameter	Normal condition	Deliberate changes	% RSD of peak area (n=3)	
			BEN	HCT
Flow rate	1.5 ml/min	0.8 ml/min	0.95	0.84
		1.2 ml/min	0.68	0.71
Mobile phase ratio	Water: Methanol (55:45)	53: 47	0.82	0.29
		57: 43	0.56	0.63
pH of mobile phase ratio	7	7.2	0.92	0.81
		6.8	0.65	0.76

Table 5: Robustness study by proposed RP–HPLC method.

Accuracy: Accuracy of an analytical method is the closeness of the test results to the true value. It was determined by the application of analytical procedure to recovery studies, where a known amount of standard is spiked into pre-analysed sample solutions. The recoveries were found to be 99.2-99.32%, and 99.0-99.44% for BEN and HCT respectively (Tables 3 and 4).

Limit of detection and limit of quantification: The detection limits for BEN and HCT were 0.065 µg/ml and 0.014 µg/ml respectively, while quantitation limits were 0.1 µg/ml and 0.5 µg/ml respectively. The above data shows that a nano gram quantity of both the drugs can be accurately and precisely determined.

Specificity: The specificity study was carried out to check the interference from the excipients used in the formulation by preparing synthetic mixture containing the drug and excipients. The chromatogram showed peaks for the drug without any interfering peak.

Robustness: The method was found to be robust, as small but deliberate changes in the method parameters have no detrimental effect on the method performance as shown in table. The low value of relative standard deviation was indicating that the method was robust (Table 5).

Forced degradation study: Chromatogram of base hydrolysis performed at 80°C for 4 hrs reflux showed degradation of BEN and HCT at retention time (RT) 9.2 min and 3.12 min respectively (Figure 3).

Chromatogram of acid hydrolysis performed at 80°C for 4 hrs reflux showed degradation of BEN and HCT at retention time (RT) 9.4 min and 3.07 min respectively (Figure 4).

The chromatogram of oxidized BEN and HCT with 6% hydrogen peroxide at 80°C for 4 hrs reflux showed degradation of BEN and HCT with at retention time 9.3 min and 3.09 min (Figure 5).

The chromatogram of BEN and HCT exposed to dry heat at 80°C

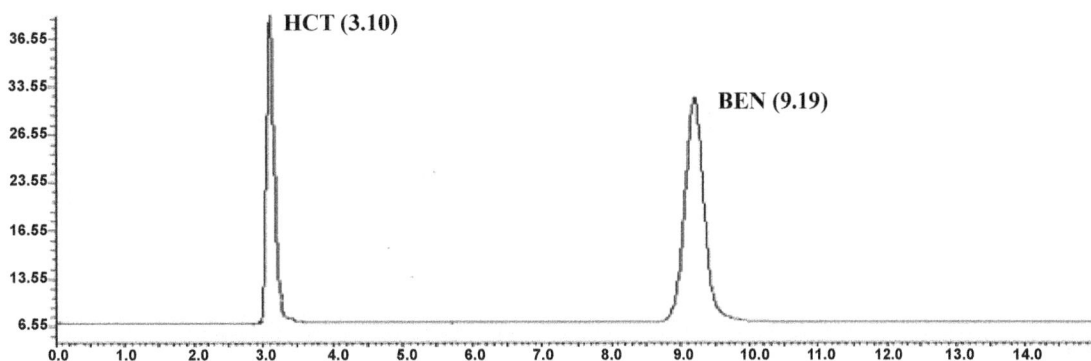

Figure 2: Chromatogram of standard BEN (10 μg/ml) and HCT (10 μg/ml) at flow rate of 1 ml/min in mix standard.

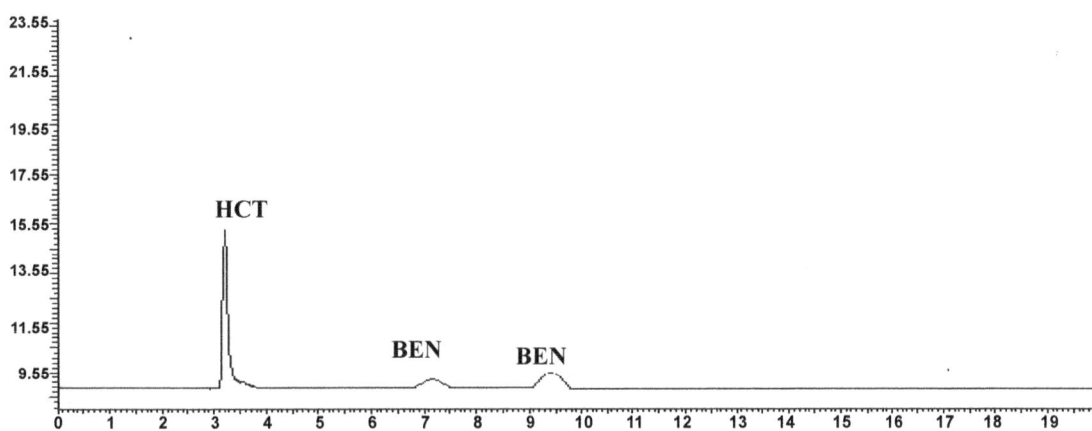

Figure 3: Chromatogram of base treated BEN (10 μg/ml) and HCT (10 μg/ml) at 80°C for 4 hrs in mixture.

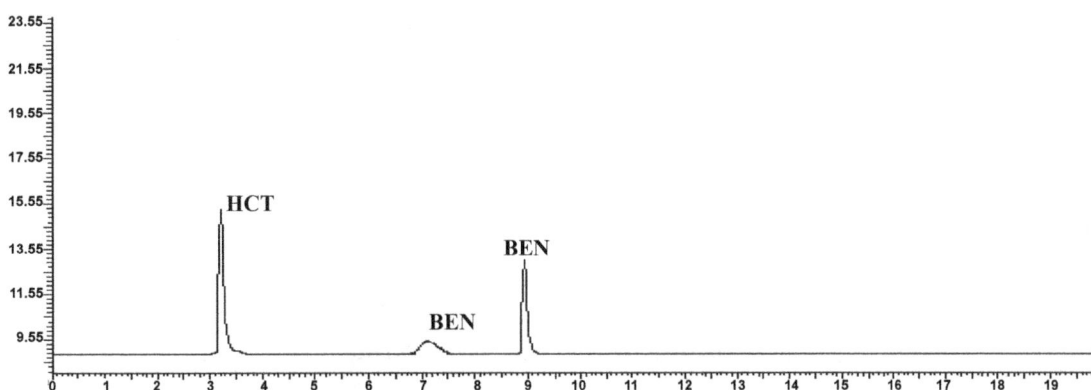

Figure 4: Chromatogram of acid treated BEN (10 μg/ml) and HCT (10 μg/ml) at 80°C for 4 hrs.

for 4 hrs showed degradation of BEN and HCT at retention time (RT) 9.10 min and 3.11 min. respectively (Figure 6).

The chromatogram of BEN and HCT expose to sun light for 72 hrs showed degradation of BEN and HCT at retention time (RT) 9.07 min and 3.14 min respectively (Figure 7 and Table 6).

Solution stability: The solution stability study showed that BEN and HCT were evaluated at room temperature for 24 hr. The relative standard deviation was found below 2.0%. It showed that solution was stable up to 24 hrs at room temperature

Analysis of marketed formulations: The proposed method was successfully applied to the determination of BEN and HCT in their combined dosage form. The % recovery was found to be more than 99.0% for all the drugs which were comparable with the corresponding labelled amounts. No interference from the excipients present in the marketed tablet formulation was observed (Tables 7 and 8).

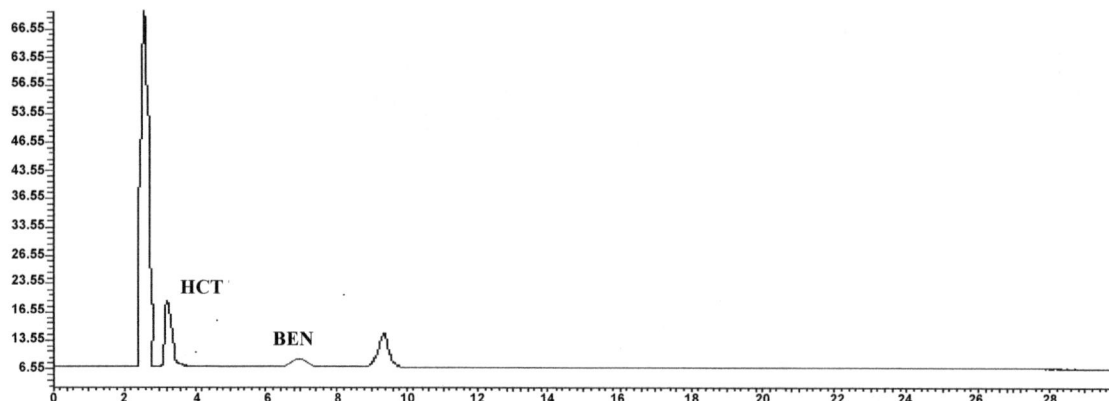

Figure 5: Chromatogram of hydrogen peroxide degraded sample of BEN (10 µg/ml) and HCT (10 µg/ml) at 80°C for 4 hrs.

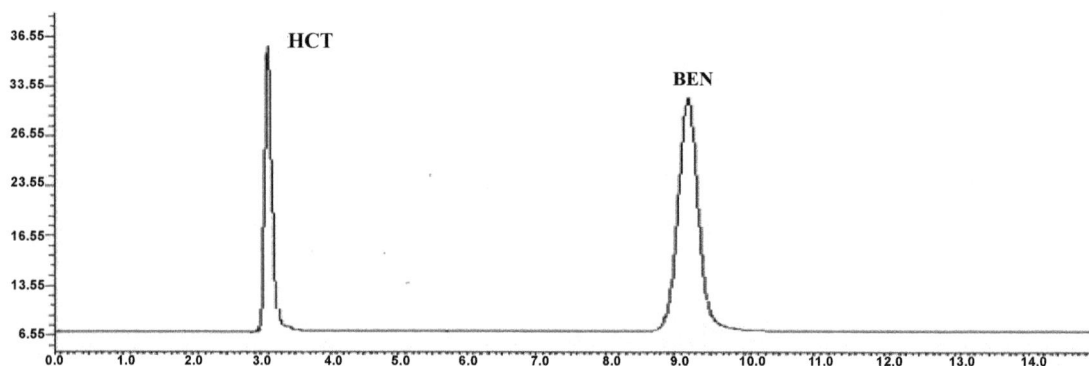

Figure 6: Chromatogram of dry Heat degraded sample of BEN (10 µg/ml) and HCT (10 µg/ml) at 80°C for 4 hrs.

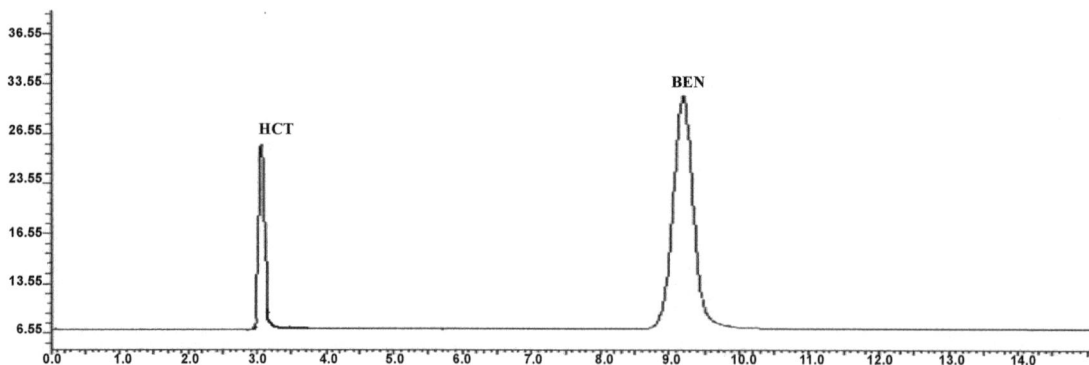

Figure 7: Chromatogram of photo degraded sample of BEN (10 µg/ml) and HCT (10 µg/ml) at 80°C for 72 hrs

Discussion

Optimization of mobile phase

To optimize the chromatographic conditions, the effect of chromatographic variables such as mobile phase pH, flow rate, and solvent ratio were studied. The resulting chromatograms were recorded and the chromatographic parameters such as capacity factor, asymmetric factor, and resolution and column efficiency were calculated. The conditions that gave the best resolution, symmetry and capacity factor were selected for estimation. The drug solutions containing BEN (10 µg/ml), HCT (10 µg/ml) and their mixture were chromatographed at a flow rate of 1 ml/min with the following mobile phases.

Various mixtures containing water and methanol were tried as mobile phases in the initial stage of method development. Methanol: Water (50:50), Methanol: Water (80:20), Methanol: Water (70:30), Methanol: Water (60:40) was tried as mobile phase but satisfactory resolution of drug and degradation peaks were not achieved.

The mobile phase methanol: water (55:45, v/v pH adjusted with 1% solution of TEA) was found to be satisfactory and gave symmetric peak for BEN and HCT. The retention time for proposed method was found

Parameter	BEN	HCT
Retention time (min)	9.19	3.10
Theoretical plates	3595.43	5133.94
Tailing factor	1.33	1.02
Base width (sec)	12.57	30.88
Resolution	9.36	

Table 6: System suitability parameter.

Formulation	Actual concentration (µg/ml)		Amount obtained (µg/ml)		% BEN Mean ± SD (n=3)	% HCT Mean ± SD (n=3)
	BEN	HCT	BEN	HCT		
Tablet (Lotensin HCT)	4	5	3.98	5.01	99.5% ± 0.41	100.2% ± 0.58

Table 7: Assay Results of Marketed Formulation.

Conditions	Time (hrs)	Recovery (%)		Retention time of degradation products	
		BEN	HCT	BEN	HCT
Base (0.1 N NaOH)	4	62.89	18.45	9.2	3.1
Acid (0.1 N HCl)	4	87.98	10.97	9.4	3.07
6% H_2O_2	4	82.42	14.03	9.3	3.09
Photo oxidation	72	97.62	12.80	9.07	3.14
Dry Heat	4	99.97	44.12	9.10	3.11

Table 8: Forced degradation study of BEN and HCT for the proposed Method.

to be 9.19 min for BEN and 3.10 min for HCT as shown in Figure 2.

Method validation

The calibration curve was found to be linear over the range of 0.1-20 µg/ml for BEN and 0.5-30 µg/ml for HCT. Instrument precision was determined by performing injection repeatability test and the %RSD value for BEN and HCT was found to be 0.30-1.68% and 0.20-1.43. The intra-day and inter-day precision studies were carried out. For BEN the intra-day study %RSD values were found to be 0.11-0.19% and for inter-day precision study %RSD values were found to be 1.11-1.72%. For HCT the intra-day study %RSD values were found to be 0.11-0.48% and for inter-day precision study %RSD values were found to be 1.22-1.70%. The low %RSD values indicate that the method is precise.

The accuracy of the method was determined by calculating recoveries of BEN and HCT by method of standard additions. The recovery of BEN and HCT was found to be 98.66-99.32% and 98.93-99.73, respectively. The values indicate that the method is accurate.

The detection limits for BEN and HCT was found to be 0.07 µg/ml and 0.02 µg/ml while quantitation limits was found to be 0.1 µg/ml and 0.5 µg/ml, respectively. The above data shows that a nanogram quantity of the drug can be accurately and precisely determined. System suitability test was carried out on freshly prepared standard stock solution of BEN and HCT.

The liquid chromatogram of the placebo used in the specificity study did not give any interfering peak in the chromatogram, which suggests that the proposed LC method is both selective and specific.

The method was found to be robust, as small but deliberate changes in the method parameters have no detrimental effect on the method performance. The low value of percentage relative standard deviation was indicating that the method was robust.

The solution stability study revealed that BEN and HCT in mixed standard solution were found to be stable for 24 h without detection of degradation. The percentage recoveries of both the drugs were found to be satisfactory.

Forced degradation study

The degradation study thereby indicated that BEN and HCT was susceptible to acid hydrolysis, base hydrolysis, oxidation (6% hydrogen peroxide), photo degradation, and dry heat. No degradation products from different stress conditions affected determination of BEN and HCT.

Conclusion

As compared with the published, the proposed method is more sensitive. Proposed study describes stability indicating LC method for the estimation of BEN and HCT in bulk and their pharmaceutical dosage form. The method was validated and found to be simple, sensitive, accurate and precise. Statistical analysis proved that method was repeatable and selective for the analysis of BEN and HCT without any interference from the excipients. The method was successfully used for the determination of drug in their pharmaceutical formulation. Also the above results indicated the suitability of the method for acid, base, oxidation, dry heat and photolytic degradation study. As the method separates the drugs from its degradation products, it can be used for analysis of stability samples. The method is suitable for the routine analysis of BEN and HCT in tablets. In addition, the HPLC procedure can be applied to the analysis of samples obtained during accelerated stability experiments to predict expiration dates of pharmaceuticals.

Acknowledgement

The authors are thankful to Dishman pharmaceutical Ltd and Cadila Pharmaceuticals Ltd., Ahmedabad for providing gratis sample of Benazepril Hydrochloride and Hydrochlorthiazide. The authors are very thankful Indukaka Ipcowala College of pharmacy, new vallabh vidyanagar, an and, for providing necessary facilities to carry out research work.

References

1. British Pharmacopoeia (2011) The department of health, social services and public safety. 2: 225-226.

2. Indian Pharmacopoeia (2007) The Indian pharmacopoeia commission, Ghaziabad, Ministry of Health & Family welfare 2: 576-578.

3. Pawar PY, Joshi RS, Sandhan V, Wagh S, Jangale K (2011) Simultaneous spectrophotometric estimation of Amlodipine Besylate and Benazepril HCl in pure and pharmaceutical dosage form, Der Pharmacia Lettre 3: 397-403.

4. Naidu KR, Kale UN, Shingare MS (2005) Stability indicating RP-HPLC method for simultaneous determination of amlodipine and benazepril hydrochloride from their combination drug product. Journal of Pharmaceutical and Biomedical Analysis 39: 147-155.

5. Sarat M, Murli PK, Rambabu C (2012) Development and validation of RP-HPLC method for simultaneous estimaton of amlodipine besyate and benazepril HCl in tablet. International Journal of Current Pharmaceutical Research 4: 80-84.

6. Safeer K, Anbarasi B, Senthil KN (2010) Analytical method development and validation of amlodipine and hydrochlorothiazide in combined dosage form by RP-HPLC. International Journal of Chem Tech Research 2: 21-25.

7. Patel G, Patel S, Prajapiti D, Mehta R (2010) RP-HPLC method for simultaneous estimation of amlodipine besylate and hydrochlorothiazide in combined dosage forms. Stamford Journal of Pharmaceutical Sciences 3: 49-35.

8. Rao LA, Bhaskara RV (2011) Simultaneous estimation of valsarten and hydrochlorthiazide in tablets by RP-HPLC method. Intentional Journal Pharmaceutical & Industrial Reserch 1: 170-174.

9. Kharoaf M, Malkieh N, Abualhasan M, Shubitah R, Jaradat N, et al. (2012) Tablet formulation and development of a validated stability indicating HPLC method for quantification of valsarten and hydrochlorthiazide combination. International Journal of Pharmacy and Pharmaceutical Sciences 4: 683-687.

10. Chabukswar AR, Jagdale SC, Kuchekar BS, Lokhande PD, Shinde SN, et al. (2010) Development and validation of a RP-HPLC method for simultaneous

estimation of hydrochlorothiazide and irbesartan. Der Pharma Chemica 2: 148-156.

11. Rosangluaia, Shanmugasundaram P, Malarkodi V (2011) Validated HPTLC method for simultaneous estimation of irbesartan and hydrochlorthiazide in a tablet dosage form. Der Pharma Chemica 3: 310-317.

12. Singh B, Patel DK, Ghosh SK (2009) Development of RP-HPLC method for simultaneous analysis of metoprolol succinate and hydrochlorothiazide in a tablet formulation. Tropical Journal of Pharmaceutical Research 8: 539-543.

13. Kumbhar ST, Chougule GK, Tegeli VS, Gajeli GB, Thorat VS, et al. (2011) A Validated HPTLC method for simultaneous quantification of nebivolol and hydrochlorothiazide in bulk and tablet formulation. International Journal of Pharmaceutical Sciences and Drug Research 3: 62-66.

14. Gupta Y, Shrivastava A, Duggal D, Patel A, Agrawal S (2009) A new RP-HPLC method for simultaneous estimation of nebivolol hydrochloride and hydrochlorthiazide in dosage forms. Journal Young Pharmacist 1: 264-269.

15. Hapse SA, Wagh VS, Kadaskar PT, Dokhe MD, Shirsath AS (2012) Spectrophotometric estimation and validation of hydrochlorothiazide in tablet dosage forms by using different solvents. Der Pharma Chemica, 4: 10-14.

16. Alaa EG, Ahmed A, Laila AF, Marwan MS (2001) Application of LC and HPTLC-densitometry for the simultaneous determination of benazepril hydrochloride and hydrochlorothiazide. Journal of Pharmaceutical and Biomedical Analysis 25: 171-179.

17. Panderi IE, Parissi-Poulou M (1999) Simultaneous determination of benazepril hydrochloride and hydrochlorothiazide by micro-bore liquid chromatography. Journal of Pharmaceutical and Biomedical Analysis 21: 1017-1024.

18. Validation of Analytical Procedures: Methodology, ICH Harmonized Tripartite Guidelines 2005.

19. ICH Harmonized tripartite Guidelines, Stability testing of New Drug Substances and products, Q1A(R2), Feb 2003.

22

Quantitative Analysis of L-Abrine and Ricinine Spiked into Selected Food Matrices by Liquid Chromatography-Tandem Mass Spectrometry

Shannon M Black*, Sadia Muneem, Deborah Miller-Tuck and Prince A Kassim

Maryland Department of Health and Mental Hygiene, 201 West Preston Street, Baltimore, MD 21201 United States

Abstract

Abrin and ricin are highly toxic and lethal proteins which have the potential to be used as bioterrorism agents.L-abrine and ricinine, molecular biomarkers for abrin and ricin, serve as useful biomarkers in assessing exposure and contamination.In this study, we developed a method for the quantitation of L-abrine and ricinine spiked into four food types: ground beef, chicken breast, hot dogs, and Eggbeaters®.This method involved sample homogenization, followed by polymeric reversed solid phase extraction, evaporation, and reconstitution.Determination and quantitation of L-abrine and ricinine was achieved using LC tandem mass spectrometry utilizing positive electrospray ionization. Quantitation was based upon fragmentation of m/z219→132 for L-abrine and m/z 165→138 for ricinine.The limit of detection achieved was 0.1 µg/L while the limit of quantitation was 0.50 µg/L for L-abrine and 0.30 µg/L for ricinine.A 7-point standard calibration curve showed good linearity for both analytes (r2 = 0.990) with a CV of <10% for replicates.Spiked fortified food matrices at 5 µg/L, 25 µg/L, and 50 µg/L yielded recoveries ranging from 93 – 119%. Validation results showed this method to be rapid, accurate, sensitive, and suitable for surveillance monitoring of food products suspected of abrin and ricin contamination.

Keywords: Abrin; ricin; L-abrine; Ricinine; Liquid chromatography; Mass spectrometry; Solid phase extraction;Quantitative analysis

Introduction

Ricin, an alkaloid of castor bean plants, is an extremely potent and deadly poison which has gained interest in the chemical emergency preparedness and response community due to its appearance in terrorism literature and its potential use as a chemical warfare agent [1]. The castor plant is widely available due to its many uses in the production of oils, lubricants, pharmaceuticals, cosmetics and engineering plastics [2].Harvested in many countries, including Asia, Africa and Europe, there is an overall production of 1 million metric tons of castor seeds which implies a total production of 10,000 metric tons of ricin each year [3]. While the most potent approaches of ricin poisoning are inhalation and injection, the most common approach has been found to be ingestion [4]. Currently there are no validated tests for the detection of ricin available to the clinical laboratories [5] and while the analysis of ricin, a large heterogeneous protein with glycosylation, has proven difficult, the identification of the ricinine has shown to be a complementary technique for the determination of castor bean extracts [1].

Abrin, another alkaloid chemical, can easily be isolated from the seed of the rosary pea plant [6]. Just like ricin and ricinine, crude abrin and abrine are easily obtained from their seeds using rather simple technology [7,8]. This simple isolation and the potential of its use to adulterate foods have made the detection of abrin and ricin in foods a top priority. However, abrin, like ricin, belongs to the large family of ribosome inactivating proteins containing two disulfide linked heterodimic chains [8], making it just as difficult as ricin to effectively isolate from contaminated matrices at low detection limits.Due to the high potential threat which abrin and ricin carry, both toxins are classified as category B Select Agents by the US Health and Human Services [8].

As it is mentioned, the primary analytes of interest to the chemical emergency preparedness community are that of abrin and ricin, however due to the difficulty of isolation and detection of these analytes from potentially tainted matrices, L-abrine and ricinine have proven to be useful markers in the detection of the primary analytes.

In addition to the easier isolation of L-abrine and ricinine, they have also demonstrated longer stability in contaminated matrices thereby allowing for a larger window between the time of contamination and detection [9].

Materials and Methods

Chemicals and food products

L-abrine was purchased from Sigma Aldrich (St. Louis, MO) and ricinine was purchased from Accurate Chemical and Scientific Corporation (Westbury, NY). Isotopically-labeled L-abrineand ricinine were purchased from Cerilliant (Round Rock, TX) and used as internal standards for instrumental quantitation.Formic acid and HPLC grade solvents including acetonitrile and methanolwere purchased from Sigma Aldrich (St. Louis, MO), and used throughout the study.High purity water was supplied from within the laboratory using a Millipore Elix Water Purification System (Billerica, MA).Raw food products, including ground beef (15% fat), chicken breast, hot dog, and EggBeaters˙, were purchased from a local grocery store.

Preparation of standards

Assessment of standard purity: The sources of L-abrine and ricinine were analyzed on an Agilent Technologies 1100 HPLC coupled with an Applied Biosystems API 4000 quadrupole linear ion trap mass spectrometer (LC/MS/MS) controlled by Analyst software.

*Corresponding author: Shannon M Black, Maryland Department of Health and Mental Hygiene, 201 West Preston Street, Baltimore, MD 21201, United States, E-mail: ShannonBlack2012@gmail.com

Stock standard solutions of abrine and ricinine were prepared by dissolving 50 mg of each analyte into 50 mL of deionized water. Working standardsolutions were prepared by diluting 0.50 mL of the stock standard solution into 100 mL deionized water.Ten μL of this 5 μg/mL solution was injected into the HPLC and chromatographed using a mobile phase gradient program.The ion transitions for L-abrine were 219/132 (quantitation) and 219/187.9 (confirmation). The ion transitions for ricinine were 165/138 (quantitation) and 165/82 (confirmation).

Sample preparation and clean-up

Preparation of standards: A working standard solution of L-abrine and ricinine with a concentration of 5.0 μg/mL was used to prepare thirteen standards to construct a linear dynamic range (LDR), seven calibration standards, and six quality control (QC) samples. The LDR for L-abrine ranged from 0.10 μg/L to 800 μg/L and the seven calibration standards ranged from 0.50 μg/L to 200 μg/L. The LDR for ricinine ranged from 0.10 μg/L to 600 μg/L and the seven calibration standards ranged from 0.30 μg/L to 100 μg/L.

Preparation of samples: For solid food samples (i.e., ground beef, hot dogs, chicken) the matrix was homogenized while the liquid sample (i.e., egg beaters) was mixed and 5.0 g or 5.0 mL of each matrix sample was aliquoted into a 15 mL centrifuge tube.Each vial was spiked with the target analytes at the equivalent to its concentration level, vortexed and kept at 4°C overnight.

Liquid extraction: Samples were prepared by allowing the sample vials to sit at room temperature for 30 min and then adding10 mL water to each tube.The sampletubes were capped and vortexed for 30 seconds. The samples were then sonicated (Branson 3510, Bransonic, Dansbury, CT) for 30 min and centrifuged at 4000 xg for 30 min (Sorvall, Thermo Scientific, Waltham, MA).The supernatant was removed with a disposable glass pipette and passed through a 0.2 μm filter.A 1 mL aliquot was transferred to a test tube and spiked with 100 μL of L-abrine and ricinine internal standard mixture.

Solid phase extraction: Strata-X SPE columns (60 mg/3 mL; Phenomenex, Torrance, CA) were conditioned with 3 mL of methanol followed by 3 mL of water.The supernatant of the sample was loaded (1 mL) and allowed to drip through by gravity.The SPE columns were then washed with 5% methanol and the L-abrine and ricinine were eluted with 3 mL of acetonitrile.The eluent was then completely evaporated under a constant flow of nitrogen at 15 psi in a Zymark Turbo Vap evaporator (Caliper Life Sciences, Waltham, MA) at 65°C for 20 minutes or until completely dry.The extracts were reconstituted in 200 μL water, vortexed and transferred to an autosampler vial for analysis by LC/MS/MS.

The optimization of the clean-up protocol for SPE was evaluated extensively.The solid phase extraction cartridges were evaluated with the elution solvents of acetonitrile, ethyl acetate, acetone and methanol. Acetonitrile as the elution solvent resulted in the highest recoveries.

Instrumental conditions

LC/MS/MS: The Applied Biosystems API 4000 quadrupole linear ion trap mass spectrometer was optimized for L-abrine and ricinine ionization and analysis from food samples.The instrument was first calibrated using positive and negative polypropylene glycol (PPG) solutions per manufacturer's specifications.After tuning the instrument with a 100 μg/L solution of L-abrine and ricinine in water + 0.1% formic acid, optimized settings were as follows: positive ion mode ESI with ion spray voltage at 5100 V, collision gas at high, curtain gas

at 20 L/hr, ion source gas 1 at 65 L/hr, ion source gas 2 at 10 L/hr and the interface temperature at 550 °C.The collision energy setting was 25 eV for Ricinine 165→138, 41 eV for Ricinine 165→82, 25 eV for labeled Ricinine 171→85, 29 eV for L-abrine 219→132, 17 eV for L-abrine 219→187.9 and 17 eV for labeled L-abrine 223→187.9.The declustering potential for all ricinine transitions was 51 V, 31 V for L-abrine 219→132 and 46 V for L-abrine 219→187.9 and 223→187.9.The collision cell exit potential for all ricinine transitions was 8 V, 16 V for L-abrine 219→132 and 12 V for L-abrine 219→187.9 and 223→187.9.Dwell time was set a 100 m/sec for all transitions.

An Agilent 1100 HPLC system consisting of a quaternary pump, in line mobile phase degasser, temperature controlled autosampler (maintained at 20°C), and column heating compartments was utilized for chromatography.Mobile phase A consisted of 10% methanol in water + 5mM formic acid and mobile phase B consisted of acetonitrile + 5mM formic acid.Twenty μL were injected onto a 2.0 mm x 100 mm i.d., 2.5 μm, Polar Reverse Phase Phenyl analytical column (Phenomenex, Torrance, CA).The column was maintained at40°C ± 5°C throughout the chromatographic run with a 3 min equilibration time between samples.The gradient mobile phase conditions were as follows: 93% A at 0 min (hold for 0.5 min) to 50% A within 2 min (hold for 1 min), return to 93% A in 0.01 min (hold for 3 min) for a total run time of 6 min with a 0.300 mL/min flow rate maintained throughout.L-abrine eluted at 3.00 min and ricinine eluted at 4.2 min.

Quantitation was calculated using a PC equipped with Analyst software by linear regression with no weighting from 0.5 μg/L to 200μg/L for L-abrine and from 0.30 μg/L to 100 μg/L forricinine with n ≥ 6 measurements per standard.A standard curve was prepared for each individual sequence run per matrix and six quality control samples were included at the beginning and end of each run.A linear dynamic range consisting of 13 standards and a blank was also included in each run.L-abrine and ricinine was analyzed using multiple reaction monitoring (MRM) and the following transitions were monitored: L-abrine, m/z 219 → 132 (quantitation ion), m/z 219 → 187.9 (confirmation ion) and 223 → 187.9 (internal standard); ricinine, m/z 165 → 138 (quantitation ion), 165 → 82 (confirmation ion) and 171 → 85 (internal standard).

Method verification

Three sets of food samples were used in the method verification procedures.Aliquots (1 g) of each matrix were spiked at three levels to obtain 5.0 μg/L, 25.0 μg/L and 50.0 μg/L concentrations (n = 6 per spike level per food sample) and were also left unfortified (n = 6) to include control samples.The food samples were extracted with water as described above in *Liquid Extraction* and using the SPE procedure described above with analysis by LC/MS/MS.

Statistical analysis

All statistical analyses were completed using SPSS Statistics software from IBM.

Results and Discussion

Method characteristics

Figure 1are example chromatograms generated from the different food spike-recovery experiments demonstrating sufficient separation/detection of native and labeled L-abrine and ricinine by LC/MS/MS in 6 minutes at a sample concentration of 40 μg/L.It also shows the three ion ratios measured for each analyte, including the quantitation, confirmation and internal standard ion ratios.Figures 2 and 3 are example chromatograms of the quantitation and confirmation ion ratios of L-abrine and ricinine respectively. Figures 4 and 5 exhibit extracted ion chromatograms of the L-abrine and ricinine and the resulting MRM.Tables 1 and 2 summarize native and labeled L-abrine and ricinine LC/MS/MS instrumental parameters and SRM configurations.L-abrine and ricinine analyzed by LC/MS/MS was quantified using a linear calibration curve with no weighting (Figures 6 and 7).The calibration curves had a minimum r^2 value of 0.9900 within the calibration range of 0.50 – 200 µg/L for L-abrine and 0.30 – 100 µg/L for ricinine for all extracted matrices demonstrating excellent method linearity.Instrumental detection limits were assigned to the molecular ions when their lowest abundance confirmation ion signal-to-noise (S/N) ≥ 3 as determined by the S/N script of the AB Sciex Analyst data analysis software, version 1.6 (Framingham, MA).Samples used in the determination of instrumental detection limits were standard solutions analyzed from several batches over several days.The instrumental detection limit was 0.10 µg/L for both analytes, L-abrine and ricinine. The analytes were considered quantitative when they calibrated with $r^2 \geq$ 0.9900, their lowest abundance confirmation ion had S/N > 3 and had reproducible and accurate quantitation (± 20% of their true value) as assessed by quality control verification standards. Figures 8 and 9 summarizes L-abrine and ricinine spike recoveries obtained for liquid and SPE extraction methods from chicken, ground beef (15% fat), hot dog and EggBeaters'.The recoveries of the analytes were tested against a plethora of different solvent combinations to determine the most effective combination in eluting the analytes from the SPE columns.Combinations tested included ethyl acetate, methanol, acetonitrile, water, and hexanes with the most effective combination being methanol, water and acetonitrile. These extraction methods used in tandem yielded decent recoveriesranging from 80-120% over all matrices analyzed.The goal of this study was to develop and validate a method for the analysis of L-abrine and ricinine in a variety of food matrices.The data presented strongly indicate that a liquid extraction coupled with a dispersive SPE sample cleanup and LC/MS/MS is a fast, selective, efficient and precise method for the determination of L-abrine and ricinine in food matrices.This method demonstrates good potential for use in monitoring exposure of abrin, L-abrine, ricin and ricinine in food.

Acknowledgements

This study was supported by the Food Safety Inspection Service, Food Emergency Response Network Cooperative Agreement (# FSIS-C-11-2011) from the U.S. Department of Agriculture to the Laboratories Administration, a Unit in the Maryland Department of Health and Mental Hygiene.

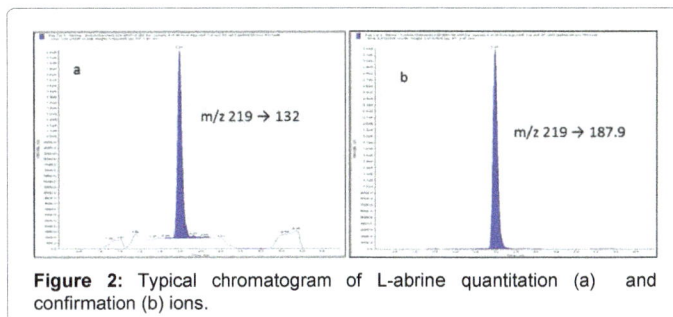

Figure 1: Total Ion Chromatograms of L-abrine (Rt = 2.97 min) , ricinine (Rt = 4.22 min) and L-abrine and ricinine labeled internal standards (Rt = 2.97 min and 4.22 min respectively) extracted from (a) raw chicken, (b) Eggbeaters®, (c) raw ground beef, and (d) hot dog. The different colors in the chromatograph are indicative of the various ion ratios (i.e. quantitation ion ratio, confirmation ion ratio, and internal standard ion ratio).

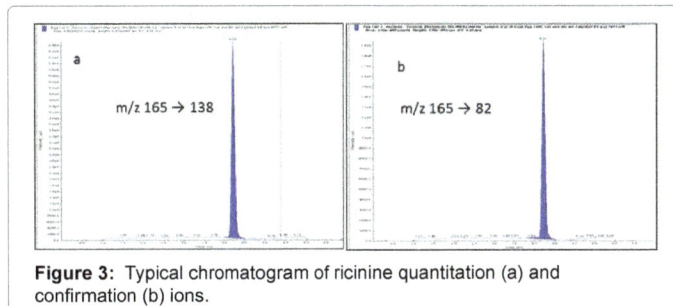

Figure 2: Typical chromatogram of L-abrine quantitation (a) and confirmation (b) ions.

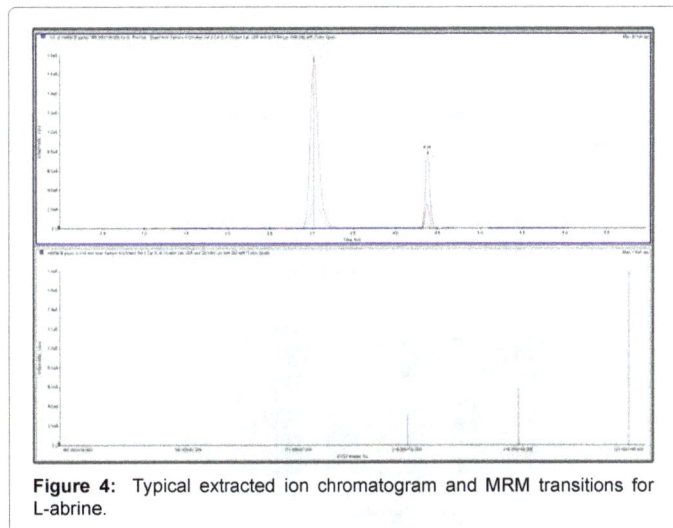

Figure 3: Typical chromatogram of ricinine quantitation (a) and confirmation (b) ions.

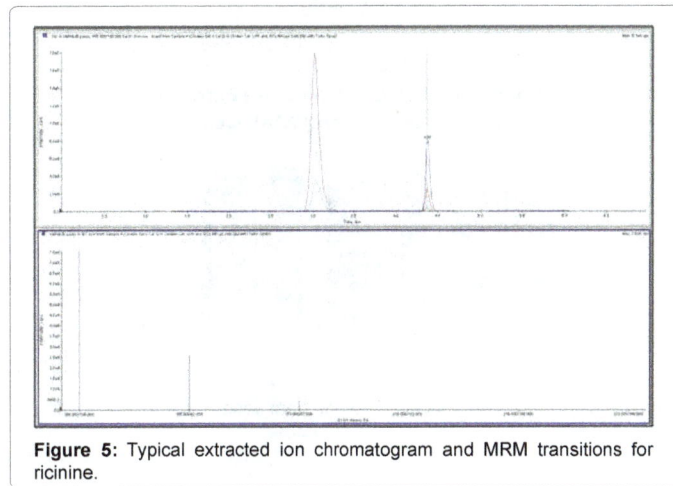

Figure 4: Typical extracted ion chromatogram and MRM transitions for L-abrine.

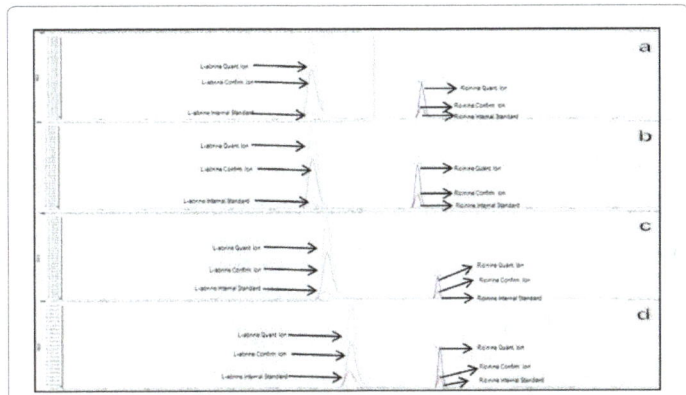

Figure 5: Typical extracted ion chromatogram and MRM transitions for ricinine.

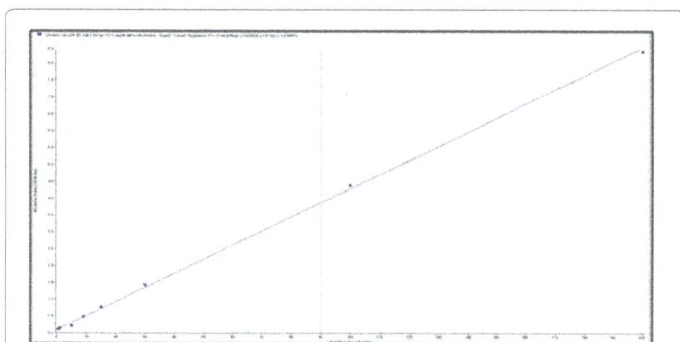

Figure 6: Typical calibration curve for L-abrine. Linear regression equation is equivalent to $y=0.0149x + 0.102$; $r^2=0.9981$

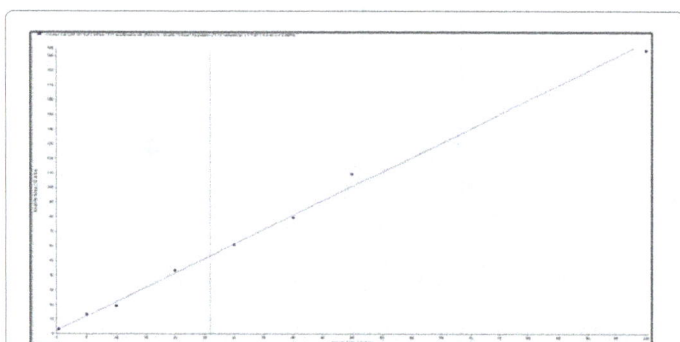

Figure 7: Typical calibration curve for ricinine. Linear regression equation is equivalent to $y=1.97x + 2.43$; $r^2=0.9984$

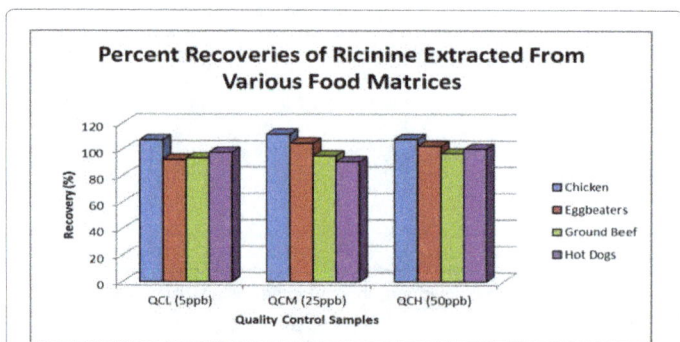

Figure 8: Average recoveries of ricinine from various food matrices. All matrices attained recoveries between 80-120%.

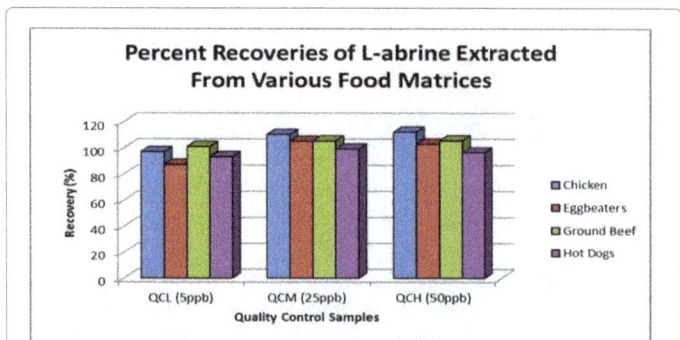

Figure 9: Average recoveries of L-abrine from various food matrices. All matrices attained recoveries between 80-120%.

Parameter	Setting			
LC Method Pump 1	Gradient			
	Reservoir A = 10% MeOH in water, 5 mM formic acid			
	Reservoir B = ACN, 5 mM formic acid			
	Time (min)	%A	%B	Flow Rate (µL/min)
	0.0	93	7	300
	0.5	93	7	300
	2.0	50	50	300
	3.0	50	50	300
	3.01	93	7	300
	6.0	93	7	300
Column Oven	Temperature ranges from 40 °C			
Column Type	Polar RP Phenyl column, 2x100 mm, 2.5µm particle size			
Injection Volume	20 µL			
Needle Rinse Time	5 - 10 seconds, using flushport			
Rinse Solvent	Mobile Phase A			
Injection Mode	Standard			
Autosampler Tray Temp.	20 °C			
Typical Retention Time	L-Abrine = 2.1 min			
	Ricinine = 3.5 min			
HPLC Degasser	Powered for continuous use (there are no settings)			
MS Scan Mode	MS/MS selected reaction monitoring (SRM)			
Ionization Type	Turbo-ionspray (similar to ESI)			
Ion Polarity Mode	Positive ion			
Scan Type	100 msec per channel			
Collision Gas (CAD)	High or "12"			
Curtain Gas (CUR)	20			
Ion Source Gas 1 (GS1)	65			
Ion Source Gas 2 (GS2)	10			
Ion Spray Voltage (IS)	5100			
Temperature (TEM)	550 (interface heater on)			
Declustering Potential (DP)	see Table 5			
Entrance Potential (EP)	10			
Collision Energy (CE)	see Table 5			

Table 1: LC/MS/MS Parameters and Settings.

Analyte	Retention Time (min)	Parent Ion (amu)	Center Ion (amu)	Collision Energy (eV)	Declustering Potential (DP)	Collision Cell Exit Potential (CXP)
Native Ricinine	4.38	165	138	25	51	8
Native Ricinine	4.38	165	82	41	51	8
Labeled Ricinine	4.38	171	85	25	51	8
Native L-abrine	3.01	219	132	29	31	16
Native L-abrine	3.01	219	187.9	17	46	12
Labeled L-abrine	3.01	223	187.9	17	46	12

Table 2: Tandem Mass Spectrometer SRM Configuration.

Disclaimer

The findings and conclusions published in this paper are those of the authors and do not represent the views of the Maryland Department of Health and Mental Hygiene.The use of trade names is for identification only and does not constitute endorsement by the Laboratories Administration, Maryland Department of Health and Mental Hygiene, or the USDA-FSIS.

References

1. Darby SM, Miller ML, Allen RO (2001) Forensic determination of ricin and the alkaloid marker ricinine from castor bean extracts. J Forensic Sci 46: 1033-1042.

2. Vignolo R, Naughton F (1991) Castor: a new sense of direction. Inform 2: 692-699.

3. Johnson RC, Lemire SW, Woolfitt AR, Ospina M, Preston KP, et al. (2005) Quantification of ricinine in rat and human urine: A biomarker for ricin exposure. J Anal Toxicol 29: 149-155.

4. Bradberry SM, Dickers KJ, Rice P, Griffiths GD, Vale JA (2003) Ricin poisoning. Toxicological reviews 22: 65-70.

5. http://emergency.cdc.gov/agent/ricin/clinicians/diagnosis.asp

6. Owens J, Koester C (2008) Quantitation of abrine, an indole alkaloid marker of the toxic glycoproteins abrin, by liquid chromatography/tandem mass spectrometry when spiked into various beverages. J Agric Food Chem 56: 11139-11143.

7. Johnson RC, Zhou Y, Jain R, Lemire SW, Fox S, et al. (2009) Quantification of l-abrine in human and rat urine: a biomarker for the toxin abrin. J Anal Toxicol 33: 77-84.

8. Felder E, Mossbrugger I, Lange M, Wölfel R (2012) Simultaneous detection of ricin and abrin DNA by real-time PCR (qPCR). Toxins 4: 633-642.

9. Knaack JS, Pittman CT, Wooten JV, Jacob JT, Magnuson M, et al. (2013) Stability of ricinine, abrine, and alpha-amanitin in finished tap water. Anal Methods 5: 5804-5811.

23

Use of Polydimethylsiloxane Preconcentration Sorbent for the Analysis of Organotins in Water Samples

Cavalheiro J[1], Tessier E[1], Baltrons O[2], Donard OFX[1] and Monperrus M[1*]

[1]LCABIE-IPREM, CNRS UMR 5254 UPPA, Université de Pau et des Pays de l'Adour, Hélioparc Pau Pyrénées, 2, av. P. Angot, 64053 Pau cedex 9, France
[2]UT2A, Bâtiment de Vinci, Hélioparc Pau Pyrénées, 2, av. P. Angot, 64053 Pau cedex 9, France

Abstract

Organotin compounds were extensively used as powerful biocides in anti-fouling paints and pesticide formulations, particularly tributyltin (TBT), and are recognized as priority pollutants by the european European Water Framework Directive (WFD). Environmental Quality Standard set by the WFD for TBT is fixed at 0.2 ng.L^{-1} for water samples which impose the use of very sensitive analytical techniques to perform such analysis. GC-ICP-MS is the only analytical technique able to achieve such low concentration level, but the expensive equipment requires also qualified personnel to operate, so not all routine laboratories can afford to use it. Preconcentration methods are therefore a good alternative to obtain a more concentrated and cleaner sample extract and later use an easier and affordable analytical technique like GC-MS. In this work, a new method to analyse organotin compounds in water samples using a preconcentration step with polydimethylsiloxane sorbents was developed and validated. The analytical method exhibited linearity higher than 0.9967 for all compounds. For butyltin compounds, LOQ were between 5.7-20.1 pg.L^{-1} with precision below 8%. For methyltin compounds, LOQ were between 26.9-106 pg.L^{-1} with precision below 22%. Matrix effects were evaluated for simulated synthetic estuarine and river waters and was then applied to real estuarine and river water samples. Very good recoveries were obtained for butyltins ranging between 87% and 131%. Methyltins, however, exhibited recoveries between 26% and 118% as a result of matrix effect no accurately corrected by the quantification by external calibration.

Keywords: Organotin compounds; Preconcentration; Polydimethylsiloxane; Water samples

Introduction

Organotin compounds are used in the industry since the 40s, when they were found to prevent discoloration and embrittlement of polyvinyl chloride (PVC) under heat and light. Nowadays, 70% of the organotin production is still used in PVC application [1]. Later, the biocidal properties of trisubstituted organotins were discovered and their introduction in anti-fouling paints began in the 1960s. These anti-fouling coatings inhibit the natural settlement of marine organisms on vessel hulls, which cause frictional drag and consequently reduce ship speed, or increase the fuel consumption in order to maintain speed [2].

The nefarious effects of organotin-based anti-fouling paints started to be clearly visible in the early 80s. In France, oyster farms registered 80%-100% of individuals with abnormalities and similar effects were suspected in the Tagus estuary and the Spanish coast [3]. This lead to the complete prohibition of the use of tributyltin (TBT) in anti-fouling paints in January 1st, 2008 [3,4].

The environmental quality standards (EQS) set by the water framework directive in water samples for TBT is 0.2 ng.L^{-1} [5] which requires very sensitive analytical methods. GC-ICPMS is often used to quantify organotins and achieve such low limits of quantification. However not all routine laboratories can afford this expensive technique which demands qualified personnel to operate [6]. An alternative solution is to improve the extraction technique used to obtain a more concentrated and cleaner sample extract and later use an easier and affordable analytical technique like GC-MS.

New pre-concentration techniques are being developed, some of which handmade and considerably low-cost. Polydimethylsiloxane was first used as a solid sorbent in solid-phase microextraction (SPME) [7] and later in stir-bar sorptive extraction (SBSE) [8]. Both these techniques were already applied to the determination of butyltin compounds [9-13]. Another PDMS-based alternative

extraction technique is the extraction using PDMS rods (silicon rods, SR, or silicon tubes, ST) which were used in the analysis of organic compounds [14,15], showing equivalent performance to SBSE. Nevertheless, this is the first time these solid sorbents are applied to the analysis of organometallic compounds. Having the same composition as stir-bars, the big advantage of the SRs or STs is their flexibility, as each user can vary their length and even divide the rod/tube in several "replicates" of the same sample. Their robustness and low cost are also important features, while stir-bars are fragile and breakable. On the other hand, since they are not yet available in a commercial format, their composition might vary depending on suppliers, and the material might not be as pure as the PDMS applied in the SBSE coatings [16].

The aim of this work was to develop and validate SRs or STs as new pre-concentration sorbents for the quantification of organotin compounds in water samples.

Materials and Methods

Reagents and standards

Tributyltin (TBT) chloride (96%), dibutyltin (DBT) dichloride

---

*Corresponding author: Mathilde Monperrus, LCABIE-IPREM, CNRS UMR 5254 UPPA, Université de Pau et des Pays de l'Adour, Hélioparc Pau Pyrénées, 2, av. P. Angot, 64053 Pau cedex 9, France, E-mail: mathilde.monperrus@univ-pau.fr

(97%), monobutyltin (MBT) trichloride (95%) were obtained from Sigma-Aldrich, whereas trimetyltin (TMT) chloride (98%), dimetyltin (DMT) dichloride (95%), monometyltin (MMT) trichloride (98%) were obtained from Strem Chemicals (Newburyport, MA, USA). The target analytes are resumed in Table 1 as well as their octanol-water partition coefficient. The isotopically enriched butyltin species used were purchased from ISC Science (Oviedo, Spain): a mix of MBT, DBT and TBT enriched in 119 Sn (82.4%) at 0.110, 0.691 and 1.046 µg.g^{-1} respectively.

All stock solutions (1000 µg.g^{-1} as Sn) were prepared by dissolving the corresponding salt in a 3:1 mixture of acetic acid/methanol and were kept in the dark at 4°C until use. Working solutions of the organotin compounds were prepared daily before analysis by dilution of the stock solutions with 1% HCl in ultrapure water.

Hydrochloric acid (HCl, 33-36%, ultrexII ultrapure reagent) and glacial acetic acid (HAc, Instra-analyzed) were purchased from J.T. Baker (Phillipsburg, NJ, USA) and sodium acetate trihydrate (NaAc, puriss p.a.) from Riedel-de-Haën (Seelze, Germany). Sodium tetraethylborate (NaBEt4) with 98% purity were purchased from Merseburger Spezialchemikalien (Germany). Sodium chloride (>99.5%) used to modify the ionic strength of samples was obtained from Avantor Materials (Deventer, The Netherlands) and humic acid sodium salt (technical mixture, 50-60%) was purchased from Alfa Aesar (supplied by VWR, Fontenay-sous-Bois, France).

Water sampling and storage

Estuarine and river water samples were collected in the Adour River (southwest of France) directly into 1 L borosilicate bottles (previously cleaned as described below). Whole-water samples were acidified to 2% with pure glacial acetic acid and placed in the storage cold room at 4°C until sample preparation.

Cleaning procedure

All glassware was carefully cleaned before use by soaking overnight into a solution 10% nitric acid followed by a solution 10% hydrochloric acid and finally in milli-Q water (18.2 MΩ). The silicon rods and tubes were previously soaked overnight in isooctane (or in the appropriate solvent, in the particular case of different solvent testing), dried immediately before use with a lint-free tissue and discarded after use.

Silicon-tube (ST) extraction method

One 0.8-cm length ST was added to 100-mL sample containing 5 mL of HAc/NaAc buffer (pH 5, 0.1 mol.L^{-1}) and previously spiked with the appropriate amount of isotopically enriched butyltin species (^{119}TBT, ^{119}DBT and ^{119}MBT). The pH was then re-adjusted with ultrapure HCl to pH 5. The organotin species were ethylated by adding 1 mL of isooctane and 1 mL of NaBEt4 1% (w/v) prepared daily, followed by 5 minutes of vigorous shaking. The ST was recovered, transferred to a 1.5 mL conic tubes (Eppendorf) were 300 µL of isooctane were added and left to complete desorption in the ultrasounds bath for 10 minutes. The organic solvent was recovered into a 300 µL glass insert in a GC-vial and the GC-ICP/MS analysis was performed within 24 hr.

GC-ICP-MS analysis

A Thermo X Series 2 inductively coupled plasma-mass spectrometer (ICP/MS) coupled to a gas chromatograph (GC) (Thermo Fisher, Waltham, MA, USA) by a commercial GC-ICP/MS interface (SilcoSteel, 0.5 m length, inner i.d. 0.28 mm and o.d. 0.53 mm, outer i.d. 1.0 mm and o.d. 1.6 mm, Thermo Fisher) was used [17]. The chromatographic

Analyte	Log K$_{ow}$	Analyte	Log K$_{ow}$
Monobutyltin (MBT)	0.18[a]	Monomethyltin (MMT)	− 2.15[a]
Dibutyltin (DBT)	1.89[a]	Dimethyltin (DMT)	−2.18 to −3.1[a]
Tributyltin (TBT)	3.9 − 4.9[b]	Trimethyltin (TMT)	n.i.

[a](Dobson et al.); [b](Bangkedphol et al.); n.i.: No information available

Table 1: Chemical structures of the target organometallic compounds and their log K$_{ow}$.

separation was performed in a (30 m × 0.32 μm i.d. × 0.25 μm coating) HP-5 capillary column (Agilent Technologies, Santa Clara, CA, USA) with the following oven program: start at 40°C for 0.5 minutes, increase at 20°C.min⁻¹ until 50°C, then increase at 50°C.min⁻¹ until 290°C and hold for 1 min. Two μL of sample were injected in the GC inlet at 280°C, the carrier gas (He) was set 1.5 ml.min⁻¹ and the interface temperature to 290°C. The ICP-MS analysis was performed with 1250 W of forward power, plasma gas flow at 15 L.min⁻¹, nebulizer gas flow at 0.6 L.min⁻¹ and finally the make-up gas flow at 0.3 L.min⁻¹. The performance of the instrument was optimized with liquid standards. Dwell time for Sn isotopes (116, 117, 118, 119, 120) was 30 ms and Sb (isotopes 121 and 123, dwell time=5 ms) was measured to check the mass bias in each chromatographic run. Butyltin species were quantified by speciated isotope dilution mass spectrometry (SIDMS) by using a mixed isotopic spike solution containing ¹¹⁹MBT, ¹¹⁹DBT and ¹¹⁹TBT, which takes into account the cross-species transformations, the loss of analytes during sample preparation and the signal drift. Methyltin compounds, on the other hand, were determined by interpolation on the external calibration curve, because no isotopic tracers are available for these species.

Results and Discussion

Several parameters were previously tested and optimized, such as the geometry of the sorbent phase, the nature and amount of solvent used for the desorption from the extracting phase (also known as back extraction), whether the use of ultrasounds helped desorption, the amount of sorbent phase and the extraction time.

Influence of the geometry of the sorbent phase: STs vs SRs

In this work, PDMS was tested in 2 different forms: rods and tubes. The rods are compact and have 1 mm of diameter, while the tubes have an internal surface, resulting in 3 mm of external diameter and 2 mm of internal diameter. The response obtained (Figure 1) is the same for both forms, while the surface area of the silicon rod is almost 2 times bigger than the tube. Previous tests performed to evaluate differences in the surface of both materials revealed that the tubes and the rods did not exhibit the typical isotherm profile of a porous material (data not shown). This was the first indication that the phenomenon behind the trap of analyte in PDMS is not adsorption into the surface pores, but migration through the polymeric net, which is normally called sorption [18]. The second evidence of this fact is the response obtained when the same mass of sorbent (rather than the same length) was compared. A tube fragment of 2 mm length and a rod fragment of 12 mm exhibited the same mass, 10 mg, and their responses were compared. Figure 1 shows that both PDMS forms presented similar sensitivities for the studied compounds, reinforcing the hypothesis that a sorption

phenomenon was occurring. Silicon tubes were therefore chosen to proceed with further optimization.

Influence of the nature and the volume of back extraction solvent

Synthetic water samples prepared with ultrapure water and all analytes at 1 ng.L⁻¹ were extracted with ST and SR and then recovered in different solvents (hexane, isooctane, acetonitrile and ethylacetate). Results are shown in Figure 1. Isooctane was the solvent exhibiting the higher responses for all the organotin compounds and was then chosen for the optimized procedure.

The use of these solvents with such different hydrophobic/hydrophilic properties had the purpose to evidence the hydrophobic/hydrophilic properties in the studied molecules. In fact, Figure 1 shows that the most hydrophilic solvent (acetonitrile) is the one that exhibits a weaker capacity to recover analytes from the organic phase, which translates in the lowest response of analytes in peak area. Peak areas increase with decreasing polarity (or increasing hydrophobicity) of the solvents, which proves the affinity of the hydrophobic molecules to the hydrophobic solvent. It is important to highlight that the sorption of the molecules into the solid sorbent occurs after derivatization with sodium tetraethylborate, which provides ethyl groups to the organotin molecules and grants them a more hydrophobic character.

The volume of solvent used for the back extraction is critical, because too little volume does not extract all the analytes from the sorbent phase and too much solvent cause's unnecessary dilution of the extract. Three separated fractions of 300 μL of isooctane were successively added for the back extraction in order to determine the percentage of analytes recovered in each fraction. The results, shown in Figure 2, show that with 2 fractions of 300 μL, 90% of the analytes are eluted from the sorbent into the solvent extract. However, with only one portion of isooctane, 80% of the analytes are recovered (except MBT) with no dilution of the extract, improving sensitivity and, consequently, analytical performances of the method.

Influence of the use of ultrasounds for the back extraction

The use of ultrasound to extract the organotin analytes from the sorbent was also evaluated and, as shown is Figure 3, better results were obtained when the back extraction was carried out in ultrasound bath for 10 minutes.

Influence of the amount of sorbent phase

Five sequential extractions were made, each with one segment of 2 mm of sorbent, in order to evaluate how many segments were needed to extract the totality of analytes from the synthetic water

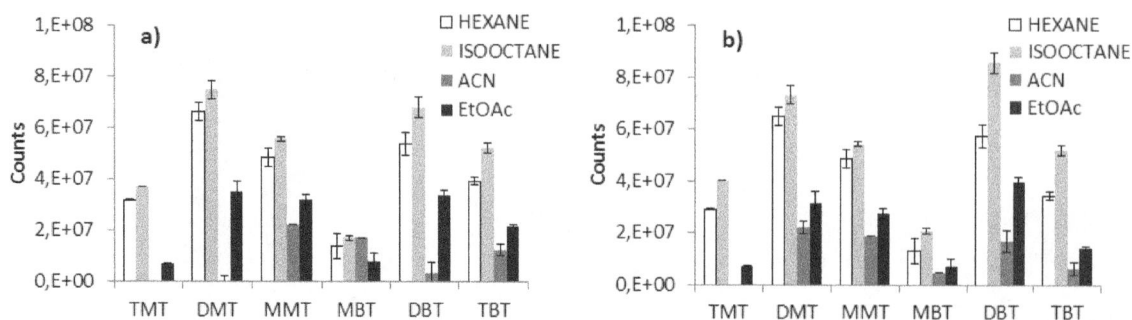

Figure 1: Response obtained for both ST (a) and SR (b) when water samples were extracted with hexane, isooctane, acetonitrile (ACN) and ethylacetate (EtOAc).

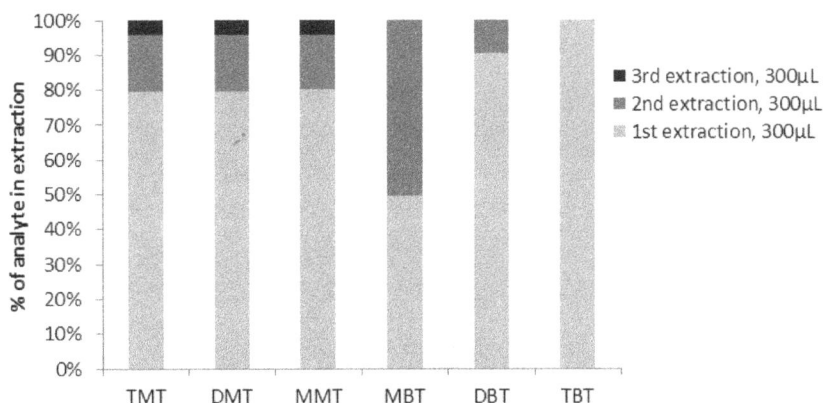

Figure 2: Percentage of analytes obtained in each fraction of solvent used for the back extraction.

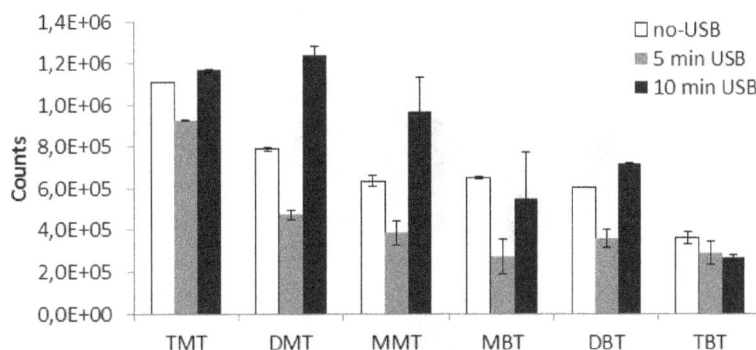

Figure 3: Results obtained when the back extraction was performed with (5 and 10 minutes) and 10 minutes without ultrasonic bath (USB).

samples. Theses synthetic water samples were prepared with ultrapure water and spiked to a final concentration of 1 ng.L^{-1}. Results (Figure 4) clearly show that more than 90% of the analytes are recovered with 4 segments, corresponding to 8 mm of sorbent phase. This amount of ST was further used for the method validation and the real samples analysis.

Influence of the extraction time

The speed and turbulence at which the simultaneous derivation/extraction step is performed has a major effect in its efficiency, so both shaking vigorously and stirring at 900 rpm were evaluated (Figure 5) in synthetic samples spiked with organotins at 1 ng.L^{-1}. The optimization of this step is crucial, because it allows the compounds to be transformed into more volatile substitutes and trap them into the organic phase, usually isooctane. Here, the organic phase was replaced by the 8 mm ST and the results obtained are represented in Figure 5.

Evaluation of the matrix effects

Matrix effects were evaluated by preparing different synthetic water samples (containing 1 ng.L^{-1} of butyl- and methyl-tins) at different salt and humic acid concentrations in order to simulate estuarine and riverine waters. Recoveries were calculated by comparing with no salt and no humic acids addition to the synthetic solutions (Figure 6).

The increasing ionic strength of the solution is usually one of the matrix modifications performed to optimize trapping the hydrophobic compounds into the organic phase. However, the presence of salt at 35 g.L^{-1} decreased the sorption of the compound into the ST. On the other hand, the presence of humic acids severely affected the sorption of these compounds no matter the concentration at which they were present. This matrix effect is higher for butyltins than for methyltins, but while for butyltins the trisubstituted molecule is the most affected (humic acid effect: TBT>DBT>MBT), for methyltins it is the opposite (humic acid effect: MMT>DMT>TMT) indicating that the derivatized molecule is more hydrophobic than the original ionic compound and therefore adsorbs more onto the organic matter.

Analytical performances

After optimization, figures of merit of the method were determined and are presented in Table 2. Linearity (represented as correlation coefficient) higher than 0.9967 was obtained for all compounds. LOD and LOQ for methyltins (between 8.1-31.9 pg.L^{-1} and 26.9-106 pg.L^{-1}, respectively) were - in average - one order of magnitude higher than for butyltins (between 1.7-6.0 pg.L^{-1} and 5.7-20.1 pg.L^{-1}, respectively) because the latter were calculated through isotope dilution analysis and species specific enriched isotopes are not available for methyltins. There are not many recent analytical publications on the determination of methyltins in water samples using GC-ICP-MS as separation and detection technique, but when compared to methods recently developed for landfill leachates, LOD obtained in this study are 150 to 800 times lower [19,20]. Butyltins also exhibit very low limits of detection, only closely achieved by techniques like solid phase extraction (SPE) and stir-bar sorptive extraction (SBSE) coupled to high performance analytical techniques like GC-ICP-MS and GC-MS-MS [21].

In what concerns precision, the quantification method performed

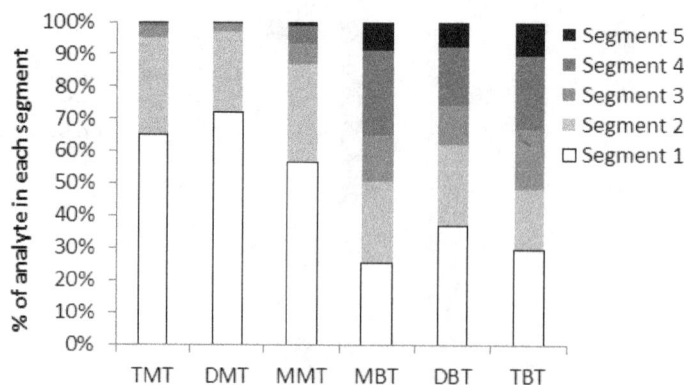

Figure 4: Percentage of analytes obtained in each 2 mm segment of sorbent.

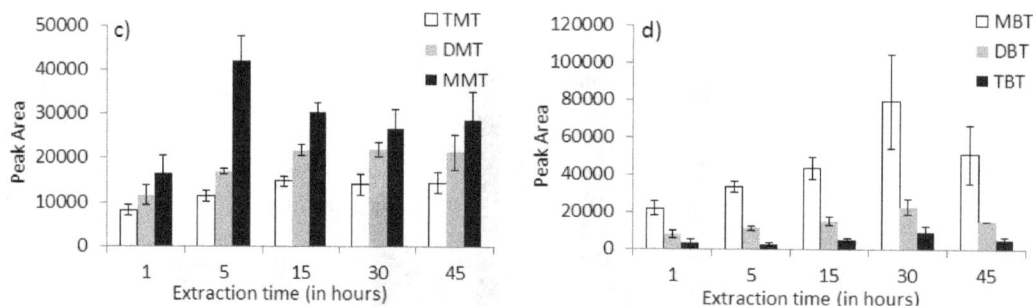

Figure 5: Peak areas obtained when using different types of agitation during extraction: shaking (a and b) and stirring (c and d).

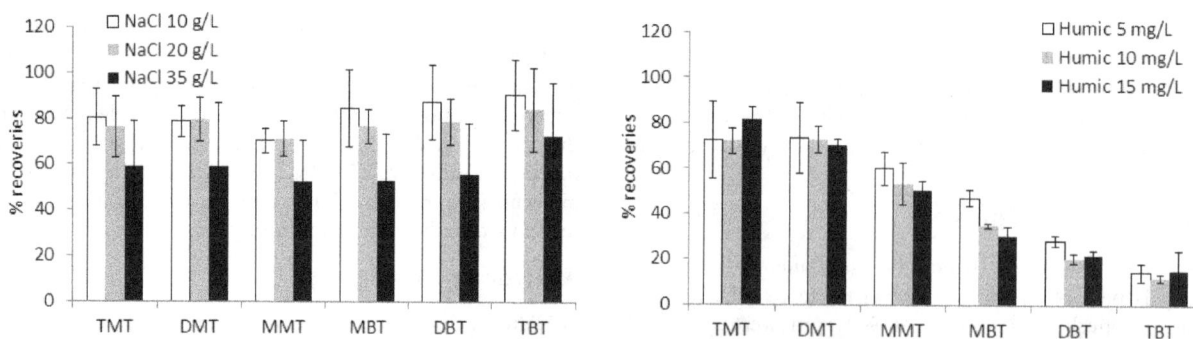

Figure 6: Recoveries determined for synthetic water samples at different salinity and humic acids concentration levels.

for the two families of compounds influenced the results obtained: for methyltins precision was below 22% while for butyltins was below 8%.

Application to real samples

Estuarine and river water samples were collected in the Adour river (southwest of France) and their physical-chemical data are described in Table 3.

Both samples were analyzed with the developed method but organotins were present in concentrations below LOQ. Therefore, both samples were spiked with a mixed solution of all organotin compounds up to a concentration of 0.1 ng.L^{-1} (an example of a chromatogram of the enriched sample is shown in Figure 7). Recoveries were calculated and are represented in Table 4.

The butyltin compounds quantified by isotope dilution analysis provided acceptable results with good recoveries ranging between 87% and 131%. Methyltins, however, were quantified by external calibration and exhibited poorer recoveries, between 26% and 118%. The method developed in this work to quantify organotins in river water samples has provided very good results when used with isotope dilution analysis.

Conclusion

In this work, a low cost preconcentration technique was applied to water analysis in order to quantify organotin compounds. Silicon tubes made with PDMS are cheap and easy to adapt to water analysis as the sample preparation is very similar to stir-bar sorptive extraction (SBSE).

	Linearity (R²)	LOD	LOQ	Precision (n=3)	
	LOQ -- 3 ng.L⁻¹	pg.L⁻¹	pg.L⁻¹	% at 0.1 ng.L⁻¹	% at 1 ng.L⁻¹
TMT	0.9985	31.9	106	17	1.8
DMT	0.9989	11.7	39.1	22	2.0
MMT	0.9999	8.1	26.9	3	1.2
MBT	0.9968	6.0	20.1	1	1.4
DBT	0.9984	1.7	5.7	5	0.1
TBT	0.9967	3.7	12.4	8	1.4

Table 2: Figures of merit of the method: linearity, LOD, LOQ and precision.

Sample	Barre	Pont Rouge
pH	7.57	7.65
Temperature	8.68°C	8.76°C
Conductivity	984 µS/cm	225 µS/cm
TDS	420 ppm	112 ppm
DOC	15.8 mg/L	16.2 mg/L

Table 3: Physical-chemical parameters of water samples collected in the Adour River.

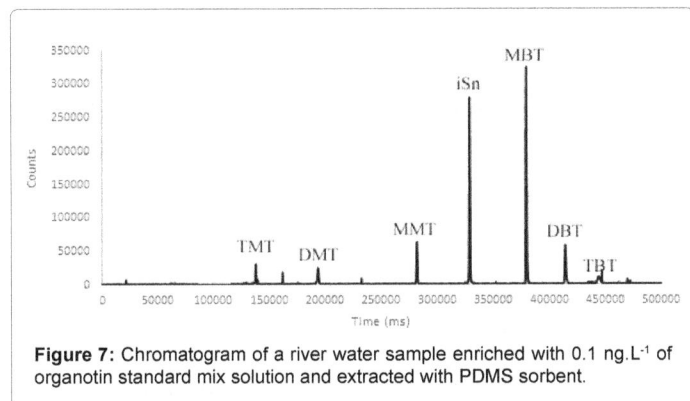

Figure 7: Chromatogram of a river water sample enriched with 0.1 ng.L⁻¹ of organotin standard mix solution and extracted with PDMS sorbent.

	TMT	DMT	MMT	MBT	DBT	TBT
Barre	26	112	48	91	87	131
Pont Rouge	26	118	40	93	90	118

Table 4: Recoveries (%) determined for estuarine and river water samples spiked with an organotin mixed solution to a final concentration of 0.1 ng.L⁻¹.

The method exhibited linearity higher than 0.9967 for all compounds. For methyltins, LOQ were between 26.9-106 pg.L⁻¹ and precision was below 22%. For butyltins, LOQ were about five times lower, between 5.7-20.1 pg.L⁻¹, and precision below 8%, as a result of using species specific isotope dilution for the quantification of these species. When applied to spiked environmental water samples, very good recoveries were obtained for butyltins: between 87% and 131%. Methyltins, however, exhibited recoveries between 26% and 118% as a result of their quantification by external calibration.

Acknowledgements

J. Cavalheiro is grateful to the French Minister of Education and Research for her PhD Grant (Doctoral School ED211, Université de Pau et des Pays de l'Adour). This work was financially supported by the Conseil Général des Pyrénées Atlantiques and the EU Interreg funding in the framework of the ORQUE SUDOE project.

References

1. Hoch M (2001) Organotin compounds in the environment: an overview. Appl Geochem 16: 719-743.

2. Dafforn KA, Lewis JA, Johnston EL (2011) Antifouling strategies: history and regulation, ecological impacts and mitigation. Mar Pollut Bull 62: 453-465.

3. Alzieu CL, Sanjuan J, Deltreil JP, Borel M (1986) Tin contamination in Arcachon Bay: Effects on oyster shell anomalies. Mar Pollut Bull 17: 494-498.

4. Comission E (2003) Regulation (EC) No 782/2003, Official Journal of the European Union, L 115/1, 09.05.2003.

5. Commission E (2013) Directive 2013/39/EU, Official Journal of the European Union, L 226/1, 24.08.2013.

6. Cavalheiro J, Preud'homme H, Amouroux D, Tessier E, Monperrus M (2014) Comparison between GC-MS and GC-ICPMS using isotope dilution for the simultaneous monitoring of inorganic and methyl mercury, butyl and phenyl tin compounds in biological tissues. Anal Bioanal Chem 406: 1253-1258.

7. Arthur CL, Pawliszyn J (1990) Solid phase microextraction with thermal desorption using fused silica optical fibers. Anal Chem 62: 2145-2148.

8. Baltussen E, Sandra P, David F, Cramers C (1999) Stir bar sorptive extraction (SBSE), a novel extraction technique for aqueous samples: Theory and principles. J Microcolumn Sep 11: 737-747.

9. Vercauteren J, Pérès C, Devos C, Sandra P, Vanhaecke F, et al. (2001) Stir bar sorptive extraction for the determination of ppq-level traces of organotin compounds in environmental samples with thermal desorption-capillary gas chromatography - ICP mass spectrometry. Anal Chem 73: 1509-1514.

10. Prieto A, Telleria O, Etxebarria N, Fernández LA, Usobiaga A, et al. (2008) Simultaneous preconcentration of a wide variety of organic pollutants in water samples: Comparison of stir bar sorptive extraction and membrane-assisted solvent extraction. J Chromatogr A 1214: 1-10.

11. Van Hoeck E, Canale F, Cordero C, Compernolle S, Bicchi C, et al. (2009) Multiresidue screening of endocrine-disrupting chemicals and pharmaceuticals in aqueous samples by multi-stir bar sorptive extraction-single desorption-capillary gas chromatography/mass spectrometry. Anal Bioanal Chem 393: 907-919.

12. Bianchi F, Careri M, Maffini M, Mangia A, et al. (2010) A fast and effective routine method based on SPME and GC/ICP-MS for the monitoring of organotin compounds in surface and sea water. Curr Anal Chem 6: 223-227.

13. Segovia-Martinez L, Bouzas-Blanco A, Campins-Falco P, Seco-Torrecillas A (2010) Improving detection limits for organotin compounds in several matrix water samples by derivatization-headspace-solid-phase microextraction and GC-MS. Talanta 80: 1888-1893.

14. Montero L, Popp P, Paschke A, Pawliszyn J (2004) Polydimethylsiloxane rod extraction, a novel technique for the determination of organic micropollutants in water samples by thermal desorption-capillary gas chromatography-mass spectrometry. J Chromatogr A 1025: 17-26.

15. Popp P, Bauer C, Paschke A, Montero L (2004) Application of a polysiloxane-based extraction method combined with column liquid chromatography to determine polycyclic aromatic hydrocarbons in environmental samples. Anal Chim Acta 504: 307-312.

16. van Pinxteren M, Paschke A, Popp P (2010) Silicone rod and silicone tube sorptive extraction. J Chromatogr A 1217: 2589-2598.

17. Monperrus M, Tessier E, Veschambre S, Amouroux D, Donard O (2005) Simultaneous speciation of mercury and butyltin compounds in natural waters and snow by propylation and species-specific isotope dilution mass spectrometry analysis. Anal Bioanal Chem 381: 854-862.

18. Mitra S (2003) Sample Preparation Techniques in Analytical Chemistry, Chemical Analysis: a series of monographs on analytical chemistry and its applications. John Wiley & Sons Inc., New Jersey, USA.

19. Cacho JI, Campillo N, Viñas P, Hernández-Córdoba (2013) Headspace sorptive extraction for the analysis of organotin compounds using thermal desorption and gas chromatography with mass spectrometry. J Chromatogr A 1279: 1-6.

20. Vahčič M, Milačič R, Sčančar J (2011) Development of analytical procedure for the determination of methyltin, butyltin, phenyltin and octyltin compounds in landfill leachates by gas chromatography–inductively coupled plasma mass spectrometry. Anal Chim Acta 694: 21-30.

21. Cole RF, Mills GA, Parker R, Bolam T, Birchenough A, et al. (2015) Trends in the analysis and monitoring of organotins in the aquatic environment. Trends Environ Anal Chem 8: 1-11.

24

Stability-Indicating HPTLC Method for Estimation of Guaifenesin in Bulk and in Pharmaceutical Formulation

Jain PS *, Kale NK and Surana SJ

RC Patel Institute of Pharmaceutical Education and Research, Karwand Naka, Shirpur Dist, Dhule 425 405 (MS) India

Abstract

A HPTLC/ Densitometric method has been developed for the determination of Guaifenesin (GFN) in bulk and pharmaceutical formulation. The estimation of drug was performed on HPTLC aluminium plates precoated with silica gel 60 RP-18 TLC F254 S using toluene: methanol: ethyl acetate: acetic acid (7:0.8:1.2:0.5 v/v/v/v) as mobile phase. The densitometric quantification for the drug was carried out at 274 nm. GFN obeyed linearity in concentration range 800 – 2800 ng/band with coefficient of correlation 0.999. The Rf for GFN was found to be 0.6 ± 0.02. The proposed method was applied for pharmaceutical formulation and % label claim for GFN was found to be 100.34 ± 1.06. The method was validated for accuracy, precision and ruggedness. Accuracy of the method was checked by recovery studies at three different levels i.e. 80 %, 100 % and 120 %. The % recovery of GFN was found to be in the range of 99.48% - 100.68 %; the % RSD value was less than 2 indicates the accuracy of the method. The method was found to be precise as indicated by the inter-day, intra-day and repeatability analysis; showing % RSD less than 2. The results did not show any statistical difference between operators showing that developed method was rugged. GFN was subjected to acid and alkali hydrolysis, oxidation and thermal degradation. The drug undergoes degradation under acid-base conditions, hydrolysis, oxidation, photo degradation except dry heat degradation. This indicates that the drug is susceptible to acid and base. The degraded product was well resolved from the pure drug with significantly different Rf value. Statistical analysis proves that the method is repeatable, selective and accurate for the estimation of investigated drug. The proposed developed HPTLC method can be applied for the identification and quantitative determination of GFN in bulk drug and pharmaceutical formulation.

Keywords: Guaifenesin; High-performance thin-layer chromatography; Validation; Stability; Degradation

Introduction

Guaifenesin, (RS)-3-(2-methoxyphenoxy)propane-1,2-diol (Figure 1), Guaifenesin may act as an irritant to gastric vagal receptors, and recruit efferent parasympathetic reflexes that cause glandular exocytosis of a less viscous mucus mixture. Cough may be provoked. This combination may flush tenacious, congealed mucopurulent material from obstructed small airways and lead to a temporary improvement in dyspnea or the work of breathing. Guaifenesin is an expectorant which increases the output of phlegm (sputum) and bronchial secretions by reducing adhesiveness and surface tension. The increased flow of less viscous secretions promotes ciliary action and changes a dry, unproductive cough to one that is more productive and less frequent. By reducing the viscosity and adhesiveness of secretions, Guaifenesin increases the efficacy of the mucociliary mechanism in removing accumulated secretions from the upper and lower airway [1,2].

Various methods such as Spectrophotometric estimation of Guaifenesin and salbutamol in pure and tablet dosage form by using different methods, Determination of Guaifenesin and Dextromethorphan in a Cough Syrup by HPLC with Flurometric Determination, HPLC determination of Guaifenesin with selected medications on underivatized silica with an aqueous-organic mobile phase, Determination of Guaifenesin in human serum by capillary gas chromatography and electron capture detection, Simultaneous, stability indicating, HPLC-DAD determination of Guaifenesin and methyl and propyl-parabens in cough syrup [3-7]. To our knowledge, no articles related to the stability-indicating chromatographic determination of guaifenesin by high-performance thin-layer chromatography (HPTLC) in pharmaceutical dosage forms have been reported in literature. The hydrolytic and the photolytic stability also are required. An ideal stability-indicating method is one that quantifies the drug and also resolves its

degradation products. Unlike HPLC, consumption of mobile phase per sample basis is quite low. This saves cost per analysis and analysis time as well. HPTLC facilitates repeated detection (scanning) of the chromatogram with the same or different parameters. HPTLC technique is most suited for impurity profile of drug substances and content uniformity test as per compendia specifications [8,9]. The aim of this work is to develop an accurate and specific, repeatable and stability-indicating method for the determination of ciprofibrate in the presence of its degradation products as per ICH guidelines [10,11].

Experimental

Chemicals and reagents

Guaifenesin was supplied as a gift sample from Macloeds Pharmaceutical, Daman, India all chemicals and reagents used were of analytical grade (MERCK Chem. Ltd., Mumbai). Methanol was selected as the solvent for sample preparation.

HPTLC instrumentation

Chromatography was performed on 20 cm × 10 cm aluminum-backed TLC plates coated with 200 µm layers of silica gel 60F$_{254}$ S (E. Merck, Darmstadt, Germany; supplied by Merck India, Mumbai, India). The plates were prewashed by methanol and activated at 100 –

*Corresponding author: Pritam Jain, RC Patel Institute of Pharmaceutical Education and Research, Karwand Naka, Shirpur Dist, Dhule 425 405 (MS) India, E-mail: pritash79@yahoo.com

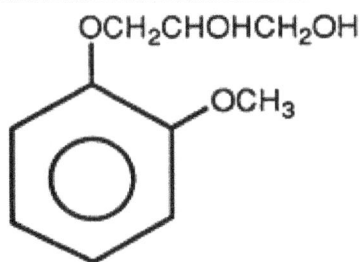

Figure 1: Chemical structure of Guaifenesin.

110 °C for 10 min prior to chromatography. The samples were applied on the plates as 6 mm wide bands, by means of a CAMAG (Muttenz, Switzerland) Linomat-5 sample applicator fitted with a 100 µL sample syringe (Hamilton, Bonaduz, Switzerland). Plate was developed to a distance of 8 cm using toluene: methanol: ethyl acetate: acetic acid (7:0.8:1.2:0.5) as mobile phase in a Camag twin-trough glass chamber previously saturated with mobile phase vapors for 10 min at ambient temperature. Densitometric scanning was performed at 274 nm using Camag TLC Scanner 3 equipped with win CATS software version 1.3.0.

Preparation of standard solution and linearity study

An accurately weighed GFN (20 mg) was transferred to 10 mL volumetric flask; dissolved in methanol and the volume was made up to mark with the same solvent to give 2000 ng/µL solution. Linearity was performed using working standard of GFN. Calibration was done by applying standard stock solution ranging from 0.4-1.4 µL on TLC Plate; which gives concentration of 800 – 2800 ng/band. The plate was developed and scanned as described under above chromatographic conditions. Calibration curve was constructed by plotting the peak area *vs.* corresponding drug concentration.

Preparation of In-House Tablet

Tablet compression

Direct compression method: MCC and talc were dispensed to the correct masses and blended together. GFN was then transferred and blended together. Powders were blended for 5 minutes and subjected for direct compression. Each tablet contains 100 mg of drug [12].

Method Validation

Precision

Intra-day and inter-day precision: Intraday precision was determined by analyzing, the three different concentrations 800 ng, 1200 ng and 1600 ng of GFN, for three times within the day. Day to day variability was assessed using above mentioned three concentrations and analyzing it for three consecutive days, which shows reproducibility of the method.

Repeatability: Repeatability of sample application was assessed by applying 0.8 µL (1600 ng) of drug solution six times on a TLC plate followed by development of plate and recording the peak height and area for 6 bands.

Limit of Detection and Limit of Quantification

Limit of Quantification (LOQ) LOD and LOQ were calculated by

the method which was based on the SD of the response and the slope (S) of the calibration curve at levels approximating the LOD and LOQ, Sensitivity of the proposed method was estimated in terms of Limit of Detection (LOD) and LOD = 3.3(SD/S) and LOQ = 10(SD/S). Stock solution of GFN was prepared and different volume of stock solution in the range 700 to 1200 ng were applied in triplicate.

Ruggedness

Ruggedness of the method was checked by analyzing 1600 ng (n = 6) of GFN, with the help of two analysts and the variations in the results were checked.

Accuracy

Recovery study was carried out by over spotting at 80, 100 and 120 % level where known amount of standard GFN was added to pre analyzed sample (800 ng of GFN) and subjected them to the proposed TLC method.

Robustness

Robustness was studied at the concentration level of 1600 ng/band. In this study, few parameters (mobile phase composition, development distance and duration of saturation) were studied and the effects on the results were examined.

Application of Proposed Method to Tablet Formulation

To determine the content of GFN in tablets, twenty tablets, each containing 250 mg GFN, were accurately weighed and finely powdered. An amount equivalent to 20 mg GFN was transferred to 10 mL volumetric flask and extracted with methanol for 20 minutes by shaking mechanically. The solution was diluted to volume with the same solvent and filtered; from it, the sample solution (0.8 µL, containing 1600 ng of GFN) was applied on TLC plate, developed and scanned.

Forced Degradation of Guaifenesin

Acid-and base-induced degradation

The 20 mg of GFN was separately dissolved in 10 ml of methanolic solution of 1N HCl and 1N NaOH. These solutions were kept for 8 h at room temperature. The 1ml of above solutions was taken and neutralized, then diluted up to 10 ml with methanol. The resultant solution were applied on TLC plate in triplicate (0.8 µl each, i.e. 1600 ng per spot).

Hydrogen peroxide-induced degradation

The 20 mg of GFN was separately dissolved in 10 ml of methanolic solution of hydrogen peroxide (10.0%, v/v). The solution was kept for 8 h at room temperature in the dark in order to exclude the possible degradative effect of light. The resultant solution was applied on TLC plate in triplicate (0.8 µl each, i.e. 1600 ng per spot).

Dry heat degradation products

The powdered drug was kept at 60°C for 8 h under dry heat condition and then 20 mg sample was taken and dissolved in 10ml methanol. Spot were applied and chromatograms were run.

Photochemical degradation

The photochemical stability of the drug was also studied by exposing the stock solution to direct sunlight for 24 h. The resultant solution (0.8 μL, i.e. 1600 ng per spot) was applied on a TLC plate and chromatograms were run.

Results and Discussion

Development of optimum mobile phase

Different ratios of toluene, methanol and ethyl acetate were tried as mobile phase was tried but, tailing of spot, less persistent spots were observed in most of the attempts. In order to overcome the problems, toluene: methanol: ethyl acetate: acetic acid (7:0.8:1.2:0.5 $v/v/v/v$) was tried and results is good resolution, sharp and symmetrical peak with R_f value of 0.6 for GFN (Figure 2).

Calibration curve

The linear regression data for the calibration curves showed good linear relationship over the concentration range 800-2800 ng/spot. Linear regression was found to be Y = 2.118 x + 428.8, Slope = 2.118, Intercept = 428.8, Correlation coefficient = 0.999

Validation of Method

Precision

The precision of the developed HPTLC method was expressed in terms of percent relative standard deviation (% RSD). The results, presented in Table 1, revealed high precision of the method.

LOD and LOQ

Detection limit and quantification limit were calculated by the method described in the section "Limit of detection and limit of quantification". The LOQ and LOD were found to be 2.24 and 6.79 ng, respectively. This indicates the adequate sensitivity of the method.

Recovery studies

The proposed method when used for extraction and subsequent estimation of ciprofibrate from the pharmaceutical dosage formed after over spotting with 80%, 100% and 120% of additional drug afforded good recovery of ciprofibrate. The amounts of drug added and determined and the percentage recovery are listed in Table 2.

Robustness of the Method

The standard deviation of peak areas was calculated for each parameter, and % RSD was found to be <2%. The low values of % RSD values, shown in Table 3, indicated the robustness of the method.

Analysis of In-House Tablet

A single spot at Rf 0.6 was observed in the chromatogram of the drug samples extracted from tablets. There was no interference from the excipients commonly present in the tablet. The % drug content and % RSD were calculated. The low % RSD value indicated the suitability of this method for the routine analysis of ciprofibrate in pharmaceutical dosage forms (Table 4 and 5).

Force Degradation

GFN was subjected to acid and alkali hydrolysis, oxidation and thermal degradation. The drug undergoes degradation under acid-base conditions, hydrolysis, oxidation, photo degradation except dry heat degradation. This indicates that the drug is susceptible to acid and base. The degraded product was well resolved from the pure drug with significantly different Rf value. Statistical analysis proves that

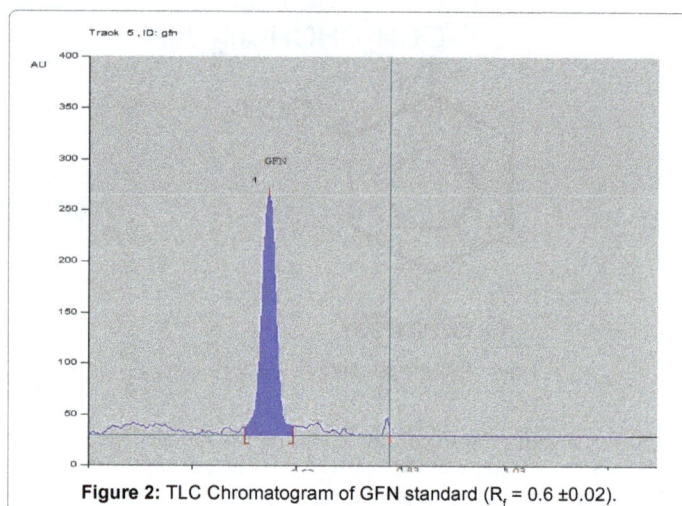

Figure 2: TLC Chromatogram of GFN standard (R_f = 0.6 ±0.02).

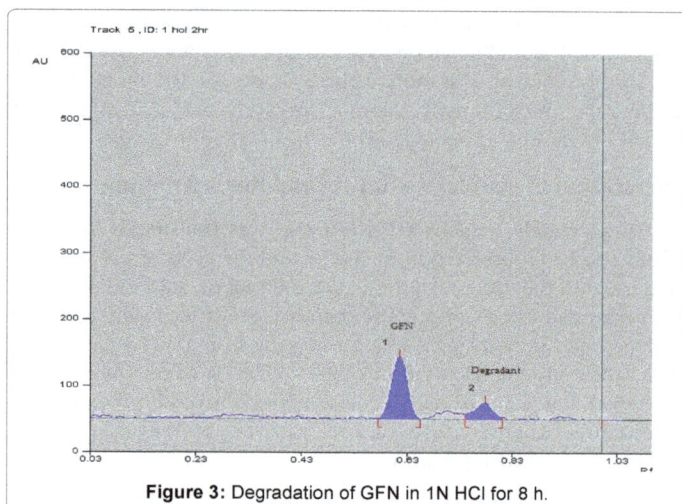

Figure 3: Degradation of GFN in 1N HCl for 8 h.

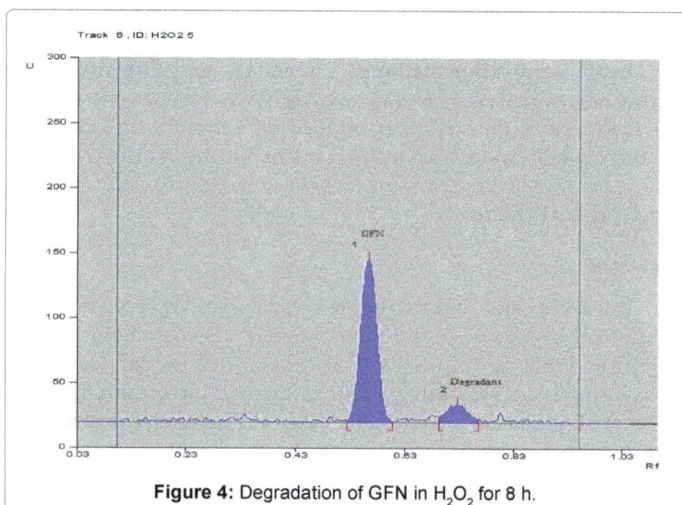

Figure 4: Degradation of GFN in H_2O_2 for 8 h.

the method is repeatable, selective and accurate for the estimation of investigated drug. The proposed developed HPTLC method can be applied for the identification and quantitative determination of GFN in bulk drug and pharmaceutical formulation (Figure 3-5).

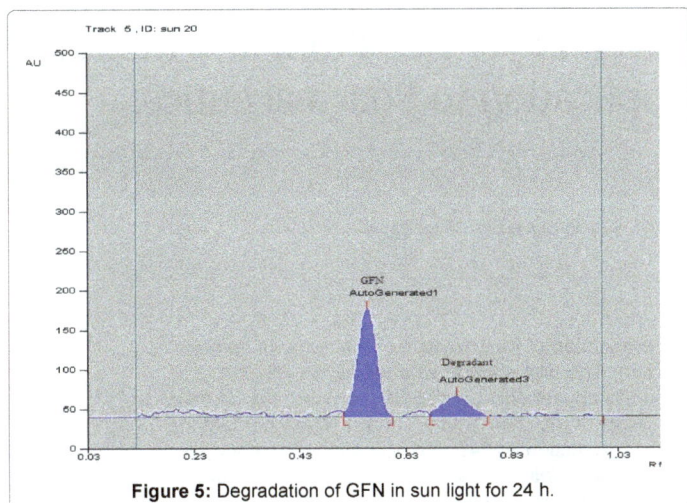

Figure 5: Degradation of GFN in sun light for 24 h.

Intra-day.

Conc. (ng/band) (n=3)	Mean ± SD	%RSD
800	1557.65 ± 4.73	0.304
1200	2510.9 ± 17.3	0.692
1600	3579.6 ± 4.80	0.134

Inter-day

Conc. (ng/band) (n=3)	Mean ± SD	%RSD
800	1663.13 ± 2.00	0.120
1200	2333.43 ± 1.24	0.053
1600	3578.23 ± 1.88	0.054

Table 1: Intra-day and Inter-day

Conc. (ng/band) (n=3)	Mean ± SD	%RSD
800	1663.13 ± 2.00	0.120
1200	2333.43 ± 1.24	0.053
1600	3578.23 ± 1.88	0.054

Table 2: Recovery Study.

Parameters	± SD of peak area (n = 3)	% RSD
Mobile phase composition (± 0.5 mL)	25.632	0.645
Mobile phase volume (± 5ml)	17.32	0.433
Duration of saturation (± 5 min.)	40.76	1.032

Table 3: Results of Robustness Study.

Parameters	Result
Linearity range	800-2800 (ng/band)
Correlation coefficient	0.999
LOD	149.23 ng
LOQ	182.65 ng
% Recovery (n = 3)	99.89
% RSD	0.759
Precision (%RSD)	
Intra-Day(n = 3)	0.134 – 0.692
Inter-Day(n = 3)	0.053 – 0.120
Repeatability (n = 6)	1.165
Ruggedness (%RSD)	
Analyst I (n = 6)	0.302
Analyst II (n = 6)	0.131

Table 4: summary of validation method.

Agent	Exposure time	Condition	Degradants peak	R_f	% Degradant
1 N HCl	8 h	Room temp	peak Found	0.75	20.25 %
1N NaOH	8 h	Room temp	peak Found	0.7	19.99 %
10% H_2O_2	8 h	Room temp	peak Found	0.69	13.84 %
Dry Heat	8 h	60° C	not Found	-	-
Light	24 h	Sunlight	peak Found	0.72	21.74 %

Table 5: Forced degradation studies.

Conclusion

The developed HPTLC method was precise, specific, and accurate and stability indicating and validated based on ICH guidelines. Statistical analysis proves that the method is repeatable and selective for the analysis of GFN as bulk drug and in pharmaceutical formulations. The method can be used to determine the purity of the drug available from the various sources by detecting the related impurities. As the method separates the drug from its degradation products, it can be employed as a stability indicating one.

References

1. Indian Pharmacopeia (2010) Ministry of Health and Family Welfare, Government of India, the Indian Pharmacopoeial Commission, Ghaziabad. 3: 558-559.

2. www.drugbank.com

3. Bankar AA, Lokhande SR, Sawant RL, Bhagat AR (2013) Spectrophotometric estimation of guaifenesin and salbutamol in pure and tablet dosage form by using different methods 3: 92-97.

4. Özdemir A, Aksoy H, Dinç E, Dermis S (2006) Determination of Guaifenesin and Dextromethorphan in a Cough Syrup by HPLC with Flurometric Determinatio. Revue Roumaine de Chimie 51: 117-122.

5. Wilcox ML, Stewart JT (2000) HPLC determination of guaifenesin with selected medications on underivatized silica with an aqueous-organic mobile phase. Journal of Pharmaceutical and Biomedical Analysis 23: 909-916.

6. Sharaf MH, Stiff DD (2004) Determination of guaifenesin in human serum by capillary gas chromatography and electron capture detection. Journal of Pharmaceutical and Biomedical Analysis 35: 801-806.

7. Grosa G, Grosso E, Russo R, Allegrone G (2006) Simultaneous stability indicating , HPLC-DAD determination of guaifenesin and methyl and propyl-parabens in cough syrup. J Pharm Biomed Anal 41: 798-803.

8. Sethi PD (1996) High Performance Thin Layer Chromatography: Quantitative Analysis of Pharmaceutical Formulations 162-165.

9. Bakshi M, Singh S, Pharm J (2002) Development of validated stability- indicating assay methods- critical review. Journal of pharmaceutical 28: 1011–1040.

10. ICH (1994) Q2A text on validation of analytical procedures, International Conference on Harmonization, October, USA.

11. ICH (1996) Q3B validation of analytical procedures: Methodology, International Conference on Harmonization, November.

12. Develop.jrs.de/Pharma/wEnglisch/handbook.../guaifenesinshtml

25

Validated Stability-indicating HPLC and Thin Layer Densitometric Methods for the Determination of Pazufloxacin: Application to Pharmaceutical Formulation and Degradation Kinetics

Nehad A Abdallah*

Experiments and Advanced Pharmaceutical Research Unit (EAPRU), Faculty of Pharmacy, Ain Shams University, Cairo, Egypt

Abstract

Two stability indicating methods were developed for the determination of pazufloxacin in presence of its acidic and alkaline degradation products, in pure form and in IV injection. The first method was based on RP-HPLC on X-terra C18 column (25 cm×4.6 mm×5 µm) using methanol and 0.5% phosphoric acid (15:85 (v/v)) with the addition of 1% Triethylamine (TEA) as a mobile phase over a concentration range 2-40 µg ml^{-1} and a mean percentage recovery of 100.467 ± 0.595. Quantification was achieved with UV detection at 249 nm. The second method was based on TLC separation of pazufloxacin from its degradation products followed by Densitometric measurement of the intact drug spot at 249 nm. The separation was carried out on silica gel 60 F_{254} aluminum sheets using chloroform, methanol and ammonia (10: 8: 2 v/v/v), as the mobile phase over a concentration range 0.1-2 µg per spot and mean percentage recovery of 99.694 ± 0.539. The two methods were simple, precise, and sensitive that could be successfully applied for the determination of pure and intravenous (IV) injection. The proposed HPLC method was used for the determination of acidic and alkaline degradation kinetics of pazufloxacin. The apparent first-order rate constants, half-life times, and activation energies of the degradation processes were calculated.

Keywords: Pazufloxacin; HPLC; TLC; Stability-indicating; kinetics

Introduction

Pazufloxacin(-)-(S)-10-(1-aminocyclopropyl)-9-fluoro-3-methyl-7-oxo-2,3-dihydro-7Hpyrido[2,3-de][1,4]benzoxazine-6-carboxylic acid monomethanesulfonate, figure 1 is a fluoroquinolone synthesized by Toyama Chemical Co. Ltd . This drug has good *in vitro* and *in vivo* activity against a broad range of bacteria, especially Gram-negative bacteria [1,2]. Clinical trials showed its intravenous injection formula was effective in treating respiratory infections [3]. Pazufloxacin is not yet the subject of a monograph in any pharmacopoeia. Reviewing the literature revealed that, few methods have been reported for the determination of pazufloxacin in raw material, pharmaceutical formulation and/or human plasma. These methods include spectrophotometric methods [4-7], spectroflurimetric methods [8,9], electrochemical method [10], capillary electrophoresis [11-13], HPLC methods [14-19]. Only one of the previous methods was used as a stability indicating method for the determination of pazufloxacin in pharmaceutical formulation and in presence of its degradation products [19]. This method was of low sensitivity (10 µg ml^{-1}) and lacks either the study of degradation kinetics of pazufloxacin or the identification of the degradation products. Also the method was validated with respect to parameters including linearity, precision and accuracy only. The ICH guideline Q1A on stability testing of new drug substances and products emphasizes that the testing of those features that are susceptible to change during storage and are unlikely to influence quality, safety and/or efficacy must be done by validated stability-indicating testing methods. It is also mentioned that forced decomposition studies (stress testing) should be performed on the drug substance to establish the inherent stability characteristics and degradation pathways to support the suitability of the proposed analytical procedures [20]. The aim of the present work is to focus on the development of efficient chromatographic methods for the determination of pazufloxacin in IV injection without interference from the excipients used and in the presence of its different degradation products in a short chromatographic run, and to study the results kinetically to prove the stability-indicating property of the method. Although HPLC method is more sensitive and less time consuming, but densitometric TLC method is technically easier and more cost effective. Also the developed densitometric TLC system can further be used in the future to separate the degradation products for their structural elucidation trying to identify the degradation pathway.

Experimental

Materials and chemicals

Pazufloxacin standard and pazufloxacin (IV) injection were purchased from Thonson technology limited, Shanghai, China.

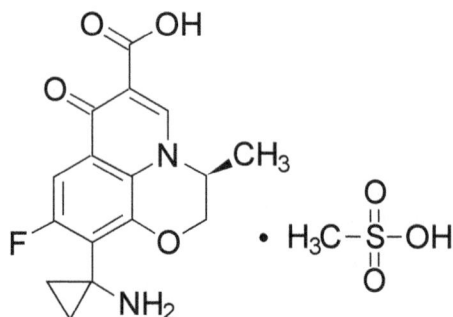

Figure 1: Chemical Structure of Pazufloxacin Mesylate.

*Corresponding author: Abdallah NA, Experiments and Advanced Pharmaceutical Research Unit (EAPRU), Faculty of Pharmacy, Ain Shams University, Cairo, Egypt; E-mail: nehad_nany@hotmail.com

HPLC grade methanol, chloroform, water, phosphoric acid and ammonia were purchased from Sigma Gmbh, Germany.

Triethylamine (TEA) analytical grade was purchased from Sigma Gmbh, Germany.

Instrumentation

For TLC densitometric method, DESAGA CD 60 HPTLC densitometer connected to IBM compatible computer fitted with Proquant evaluation software for Windows was (Sarstedt-Gruppe, Germany) was used. The spots were applied using DESAGA AS30 HPTLC Applicator (Sarstedt-Gruppe, Germany) with 25 µl Hamilton micro syringe. The used TLC plates were precoated with silica gel 60 F254 (20×20 cm) (E.Merck, Germany). The plates were pre-washed with methanol and activated at 60°C for 5 min prior to chromatography. For development, Chromatographic tank (25 cm×25 cm×8 cm) was used. UV lamp, Vilber–Lourmat (VL-6-LC, EU) 365-254 nm was used for spot detection. For HPLC, Knauer instrument (Germany) equipped with K-501 pump, Knauer injector and UV-detector was used. Data acquisition was performed on Eurochrom 2000 software. The separation was done using X-terra LC-18-DB, (25 cm×4.6 mm×5 µm) column. 0.25 µm Disposable membrane filters were used for samples filteration. A degasser, Crest Ultrasonics was used. Mettler Toledo MP 225 pH meter 3510 pH/mV was used for pH adjustment.

Chromatographic conditions

TLC densitometric method: For TLC with UV Densitometric analysis, solutions of the tested substance were applied to silica gel 60 F$_{254}$ TLC plates 20×20 using a DESAGA AS30 Applicator (Germany). Spots were applied 1.5 cm apart from each other and 2 cm from the bottom edge. The chromatographic chamber was pre-saturated with the developing mobile phase chloroform, methanol and ammonia in the ratio of (10: 8: 2 v/v/v) for 45 minutes. The spots were detected under a UV lamp at 254 nm and the drug was scanned densitometrically at 249 nm. The drug was scanned under the following instrumental conditions, photo Mode: reflection, scan Mode: Linear slit scanning, result Output: densitogram and integrated peak list, slit width: 6mm, slit height: 2 mm.

HPLC method: The mobile phase used was prepared by mixing methanol and 0.5% phosphoric acid in a ratio 15:85 with the addition of 1% (TEA). The mobile phases were filtered by vacuum filtration through 0.45 µm filter and degassed by ultrasound sonication for 50 minutes just prior to use. The column was equilibrated with the mobile phase. The analysis was done under isocratic conditions at a flow rate 1 ml.min-1 and at ambient temperature using UV detector at 249 nm.

Standard solutions

Standard solutions preparation was conducted at ambient temperature. The solutions were protected from light with aluminum foil wrapping and stored at -20°C.

Stock standard solution: It was prepared by accurately weighing 25 mg of pazufloxacin into 25-ml volumetric flask, dissolved in and diluted to volume with methanol.

Working standard solution: For TLC, pazufloxacin (200 µg ml-1) was prepared by transferring 5 ml of the previously prepared stock standard solution into 25 ml volumetric flask; the volume was then completed with methanol. For HPLC, pazufloxacin (100 µg ml^{-1}) was prepared by transferring 2.5 ml of each of the previously prepared stock standard solutions into 25 ml volumetric flasks; the volume was then completed with mobile phase.

Calibration curves

For TLC densitometric method: Accurately measured aliquots (0.5-10.00 µl) of working standard solution (200 µg ml^{-1}) were applied separately to the TLC plates, in triplicates. The specified chromatographic conditions were set. After development and scanning, the average peak areas were calculated, the calibration curve, relating the integrated peak area and its corresponding concentration, was constructed and the regression equation was computed.

For HPLC method: Accurately measured aliquots (0.2-4 ml) of working standard solution were transferred into a series of 10 ml-volumetric flasks, diluted to volume with mobile phase to obtain a concentration range of (2-40 µg ml^{-1}). 20 µls volume of each solution was injected, in triplicates, and the drug was analyzed using the chromatographic conditions described above and average peak areas were calculated. The calibration curve, representing the relationship between the average peak area and corresponding concentration was plotted and the regression equation was computed.

Method validation

Linearity: The linearity of response for pazufloxacin mesylate was assessed in the range of 0.1-2 µg per spot and 2-40 µg ml^{-1} for standard drug for TLC and HPLC methods, respectively.

Accuracy and precision: The accuracy of an analytical method is defined as the similarity of the results obtained by this method to the true values. To test the validity of the method it was applied to the determination of pure samples of pazufloxacin over the concentration range of 0.1-2 µg per spot and 2-40 µg ml^{-1} for TLC and HPLC methods, respectively. It was expressed as percent recovery [mean back-calculated concentration/theoretical concentration×100].

The intra-day precision was evaluated through replicate analysis of three concentrations of pazufloxacin in pure form on three successive times. The inter-day precision was also evaluated through replicate analysis of three concentrations for a period of three successive days. The precision of the methods was expressed in terms of S.D. and CV%.

Limit of detection and limit of quantitation: The limit of detection (LOD) and limit of quantitation (LOQ) of the studied drugs by the proposed methods were determined using calibration standards close to the expected LOD and LOQ. LOD and LOQ were calculated as 3.3 σ/s and 10 σ/s, respectively, where σ is the standard deviation of y-intercept of regression equation and s is the slope of the calibration curve [20].

Specificity and system suitability tests: Specificity of the method was also confirmed by its ability to measure unequivocally the drug in the presence of degradation products.

For TLC, Specificity of the method was established through the study of the resolution of the drug peak from the nearest resolving peak in stressed degradation samples. System suitability parameters as capacity factor (K), selectivity factor (α) and resolution (R$_s$) were calculated.

For HPLC, Specificity of the method was studied by determination of peak purity in stress acidic and alkaline degradation samples using PDA detector. The resolution factor of the drug peak from the nearest degradant peak was also calculated. The capacity factor (K), tailing factor (T) and theoretical plates number (N) were also tested.

Robustness: The robustness is a measure of method capacity to remain unaffected by small but deliberate variations in method parameters.

For TLC, it was studied through testing the influences of small changes in mobile phase composition. Chloroform: methanol: ammonia in varying ratios (10: 6: 2 v/v/v), (11: 6: 2.5 v/v/v), (10: 7: 3 v/v/v) were tried. Changes in mobile phase volume (± 10%), duration of saturation (± 10 minutes), time from spotting to chromatography and time from chromatography to scanning (± 10 minutes) were also studied.

For HPLC, the influences of small changes in mobile phase composition (methanol ± 3.00%.), pH (± 0.20), wavelength of detection (± 2.00), flow rate (± 0.1) and column supplier. [Pheonomenex-C18 (4.6×250 mm)] column was used.

Application to pazufloxacin IV injection

The proposed methods were applied to the assay of pazufloxacin in IV injection. The average percent recoveries of different concentrations were calculated. It was based on the average of three replicate determinations.

The accuracy of the proposed methods was assessed by applying the standard addition technique. Known amounts of the drug were added to the pharmaceutical product. The procedure stated under linearity was then applied. The concentrations, mean recoveries and the standard deviations were calculated for each added concentration.

Accelerated acidic and alkaline degradation of pazufloxacin mesylate

The accelerated degradation in acidic and basic media was performed in the dark in order to exclude the possible degradative effect of light on the drug.

Preparation of acidic induced degradation product: Ten ml pazufloxacin stock solution were transferred into a conical flask and mixed with 10 ml 1M HCl. It was heated in a thermostatically controlled water bath at 70°C for 2 hours while fitting an air condenser. The solution was cooled and neutralized with 5M NaOH to pH (6-6.5). Then the solution was transferred quantitatively into 25 ml volumetric flask. For TLC, the volume was completed with methanol and 10 and 5 µls of pazufloxacin were spotted on TLC plates in triplicates and analyzed using the chromatographic conditions described above. For HPLC, the volume was completed with methanol. Further dilution was carried out by transferring 10 ml to 100 ml volumetric flask and the volume was completed with mobile phase to obtain a solution containing 40 µg ml^{-1} of intact drug. 20 µls were injected, in triplicates, and analyzed using the chromatographic conditions described above.

Preparation of alkaline induced degradation product: Ten ml pazufloxacin stock solution were transferred into a conical flask and mixed with 10 ml 1M NaOH. It was heated in a thermostatically controlled water bath at 70°C for 2 hours while fitting an air condenser. The solution was cooled and neutralized with 5M HCl to pH (6-6.5). Then the solution was transferred quantitatively into 25 ml volumetric flask. For TLC, the volume was completed with methanol and 10 and 5 µls of pazufloxacin were spotted on TLC plates in triplicates and analyzed using the chromatographic conditions described above. For HPLC, the volume was completed with methanol. Further dilution was carried out by transferring 10 ml to 100 ml volumetric flask and the volume was completed with mobile phase to obtain a solution containing 40 µg ml-1 of intact drug. 20 µls were injected, in triplicates, and analyzed using the chromatographic conditions described above.

Study of the acidic and alkaline degradation kinetics using the proposed HPLC method

Study the kinetic order of the reactions: Two sets of two different concentrations each was formed of nine replicates were prepared and passed through stressed acidic and alkaline degradation reactions. The proposed HPLC method was used to determine the remaining concentration of the drug in each solution. The degradation reactions were terminated at 15 minutes intervals by adding calculated volume of either 5M HCl or 5M NaOH to adjust the pH in the range 5.5-6.5. The study time was 120 minutes. Triplicate 20 µL injections were made for each sample at each time interval.

Study the effect of reagent concentration on the reactions rate: The proposed HPLC method was followed using 0.5 and 1.5M HCl and 0.7 and 1.3M NaOH. The reactions were terminated by appropriate volumes of either 3M NaOH or 3M HCl. Triplicate 20 µL injections were made for each sample at each time interval.

Study the effect of temperature on the reactions rate: The stressed degradation reactions were followed using 1M HCl and 1M NaOH at 60°C and 80°C, separately. The reactions were terminated by appropriate volumes of either 5M NaOH or 5 M HCl.

Results and Discussion

Optimization of chromatographic conditions

TLC Densitometric method: A TLC densitometric technique is suggested for the determination of pazufloxacin in the presence of its acidic and alkaline degradation products based on the difference in Rf values. Several mobile phases were tried to accomplish complete separation of pazufloxacin from its degradation products. Using the mobile phase ethyl acetate: ethanol: ammonia system in varying ratios. Significant tailing was observed in the intact drug spot which caused interference with the degradation products spots. Good resolution and complete separation of the intact drug from the degradation products peaks were achieved by using chloroform: methanol: ammonia (10: 8: 2 v/v/v). It gave a sharp and symmetric peak of pazufloxacin at R$_f$=0.33 with good separation of the drug peak from the degradants peaks of R=0.54, 0.64 and 0.39 for acidic degradation product (PD1) and alkaline degradation products (PD2 and PD3). A wavelength of 249 nm was used for the quantification of the drug. Representative Densitograms are shown in figure 2 and 3.

HPLC method: The developed HPLC method has been applied for the determination of pazufloxacin in presence of its acidic and alkaline degradation products. To optimize the HPLC parameters, several mobile phases' composition were tried. It is apparent that the retention increases when the pH of the mobile phase is increased. This is because the compounds are less ionized at high pH and thus have more affinity for the stationary phase. By the use of acetonitrile in concentration above 18% at pH 2.5 the retention time decreased to 2.4 minutes but the resolution of pazufloxacin from its alkaline degradation products decreased. The use of methanol instead of acetonitrile as organic modifier led to further increase in the retention of pazufloxacin to 5 minutes.

The chosen mobile phase was methanol: 0.5% phosphoric acid (15:85) with the addition of 1% TEA to the aqueous phase with a flow rate 1ml.min^{-1}.

Figure 4 and 5 show the chromatograms of pazufloxacin with PD1, PD2 and PD3. The average retention times ± SD, for 3 replicate

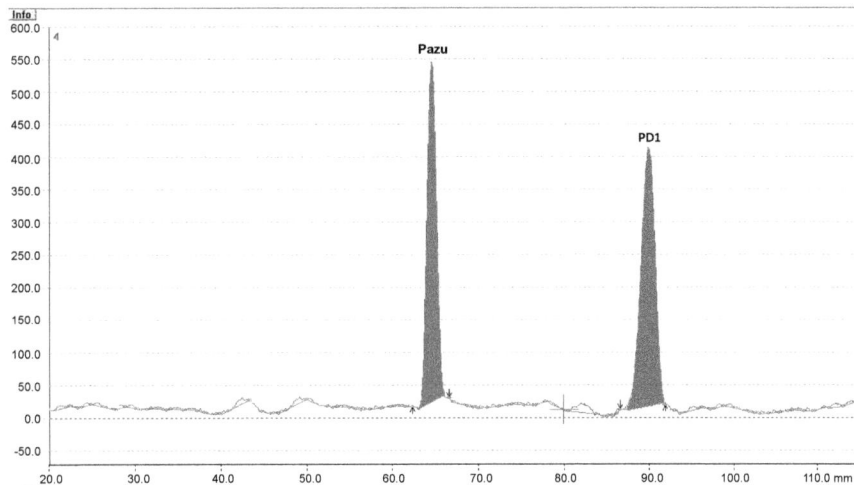

Figure 2: TLC densitogram of pazufloxacin stressed acidic degradation sample.

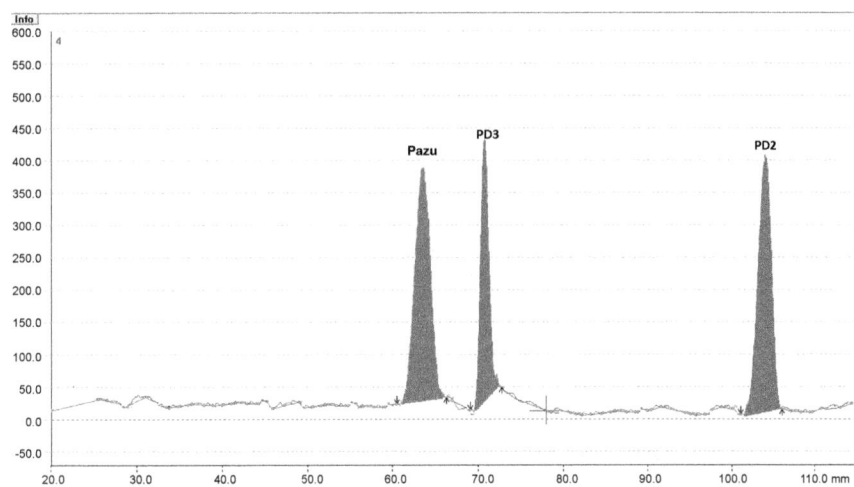

Figure 3: TLC densitogram of pazufloxacin stressed alkaline degradation sample.

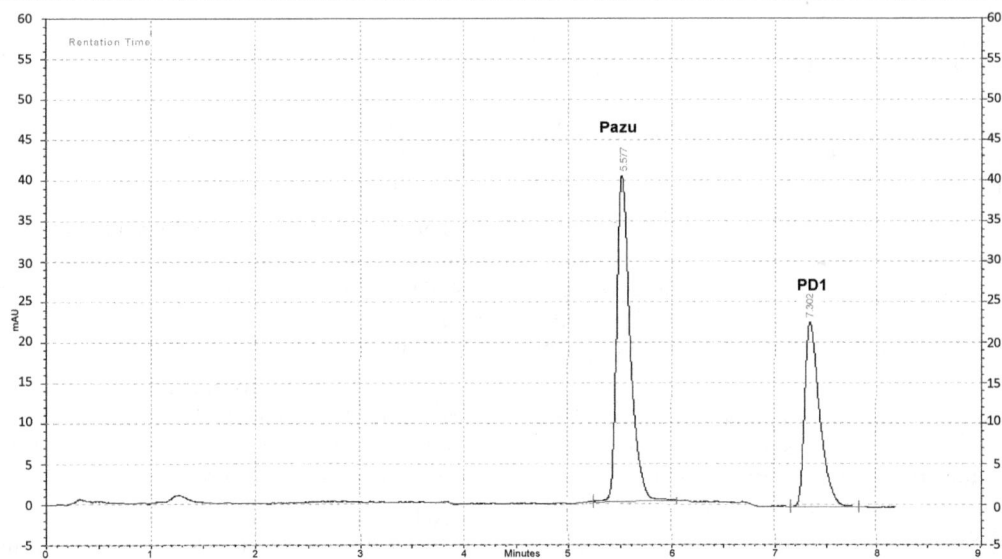

Figure 4: HPLC chromatogram of pazufloxacin stressed acidic degradation sample.

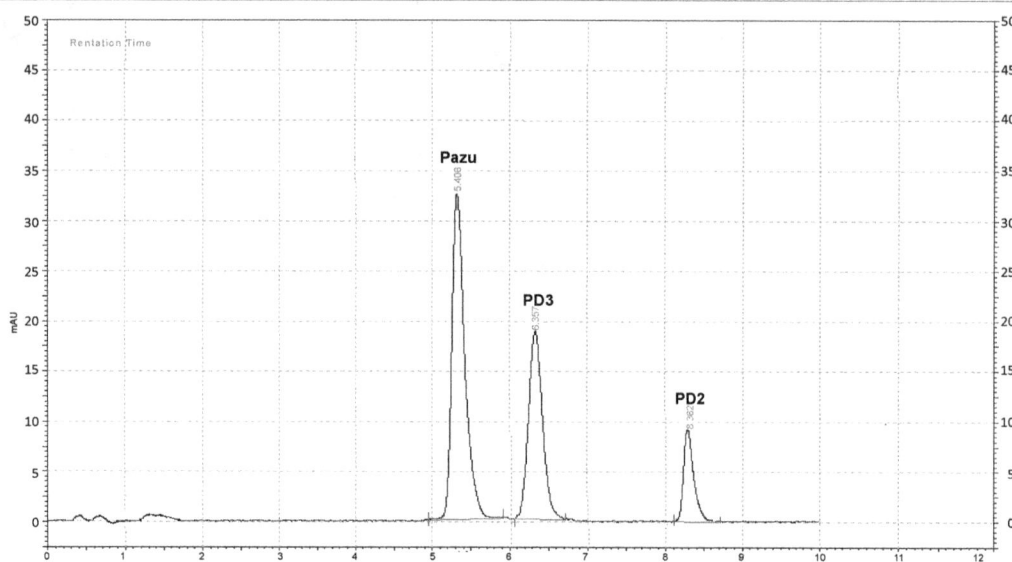

Figure 5: HPLC chromatogram of pazufloxacin stressed alkaline degradation sample.

Parameters	TLC densitometric method [n=11]	HPLC method [n=10]
Linearity range	0.100-2.00 µg per spot	2-40 µg.ml^{-1}
Correlation coefficient	0.9996	0.9998
Detection limit (LOD)	0.033	0.593
Quantitation limit (LOQ)	0.100	1.790
Slope ± SD	1902.5 ± 13.555	3.4813 ± 0.0275
Confidence limit of slope[a]	1900.3-1905.7	3.461-3.500
S.E. of slope	4.087	0.0087
Intercept ± SD	86.803 ± 19.025	0.2747 ± 0.626
Confidence limit of intercept[a]	85.765-87.435	-0.183-0.732
S.E. of intercept	5.736	0.198

a: 95% confidence limit

Table 1: Linear regression data for the calibration curves.

injections of Pazufloxacin and µµ (PD1) were found to be 5.553 ± 0.131 and 7.266 ± 0.083; respectively.

The average retention times ± SD, for 3 replicate injections of Pazufloxacin and its alkaline degradation products (PD2, PD3) were found to be 5.452 ± 0.086, 6.231 ± 0.056 and 8.311 ± 0.097; respectively.

Method validation

Linearity: The linear regression data for the calibration curves (n=6) showed good linear relationship over the concentration range of 0.1-2 µg per spot and 2-40 µg.ml^{-1} for standard drug for TLC and HPLC methods, respectively.

Characteristic parameters for regression equations and correlation coefficients were given in table 1. The linearity of the calibration graphs were validated by the high value of correlation coefficients of the regression.

Accuracy and precision: The accuracy and precision of the developed methods were expressed in terms of recovery% and %RSD, respectively. Table 2 summarizes the accuracy, intra and inter-day precision of pazufloxacin. The low values of SD and % RSD demonstrate excellent precision of the methods.

Limit of detection and limit of quantitation: For calculation

of LOD and LOQ, the standard deviation of response method based on the standard deviation of intercept was used. They were found to be 0.033 and 0.100 µg per spot, respectively, for TLC densitometric method and 0.593 and 1.790 µg ml^{-1}, respectively for HPLC method which indicates the adequate sensitivity of the methods.

Specificity and system suitability tests: For TLC, the densitograms of stressed samples of pazufloxacin showed separated spots of the drug from the nearest degradants. This indicates that the method is sufficiently specific to the drug.

For HPLC, Peak purity test results determined by PDA detector under the optimized chromatographic conditions, confirmed that no additional peaks were co-eluted with the drug in stressed degradation samples and evidencing the ability of the method to assess the drugs of interest in the presence of stressed samples.

The results of the system suitability tests in table 3 assured the adequacy of the proposed TLC and HPLC methods for the routine analysis of pazufloxacin and the methods capacity remains unaffected by small variations in methods parameters.

Robustness: The S.D, %RSD and S.E. of the peak areas for each parameter at a concentration level 1 µg per spot and 20 µg ml^{-1} for TLC and HPLC, respectively, are summarized in table 4. The low values

Intra-day		TLC densitometric method[a]				HPLC method[a]				
		Concentration µg per spot	Recovery%	SD	%R.S.D.	Concentration µg.ml⁻¹		Recovery%	SD	%R.S.D.
		0.300	100.667	0.004	1.325	6.000		100.500	0.065	1.078
		1.200	99.667	0.031	2.592	20.000		99.450	0.211	1.061
		2.000	99.250	0.042	2.116	35.000		100.886	0.432	1.223
	0.300	1st day	99.000	0.005	1.684		1st day	99.670	0.068	1.137
		2nd day	101.333	0.004	1.316	6.000	2nd day	98.500	0.072	1.218
		3rd day	99.667	0.006	2.007		3rd day	98.000	0.075	1.276
	1.200	1st day	100.917	0.032	2.642		1st day	99.550	0.266	1.336
		2nd day	100.500	0.027	2.239	20.000	2nd day	100.550	0.301	1.497
		3rd day	99.083	0.029	2.439		3rd day	99.850	0.289	1.447
	2.000	1st day	97.700	0.042	2.149		1st day	98.770	0.511	1.478
		2nd day	101.050	0.038	1.880	35.000	2nd day	100.340	0.587	1.671
Inter-day		3rd day	102.250	0.045	2.200		3rd day	98.91	0.603	1.742

[a]Each result is an average of three determinations.

Table 2: Accuracy and precision of TLC densitometric and HPLC methods.

TLC densitometric method			HPLC method		
Parameters	Acidic degradation	Alkaline degradation	Parameters	Acidic degradation	Alkaline degradation
Capacity factor [K']	2.211	2.015	Capacity factor [K']	4.567	4.411
Selectivity factor [α]	2.388	1.301	Selectivity factor [α]	1.377	1.215
Resolution [R]	4.714	1.667	Resolution [R]	3.833	1.898
			Theoretical plates [N]	3110.29	2924.65
			Tailing factor [t]	1.25	1.25
			Peak purity	0.996	0.997

Table 3: System Suitability results of TLC densitometric and HPLC methods.

	Parameters	S.D. [peak area]	% R.S.D.	S.E.
TLC densitometric method	Mobile phase composition	0.415	0.021	0.169
	Mobile phase volume [± 10%]	0.799	0.040	0.326
	Time of saturation [± 10 minutes]	0.225	0.011	0.092
	Time from spotting to development [±10 min]	0.309	0.015	0.126
	Time from chromatography to scanning [±10 min]	0.319	0.016	0.130
HPLC method	Change in organic phase percent in mobile phase [± 3.00%]	0.383	0.542	0.156
	Change in mobile phase pH [± 0.20]	0.187	0.265	0.076
	Change in wavelength of detection [± 2.00]	0.137	0.194	0.056
	Change in flow rate [± 0.1]	0.375	0.531	0.153
	Change column supplier	0.467	0.661	0.191

Table 4: Robustness of TLC densitometric and HPLC methods [n=6].

obtained after introducing small deliberate changes in the developed methods.

Application to pazufloxacin IV injection

The proposed methods were successfully applied for the determination of pazufloxacin in its pharmaceutical formulation. No interaction was observed between the drug and excipients present in the formulation. A single spot at R_f=0.33 was observed in the densitogram of pazufloxacin samples extracted from the pharmaceutical formulation.

TLC densitometric method		HPLC method	
Added standard (µg per spot)	Recovery %	Added standard (µg/ml)	Recovery %
0.2	100.450	5.000	101.740
0.6	98.717	10.000	99.210
1	99.650	20.000	99.490
1.4	99.114	30.000	100.040
1.8	99.500	40.000	99.160

Table 5: Application of standard addition technique on pazufloxacin IV injection using TLC densitometric and HPLC methods [n=3].

	TLC densitometric method	HPLC method	Reported method [5]
Mean	98.654	100.125	99.651
SD	1.0221	0.8765	1.0881
n	6	6	6
Variance	1.0447	0.7683	1.1839
t [2.228]*	1.64	0.83	
F [5.05]*	1.1332	1.541	

Table 6: Statistical comparison between the results obtained by the proposed TLC densitometric and HPLC methods with the reported method [5] for the analysis of pazufloxacin in IV injection.

The standard addition technique was applied to assess the accuracy of the proposed methods as shown in table 5. Statistical comparison showed that there was no significant difference between the results obtained from the proposed methods and those obtained from the reported method [5] as shown in table 6, since the calculated t and F values are less than the tabulated ones.

Study of stressed acidic and alkaline degradation of pazufloxacin

Kinetic studies of pazufloxacin have not been previously investigated. It was investigated in 1M HCl and 1M NaOH using the proposed HPLC method for a period of 2 hours at 15 minutes interval. A decrease in concentration of drug with increasing time was observed. To study the effect of the reagent concentration on the reaction rate, experiments were performed using 3 different concentrations of HCl and NaOH that are always in large excess with respect to the drug. (Figure 6 and Figure 7) and table 7 show that the reaction rate was increased by uplifting the reagent concentration. For each experiment k_{obs}, $t_{1/2}$ and t_{90} (time where 90% of original concentration of the drug is left) were determined. The influence of temperature on the acid and alkaline degradation process of pazufloxacin was investigated at 60,

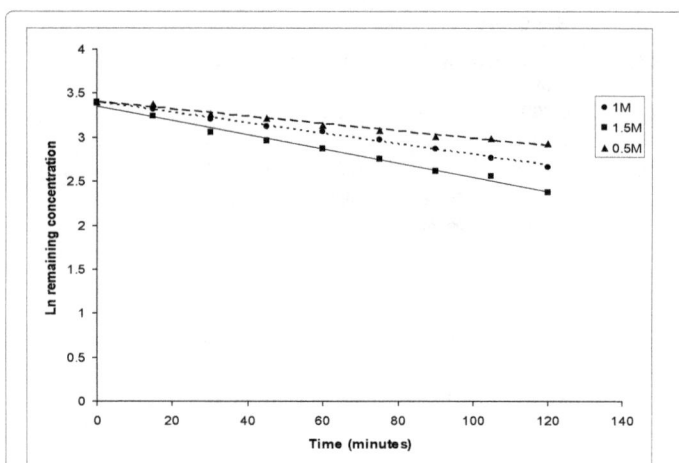

Figure 6: Effect of HCl concentration on the degradation of Pazufloxacin using 0.5, 1 and 1.5M HCl.

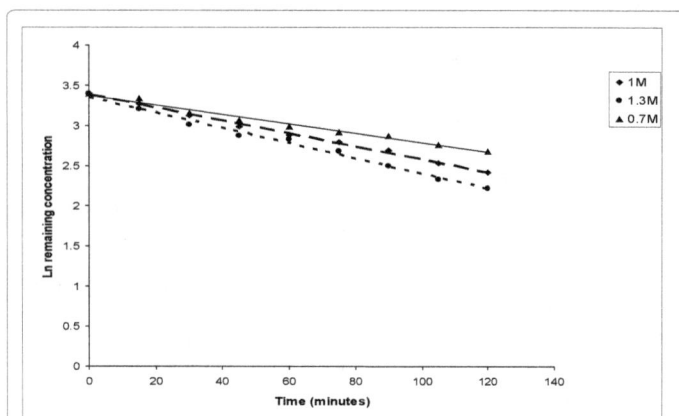

Figure 7: Effect of NaOH concentration on the degradation of Pazufloxacin using 0.7, 1 and 1.3M NaOH.

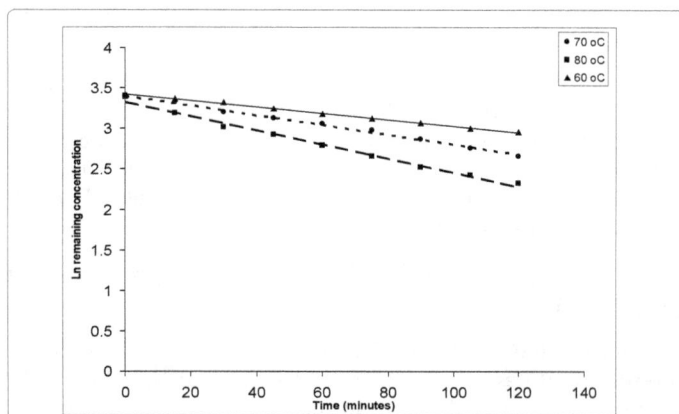

Figure 8: Kinetic plot for degradation of Pazufloxacin in 1M HCl at 60°C, 70°C and 80°C

70, 80°C below 60°C the reaction is too slow to be monitored. At the selected temperatures the degradation process followed pseudo first order kinetics (Figures 8 and 9). Data obtained from first order kinetics treatment were further subjected to fitting in Arrhenius equation

$$Ln\ K = Ln\ A - E_a/RT$$

Where, K specific reaction rate (min[-1]), A constant known as Arrhenius factor or frequency factor (min[-1]), E_a energy of activation (Kcal/mol), R gas constant (1.987 cal/deg.mole) and T absolute temperature

The plot of ln k_{obs} versus 1/T gave the Arrhenius plots (Figures 10 and 11) which were found to be linear in the temperature range 60-80°C. From the regression equations, the activation energy and the frequency factor were calculated for the acidic and alkaline degradation of the drug. An extrapolation to Arrhenius plot is used to calculate the degradation rate constants at room temperature (25°C ± 2). The half-life and t_{90} were also calculated at room temperature as shown in tables 8 and 9. The obtained data suggests that the drug is susceptible to acidic and alkaline degradation.

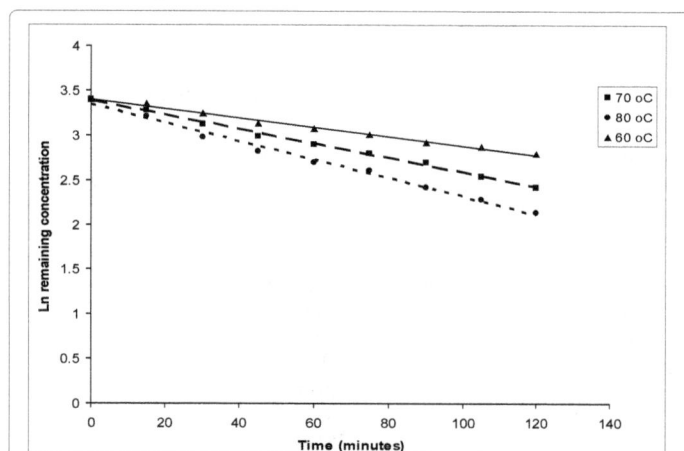

Figure 9: Kinetic plot for degradation of Pazufloxacin in 1M NaOH at 60°C, 70°C and 80°C

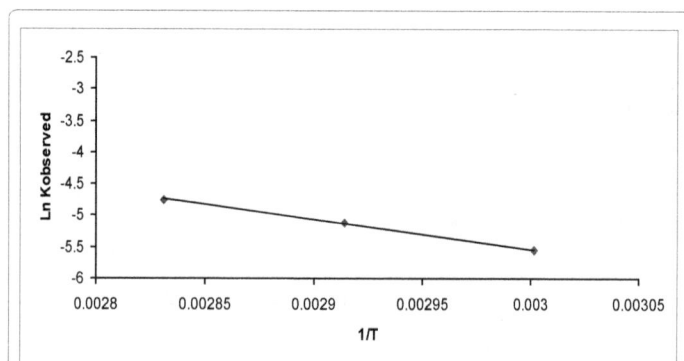

Figure 10: Arrhenius plot for acidic degradation of Pazufloxacin in 1M HCl.

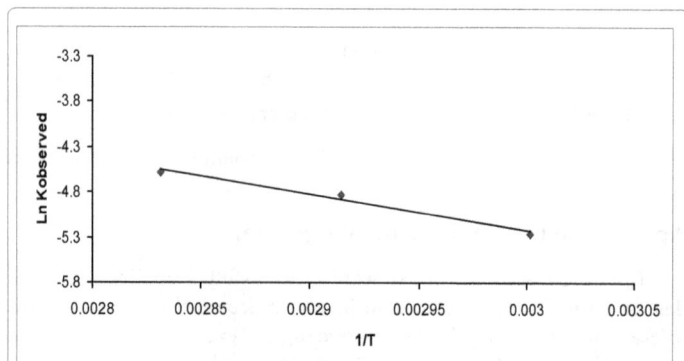

Figure 11: Arrhenius plot for alkaline degradation of Pazufloxacin in 1 M NaOH.

Parameter		$K_{observed}$ (min^{-1})	$t_{1/2}$ (min)	t_{90} (min)
Acid-induced degradation	0.5 M	0.0041	169.024	25.609
	1 M	0.006	115.500	17.500
	1.5 M	0.0081	85.556	12.963
Alkaline-induced degradation	0.7 M	0.0059	117.458	17.797
	1 M	0.008	86.625	13.125
	1.3 M	0.0095	72.947	11.053

Table 7: Kinetic data of pazufloxacin acidic and alkaline degradation at 70°C.

Parameter		$K_{observed}$ (min^{-1})	$t_{1/2}$ (min)	t_{90} (min)
Acid-induced degradation	60°C	0.0039	177.692	26.923
	70°C	0.006	115.500	17.500
	80°C	0.0087	79.655	12.069
Alkaline-induced degradation	60°C	0.0052	133.269	20.192
	70°C	0.008	86.625	13.125
	80°C	0.0102	67.941	10.294

Table 8: Kinetic data of Pazufloxacin acidic and alkaline degradation in the presence of 1M HCl and 1M NaOH at different temperatures.

Parameters	1M HCl	1M NaOH
Activation energy, Ea(Kcal mol^{-1})	9.382	7.894
Degradation rate constant k_{25} (min^{-1})	0.00074	0.00132
Half life, $t_{1/2}$ (min)	936.487	525.000
Shelf life, t_{90} (min)	141.892	79.545
Arrhenius frequency factor, A (min^{-1})	5611.649	806.981

Table 9: Kinetic data of Pazufloxacin degradation at 25°C.

Conclusion

The goal of this work was achieved by separating and quantitation the new fluoroquinolone pazufloxacin in presence of its acidic and alkaline degradation products in bulk powder and in IV infusion. HPLC and TLC densitometric methods have been developed and validated for the determination of the drug without any interference from excipients and in presence of degradation products. They provide significant sensitivity as well as significant decrease in sample preparation, instrument run time over other separation methods.

References

1. Muratani T, Inoue M, Mitsuhashi S (1992) In vitro activity of T-3761, a new fluoroquinolone. Antimicrob Agents Chemother 36: 2293-2303.

2. Fukuoka Y, Ikeda Y, Yamashiro Y, Takahata M, Todo Y, et al. (1993) In vitro and in vivo antibacterial activities of T-3761, a new quinolone derivative. Antimicrob Agents Chemother 37: 384-392.

3. Higa F, Akamine M, Haranaga S, Tohyama M, Shinzato T, et al. (2005) In vitro activity of pazufloxacin, tosufloxacin and other quinolones against Legionella species. J Antimicrob Chemother 56: 1053-1057.

4. Rewatkar NS, Dinesh R, Mahavir C, Ghante H, Umarkar AR (2011) Development of UV spectrophotometric methods for estimation of Pazufloxacin in infusion form using absorbance ratio method Research. Journal of Pharmaceutical, Biological and Chemical Sciences 2: 456-463.

5. Rewatkar NS, Umarkar AR, Chaple D, Ghante H (2011) Validated spectrophotometric determination of pazufloxacin in formulation. J Pharm Res 4: 1394-1395.

6. Wang X, Chen S, Zhao H, Jin L (2005) Europium sensitized chemiluminescence determination of pazufloxacin mesylate in urine and serum. Anal Lett 38: 971-979.

7. Yang C, Zhang Z, Chen S, Yang F (2007) Molecularly imprinted on-line solid-phase extraction combined with chemiluminescence for the determination of pazufloxacin mesylate Mikrochim Acta 159: 299-304.

8. Jin J, Zhang X (2008) Spectrophotometric studies on the interaction between pazufloxacin mesilate and human serum albumin or lysozyme. J Lumin 128: 81-86.

9. Chen S, Ma H, Zhao H, Feng R, Jin L (2004) Terbium-sensitized fluorescence method for the determination of pazufloxacin mesilate and its application. Anal Sci 20: 1075-1078.

10. Lu Z, Rong YG, Ruo Y, YaQin C, SuMing C. (2007) A capacitive sensor based on molecularly imprinted polymers and poly(p-aminobenzene sulfonic acid) film for detection of pazufloxacin mesylate Sci China Ser B-Chem. 50: 547-553.

11. Zu LZ, Jian JL, Ling BQ, Ran Y (2008) Determination of Pazufloxacin Mesylas by Capillary Electrophoresis with Electrochemiluminescence Detection. Chinese J Anal Chem 36: 941-946.

12. Zhou S, Ouyang J, Baeyens WR, Zhao H, Yang Y (2006) Chiral separation of four fluoroquinolone compounds using capillary electrophoresis with hydroxypropyl-beta-cyclodextrin as chiral selector. J Chromatogr A 1130: 296-301.

13. Wang Y, Baeyens WR, Huang C, Fei G, He L, et al. (2009) Enhanced separation of seven quinolones by capillary electrophoresis with silica nanoparticles as additive. Talanta 77: 1667-1674.

14. Li Q, Wang R, Pei F (2004) Determination of pazufloxacin mesylate in human plasma and urine by high performance liquid chromatography. Asian Journal of Drug Metabolism and Pharmacokinetics Asian Journal of Dug Metabolism and Pharmacokinetics 4: 289-294.

15. Zhang Z, Yang G, Wang D, Liang G, Chen Y (2004) Chiral Separation and Enantiomeric Purity Determination of Pazufloxacin Mesilate by HPLC Using Chiral Mobile Phase Additives. J Liq Chromatogra Relat Technol 27: 813-827.

16. Zhang H, Ren Y, Bao X (2009) Simultaneous determination of (fluoro) quinolones antibacterials residues in bovine milk using ultra performance liquid chromatography-tandem mass spectrometry. J Pharm Biomed Anal 49: 367-374.

17. Watabe S, Yokoyama Y, Nakazawa K, Shinozaki K, Hiraoka R, et al. (2010) Simultaneous measurement of pazufloxacin, ciprofloxacin, and levofloxacin in human serum by high-performance liquid chromatography with fluorescence detection. J Chromatogr B Analyt Technol Biomed Life Sci 878: 1555-1561.

18. Phapale PB, Lee HW, Kim SD, Lim MS, Kale DD, et al. (2010) Analysis of Pazufloxacin Mesilate in Human Plasma and Urine by LC with Fluorescence and UV Detection, and Its Application to Pharmacokinetic Study Chromatographia 71: 101-106.

19. Rewatkar NS, Umarkar AR, Chaple D, Ghante H (2011) A validated stability-indicating RP-HPLC method for the estimation of pazufloxacin in presence of its degradation products. J Pharm Res 4: 3060-3062.

20. http://www.ich.org/fileadmin/Public_Web_Site/ICH_Products/Guidelines/Efficacy/E6_R1/Step4/E6_R1__Guideline.pdf

26

Simultaneous Determination of Omeprazole, Tinidazole and Clarithromycin in Bulk Powder and Helicure® Tablets by HPLC

Hesham Salem[1], Safa M Riad[2], Mamdouh R Rezk[2], Kholoud Ahmed[1]*

[1]*Pharmaceutical Analytical Chemistry Department, Faculty of Pharmacy, October University for Modern Sciences and Arts, Egypt*
[2]*Analytical Chemistry Department, Faculty of Pharmacy-Cairo University, Kasr El-Aini Street, Egypt*

Abstract

Sensitive and precise chromatographic method was developed and validated for simultaneous determination of omeprazole (OMP), tinidazole (TND) and clarithromycin (CLR) in bulk powder, laboratory prepared mixture and pharmaceutical preparation. The technique adopted for quantification is HPLC. A mixture of acetonitrile, methanol, phosphate buffer at pH 3.5 (33: 17: 50, v/v/v) was used as a mobile phase. The stationary phase used was (150 mm×4.6 mm, 10µm) C8 LichrosorbTM analytical column. The method was linear in the range of 0.2-250 µg mL^{-1}, 0.5-250 µg mL^{-1} and 75-2000 µg mL^{-1} for OMP, TND and CLR respectively. The selectivity of the proposed method was checked using laboratory prepared mixtures. The proposed method was successfully applied to the analysis of OMP, TND and CLR in their mixture and in pharmaceutical dosage form without interference from other additives.

Keywords: Omeprazole; Tinidazole; Clarithromycin; HPLC.

Introduction

Omeprazole (OMP), is 6-methoxy-2-[[(4-methoxy-3,5-dimethyl-2-pyridinyl) methyl] sulphinyl]-1H-benzimidazole [1], (Figure 1). It is the first member of the "proton pump inhibitors" that are widely used for the prophylaxis and treatment of both gastro-duodenal ulcers and symptomatic gastro-esophageal reflux. It is highly effective in the treatment of Zollinger-Ellison syndrome [2]. Tinidazole (TND) is 1-[2-(ethyl sulphonyl) ethyl]-2-methyl-5-nitro-1H-imidazole, [1] (Figure 2). It is used as antiprotozoal agent. Clarithromycin (CLR), is (3R,4S,5S,6R,7R,9R,11R,12R,13S,14R)-4-[(2,6-Dideoxy-3-C-methyl-3-O-methyl-a-L-ribohexopyranosyl) oxy]-14-ethyl-12,13-dihydroxy-7-methoxy-3,5,7,9,11,13-hexamethyl-6-[[3,4,6-trideoxy-3-(dimethylamino)-b-D-xylo hexopyranosyl] oxy] oxacyclotetradecane-2,10-dione (6-O-methylerythromycin A), [1] (Figure 3). CLR is semi-synthetic macrolide antibacterial agent [1].

The literature survey reveals several analytical methods for quantitative estimation of OMP alone in body fluids and in pharmaceutical formulations these methods include spectrophotometry [3-14], electrochemical methods [15], HPLC [16-21], liquid chromatography-electrospray ionization tandem mass spectrometry [22] and electrophoresis [23]. Tinidazole was estimated in body fluids and in pharmaceutical formulations by spectrophotometry [24-29], potentiometry [29], HPLC methods [29-31], polarography [32,33] and resonance light scattering technique [34]. Clarithromycin has been reported to be estimated in body fluids and in pharmaceutical formulations by spectrophotometry [35], HPLC methods [36-44]. Omeprazole, Tinidazole and Clarithromycin were simultaneously determined by spectrophotometry [45,46].

Up to our knowledge, there is no isocratic HPLC method was described for the simultaneous determination of the three studied drugs in their laboratory prepared mixtures and in the pharmaceutical dosage form without prior derivatisation. The present work aimed to develop an isocratic HPLC method for simultaneous determination of OMP, TND and CLR in laboratory prepared mixtures and pharmaceutical dosage form. The proposed method has advantage of being cheap, simple, rapid and time saving (one run in less than 7 minutes).

Experimental

Instruments

A liquid chromatograph consisted of an quaternary pump (Agilent model G1316 A/G1316 B), a diode array multiple wavelength detector (model G1316 C/D and G1365C/D, Agilent 1200 Series), Standard and preparation autosamplers (Agilent 1200 series) equipped vacuum degasser, Agilent. Stationary phase (150 mm×4.6 mm, 10 µm)

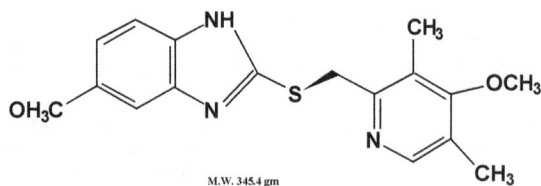

M.W. 345.4 gm

Figure 1: Chemical structure of Omeprazole (OMP).

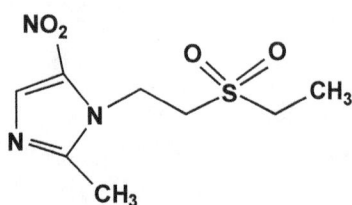

M.W. 247.3 gm

Figure 2: Chemical structure of Tinidazole (TND).

***Corresponding author:** Kholoud Ahmed, Pharmaceutical Analytical Chemistry Department, Faculty of Pharmacy, October University for Modern Sciences and Arts, Egypt; E-mail: m.habashyyy@hotmail.com

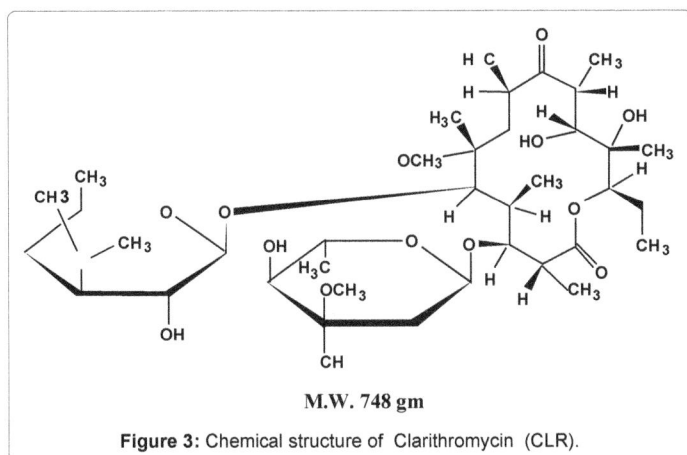

M.W. 748 gm

Figure 3: Chemical structure of Clarithromycin (CLR).

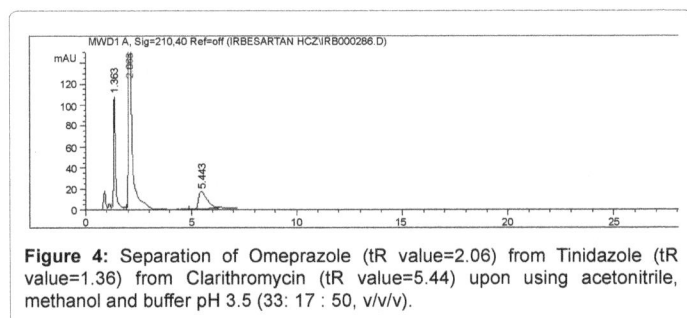

Figure 4: Separation of Omeprazole (tR value=2.06) from Tinidazole (tR value=1.36) from Clarithromycin (tR value=5.44) upon using acetonitrile, methanol and buffer pH 3.5 (33: 17 : 50, v/v/v).

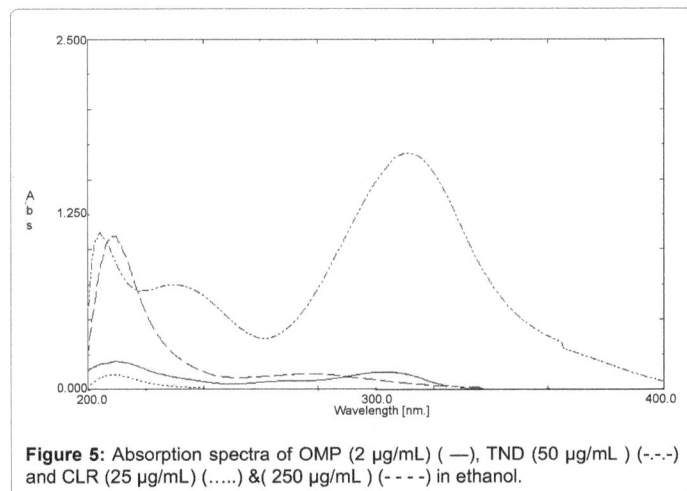

Figure 5: Absorption spectra of OMP (2 µg/mL) (—), TND (50 µg/mL) (-.-.-) and CLR (25 µg/mL) (.....) &(250 µg/mL) (- - - -) in ethanol.

C8 Lichrosorb TM analytical column. Mobile phase; acetonitrile, methanol, buffer at pH 3.5 (33: 17 : 50, v/v/v). The mobile phase was filtered through a 0.45 µm Millipore membrane filter and was degassed for 15 min in an ultrasonic bath prior to use. UV-detection was done at 210 nm. The samples were filtered also through a 0.45 µm membrane filter.

Standards, solvents, and pharmaceutical preparation

Reference omeprazole (OMP), reference tinidazole (TND) and reference clarithromycin (CLR) were kindly donated by EGYPHAR Pharmaceuticals Co. The potency was found to be 100.30%, 100.13% and 100.16% for OMP, TND and CLR, respectively. Pharmaceutical dosage form (Heli-cure tablets were kindly supplied by EGYPHAR and were claimed to contain 20 mg of OMP, 500 mg TND and 250 mg of

CLR per tablet. Acetonitrile, methanol (HPLC grade) and phosphate buffer adjusted to pH 3.5.

Standard solutions

OMP, TND standard solutions (each 0.5 mg mL^{-1}) and CLR standard solution (2 mg mL^{-1}) were prepared in mobile phase for the suggested HPLC method. The standards solutions were freshly prepared on the day of analysis and stored in a refrigerator to be used within 24 hr.

Procedures

Linearity: Portions of OMP, TND standard solutions (each 0.5 mg mL^{-1}) and CLR standard solution (2 mg mL^{-1}) were transferred separately into a series of 10-mL volumetric flasks and completed with mobile phase. Several dilutions were done and the content of each flask was completed to volume with the mobile phase to get the concentrations of 0.2-250 µg mL^{-1} OMP, 0.5-250 µg mL^{-1} TND and 75-2000 µg mL^{-1} CLR. The samples were then chromatographed using the following chromatographic condition. Stationary phase (150 mm×4.6 mm,10 µm) C8 lichrosorb^{TM}. Many mobile phases such as methanol and acetonitrile (50:50, 60:40, 65:35, by volume), methanol, acetonitrile and phosphate buffer adjusted at pH 3.5 (30: 20:50, 40: 30: 30) by volume and different other ratios but the mobile phase which give the best separation and peaks shape was found to be a mixture of acetonitrile, methanol, buffer at pH 3.5 (33: 17: 50, v/v/v). The mobile phase was filtered through a 0.45 µm millipore membrane filter and was degassed for about 15 min in an ultrasonic bath prior to use, flow rate; 0.7 mL min^{-1} [isocratically at temperature (35°C)], with UV-detection at 210 nm, the detection wavelength was set regarding the UV absorption spectra of the drugs (Figure 5) and their relative concentrations within the pharmaceutical formulation. Whereas TNZ is nominally 2 and 25 times more concentrated than CLR and OMP, respectively. The drugs have strong contributions in the overall UV region (200-375 nm). This is why an optimum detection wavelength was set at 210 nm during the chromatographic separation, favoring the quantification of both CLR and OMP, which represent the less concentrated components of this ternary mixture. In addition, this chosen detection wavelength can greatly improve the sensitivity of the proposed method for the CLR determination because it exhibits absorption maxima (at 210 nm). The samples were filtered also through a 0.45 µm membrane filter. To reach good equilibrium, the analysis was usually performed after passing 50-60 mL of the mobile phase, just for conditioning and pre-washing of the stationary phase. The relative peak area ratios were then plotted versus the corresponding concentrations of OMP, TND and CLR to get the calibration graphs and to compute the corresponding regression equations.

Analysis of laboratory prepared mixtures containing different ratios of OMP, TND and CLR: Aliquots of each standard solution were mixed to prepare different mixtures containing different ratios (3: 4:90, 1:0.2:26, 7:4:130, 2:0.2:23, 0.5:50:336, 30:12.5:168, 6:2.5:124, 0.5:1:144) of OMP, TND and CLR, respectively. The concentrations were calculated from the corresponding regression equations.

Assay of pharmaceutical formulations (Heli-cure tablets): Twenty tablets were powdered well and homogeneously mixed in a morter. A mass of the powdered tablets equivalent to 20 mg of OMP, 250 mg of CLR and 500 mg of TND was weighed and transferred to a 100-ml volumetric flask. The powder was extracted by shaking with 3×30 mL mobile phase with vigorous shaking for 15 minutes then filtered. The volume was completed to the mark with the mobile phase. Several portions 0.5-2 mL of aliquot were transferred separately to 10-

mL volumetric flasks, the volumes were completed to the mark with mobile phase and chromatographed under the previous mentioned conditions.

Results and Discussion

High-performance liquid chromatography

A simple isocratic high-performance liquid chromatographic method was developed for the determination of OMP, TND and CLR in pure form and in pharmaceutical preparation using (150 mm×4.6 mm, 10 µm) C8 lichrosorb™ analytical column. The mobile phase consisted of acetonitrile, methanol, buffer at pH 3.5 (33: 17: 50, v/v/v). The mobile phase was chosen after several trials to reach the optimum stationary/mobile-phase matching. The average retention times under the conditions described are 2.06 min for OMP, 1.36 min for TND and 5.44 for CLR (Figure 4). One sample can be chromatographed in less than 6 min.

Peak purity was confirmed for the HPLC peaks of OMP, TND and CLR by a pilot run using a photodiode array detector. Calibration graph was obtained by plotting the relative peak area ratios against concentrations. Linearity range was found to be 0.2-250 µg mL^{-1} for OMP, 0.5-250 µg mL^{-1} TND and 75-2000 µg mL^{-1} CLR. The regression equation for OMP: A=0.1832C+0.1946 (r=0.9999), for TND: A=0.0241C+0.0513 (r=0.9999) and for CLR: A=0.0021C+0.0280 (r=0.9999) where A is the relative peak area ratio, C is the concentration in µg mL^{-1} and r is the correlation coefficient. The mean percentage recovery was found to be 100.08 ± 0.454 for OMP, 100.40 ± 0.535 for TND and 100.65 ± 0.862 for CLR (Tables 1 and 2).

Analysis of laboratory prepared mixtures containing different ratios of OMP, TND and CLR

The suggested HPLC method was successfully applied for the determination of the studied drugs in their laboratory prepared mixtures. The precision of the proposed method was checked by the analysis of different concentrations (Table 2).

The mean percentage recovery was found to be:

99.96 ± 0.407   for OMP

100.14 ± 0.332   for TND

99.97 ± 0.216   for CLR.

Analysis of dosage form (Heli-cure tablets)

The suggested HPLC method was successfully applied for the

Parameter	OMP	TND	CLR
Range (µg mL^{-1})	0.2-250	0.5-250	2000-75
Slope	0.183	0.024	0.002
Intercept	0.195	0.051	0.028
Variance	0.206	0.286	0.743
Coefficient of variation	0.454	0.535	0.862
Correlation coefficient (r)	1	1	1
Accuracy mean	100.08	100.40	100.65
RSD%	0.454	0.535	0.862
Precision (RSD%	0.201	0.184	0.237
Repeatability Intermediate precision	0.332	0.409	0.294
Specificity mean	99.96	100.14	99.97
RSD%	0.407	0.332	0.216

Table 1: Validation and regression parameters for the determination of OMP, TND & CLR by the proposed HPLC method.

OMP : TND : CLR	OMP			TND			CLR		
	Taken ug mL^{-1}	Found ug mL^{-1}	R (%)	Taken ug mL^{-1}	Found ug mL^{-1}	R (%)	Taken ug mL^{-1}	Found ug mL^{-1}	R (%)
3 : 4 : 90	30	29.75	99.17	40	40.15	100.38	900	900.14	100.02
0.5 : 1 : 144	2.5	2.503	100.12	5	4.99	9980	720	721.33	100.18
1 : 0.2 : 26	50	49.92	99.84	10	10.02	100.20	1300	1299.05	99.93
7 : 4 : 130	7	6.98	99.71	4	3.98	99.50	130	130.1	100.08
0.5 : 50 : 336	0.5	0.502	100.40	50	50.11	100.22	336	334.82	99.65
30 : 12.5 : 168	60	60.23	100.38	25	25.09	100.36	336	336.71	100.21
6 : 2.5 : 124	30	29.96	99.87	12.5	12.52	100.16	620	620.14	100.02
2 : 0.2 : 23	20	20.04	100.20	2	2.01	100.50	230	229.19	99.65
Mean ± SD	99.96 ± 0.407			100.14± 0.332			99.968 ± 0.216		

Table 2: Determination of OMP, TND and CLR in laboratory prepared mixtures containing different ratios (3: 4: 90, 0.5: 1: 144, 1: 0.2: 26, 7: 4: 130, 0.5: 50: 336, 30: 12.5: 168, 6: 2.5: 124, 2:0.2:23) of OMP, TND and CLR, respectively by the proposed method.

OMP			TND			CLR		
Taken ug mL^{-1}	Found ug mL^{-1}	R (%)	Taken ug mL^{-1}	Found ug mL^{-1}	R (%)	Taken ug mL^{-1}	Found ug mL^{-1}	R (%)
10	10.03	100.30	4	4.01	100.25	125	124.67	99.74
15	14.99	99.93	5	4.98	99.60	250	250.71	100.28
20	20.09	100.45	30	30.05	100.17	500	501.33	100.27
100.227 ± 0.268			100.01 ± 0.354			100.097 ± 0.216		

Table 3: Determination of OMP, TND and CLR in Helicure® tablets by the proposed method

OMP			
Claimed amount taken (ug mL^{-1})	Authentic added (ug mL^{-1})	Authentic found (ug mL^{-1})	Recovery (%)
40.00	20.00	20.11	100.55
	40.00	40.11	100.28
	60	60.14	100.23
Mean ± SD			100.35 ± 0.172
TND			
Claimed amount taken (ug mL^{-1})	Authentic added (ug mL^{-1})	Authentic found (ug mL^{-1})	Recovery (%)
25.00	10.00	10.04	100.40
	25.00	24.94	99.76
	50.00	49.95	99.90
Mean ± SD			100.02 ± 0.336
CLR			
Claimed amount taken (ug mL^{-1})	Authentic added (ug mL^{-1})	Authentic found (ug mL^{-1})	Recovery (%)
1000.00	2000.00	1987.62	99.38
	1500.00	1499.23	99.95
	4000.00	3998.60	99.97
Mean ± SD			99.78 ± 0.335

Table 4: Application of the standard addition technique to the proposed HPLC method of OMP, TND & CLR in their pharmaceutical formulation.

determination of the studied drugs in their pharmaceutical formulation which is Heli-cure tablets. The precision of the proposed method was checked by the analysis of different concentrations (Table 3).The mean percentage recovery was found to be:

100.23 ± 0.268   for OMP

100.01 ± 0.354   for TND

100.097 ± 0.216   for CLR.

Conclusion

Validation of the accuracy of the proposed HPLC method was

Parameters	HPLC			B.P official method[2] OMP	B.P official method[2] TND	B.P official method[2] CLR
	OMP	TND	CLR			
Mean	100.076	100.403	100.647	100.30	100.13	100.16
± S.D	0.454	0.535	0.862	0.401	0.722	0.531
Variance	0.206	0.286	0.743	0.161	0.521	0.282
F-test	1.28 (4.95)a	1.82 (4.95)a	2.63 (5.05)a			
Student,s t-test	0.723 (2.201)a	0.763 (2.201)a	1.176 (2.23)a			
N	7	7	6	6	6	6

Table 5: Statistical comparison for the results obtained by the proposed method and the official method for analysis of OMP, TND and CLR.

Parameters	HPLC			Reported method[48] OMP	Reported method[48] TND	Reported method[48] CLR
	OMP	TND	CLR			
Mean	100.227	100.01	100.097	100.20	100.12	100.12
± S.D	0.268	0.354	0.216	0.283	0.167	0.281
Variance	0.072	0.125	0.047	0.080	0.028	0.079
F-test	1.14 (5.79)a	4.46 (5.79)a	1.68 (5.79)a			
Student,s t-test	0.13 (2.57)a	0.497 (2.57)a	0.123 (2.57)a			
N	3	3	3	4	4	4

aThe values in the parenthesis are corresponding theoretical t- and F-values at P=0.05 [44]

Table 6: Statistical comparison for the results obtained by the proposed HPLC method and the reported TLC method for analysis of OMP, TND and CLR in dosage form.

Parameter	OMP	TND	CLR	Limit [49,50]
Retention time (t_R)	2.063	1.353	5.443	
Resolution (R_s)		3.95		$R_s > 2$
Tailing factor (T)	0.80	0.88	0.40	T=1 for a typical symmetric peak
Capacity factor (K')	1.58	1.48	5.06	1-10 acceptable
Selectivity factor (α)		3.27		α>1
Column efficiency (N)	9811	6991	2451	N>2000
Height equivalent to theoretical plate (HETP)	0.003	0.004	0.010	The smaller the value, the higher the column efficiency

Table 7: System suitability parameters of the proposed HPLC method.

confirmed using standard addition technique (Table 4). Statistical comparison with the official and reported methods showed that the proposed HPLC is sensitive and precise (Table 5 and 6). Application of the proposed methods to the analysis of OMP, TND and CLR in their pharmaceutical formulation (Table 3) shows that excipients do not interfere with the determination. The system suitability parameters of the proposed HPLC method (Table 7). The proposed method has advantage of being sensitive and applicable over wide range. The proposed method can be used for routine analysis of omeprazole, tinidazole and clarithromycin in quality control laboratories.

References

1. The British pharmacopoeia I & II (2014) Bp commission (7th edn) Great Britain, her Majesty's stationery office, Dublin.

2. Souney PF, Matthews SJ (1994) Comprehensive Pharmacy Review, (2nd edn) Harwal, 765-777.

3. Mahmoud AM (2009) New sensitive kinetic spectrophotometric methods for determination of omeprazole in dosage forms. Int J Anal Chem 2009: 307045.

4. Bhandage A, Bhosale A, Kasture A, Godse VP (2009) Extractive Spectrophotometric determination of omeprazole in pharmaceutical preparations. Trop J Pharm Res 8: 449-454.

5. Wahbi AM, Abdel-Razak O, Gazy AA, Mahgoub H, Moneeb MS (2009) Spectrophotometric determination of omeprazole, lansoprazole and pantoprazole in pharmaceutical formulations. J Pharm Dev technol 14: 516-523.

6. Dhumal SN, Dikshit P M, Ubharay II, Mascarenhas BM, Gaitonde CD (1991) Individual UV-spectrophotometric assays of trazodone hydrochloride and omeprazole from separate pharmacetical dosages. J Indian Drugs 28: 565-567.

7. Ozaltin N, Koçer A (1997) Determination of omeprazole in pharmaceuticals by derivative spectroscopy. J Pharm Biomed Anal 16: 337-342.

8. Castro D, Moreno MA, Torrado S, Lastres JL (1999) Comparison of derivative spectrophotometric and liquid chromatograpic methods for the determination of omeprazole in aqueous solutions during stability studies. J Pharm Biomed Anal 21: 291-298.

9. El-Kousy NM, Bebawy LI (1999) Stability-indicating methods for determining omeprazole and octylonium bromide in the presence of their degradation products. J AOAC Int 82: 599-606.

10. Karljikovic-Rajic K1, Novovic D, Marinkovic V, Agbaba D (2003) First-order UV-derivative spectrophotometry in the analysis of omeprazole and pantoprazole sodium salt and corresponding impurities. J Pharm Biomed Anal 32: 1019-1027.

11. Salama F1, El-Abasawy N, Abdel Razeq SA, Ismail MM, Fouad MM (2003) Validation of the spectrophotometric determination of omeprazole and pantoprazole sodium via their metal chelates. J Pharm Biomed Anal 33: 411-421.

12. Wahbi AA, Abdel-Razak O, Gazy AA, Mahgoub H, Moneeb MS (2002) Spectrophotometric determination of omeprazole, lansoprazole and pantoprazole in pharmaceutical formulations. J Pharm Biomed Anal 30: 1133-1142.

13. el-Kousy NM, Bebawy LI (1999) Stability-indicating methods for determining omeprazole and octylonium bromide in the presence of their degradation products. J AOAC Int 82: 599-606.

14. Li HK, Zhou WB, Li GSY (2004) Spectrophotometric Determination of Omeprazole Based on the Charge Transfer Reaction between Omeprazole and Chloranilic Acid. Chin J spectroscop Lab 21: 646-649.

15. Qaisi AM, Tutunji MF, Tutunji LF (2006) Acid decomposition of omeprazole in the absence of thiol: a differential pulse polarographic study at the static mercury drop electrode (SMDE). J Pharm Sci 95: 384-391.

16. Rezk NL, Brown KC, Kashuba ADM (2006) A Simple and sensitive bioanalytical assay for simultaneous determination of omeprazole and its three major metabolites in human blood plasma using RP-HPLC after a simple liquid-liquid extraction procedure. J Chromatogr B 844: 314-321.

17. Hofmann U, Schwab M, Treiber G, Klotz U (2006) Sensitive quantification of omeprazole and its metabolites in human plasma by liquid chromatography-mass spectrometry. J Chromatogr B Analyt Technol Biomed Life Sci 831: 85-90.

18. Linden R, Ziulkoskil AL, Wingert M, Tonello P, Souto AA (2007) Simultneous determination of omeprazole, hydroxyl omeprazole and omeprazole sulphone in human plasma by isocratic HPLC-DAD: application to the phenotyping of CYP2C19 and CYP3A4 in Brazilian volunteers. J Braz Chem Soc 18: 733-740.

19. Sivasubramanian L, Anilkumar V (2007) Simultaneous HPLC estimation of omeprazole and domperidone from tablets. Ind J Pharm Sci 69: 674-676.

20. Yeung PKF, Little R, Jiang Y, Buckley SJ, Pollak PT, et al. (1998) A simple high performance liquid chromatography assay for simultaneous determination of omeprazole and metronidazole in human plasma and gastric fluid. J Pharm Biomed Anal 17: 1393-1398.

21. Schubert A, Werle AL, Schmidt CA, Codevilla C, Bajerski L, et al. (2003) Determination of omeprazole in bulk and injectable preparations by liquid chromatography. J AOAC Int 86: 501-504.

22. Vittal S, Ganneboina R, Layek B, Trivedi RK, Hotha KK, Bharathi DV, et al. (2009) Highly sensitive method for the determination of omeprazole in human plasma by liquid chromatograpy-electrospray ionization tandem mass spectrometry: application to a clinical pharmacokinetics study. J Biomed Chromatogr 23: 390-396.

23. Perez-Ruiz T, Martínez-Lozano C, Sanz, E Bravo A, Galera R (2006) Determination of omeprazole, hydroxyl omeprazole and omeprazole sulfone

using automated solid phase extraction and micellar electrokinetic capillary chromatography. J Pharm Biomed Anal 42: 100-106.

24. Singh L (2010) Validated method for determination of tinidazole using direct colorimetry, UV-visible spectrophotometry and difference spectrophotometry in pure and tablet dosage form. J Pharmacy Research 3: 1211.

25. Okunrobo LO (2007) Titrimetric and spectrophotometric determination of tinidazole tablets. World J Chem 2: 63-66.

26. Prasad CV, Sripriya V, Saha RN, Parimoo P (1999) Simultaneous determination of tinidazole, furazolidone and diloxanide furoate in a combined tablet preparation by second-derivative spectrophotometry. J Pharm Biomed Anal 21: 961-968.

27. Adegoke OA, Umoh OE (2009) A new approach to the spectrophotometric determination of metronidazole and tinidazole using p-dimethylaminobenzaldehyde. Acta Pharm 59: 407-419.

28. Prasad CV, Parihar C, Sunil K, Parimoo P (1997) Simultaneous determination of tinidazole-clotrimazole and tinidazole-norfloxacin in combined tablet preparations by derivative spectroscopy. J Pharm Pharmacol Commun 3: 337-341.

29. Basavaiah K, Nagegowda P, Chandrashekar U (2005) Determination of tinidazole by potentiometry, spectrophotometry and high performance liquid chromatography. Indian J Chemical Technology 12: 273-280.

30. Feng MX, Gao H, Yu FY (1997) Determination of tinidazole and its degradation products in tinidazole injections by reverse-phase HPLC. Fenxi Zazhi 17: 247-249.

31. Pai PN, Rao GK, Srinivas B, Puranik S (2008) RPLC Determination of Tinidazole and Diloxanide Furoate in Tablets. Indian J Pharm Sci 70: 670-672.

32. Abu Zuhri AZ, Al-Khalil S, Shubietah RM, El-Hroub I (1999) Electrochemical study on the determination of tinidazole in tablets. J Pharm Biomed Anal 21: 881-886.

33. Rege PV, Sathe PA, Salvi VS, Trivedi ST (2011) Simultaneous determination of norfloxacin and tinidazole in combined drug formulation by a simple electroanalytical technique. Int J Pharm Research Development 3.

34. Jiang XY, Chen XQ, Dong Z, Xu M (2007) The application of resonance light scattering technique for the determination of tinidazole in drugs. J Autom Methods Manag Chem 2007: 86857.

35. Shah J, Rasul JM, Suraya M (2001) Extractive Spectrophotometric methods for determination of clarithromycin in pharmaceutical formulations using bromothymol blue and cresol red. J Biomed Chromatography 15: 507-508.

36. Niopas I, Daftsios AC (2001) Determination of clarithromycin in human plasma by HPLC with electrochemical detection: validation and application in pharmacokinetic study. Biomed Chromatogr 15: 507-508.

37. De Velde F, Alffenaar JW, Wessels AM, Greijdanus B, Uges DR () Simultaneous determination of clarithromycin, rifampicin and their main metabolites in human plasma by liquid chromatography-tandem mass spectrometry, J Chromatogr B 877: 1771-1777.

38. Amini H, Ahmadiani A (2005) Sensitive determination of clarithromycin in human plasma by high-performance liquid chromatography with spectrophotometric detection. J Chromatogr B Analyt Technol Biomed Life Sci 817: 193-197.

39. Jiang Y, Wang J, Li H, Wang Y, Gu J (2007) Determination of clarithromycin in human plasma by liquid chromatography-electrospray ionization tandem mass spectrometry. J Pharm Biomed Anal 43: 1460-1464.

40. Gurule S, Verma PRP, Monif T, Khuroo A, Partani P (2008) Sensitive liquid chromatographic determination of clarithromycin and 14-hydroxy clarithromycin in human plasma with tandem mass spectrometry. J Liquid Chromatogr & Related Technol 31: 2955-2973.

41. Oswald S, Peters J, Venner M, Siegmund W (2011) LC-MS/ MS method for the simultaneous determination of clarithromycin, rifampicin and their main metabolites in horse plasma, epithelial lining fluid and broncho-alveolar cells, J Pharm Biomed Anal 55: 194-201.

42. Elkhoudary MM , Abdel Salam RA, Hadad GM (2014) Current Pharm Anal 10: 58-70.

43. Darwish KM, Salama I, Mostafa S, El-Sadek M (2013) RP-HPLC/pre-column derivatization for analysis of omeprazole, tinidazole, doxycycline and clarithromycin. J Chromatogr Sci 51: 566-576.

44. Kasnia V, Kumar M S, Mahadevan N (2012) International Journal of Recent Advances in Pharmaceutical Research 2: 78-83

45. Lotfy HM (2006) Simultaneous determination of omeprazole, tinidazole and clarithromycin in combination. Bull Fac Pharm Cairo Univ 44: 27-39.

46. Lotfy HM, Abdel-Monem Hagazy M (2012) Comparative study of novel spectrophotometric methods manipulating ratio spectra: An application on pharmaceutical ternary mixture of omeprazole, tinidazole and clarithromycin. J Spectrochim Acta A Mol Biomol Spectrosc 96: 259-270.

47. Spiege MR, Stephns LJ (1999) Schaumoutline of theory and problems of statistics, Schaum Outline Series.

48. Salem H, Riad SM, Rezk MR, Ahmed K (2013) Simultaneous determination of omeprazole, tinidazole and clarithromycin in bulk powders and Helicure tablets by TLC densitometric technique. J Pharm Educ Res 4: 34.

49. Andrea W, Phyllis R (1997) HPLC and CE principles and practice, London: Academic press. 7-15.

50. Adamovics AJ (1997) Chromatographic analysis of pharmaceuticals, New York: Marcel Dekker Inc.

27

Quantification of Cilostazol and Telmisartan in Combination using Risk Profile and Uncertainty Contour

Dharmendra D, Karan M, Bhoomi P and Rajshree CM*

Quality Assurance Laboratory, Centre of Relevance and Excellence in Novel Drug Delivery system, Department of Pharmacy, Shree G.H.Patel Pharmacy building, Donor's Plaza, The Maharaja Sayajirao University of Baroda, Fatehgunj, Vadodara-390002, Gujarat, India

Abstract

Risk profile and uncertainty estimation are the two major and important parameter that need to be carried out during the development of pharmaceutical process, to obtain reliable results. The conventional method validation schedule needs to be improvised so as to certify extraordinary method reliability to measure quality feature of a drug product. Risk profile assessment, expanded uncertainty and combined standard uncertainty in the analysis of cilostazol and telmisartan in combined tablet dosage form were studied in this research work. RP-HPLC method was validated in our laboratory as per ICH guideline and risk profile assessment has been outlined including uncertainty estimation using the cause-effect approach. In the course of validation, the calibration model found to be impregnable when encountered with lack of fit test and Levene's test. In uncertainty major contribution is due to sample concentration and mass. The proposed research work clearly demonstrate the application of theoretical concept of calibration model tests, relative bias, risk profile and uncertainty in the methods used for analysis in drug discovery process.

Keywords: Cilostazol (CLZ); Telmisartan (TLM); RP-HPLC; Risk profile; Relative bias; Combined standard uncertainty; Expanded uncertainty

Introduction

Decisions are made by measurement. When we use number to make decision every time we run with the risk of making fault, because all number are more or less reliable. Since every measurement is suspected, it is necessary to know how and why it is so. So to avoid this risk of making mistake with every measurement and to clear worries associated with measurement recently a new approach has been familiarized known as measurement of uncertainties in analysis of sample. The quality of method is expressed in terms of its uncertainty and assessment of uncertainty becomes key parameter for method validation to get certification [1,2]. Analytical method for evaluation of drug component in pharmaceutical formulation has been introduced and available worldwide. But the validation by total error approach and quantification of causes of uncertainties in this developed and published method has been omitted. In this study simple method for quantification of uncertainty components and combined standard uncertainty (CSU) is presented by assessing this computation for RP-HPLC measurement of cilostazol and telmisartan in tablet dosage form. Some article recommends the conventional estimation of analytical measurement and uncertainty as well. For perfect and complete study of uncertainty components, a new RP-HPLC method for cilostazol and telmisartan in combined tablet dosage form has been established and validated as per the ICH guideline [3,4].

A simple, rapid, accurate, precise, reliable and economical RP-HPLC method with UV detection was optimized, developed and validated as per ICH-Q2 guideline for the simultaneous estimation of cilostazol (CLZ) and telmisartan (TLM) in tablet dosage form. The retention behavior of CLZ and TLM as a function of mobile phase pH, composition and flow rate was inspected. Separation was developed on a reverse-phase C_{18} column (250 mm × 4.6 mm i.d, 5 μ particle size), using a mobile phase consisting of Acetonitrile: Phosphate Buffer (pH-3.5 adjusted with Orthophosphoric acid) in the ratio of 60:40 (%v/v) at a flow rate of 1.0 ml/min with UV detection at 255 nm within 10 min with retention time of 4.23 and 6.34 for CLZ and TLM respectively. The standard curves were linear over the concentration range of 12-32 μg/mL and 3-8 μg/mL for CLZ and TLM. The developed method was validated in terms of accuracy, precision, and linearity, limit of detection and limit of quantification. From the validation outcomes it was established that proposed method can be used for the approximation of both drugs in combined pharmaceutical tablet dosage form.

Cilostazol (CLZ) (Figure 1) is chemically known as 6-[4-(1-cyclohexyl-1H-tetrazol-5-yl) butoxy]-3,4-dihydro-2 (1H) – quinolinone and is a quinolinone derivative that inhibits cellular phosphodiesterase III, and is used for the inhibition of platelet aggregation and as a vasodilator [5-6]. Telmisartan (TLM) (Figure 2) is chemically known as 4'-([4-methyl-6-(1-methyl-lH-benzimidazol-2yl)-2-propyl-lH-benzimidazol-1-yl]methyl}-2-biphenylcarboxylic acid. Telmisartan is a new angiotensin II receptor antagonist for the treatment of essential hypertension and useful in the treatment of mild to moderate hypertension, well tolerated with a lower incidence of cough than ACE inhibitors [7,8]. Several method for determination of cilostazol and telmisartan individual and in combination with other drug has been reported in the past, such as spectrophotometric [9-12], high-performance liquid chromatography (HPLC) [13-15]. Although the RP-HPLC method for simultaneous estimation of cilostazol and telmisartan is already developed but this method include the uncertainty measurement and [16-21] risk profiling using total error approach and developed method is more economical as compare to previous published method [22].

*Corresponding author: Rajshree C Mashru, Quality Assurance Laboratory, Centre of Relevance and Excellence in Novel Drug Delivery system, Pharmacy Department, Shree G.H.Patel Pharmacy building, Donor's Plaza, The Maharaja Sayajirao University of Baroda, Fatehgunj, Vadodara-390002, Gujarat, India E-mail: rajshreemashru@yahoo.com ; damor03.dd@gmail.com

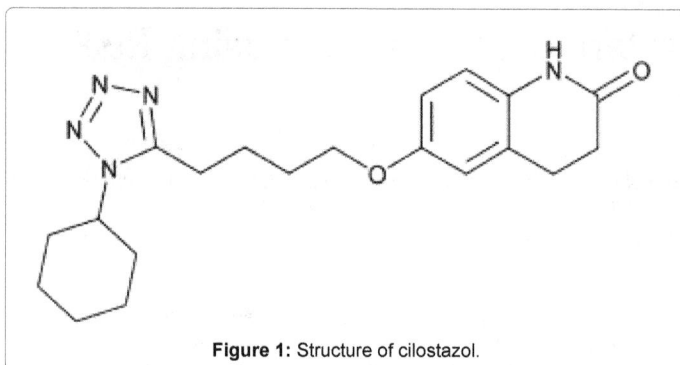

Figure 1: Structure of cilostazol.

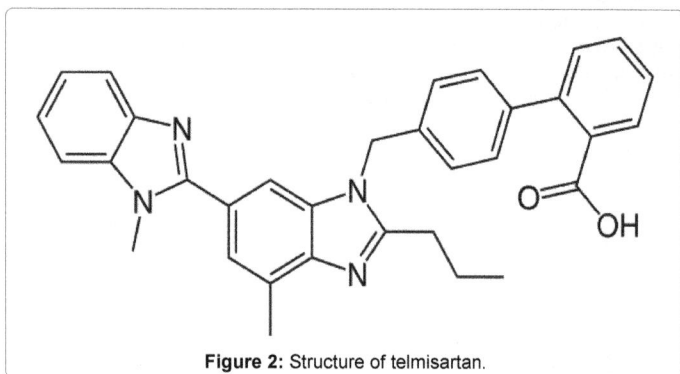

Figure 2: Structure of telmisartan.

The present research paper describes a validated RP-HPLC method for quantification of cilostazol and telmisartan in combined dosage formulation with total error approach and uncertainty measurement.

Experimental Section

HPLC Instrumentation

Chromatography was performed on Shimadzu (Shimadzu Corporation, Kyoto, Japan) chromatographic system equipped with Shimadzu LC-20AT pump and Shimadzu SPD-20AV absorbance detector. Samples were injected through a Rheodyne 7725 injector valve with fixed loop at 20 µl. Data acquisition and integration was performed using Spinchrome software (Spincho biotech, Vadodara). The chromatographic elution of analytes was obtained by using Kromasil C_{18} 5 µm (250 × 4.6) mm column at flow rate of 1 ml/min was used.

Reagents and chemicals

Telmisartan was produced as gift sample from Alembic pharmaceutical, Vadodara and cilostazol was purchased from Swapnroop drug Pvt.. Ltd Bombay. HPLC grade Methanol, Acetonitrile and Ortho Phosphoric Acid were supplied by Spectrochem Pvt. ltd, Mumbai. Water used throughout the experiment was Purified HPLC grade water obtained by filtering double distilled water through nylon filter paper 0.2 µm pore size and 47 diameters (Pall Life sciences, Mumbai, India).

Preparation of mobile phase

Phosphate Buffer of pH 3.5 (pH adjusted with Ortho phosphoric acid) was prepared and this phosphate buffer was filtered through 0.20 µm filter paper and then proper mixing of Acetonitrile and Phosphate Buffer (60:40) was done. Then the prepared mixture were sonicated for 5 min in ultrasonic bath, and then used as mobile phase.

Preparation of standard solutions of CLZ and TLM

Stock solution of (1000 µg mL^{-1}) for both cilostazol and telmisartan was prepared by accurately weighing 25 mg of both drugs in 25 ml volumetric flask and then diluted with Methanolup to the mark. From this stock 10 ml solution was withdrawn and transferred to 100 ml of volumetric flask and then diluted up to the mark with Mobile phase to get working standard solution of (100 µg mL^{-1}) of cilostazol and telmisartan.

Preparation of calibration curve

From the working standard solution of cilostazol (100 µg mL^{-1}) aliquots ranging from 1.2 ml to 3.2 ml were taken, in 10 ml volumetric flask and diluted to volume with mobile phase to give final concentrations of 12, 16, 20, 24, 28, 32 µg mL^{-1} of CLZ. And aliquots ranging from 0.3 ml to 0.8 ml were taken, from the working standard telmisartan (100 µg mL^{-1}) in 10 ml volumetric flask and diluted to volume with mobile phase to give final concentrations of 3, 4, 5,6 7, 8 µg mL^{-1} of TLM.

Injections of 20 µl were made for each concentration and chromatogram was obtained under the condition described in. Calibration graph was constructed by plotting peak area versus concentration of each drug and the regression equation was calculated.

Method Development

Selection of common solvent

CLZ and TLM were found to be soluble in methanol and hence, they were first dissolved in methanol and then, to avoid interference of methanol in chromatograms, the dilutions were made with mobile phase.

Selection of detection wavelength

The detection wavelength should be the one where both the drugs show considerable absorbance for the purpose of obtaining good sensitivity. Both the drugs are having appreciable absorbance at 255 nm hence 255 nm selected for detection.

Selection and optimization of chromatographic conditions

To optimize the chromatographic conditions, the effect of chromatographic variables such as composition of mobile phase, pH of mobile phase and flow rate were studied. The resulting chromatograms were recorded and the chromatographic parameters such as capacity factor, asymmetric factor, resolution and theoretical plates were calculated plates were calculated. The conditions that gave the best resolution, symmetry and theoretical plate were selected for estimation. Several mobile phases with different pH and ratio were tried and good and well resolved peak was obtained in Acetonitrile: Phosphate Buffer (60:40 pH 3.5).

Validation procedure using total error approach

Current ad-hoc approaches to method validation are inconsistent with ensuring method suitability. A total error approach based on the use of two-sided b-content tolerance intervals was developed. The total error approach offers a formal statistical framework for assessing analytical method performance. The approach is consistent with the concept of method suitability and controls the risk of incorrectly accepting unsuitable analytical methods. Risk profiling is a process for finding the optimal level of risk for developed method and risk associated with method. The present method was validated as per ICH guideline [3,4] and ISO guideline which were grounded upon "total error" approach [1]. In this approach "total error" was estimated by merging the systemic error and random error to recognize the difference between observed and true value. In the proposed method sensitivity of the method and effect of sample matrix were also studied. The selectivity of the studied method was investigated by comparing chromatogram of blank without CLZ and TLM, blank mobile phase and sample with CLZ and TLM and sample of formulation. Response function in proposed method three sets of calibration curve were plotted between area and different concentration of CLZ and TLM and on these three different series regression analysis was performed and series with best coefficient of determination was selected and the selected series has been further diagnosed by Levene's test (Table 1) and standardized residual plot (Figures 3 and 4). Trueness of calibration curve was calculated by back calculation of concentration to justify the calibration line. The result of trueness was expressed in terms of absolute and relative bias (Tables 2.1 and 2.2). In order to confirm the reproducibility of standard precision at two different levels were studied, first one is repeatability under same operating condition over short time interval and second is intermediate precision assessed on different days. The precision result expressed in terms of % relative standard deviation (RSD) (Tables 3.1 and 3.2) (Figures 5 and 6). Relative and absolute precision at both level were calculated with 95% Upper confidence limit (Tables 3.1 and 3.2). The recovery study which is the most critical parameter in method validation requires an extra precaution during study and interpretation of recovery results. Therefore, the results of accuracy studies were interpreted and represented in the β- expectation tolerance limits (Tables 4.1 and 4.2). In addition to this parameter, risk profile has also been studied to know the future application of the method. Linearity profile was also studied demonstrate the relationship between nominal and observed concentration in matrix and furthermore, residual plot was generated to know the outliers in the determination of CLZ and TLM in sample matrix (Figure 7). Limit of detection and quantification represents the sensitivity of the method which has been calculated as per ICH guidelines. Limit of detection LOD and limit of quantification are two important parameter which show the application of method in quantification and detection of different sample. These are calculated according to the procedure mentioned in the ICH guideline [4,5].

Limit of detection (LOD) and limit of quantification (LOQ) are two important parameter which show the application of method in quantification and detection of different sample. These are calculated according to the procedure mentioned in the ICH guideline [4,5].

Uncertainty Estimation

Cause-effect diagram: Even though estimation method was validated as per guidelines but still doubt was there in result as during the validation of method small influence which can affect the results has not been studied, Such as error during sample weighing, discharge of volumetric flask etc. Therefore, to overwhelm such doubt during result collation were clarified by estimation of uncertainty in result obtained from validation. The protocol for uncertainty estimation starts with identification of sources of uncertainty. The best way of listing uncertainty sources is to use the cause-effect diagram plan, as it outlines the sources connection to each other demonstrating their impact on the result. Thus a cause-effect diagram was assembled as presented in Figure 8. The parameter taken in consideration was volume of volumetric flask V_{10}, and mass of sample, recovery of method R_m and precision of method. These all parameter contribute to uncertainty in the interpreted results. This diagram also helps in resolving any repeatability of component in uncertainty. The parameter comes in consideration after constructing cause-effect diagram was illustrated in Equation 1.

$$CLZ / TLM_{sample} = C_{10}V_{10}10^{-3} / m_{sample} R_m \qquad (1)$$

Where, CLZ_{sample} and TLM_{sample}, CLZ and TLM quantity in (mol/kg); C_{10}, CLZ and TLM concentration in 10 mL volumetric flask (M); V_{10} volume of 10 mL volumetric flask (mL); m_{sample}, CLZ and TLM sample mass taken (kg); Rm, Recovery of method.

These identified sources were quantified and their discrete effect of on inclusive uncertainty was assembled as CSU and EU.

Individual parameter showing effect on overall uncertainty

Liberation of solution from volumetric flask: The uncertainty due to liberation of volumetric flask was evaluated by performing experiment involving filling up and weighing of 10 mL volumetric flask

	Source	SS	df	MS	Fcalc	Fcrit, 95%	p-value
CLZ	Model	12.88	5	2.576	2.299	2.409	0.05947
	Error	53.77	48	1.12			
TLM	Model	5.213	5	1.043	4.324	2.409	0.002501
	Error	11.57	48	0.2411			

Table 1: Evaluation of homogeneity of variances (Levene's test) for CLZ and TLM.

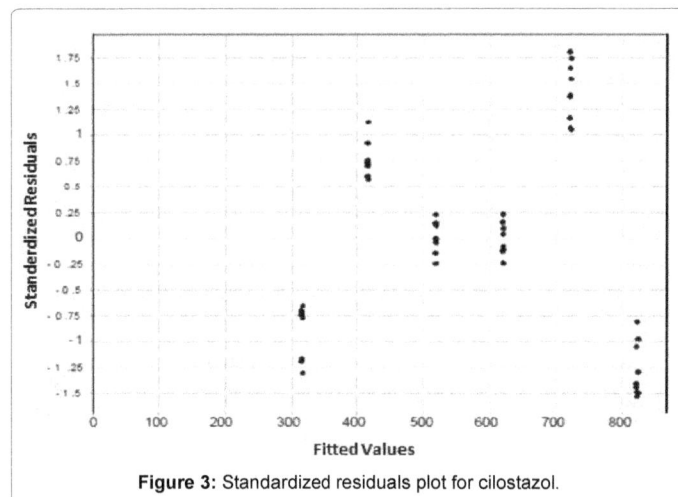

Figure 3: Standardized residuals plot for cilostazol.

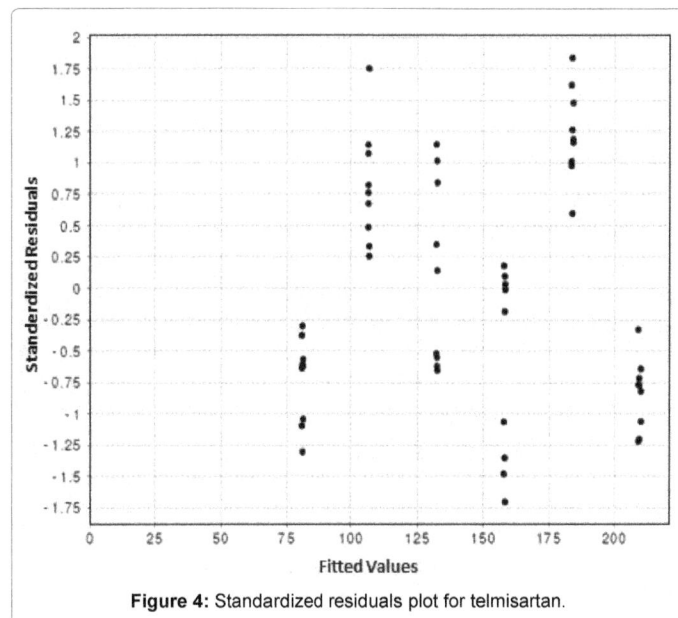

Figure 4: Standardized residuals plot for telmisartan.

Concentration level (mcg/ml)	Mean introduced concentration (mcg/ml)	Mean Back calculated concentration (mcg/ml)	Absolute bias (mcg/ml)	Relative Bias (%)	Recovery (%)	95% Confidence Interval of Recovery (%)
1.0	12.00	11.63	-0.3668	-3.057	96.94	[96.27 , 97.61]
2.0	16.00	16.32	0.3215	2.009	102.0	[101.7 , 102.4]
3.0	20.00	19.99	-0.008377	-0.04188	99.96	[99.69 , 100.2]
4.0	24.00	23.99	-0.005115	-0.02131	99.98	[99.77 , 100.2]
5.0	28.00	28.58	0.5838	2.085	102.1	[101.8 , 102.4]
6.0	32.00	31.48	-0.5249	-1.640	98.36	[98.10 , 98.62]

Table 2.1: Result of Trueness in terms of relative bias (%) for CLZ.

Concentration level (mcg/ml)	Mean introduced concentration (mcg/ml)	Mean back calculated concentration (mcg/ml)	Absolute bias (mcg/ml)	Relative bias (%)	Recovery (%)	95% Confidence interval of recovery (%)
1.0	3.000	2.947	-0.05291	-1.764	98.24	(97.60, 98.87)
2.0	4.000	4.059	0.05864	1.466	101.5	(100.8, 102.2)
3.0	5.000	5.009	0.009405	0.1881	100.2	(99.37, 101.0)
4.0	6.000	5.956	-0.04397	-0.7328	99.27	(98.56, 99.98)
5.0	7.000	7.090	0.08972	1.282	101.3	(101.0, 101.6)
6.0	8.000	7.939	-0.06088	-0.7610	99.24	(99.03, 99.45)

Table 2.2: Result of Trueness in terms of relative bias (%) for TLM.

Nominal conc (mcg/ml)	Repeatability (RSD%)	Intermediate precision (RSD%)	Repeatability (SD - mcg/ml)	Intermediate precision (SD - mcg/ml)	95% Upper confidence limit repeatability (SD - mcg/ml)	95% Upper confidence limit intermediate precision (SD - mcg/ml)
12.00	0.8732	0.8732	0.1048	0.1048	0.1793	0.1793
16.00	0.4602	0.4602	0.07364	0.07364	0.1260	0.1260
20.00	0.3487	0.3487	0.06973	0.06973	0.1193	0.1193
24.00	0.2665	0.2665	0.06397	0.06397	0.1094	0.1094
28.00	0.4125	0.4125	0.1155	0.1155	0.1976	0.1976
32.00	0.3444	0.3444	0.1102	0.1102	0.1886	0.1886

Table 3.1: Result of relative and absolute Intermediate Precision and Repeatability in terms of (%RSD) CLZ.

Nominal conc (mcg/ml)	Repeatability (RSD%)	Intermediate precision (RSD%)	Repeatability (SD-mcg/ml)	Intermediate precision (SD – mcg/ml)	95% Upper confidence limit repeatability (SD – mcg/ml)	95% Upper confidence limit intermediate precision (SD - mcg/ml)
3.000	0.8265	0.8265	0.02479	0.02479	0.04242	0.04242
4.000	0.8951	0.8951	0.03580	0.03580	0.06126	0.06126
5.000	1.069	1.069	0.05346	0.05346	0.09147	0.09147
6.000	0.9230	0.9230	0.05538	0.05538	0.09475	0.09475
7.000	0.4117	0.4117	0.02882	0.02882	0.04930	0.04930
8.000	0.2701	0.2701	0.02161	0.02161	0.03697	0.03697

Table 3.2: Result of relative and absolute intermediate precision and repeatability in terms of (%RSD) TLM.

with standard solution for 10 times.

Mass (m_{sample}): Difference between weighing glass with and without the sample provide the sample mass.

Concentration, C_{10}: The uncertainty in concentration of drug obtained from calibration curve is expressed as uncertainty due to concentration C_{10}. This is estimated using Equation 2.

$$U(c) = \frac{sr}{b} \sqrt{\frac{1}{n} + \frac{1}{p} + \frac{(c-\overline{c})^2}{sxx}} \qquad (2)$$

Where $Sr = \sqrt{\sum_{j=1}^{n} \frac{[yj - (bxi + a)]}{n-2}}$ (3)

$$sxx = \sum (ci - \overline{c})^2 \qquad (4)$$

S_r standard deviation of residual; n, number of measurement used for calibration curve; p, number of measurement used to obtain

concentration of sample; c, concentration in sample (M); c, average of standard solution (M); Y_j, response obtained from the measurement; j, index for number of measurement made in order to obtain the calibration curve; i, index for number of solution for calibration; b, slope of calibration curve (L mol^{-1}); a, calibration curve intercept;

Recovery of method: Uncertainty associated with recovery of method was evaluated using Equation 5 and it depends upon spiked and recovered concentration of standard in sample matrix.

$$U(Rm) = Rm \times \sqrt{\left(\frac{Sobs^2}{n \times Cobs^2}\right) + \left(\frac{U(Cspike)^2}{Cspike}\right)} \qquad (5)$$

Where, C_{obs}, mean of concentration observed from replicate analysis of spiked sample; C_{spike}, nominal concentration of drug in spiked sample. S_{obs}, means standard deviation of result from the replicate analyses of spiked sample; n, number of replicates; U (C_{spike}), standard uncertainty in concentration of spiked sample.

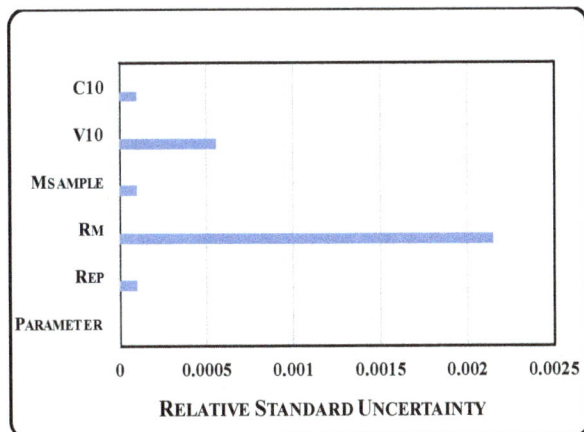

Figure 5: Uncertainty profile for cilostazol determination.

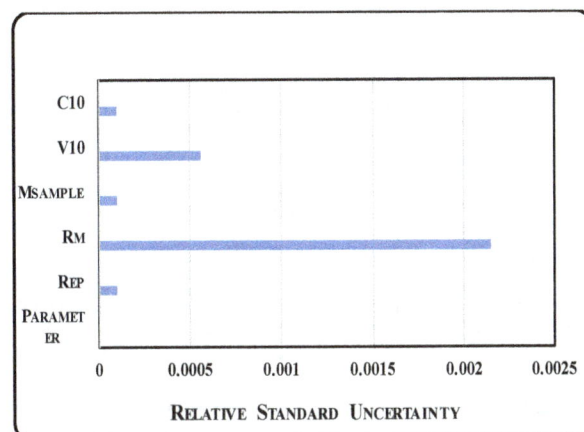

Figure 6: Uncertainty profile for telmisartan determination.

Precision (P): During the validation of method precision studies were carried out. In this study repeatability and variability associated with the measurement were included in overall precision uncertainty estimation.

Results and Discussions

Validation parameters

In this method calibration curves from the response of different concentration were prepared using linear regression model. The four different sets were prepared for response function studies with range of 12-32 µg/ml for CLZ and 3-8 µg/ml for TLM, from their regression analysis studies series 3 show the best results with coefficient of determination r^2 0.997 and 0.9986 for CLZ and TLM respectively, so this series was selected for further competition for validation and sample analysis. Moreover, the selected series and regression model was diagnosed and confirmed using lack of fit test (LOF). The p-value were calculated and found to be greater than 0.05 and further to demonstrate that no outliers were found in calibration curve standard residual plot were also plotted as represented in Figures 3 and 4. As the model was established now in order to authenticate the regression equation back calculation were done and linear plot using absolute β-expectation limit was constructed between nominal and back calculated concentration which showing the 0.9968 and 0.9984 for CLZ and TLM respectively and confirming the authenticity of

regression equation. Trueness of method justified by calculation of % relative bias which was found to be limited between [-3.057 - 2.085] for CLZ and [1.282- 1.764] for TLM as illustrated in Tables 2.1 and 2.2 from which it has been concluded that the trueness of method is adequate. The method precision and reproducibility was authenticated by result obtained from precision studies which were found to be <2% in terms of RSD for both repeatability and intermediate level as illustrated with 95% confidence upper limit in Tables 3.1 and 3.2. After the confirmation of accuracy of all the parameters related to the system and developed method, sample matrix was incorporated in validation process which includes recovery studies. Recovery studies were carried out using standard addition method in sample matrix. These recovery studies receipts into account total error of test result and is represented by the β-expectation tolerance limit. The result of accuracy studies has been illustrated in Table.4.1 and 4.2. The β-expectation tolerance limit was also found to be in the acceptance as accuracy profile illustrated in Figures 9-12. Further, these recovery studies of model was justified by plotting risk profile keeping maximum risk level at 5.0% from which it was concluded that risk of outliers are within limits and in future analysis of the sample using this developed and validated method will fall within range. The results of LOD show that this method is sensitive enough to analyze marketed formulations; LOD was found to be 1.232 and 0.2174 for CLZ and TLM respectively.

Application of the Developed Method

Analysis of formulation

It is evident from the aforementioned results that proposed method gave satisfactory results with the CLZ and TLM in bulk drug. The dosage forms were subjected to analysis for their content of active drug material by the proposed method. The percentage purity for tablet was found to be 102.475% for CLZ and 99.50% for TLM (Table 5). It is evident from the above mentioned results that proposed method is applicable to the analysis of drug in its bulk drug as well as synthetic forms with comparable analytical performance.

Measurement of uncertainty

Once uncertainty sources have been identified, they were evaluated and their magnitude was determined. In order to assure the traceability for uncertainty results all the computation were done in International System of Unit as concentration in M and weight in kg.

Uncertainty of volumetric flask

The uncertainty of volumetric flask is mainly influenced by the three parameter i.e. calibration of the volumetric flask at the time of manufacturing, repeatability and temperature.

Calibration of volumetric flask: Deviance from nominal volume of 10 mL volumetric flask is ±0.006 mL (at 27°C) as given by manufacturer. Standard value of uncertainty can be calculated with triangular distribution. So, uncertainty related to the liberation of volume by volumetric flask u (V_{10cal}) is 0.00245.

Repeatability, u (V_{10rep}): In experiment repeatedly weighing and filling of volumetric flask standard uncertainty established was 0.0014 mL.

Temperature

The manufacturer has calibrated volumetric flask at time of manufacturing at temperature of 27°C, while temperature at laboratory varied with $\Delta t=\pm4°C$. This difference can be overcome by calculating uncertainty value with estimation of temperature range and volume dilation coefficient. Volume expansion of liquid was taken

Concentration level (mcg/ml)	Mean introduced concentration (mcg/ml)	Beta-expectation tolerance limits (mcg/ml)	Relative beta-expectation tolerance limits (%)	Risk² (%)
1.0	12.00	(11.38, 11.89)	(-5.193, -0.9208)	3.454
2.0	16.00	(16.14, 16.50)	(0.8831, 3.135)	0.01566
3.0	20.00	(19.82, 20.16)	(-0.8949, 0.8111)	0.0001150
4.0	24.00	(23.84, 24.15)	(-0.6734, 0.6307)	0.00001532
5.0	28.00	(28.30, 28.87)	(1.076, 3.094)	0.008976
6.0	32.00	(31.21, 31.74)	(-2.483, -0.7977)	0.0009638

Table 4.1: Method accuracy obtained by considering linear regression for cilostazol.

Concentration level (mcg/ml)	Mean introduced concentration (mcg/ml)	Beta-expectation tolerance limits (mcg/ml)	Relative beta-expectation tolerance limits (%)	Risk² (%)
1.0	3.000	(2.886, 3.008)	(-3.786, 0.2582)	0.3188
2.0	4.000	(3.971, 4.146)	(-0.7237, 3.656)	0.3098
3.0	5.000	(4.879, 5.140)	(-2.428, 2.804)	0.2442
4.0	6.000	(5.821, 6.092)	(-2.991, 1.525)	0.1481
5.0	7.000	(7.019, 7.160)	(0.2745, 2.289)	0.001690
6.0	8.000	(7.886, 7.992)	(-1.422, -0.1002)	0.00003217

Table 4.2: Method accuracy obtained by considering linear regression for telmisartan.

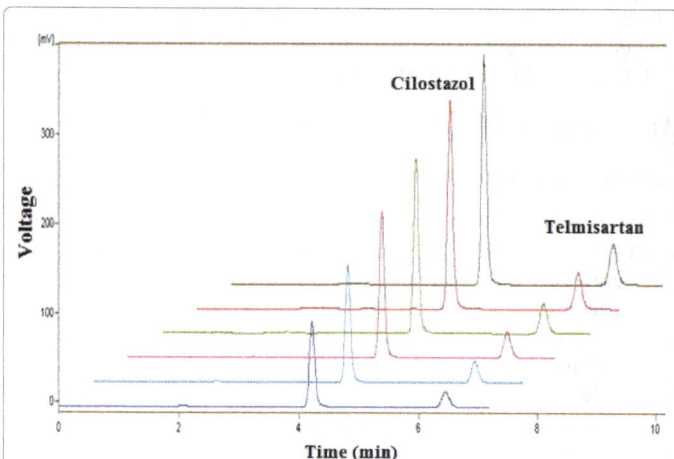

Figure 7: Linearity, overlain chromatogram of CLZ (12-28 µg/ml) and TLM (3.0-8.0 µg/ml).

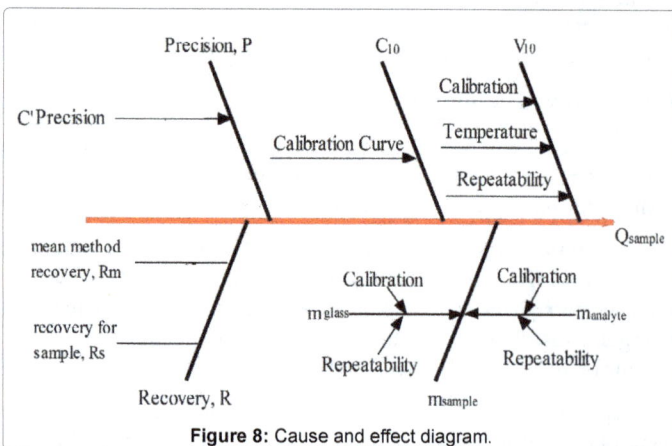

Figure 8: Cause and effect diagram.

into consideration as it is quite higher than expansion of volumetric flask. The volume expansion coefficient, λ, of water is 2.1×10^{-4} /°C. Uncertainty for 10 mL volumetric flask ΔV_{10} was calculated by Equation 6.

$$\Delta V_{10} = V_{10} \times \gamma \times \Delta t \tag{6}$$

Where $\Delta V10$, uncertainty of the 10 mL volumetric flask; $V10$, volume of the 10 mL volumetric flask; γ, volume dilation coefficient; Δt, temperature variation in the laboratory.

Thus, we obtain uncertainty for volumetric flask of 10 mL is 0.0084 mL, standard uncertainty due to temperature on liberation of volumetric flask was found to be 0.0048 mL.

Uncertainty associated with the sample mass m_{sample}

Sample mass has three types of uncertainty sources sensitivity, linearity, and repeatability. Mass of the sample was expressed in kg to convince traceability of results.

Sensitivity: The difference in weighed mass was in very less range and it was measured on the same weighing balance. Thus uncertainty due to sensitivity of balance can be neglected.

Linearity: A rectangular distribution was assumed to convert contribution of linearity. It was calculated as Equation 7.

$$u = \frac{1.06 \times 10^{-5}}{\sqrt{3}} = 6.12 \times 10^{-8} \, Kg \tag{7}$$

Repeatability: Uncertainty associated with repeatability is found to be 0.0001 kg.

Computation of relative uncertainty due to sample mass: Using the uncertainty due to linearity and repeatability the uncertainty due to sample mass u (m_{sample}) was calculated using Equation 8.

$$u = (m_{sample}) = \sqrt{2 \times (6.11 \times 10^{-6})^2 + (2.8 \times 10^{-5})^2} = 2.25 \times 10^{-7} \, Kg \tag{8}$$

Uncertainty associated with concentration, (C_{10})

Analytical response was collected after each injection in HPLC system of standard solution of different concentration. These responses were used to construct calibration curve. Regression equation of calibration curve was identified such as, slope 9364221442.3334 and intercept 8.8305 for CLZ and slope 13348714077.4857 and intercept 2.4076 for TLM. Uncertainty involved in the construction calibration curve was estimated by injecting 6 different concentration solution each measured three times and sample solution was measured ten times from which Sr and Sxx value were computed as shown in Equation 3 and 4, which were further used to calculate standard relative uncertainty, due to concentration.

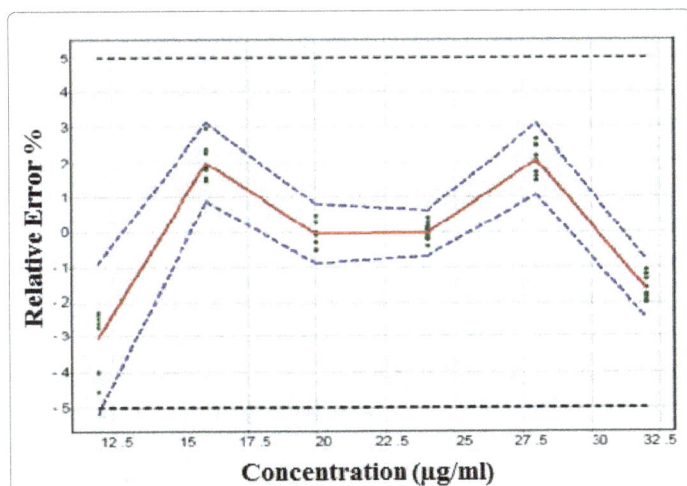

Figure 9: Accuracy profile of cilostazol obtained by considering linear regression the plain red line is the relative bias, the dashed lines are the β-expectation tolerance limits and the dotted curves represent the acceptance limits.

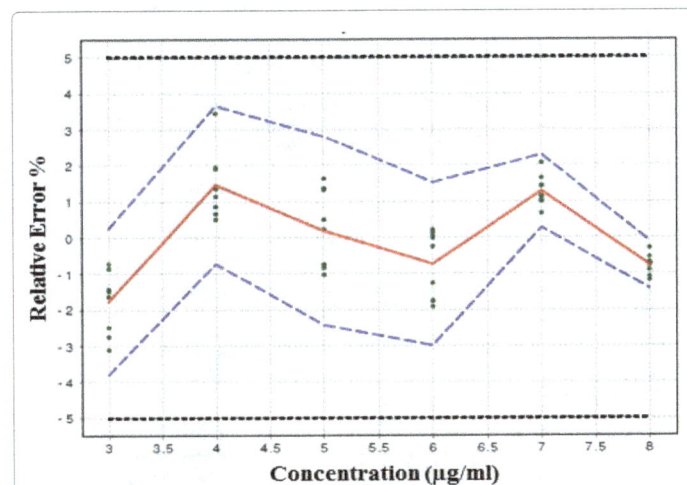

Figure 10: Accuracy profile of telmisartan obtained by considering linear regression the plain red line is the relative bias, the dashed lines are the β-expectation tolerance limits and the dotted curves represent the acceptance limits.

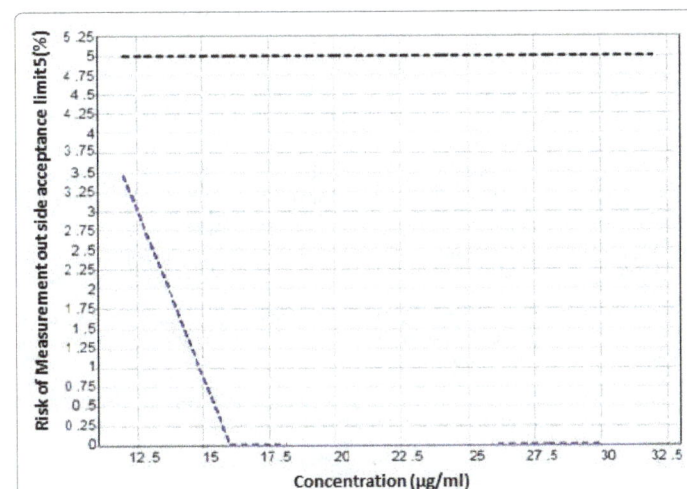

Figure 11: Risk profile obtained by considering linear regression cilostazol, the dotted line represents the maximum risk level chosen: 5.0%.

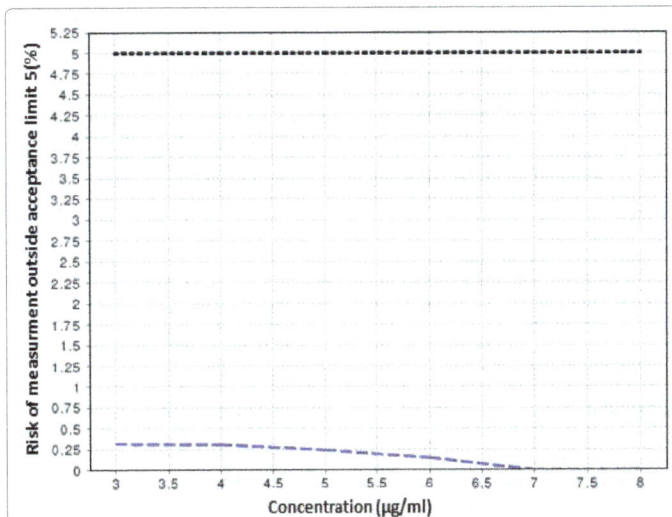

Figure 12: Risk profile obtained by considering linear regression telmisartan, the dotted line represents the maximum risk level chosen: 5.0%.

For CLZ, $Sxx=2.05127 \times 10^{-15}$

$Sr=0.042414$

For TLM, $Sxx=6.60794 \times 10^{-17}$

$Sr=0.011109$

Uncertainty due to recovery of method

Results of recovery are evaluated as percentage recovery from sample matrix after spiking a known amount. When team 'spike' is used to estimate recovery, the recovery of analytes from the sample may differ from recovery of spike so that an uncertainty needs to be evaluated. Uncertainty due to spiking is found to be 1.22471×10^{-7} for CLZ and 2.19796×10^{-8}. Standard relative uncertainty of method recovery was calculated using uncertainty due to mass of CLZ and TLM (from balance), calibration of pipette, calibration of flask and temperature effect, which was found to be 7.07107×10^{-5}, 0.0058, 0.00245 and 0.0048 respectively. Combined uncertainty due to these factor were found to be U $(R_f)=0.007917$.

Uncertainty due to precision

Method validation results show the repeatability for determination of CLZ and TLM in terms of %RSD (0.08493) and (0.074081) respectively. This equation can be used directly for calculation of CSU.

U (Rep)=RSD

U (Rep)$_{CLZ}=0.08493$

U (Rep)$_{TLM}=0.07408$

Combined standard uncertainty (CSU)

The values of all the parameters having effect on CLZ and TLM determination, these are compiled up in Tables 6 and 7 respectively. These values of parameter were further used to calculate CLZ and TLM quantity by using Equation 1 and thus, we obtained a quantity of 4.96×10^{-3} for CLZ and 6.17×10^{-6} mol/kg.

Expanded standard uncertainty (EU)

Expanded uncertainty of CLZ and TLM in sample matrices was obtained by multiplying the combined standard uncertainty by coverage factor k=2 at confidence level of 95%, and, the EU (CLZ/TLM$_{sample}$) is as shown.

Formulation: Synthetic mixture		
Drug	Composition of mixture	%Drug found* ± SD
Cilostazol	40 mg	102.475 ± 0.1150
Telmisartan	10 mg	99.50 ± 0.0709

Table 5: Application of the developed method in synthetic mixture.

Formulation	Parameter	Volume, V_{10}(ml)	Sample conc, C_{10} (M)	Mass sample m_{sample}(kg)	Recovery method	Repeatability
Tablet	Value	10	4.376×10^{-8}	1.00×10^{-4}	101×10^{-2}	----
	Standard uncertainty, u(x)	5.56×10^{-3}	3.99×10^{-9}	4.49×10^{-5}	2.85×10^{-2}	8.4×10^{-2}
	RSU*, u(x)/x	5.56×10^{-4}	9.98×10^{-5}	0.277	2.82×10^{-2}	8.4×10^{-2}

Table 6: Summary of contribution to the measurement uncertainty for determination of cilostazol through RP-HPLC in tablet dosage form.

Formulation	Parameter	Volume, V_{10}(ml)	Sample conc, C_{10} (M)	Mass sample m_{sample}(kg)	Recovery method	Repeatability
Tablet	Value	10	7.83×10^{-9}	1.00×10^{-4}	100.7×10^{-2}	----
	Standard uncertainty, u(x)	5.56×10^{-3}	1.021×10^{-9}	2.91×10^{-7}	2.85×10^{-2}	7.4×10^{-2}
	RSU*, u(x)/x	5.56×10^{-4}	1.021×10^{-4}	0.015	2.82×10^{-2}	7.4×10^{-2}

Table 7: Summary of contribution to the measurement uncertainty for determination of telmisartan through RP-HPLC in tablet dosage form.

EU (CLZ_{sample}) tab=1.73×10^{-3} mol/kg

EU (TLM_{sample}) tab=7.77×10^{-7} mol/kg

The contribution of different parameter in uncertainty is shown individually for sample matrix has been illustrated in Figure 8.

Conclusion

All analytical endeavors generate measurement data and hence, should necessarily employ appropriate statistical techniques and method of inference, to present and interpret the data. The accurate estimation of variability is challenging. Bayesian approaches offers a different path to the assessment of variability by combining probabilities estimated from detailed study of sub-processes. Developing a new pharmaceutical product requires the designing and testing of manufacturing and measurement process. The resulting process produces quality products when measurements indicative of product quality are on target with minimum variance. In the present study, error propagation break up statistical methods are successfully applied. In this validation was based on the "total error" approach and it can be seen that the method is suited for routine analysis of CLZ and TLM in tablet dosage form with minimum error. In addition, it also illustrates the application of cause-effect analysis in order to estimate the uncertainty in the measuring of CLZ and TLM from pharmaceutical formulation through RP-HPLC. The estimation of uncertainty components proved to be a good way for the experimental model to obtain contribution of the uncertainty in the present experiment, concentration of sample is the major contribution towards uncertainty.

Reference

1. International organization for standardization, ISO-IEC (1999) General Requirements for the competence of Testing and Calibrating Laboratories, 17025, Geneva.

2. Amanatidou E, Trikoilidou E, Tsikritzis L, Foteini K (2011) Uncertainty in spectrophotometric analysis-"error propagation break up", a novel statistical method for uncertainty management. Talanta 85: 2385-2390.

3. International Conference on Harmonization (1996) Q2B Validation of Analytical Procedure Methodology, Consensus Guideline, ICH Harmonized Tripartite Guidelines.

4. International Conference on Harmonization (1994) Validation of Analytical Procedure Methodology, Consensus Guideline, ICH Harmonized Tripartite Guidelines.

5. Liu Y, Shakur Y, Yoshitake M, Kambayashi J (2001) Cilostazol (pletal): a dual inhibitor of cyclic nucleotide phosphodiesterase type 3 and adenosine uptake. Cardiovasc Drug Rev 19: 369-386.

6. http://www.otsuka.pk/Left/Products/pletaal.pdf

7. Wolfgang W, Michael E, Meel JCA, Joachim S, Ulrich B, et al. (2000) A Review on Telmisartan: A Novel, Long-Acting Angiotensin II-Receptor Antagonist. Cardiovascular Drug Reviews 18: 127-154.

8. http://www.accessdata.fda.gov/drugsatfda_docs/label/2014/021162s037lbl.pdf

9. Padmavathi M, Reshma MSR, Sindhuja YV, Venkateshwararao KCh, NagaRaju K (2013) Spectrophotometric Methods for Estimation of Telmisartan Bulk Drug and Its Dosage Form. IJRPC 3: 320-325.

10. Sujit P, Deeomala M (2012) Simultaneous Spectrophotometric Estimation of Telmisartan and Indapamide in Capsule Dosage Form. Int J Pharm Pharm Sci 4: 163-166.

11. Jat RK, Sharma S, Chippa RC, Singh R, Alam I (2012) Quantitative Estimation of Telmisartan in Bulk Drug and Tablet Dosage Form. Int J Drug Res Tech 2: 268-272.

12. Pawan KB, Vinesh K, Prabhat K, Shrivastava, Deepti J (2010) Spectrophotometric Determination of Cilostazol in Tablet Dosage Form. Trop J Pharm Res 9: 499-503.

13. Sunil J, Jeyalakshmi K, Krishnamurthy T, Kumar Y (2010) Development and Validation of Simultaneous HPLC method for Estimation of Telmisartan and Ramipril in Pharmaceutical Formulations. Int J PharmTech Res 2: 1625-1633.

14. Seshagiri RJVLN, Vijayasree V, Palawan C (2013) A validated RP-HPLC Method for the Estimation of Telmisartan in Tablet Dosage Form Am. J. PharmTech Res 3: 763-769.

15. Jose K, Jaysekhar P, Jinu J (2014) HPTLC Determination of Cilostazol in Pharmaceutical Dosage Forms. International Journal of Advanced Research 2: 952-957.

16. Ulrich JC, Shihomastu HM, Hortellani MA, Sarkis J (2006) Estimate of uncertainty of measurement in the determination of methyl mercury in fish sample by FIA-CV-AAS. XVIII Imeko World Congress metrology for a sustainable development.

17. Kaur N, Mittal K, Nagar R, Nepali K, Thakkar A (2010) A sensitive spectrophotometric method for the determination of pregabalin in pure and pharmaceutical formulation through benzoylation. IJRP 1: 175-180.

18. Mittal K, Kaushal R, Mashru RC (2010) Estimation of the galanthamine using derivative spectroscopy in bulk drug and formulation. J Biomed Sci Eng 3: 439-441.

19. Mittal K, Dhingra T, Upadhyay A (2013) Estimation of uncertainty for measuring galanthamine hydrobromide in pharmaceutical formulation using ultraviolet spectroscopy. J Pharm Res 12: 34-38.

20. Eurachem Guide: (2012) Quantifying Uncertainty in Analytical Measurement (3rdedn) Laboratory of the Government Chemist London.

21. (2002) Martindale, 'The Complete Drug Reference' (33thedn) London: Pharmaceutical Press. Pp. 612, 613, 1161, 1162.

22. Patel RR, Dilip GM (2015) A new RP-HPLC method for simultaneous estimation of telmisartan and cilostazol in synthetic mixture. IJRSR 6: 3306-3310.

28

Value the Unique Merit of HPTLC Image Analysis and Extending its Performance by Digitalization for Herbal Medicines Quality Control

Pei-Shan Xie[1]*, Shuai Sun[2], Shunjun Xu[2] and Longgang Guo[1]

[1]*Guangdong UNION Biochemical Development Ltd. Co. Guangzhou, China*
[2]*Guangdong ImVin Pharmaceutical Corporation, Guangzhou, China*

Abstract

It is well known that High-Performance Thin Layer-Chromatography (HPTLC), the off-line planar chromatographic technique, has been employed in the pharmacopeias of many countries for identification by virtue of its low-cost, less dependent on expensive equipment, flexible mobile phase composition and easy post-derivatization. In particular, the unique merit of HPTLC is able to provide the picture-like images of multiple samples in parallel on the same plate for instant viewing. In addition to the two dimension parameters (migration distance (R_f value), integration peak value), the attractive color image, as the third dimension parameter, enhances the potency of HPTLC identification. The bioactivity of herbal medicines released from its compound bioactive ingredients in a holistic manner. Therefore the detected total chemical composition in the image should be reasonably more meaningful to assessing the inherent quality than only any selected single marker constituent of the given herbal drug. A vivid colorful picture-like HPTLC image could be easily recognized but not described properly by words. Hence extending the plate spot/band capacity on the plate for involving the bioactive components as much as possible. Combining digitizing the HPTLC image, the acquired infographic (HPTLC image, the digital scanning profile and the integrated parameters) can be used for establishing the fingerprint common pattern of the given species. It can be coupled with chemometrics analysis for more effectively quantifiable assessing the inherit quality of the herbal drugs. In this paper we reported methodically some Chinese herbal drugs analysis via various levels for demonstration of the practical application in QC.

Keywords: Unique merit of HPTLC; Herbal drug image analysis; Digital fingerprinting; Meaningful quality control

Introduction

High-Performance Thin-Layer Chromatography (HPTLC) had its glorious past in the second half of last century [1-10]. As identification of herbal medicines, HPTLC is adopted widely in the monographs of the pharmacopeias in various countries. This open system of planar chromatographic technique provides picture-like chromatographic images of herbal medicines for identification analysis. Taking the advantage of its low-cost, less dependent on expensive equipment, flexible mobile phase composition, easy post-derivatization and the intact images of the multiple herbal samples in parallel on the same plate for being instant comparative viewing. Therefore it is still keeping active and alive in herbal medicines quality control nowadays [11-15]. In addition to the two dimension parameters of migration distance (retention time for HPLC, R_f value for HPTLC) and integration peaks area/height, the various visible or fluorescence images is uniquely the third dimension parameter for validation of the chromatograms. The specific chemical composition of the given species is expressed by the colorful image and often recognized impressively at the first sight, but is hard described by words properly [16]. Particularly herbal's bioactive ingredients are always a complex composition, opposite to the single chemical medicine, the herbal medicines play the role mainly of adjusting the unbalance of human body's functionality based on the holistic activities of multiple bioactive substances in the herbals [17]. It would be meaningful that assessing the inherent quality of the herbals through the spectrum of the total detectable ingredients by means of HPLC, HPTLC and GC etc. Amongst them, HPTLC image analysis is the most inexpensive, and being able to compare the multiple samples on the same plate for parallel quick discrimination.

Many papers have successively been published on the perspectives of TLC/HPTLC and its widely applications [13,14,18-30]. However, ignoring the standardized operation procedure and lack of fundamental knowledge on this open system chromatographic technique is not rare in general herbal medicine analysis laboratories So the inferior quality of HPTLC image of herbal medicines is still quite common which either reduces the TLC/HPTLC's due performance or diminishes the operator's interest, to say nothing about exploring the potential merit of HPTLC. From pragmatic view, in addition to strengthening the SOP of TLC/HPTLC, how to exploring its potential capability to enhance the performance in routine quality herbal medicines control is now still significant.

Exploring the potential merit of the image analysis would be the most concern for strengthening the power of HPTLC image analysis. There are some directions need to be solved. (1) How to extend the spot/band capacity on the HPTLC silica gel plate so as to disclose the composition detail as possible as can be done; (2) how to recognize and reckon the HPTLC image to meet the special requirement of the herbals identification and (3) how to use chemometric algorism to clarify logically the HPTLC images during bulk samples tested [31]. Through the selected Chinese herbal drugs the practical fruition of HPTLC image analysis are demonstrated as the following.

*Corresponding author: Pei-Shan Xie, Guangdong UNION Biochemical Development Ltd. Co. Guangzhou, China, E-mail: psxie163@163.com

Experimental

Samples

Bupleuri radix (*Chai Hu*), Forsythiae frunctus (*Lian-Qiao*), Rehmanniae radix (*Di-Huang*), Salviae miltiorrhizae radix (*Dan-Shen*).

Instrument and reagent

Sample applicator (Automatic or semi-automatic), Twin-rough developing chamber, Reagent sprayer or HPTLC Reagent merging device, TLC visualizer (all from CAMAG, Muttenz, Switzerland); Pre-coated Silica gel 60 HPTLC plate (Merck, Germany); Dichloromethane, Chloroform, toluene, hexane, ethyl acetate, methanol, formic acid; p-Dimethylamino-benzaldehyde (DMAB), all are analytical grade).

Sample solution preparation

Generally take appropriate amount powdered herbal drugs, ultrasonicate twice with methanol/water for 30 min. Evaporated the filtrate or supernatant liquid to dryness, dissolve the residue with moderate solvent (e.g., methanol) and make it as sample solution (Table 1).

Chromatographic procedure

Unless otherwise stated, the total operation procedure followed the appendix of Chinese Pharmacopoeia Volume I and the operation guidance in the first chapter of The TLC Atlas of Chinese crude drugs in Chinese Pharmacopoeia [16]. The environment conditions in the lab is controlled, the temperature kept around at 20-25°C, relative humidity < 50%.

Sample application: appropriate amount of the sample solution and herbal reference substance (HRS) / chemical reference substance (CRS) / extractive reference substance (ERS) apply band-wise on the HPTLC silica gel plate. Band width: 8 mm, interval between bands: 4 mm; the sample-loaded plate is placed in a vacuum desiccator over P_2O_5 (anhydrous) till development for keeping the silica gel's activity.

Development

Ascending develops to 8 cm. T: 20-28°C, RH: 40% - 50%. (The developing chamber should be previous equilibration by the vapor of the mobile phase (solvent system) for 15 – 20 min, (put appropriate amount of solvent system in the one of the twin trough of the chamber beforehand).

Visualization

Dry the developed plate with hair dryer to remove the remnant solvent and visualize by means of spraying or dipping the corresponding reagents. Documentation is followed by using the appropriate facility such as TLC visualizer (CAMAG) or Digital camera (visual color image under white light).

Digital scanning

The profile of the HPTLC image can be digital scanned with corresponding software (the associated software installed in the TLC visualizer or other commercial independent software (e.g., Origin Pro) coupled with MS Excel or self-developed software for acquiring the image profile as well as the integrated of peaks area/height value.

The brief testing conditions of the species respectively listed in Table 1.

Documentation processing of post-chromatographic experiment

The herbal chromatographic image on the HPTLC plate can be instantly observed for identification. The general ensuing practice is to take photograph of the image via TLC visualizer or similar device for archiving. On the other hand, the HPTLC image can be scanned by means of TLC scanner to produce a curve profile with corresponding integration data (peak height/peak area) for quantitation. Alternatively the picture-like image photo, the permanent record which equals as viewing the HPTLC image on the plate, can do the same thing more rapid, low cost and convenient via software without any hardware. For ensuring the quality of the image photo, in addition to standardizing operation procedure of the experiment, the optical parameters for shot work should be carefully adjusted and recorded to ensure the photo as the original image as possible. Since the total chromatographic process goes under an open environment, some unexpected contamination possibly occurred when development and visualization, some requisite post-correction of the photo should be made appropriately through software. No matter how the image photo was processed, the resultant must closely resemble that of the direct real sight of the image.

Results

Bupleuri radix (*Chai- Hu*)

Visual differentiation of various species of *Bupleurum* [31,32]: The HPTLC fluorescence images of the roots of *Bupleurum chinense* DC (*Bei-Chai-Hu*), *B. scorzonerifolium* Willd (*Nan-Chai-Hu*), *B. falcatum* (San-Dao-Chai-Hu), *B. tenue* Huch. Ham. ex D. Don. (Xiao-Chai-Hu); *B. marginatum* Wall. ex DC. var. *stenophyllum* (Wolff) Shan et Y. Li (Xi-Zang-Chai-Hu), *B. marginatum* Wall. ex DC. (*Zhu-Ye-Chai-*

Chinese herbal drug	Sample solution	Developing solvent system	Visualization
Bupleuri radix (*Chai Hu*) [34,35]	ethanol extract (containing trace amount pyridine) – C18 cartridge clean-up – 80% eluate – methanol solution	dichloromethane-acetyl acetate- methanol-water (30:40:15:3)	Dimethylamino-benzaldehyde (DMAB) (2%) in sulfuric acid alcohol solution (10%), heat at 105°C for 1 min, observe fluorescence image
Forsythiae fructus (*Lian Qiao*) Unpublished data	ultrasonicate with acetone – methanol solution	(1) for higher polarity fraction:: chloroform – ethyl acetate – methanol – formic acid (30 : 5 : 10 : 1) (2) for lower polarity fraction: Toluene – ethyl acetate – 36% acetic acid (90 : 25 : 2)	10% sulphuric acid in ethanol, heated at 105°C until the bands color are distinct enough, examine the visible color images under day light and the fluorescence images under UV light at 366 nm.
Rehmanniae radix (*Di-Huang*) [37]	ultrosonicate at 40°C - centrifuge for 10 minutes (4000 r/min), use the supernatant as sample solution.	dichloromethane-ethyl acetate-methanol-water-glacial acetic acid (8:4:7:1:1)	2% anisaldehyde in 5% sulphuric acid ethanol solution, heat at 105°C until the color of the bands are clearly visible.
Salviae miltiorrhizae radix (*Dan Shen*) Unpublished data	• Lipophilic fraction: ultrasonicate with ethyl acetate extract – ethyl acetate solution; • Hydrophilic fraction: ultrasonic extract at 60°C with water, the filtrate added small amount 10% HCl, extracted by ethyl acetate – ethyl acetate solution	• Lipophilic fraction: n-Hexane - ethyl acetate - formic acid (30:10:0.5); • Hydrophilic fraction: toluene- ethyl acetate – methanol – formic acid (5:4:0.5 : 2)	• Lipophilic fraction: visible color image under day light; fluorescence-quenching image under UV 254 nm light • Hydrophilic fraction: 5%vanillin/conc. H_2SO_4 reagent; visible color image under day light;

Table 1: HPTLC experiment conditions of the species exemplified in this paper.

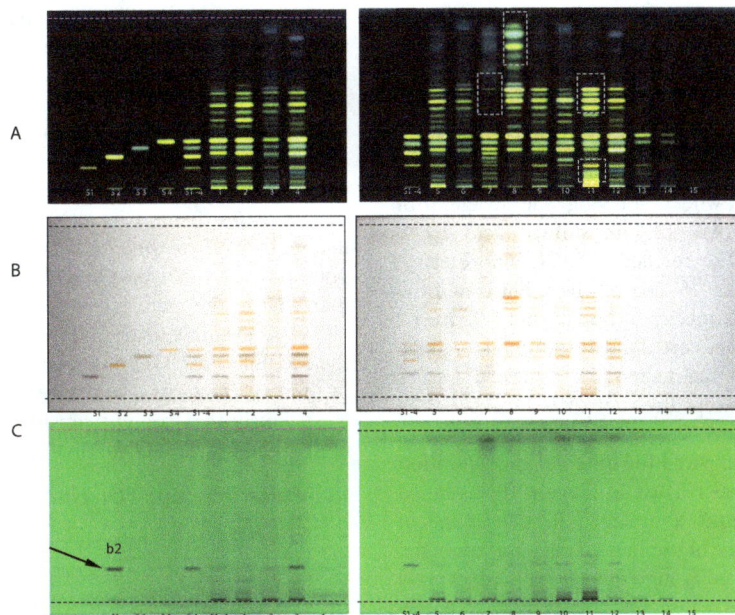

Figure 1: HPTLC image of Bupleuri radix (Chai-Hu)*. (A) HPTLC fluorescence image (365nm); (B) HPTLC visible color image; (C) HPTLC fluorescence-quenching image (254 nm), (saikosaponin b2 can be detected under UV 254nm as a fluorescence-quenching band). (S1): saikosaponin f; (S2) saikosaponin b2; (S3) saikosaponin a; (S4) saikosaponin d; (1)(2)(5): Bupleurum chinese, (3) (6):B. scorzonerifolium, (4) B. falcatum, (7) B. longiradiatum, (8) B.marginatum, (9) B. poly clonum, (10) B. wenchuanese, (11) B. marginatum var. stenophyllum, (12) B. falcatum, (13) B. yinchowense, (14) B. simithii var. parvifolium, (15) B. tenue.

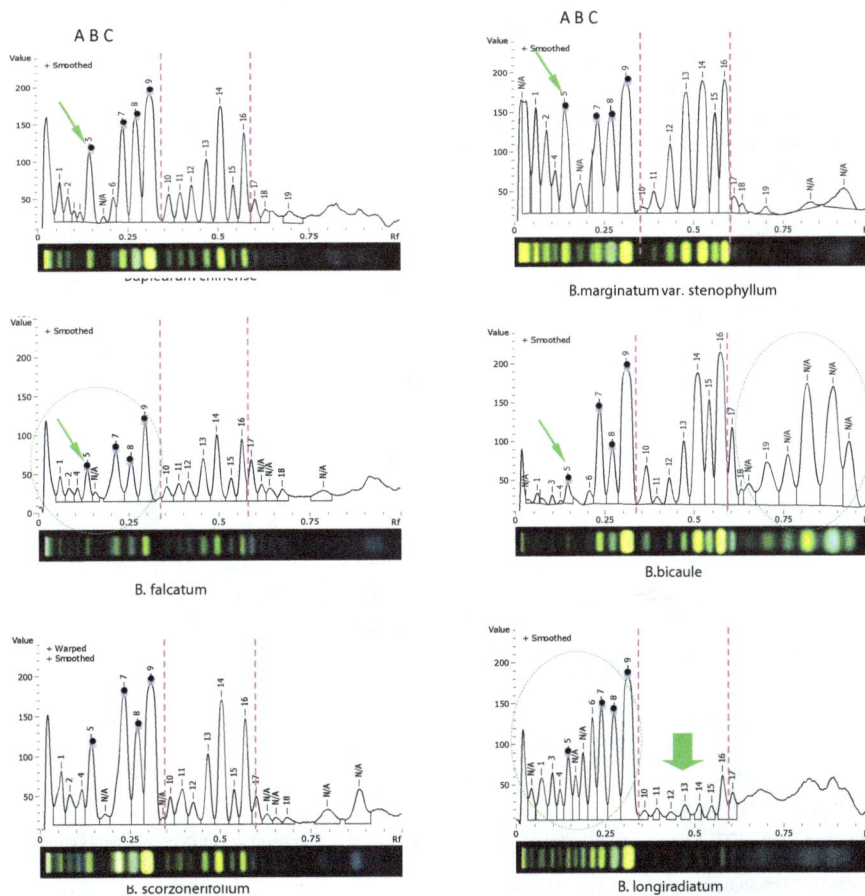

Figure 2: HPTLC fluorescence images and corresponding digital scanning profiles of 6 species of Bupleuri radix. (A): main saikosaponins region; (B): inter-species identifier region; (C) low-polarity ingredients region (cf. Figure 1).

Hu), *B. longiradiatum* Turcz (*Da-Ye-Chai-Hu*), *B. polyclonum* Y. Li et S. L. Pan (*Duo-Zhi-Chai-Hu*), *B. Wenchuanense* Shan et Y. Li (*Wen-Chuan-Chai-Hu*), *B. yinchowense* Shan et Y. Li (*Yin-Zhou-Chai-Hu*), *B. simithii* Wolff. *var. parvifolia* Shan et Y. Li (*Xiao-Ye-Yin-Chai-Hu*) are demonstrated as Figures 1 and 2.

The major hydrophilic bioactive ingredient in *Chai-Hu* roots is triterpenoid saponin named saikosaponin which can generate fluorescence under UV 365 nm whilst spraying the Reagent of Dimethylamino-benzoic acid (DMAB) (2% in sulfuric acid 10 % alcohol solution on the Plate). The formed HPTLC fluorescence color images of saikosaponins under UV 365 nm light illustrated the components composition of various species of Bupleurum (Figure 1). For easy recognition, the whole images are divided into three regions. Region (A) includes 9 peaks of the main saponins, all tested species share the common features, and except the root of *B. longiradiatum* contains more 2 peaks than the others. The digital scanning profiles of those images can be quantifiably compared the different concentration distribution among the tested species (Figure 2). Region (B) assigned as inter-species identifier. The peaks abundance in the image of *B.marginatum* var. *stenophyllum* is the highest and *B. longiradiatum* the lowest. Region (C) contains the low-polar ingredients, *B. bicauli* express specifically six stronger peaks in this region, and the species had barely some small irregular heaves in the profiles. All the features disclosed in the image and the corresponding digital scanning profiles.

Forthysiae fructus (*Lian-Qiao*)

Extended view of the detectable bioactive ingredients of *Lian-Qiao* by means of the seamless association of two fractions of HPTLC images: Forsythoside A and Forsythin (phyllyrin) are the main bioactive components of phenylethanol and lignan derivatives in Forsythiae fructus ('*Lian-Qiao*'). A HPTLC with the solvent system (A). Table 1 distinguishes clearly the two components in the fingerprint (image plus scanning curve profile) of phenylethanols fraction, but the low polar bioactive compounds crowd on the front edge of the

image. As conventional practice, that part on the front edge would be neglected. But it involves bioactive ingredients too. For separation of such crowded constituents, the sample was dedicatedly developed on another plate by using solvent system (B) (Table 1). Connecting the two HPTLC images and their profiles constitutes the intact HPTLC fingerprint of '*Lian-Qiao*'. So the assessment of quality of *Lian-Qiao* is obviously far meaningful than the single image let alone the selecting one or two marker for identification. There are two grades of *Lian-Qiao*: the immature fruits (*Qing-Qiao*; "*Qing*" means immature here) and mature fruits (*Lao-Qiao*; "*Lao*" means aged, mature). The images of both polarities show that the *Qing-Qiao* contains consistent higher contents than that in the mature fruits (*Lao-Qiao*) (Figure 3). In methodology view, this example focus on how to select the boundary line to divided the two polarities fractions to ensure no overlapping of the components in the two fractions. From view point of herbal drugs application, it is obviously that content-consistent '*Qing-Qaio*' (immature fruits) is better than the '*Lao-Qiao*' (mature fruits).

Rehmanniae radix (*Di-Huang*)

Reveal the relationship between the dynamic change of the bioactive constituents and the transformation of "drug property" between the sun-dried drug and the steaming-processed drug.

Rehmanniae radix (*Di-Huang*) is the tuberous roots of *Rehmannia glutinosa* Libosch, belonging to the family Scrophulariaceae. There are two grades of *Di-Huang* available in the market: Sun-dried or mediate-heating dried drug (*Sheng-Di-Huang*, '*Sheng*' means raw) and steam-heating processed drug (*Shu-Di-Huang*, '*Shu*' means processed). The two kinds of *Di-Huang* have different clinical efficacy according to the traditional Chinese medicine's experience due to their different 'medical properties'. It was described that the '*Sheng-Di-Huang*' has 'cold' property and the *Shu-Di-Huang* has a 'warm' property. The change in herbal 'property' from 'cold' to 'warm' implies that some significant alteration of the inherent secondary metabolites in *Di-Huang* roots occurs before and after the steam-heating processing

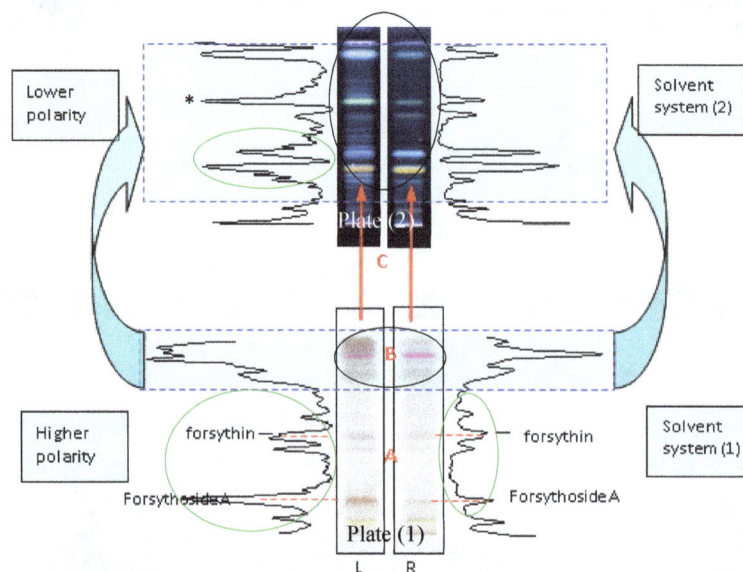

Figure 3: Extending separation capacity of HPTLC image and digital scanning profile of Forthysiae fructus with two plates associated seamlessly.
(**L**) immature fruits ('Qing-Qiao') (**R**) mature fruits ('Lao-Qiao')
A: visible color image; higher polarity fraction developed with solvent system (1);
B: lower-polarity fraction, squeezed on the front of the image (A);
C: fluorescence image of **B** (lower polarity fraction) developed with solvent system (2

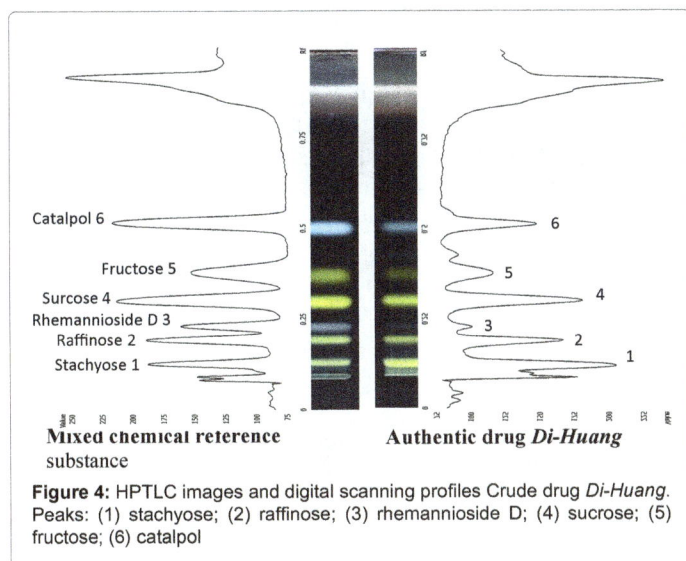

Figure 4: HPTLC images and digital scanning profiles Crude drug *Di-Huang*. Peaks: (1) stachyose; (2) raffinose; (3) rhemannioside D; (4) sucrose; (5) fructose; (6) catalpol

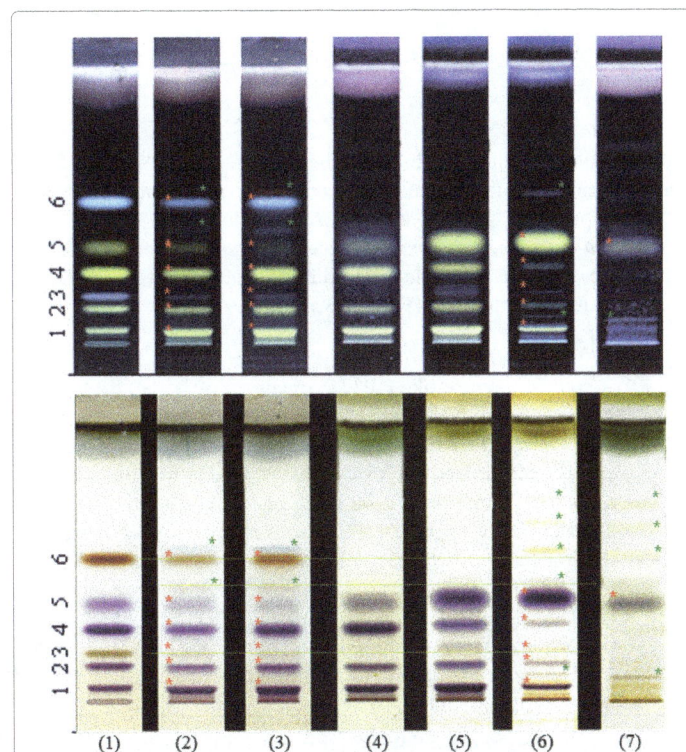

Figure 5: The HPTLC images of 'Di-Huang' are divided four grades. (The figure below is the inverted color image from the upper fluorescence figure for more easy comparison), Track (1) Chemical reference substances: (1) stachyose, (2) raffinose, (3) rehmannioside D, (4) sucrose, (5) fructose, (6) catalpol Track (2) authentic sample of 'Sheng-Di-Huang'; Track (3) commercial sample of 'Sheng-Di-Huang' Track (4) similar appearance of the crude drug with the authentic sample (2) pattern, but catalpol exceptionally disappeared and the rehmanioside D was very weak. (cf. Figure 4) Track (5) – (6) the crude drugs appearances and the HPTLC pattern of the commercial samples were up to the description of 'Shu-Di-Huang' in Chinese Pharmacopoeia, but the extent of hydrolysis of the saccharides differ, the monosaccharide – fructose increased drastically. Note that the stachyose (band 1) was relatively more stable than the others. Track (7) an over-steaming-processed sample of 'Shu-Di-Huang', almost all the major ingredients disappeared or only residual amounts remained, but fructose being abundant. The blue stars marked in the invert color images (the figure below) showed some weak chemicals which are almost hard to be recognized under deep color background in the original fluorescence image.

cycles. The HPTLC images of *Di-Huang* unfolded the dynamic change of the iridoid glycosides and oligosaccharides, the two major bioactive constituents, from *Sheng-Di-Huang* to *Shu-Di-Huang*. The HPTLC fingerprints showed that the iridoids represented by catalpol in *Sheng-Di-Huang* could be easily decomposed by steam-heating, even disappeared in *Shu-Di-Huang*. The oligosaccharides including stachyose (tetrasaccharide), raffinose (trisaccharide) were gradually hydrolyzed into sucrose (disaccharide) and glucose and fructose (monosaccharide) by successive steam-heating cycles. But the stachyose was hydrolyzed steadily, hence it still existed in the proper processed roots, nevertheless the others disappeared from the fingerprint (Figures 4 and 5). Referring to bioactive research as reported in the literature, it could be understood the iridoids ingredients in *Sheng-Di-Huang* expressed mainly the 'cold' property, and the stachyose is the prime component in *Shu-Di-Huang* exerts seemingly the role of the 'warm' property [33].

Observation in depth will find the traditional processing method being uncontrollable so as to be difficult ensuring the consistent composition. The Principal Component Analysis (PCA) of the samples tested showed the dynamic change trends (Figure 6). It hints that the steam-heating process of *Di-Huang* should be improved to keep the oligosaccharides pattern consistent in the final processed entities (*Shu-Di-Huang*).

Salviae miltiorrhizae radix (*Dan-Shen*)

Comprehensive quality assessment of by multi-levels HPTLC fingerprint analysis: There are two kinds of bioactive compounds, diterpene quinones derivatives (lipophilic) and mono- and poly-phenols (hydrophilic) in the roots (Figure 7). As a cost/effective offline HPTLC technique plays the role for comprehensive control the quality in a holistic manner. The common patterns of the images both lipophilic and hydrophilic ingredients in *Dan-Shen* serve as the chemical fingerprint. Comparative observation disclosed The image pattern are very similar within the same species (Salvia *miltiorrhizae*),

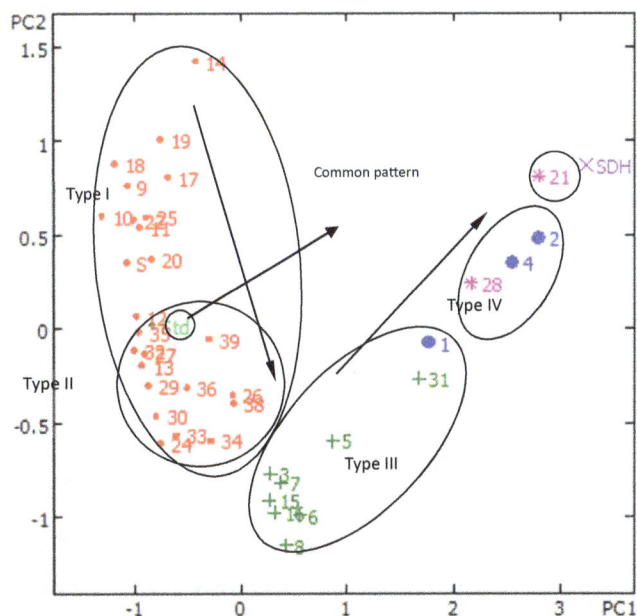

Figure 6: PCA projection plot of HPTLC fingerprint of *Di-Huang*. The dynamic change of the oligosaccharides accompanying the steam-processing caused some samples distance from the common pattern. The more lasting the steam-heating times, the more severe of hydrolysis of raffinose and sucrose into fructose will be. But the stachyose relatively stable (cf. Figure 5).

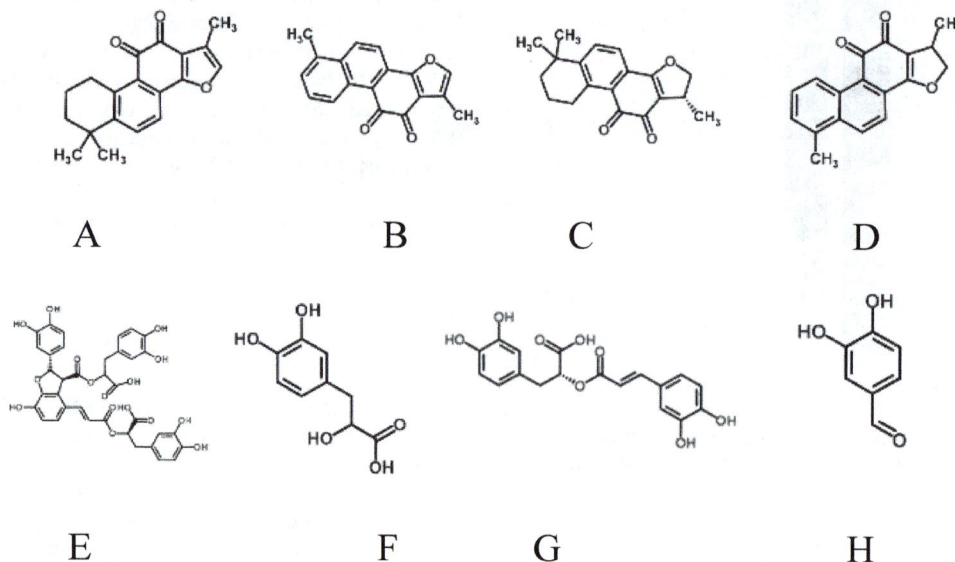

Figure 7: Chemical structure of major bioactive components in *Dan-Shen*. (A) Danshinone IIA; (B) Danshinone I; (C) Cryptotansheninone; (D) Dihydrotanshenone; (E) Salvanolic acid B; (F) Danshensu; (G) Rosmarinic acid; (H) protocatechualdehyde.

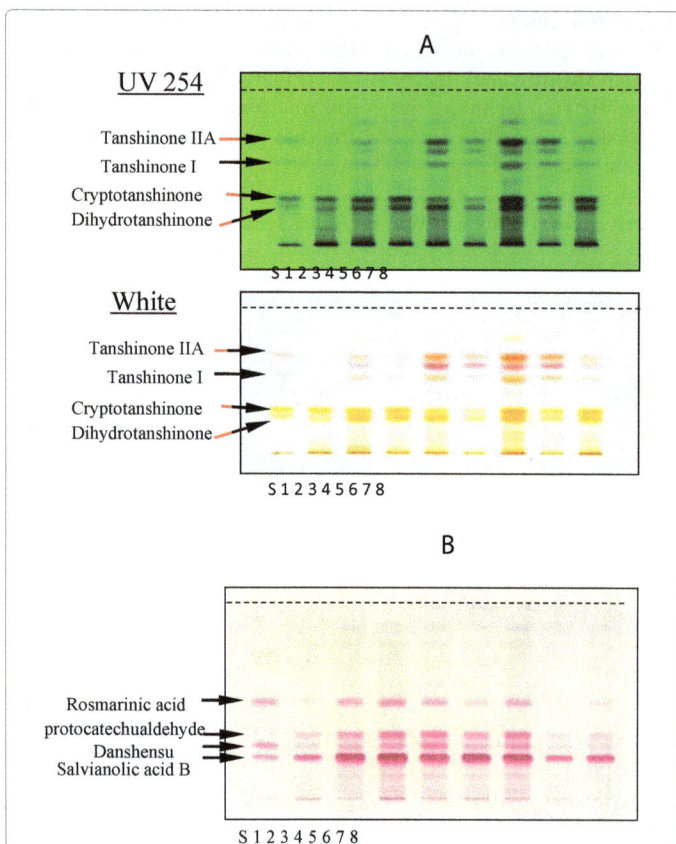

Figure 8: HPTLC fluorescence-quenching images and visible color images of *Dan-Shen*.
(A): Lipophilic fraction (fluorescence-quenching images and visible color images); (B): Hydrophilic fraction (visible color image); Track (S): CRS for lipophilic (upward): dihydrotanshenone. Cryptotanshenone, tanshenone I, tanshenone IIA; CRS for hydrophilic (upward): salvianolic acid B, danshensu, protocatechualdehyde, rosmarinic acid. Sample: (1) – (7) commercial samples of *Dan-Shen*; (8) authenticated Salvia multiorrhizae Radix (*Dan-Shen*).

but the concentration of the ingredients differ. That showed lower concentration should be inferior commerce. The patterns among different species (*S. militirrhizae, S.militirrhizae f. alba, S.prezewalski, S. castanea f. tomentosa*) are diverse each other. The associated two (lipophilic and hydrophilic ingredients) images strengthen the specificity for identification (Figures 8 and 9).

Discussion

1. A colorful picture-like HPTLC image of the herbal drugs is the unique advantage compared with the column chromatography. The examples in this paper demonstrated the vivid image is visually recognized and evaluated the quality in parallel among the various samples on the same plate. Particularly the subtle but significant specific part in some kindred species such as differenciation between the roots *Astraglus membranaceus* (AM; *Mo-jia-Huang-Qi*) and *A. membranaceus var. mongholicus* (AMM; *Meng-Gu-Huang-Qi*) in the HPTLC images counts on the subtle difference in the middle part of the images, which is hard to be literal described (Figure 10). Hence the elaborated high quality of HPTLC image is the vital prerequisite.

2. Offline operation of HPTLC is often daunting to the inadequate-trained practitioners. But the other side of offline makes flexible selection of the solvent systems (mobile phase), thus the separation would be much more selective and specific (Figure 11). Opposite to Reverse-phase HPLC, medium-polar to non-polar compounds are separated more effectively by HPTLC, so both HPTLC and HPLC are complementary separation techniques.

3. In HPTLC, the vapors of the solvent systems in the chamber also participates the chromatographic behavior and afford unique influence on the separation. Hence the solvent vapor is the 'third phase (vapor phase)' in addition to the mobile phase (solvent system) and the stationary phase (silica gel). That is the reason why the chamber often needs to be equilibrated with the solvent mixture for a certain time before developing in most cases [2].

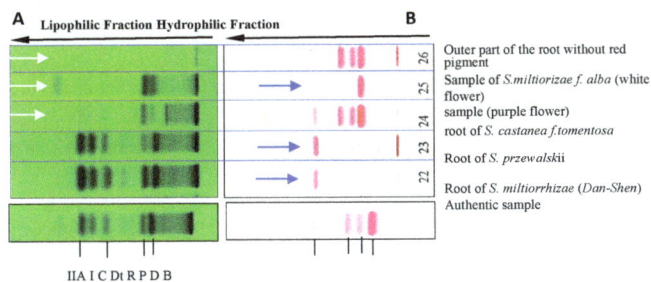

Figure 9: Differentiation among various species of *Dan-Shen* (Salvia spp.).
(A): Lipophilic fraction (fluorescence-quenching images);
(B): Hydrophilic fraction (visible color images)
CRS for lipophilic: (IIA) tanshenone IIA; (I) tanshenone I; (Dt) dihydrotanshenone; (C) Cryptotanshenone
CRS for hydrophilic: (B) salvianolic acid B; (D) danshensu; (P) protocatechualdehyde; (R) rosmarinic acid
Sample: (1) authenticated Salvia multiorrhizae Radix (*Dan-Shen*); (22)-(26) different species of *Salvia*.

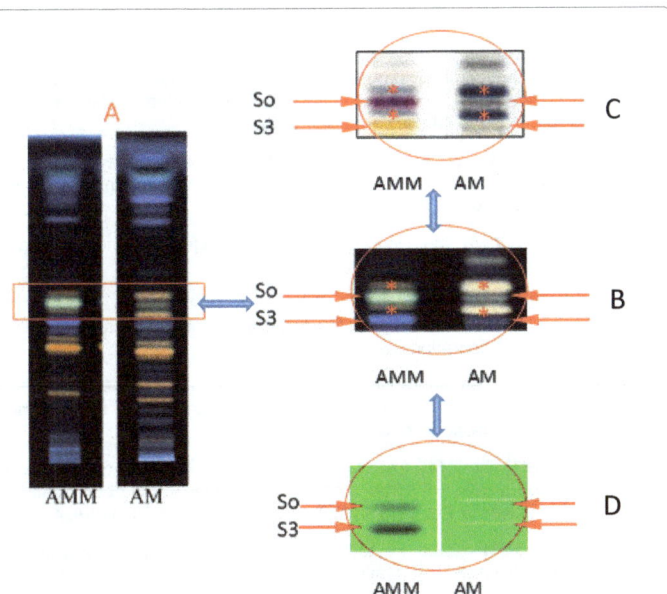

Figure 10: Partial detail of HPTLC image of the roots of two species of Astragalus (*Huang-Qi*) to show distinction between the two species: AMM = Astragalus membranaceus var. mongholicus; AM = Astragalus membranaceus (S₃) calyosin-7-O-glucoside; (So) ononin.
A: HPTLC fluorescence images of (L) Astragulus membranaceus (AM) and (R) A. membranaceus var. mongholicus (AMM)
B: Partial images of (A): The partial fraction of (A): AMM showed distinct light greyish-green fluorescence band of ononin (So), and greyish blue fluorescence band of clyosin-7-glucoside (S3); but AM showed two obvious light orange fluorescence bands of saponins (*), (So) and (S3) are very weak.
C: The invert color of image transferred from image (B) for clearer discern. AMM: the band (So) turned to dark purple color, (S3) turned to weak brownish yellow; AM: two saponin bands (*) turn to greyish blue color, (So) and (S3) are shallow.
D: The fluorescence-quenching image on silica gel 254 plate. The distinct fluorescence quenching bands of calyosin-7-glucoside (S3) and ononin (So) under UV 254 light in AMM are obvious but hard observed in AM.

4. The routine HPTLC can generally provide a better image of medium polarity components, but the low polarity components is often jam-packed on the front edge of the solvent on the plate by using the same solvent system. For disclosing the low polarity substances squeezed on the front edge on the first plate, the second development with suitable solvent system on another plate should be done to achieve the purpose. Hence the boundary of the both higher and lower polarity fractions must be defined in case unexpected overlap. The two images together virtually doubled the plate length. Seamless association of the two images constitutes the 'whole view' of the detectable bioactive compounds (Fig. 3). Generally, the bigger the gap between the two polarities fractions, the more easily affirming the boundary between both, then the more clarity of the whole image is recognized.

5. Multiple levels of visualization of the HPTLC image is another benefit for chromatographic identification. There are a lot of chemical reagents which react with the various metabolic compounds in the images of the herbals, various visible color, fluorescence as well as fluorescence-quenching HPTLC images can then be obtained. Two kinds of chemical reagents, the general reagents and special reagents, can fit to react with various categories of compounds [34]. For pilot screening test, multiple levels of visualized image can be produced on the same plate by means of spraying several reagents one-by-one sequentially if only the used reagents without interference each other. For example, it can be observed sequentially multiple levels of images from the same plate: (1) the compounds' original color image (e.g., some pigments), (2) fluorescence-quenching image (Silica gel 254 plate for aromatic compounds, observed under UV 254 nm), (3) the compounds original fluorescence image (UV 366nm, e.g., berberine, quinine, coumarins), (4) fumigating with ammonia vapor (some flavonoids, coumarins, anthroquinones), (5) spraying $AlCl_3$ reagent (some flavonoids, phenols), (5) then spraying selected $FeCl_3$ reagent (for mono- or poly-phenols) or Dragendorff reagent (alkaloids) or sulfuric acid ethanol/water solution (general reagent, generating fluorescence image of some of triterpenoids, steroids, saponins, some flavonoids, coumarins, saccharides observed under UV

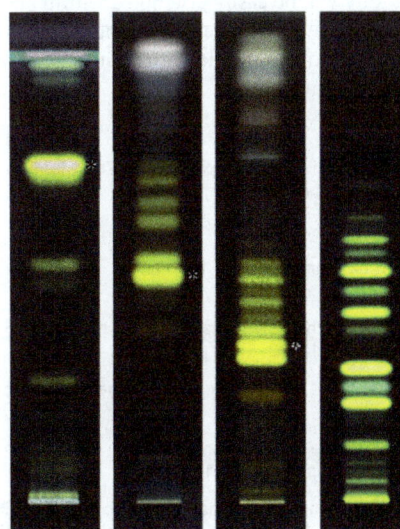

Figure 11: Comparison of different mobile phases used in the TLC analysis of Bupleurum. (The asterisk symbol shows the location of **saikosaponin a**).
1: chloroform-methanol-water (7:2:0.2);
2: ethyl acetate-methanol-water (8:2:1);
3: chloroform-ethyl acetate- methanol-water-acetic acid (20:40:22:9.5:0.5); lower layer after layering at 10°C
4: chloroform-ethyl acetate-methanol-water (20:40:20:10); lower layer after layering at 10°C.

366 nm) or vanillin or anisaldehyde/sulfuric acid (essential oils, saponins, saccharides) etc. [33]. Then the acquired successively various images can be investigated and documented sequentially on one plate. Thereby the multiple levels information (the possible compounds detected, the suitability of the mobile phase), positive or negative, in a very rapid, cost/effective way. It is most useful in screening pilot test for herbal drugs samples.

6. The obvious benefit is of easy-doing, low cost, no time bound, and suitable in general herbal laboratories, in-process quality control of the herbal industry and surveillance in the market. If using calibrated Extractive Reference Substance (ERS) replace the pure chemical standard substance as the external standard, it can be used for not only quantitative assay, but also for integrative quantifiable fingerprint analysis [35]. Furthermore, the infographic obtained from all the image and the corresponding parameters is aided by chemometric analysis makes the quality evaluation more legible (Figure 6).

Conclusion

By comparison, HPLC is mainly reversed phase chromatography suitable for analyzing higher or medium polarity compounds. The opposite, HPTLC is mainly normal phase chromatography in an open system. It is adept at separation of lower and medium polar compounds. Particularly, facing the complicated chemical composition in the herbal drugs, there are too many unknown ingredients to be separated satisfactory with any single separation technique as routine analysis. The both chromatographic techniques are complementary. Any undervaluing the merits of HPTLC should be unadvisable.

A satisfactory and reproducible experiment relies on the standardized operation, qualified instrument and relative consistent environment condition. In addition to the TLC ADC apparatus (CAMAG, Mutenz, Switzerland) for controlling the relative humidity, A well-designed small customized lab with controllable temperature and relative humidity will be easily assembled within about 10 m² area for keeping a consistent environmental condition. Anyway, the most important is the practitioners should participate a short-term pre-job-training by the eligible tutors.

Acknowledgement

The author thanks Yu-zhen Yan, He-ping Liu, Yu Zhao, for their contribution on the experimental works cited in this paper.

References

1. Geiss F (1987) Fundamentals of Thin Layer Chromatography. Huethig, Heidelberg.

2. Geiss F (1988) Role of the Vapor Phase in Planar Chromatography. J Planar Chromatogr - Mod TLC 1: 102.

3. Sherma J, Fried B (1991) Handbook of Thin Layer Chromatography. Marcel Dekker Inc., New York.

4. Kaiser RE (1988) Scope & Limitation of Modern Planar Chromatography. J Planar Chromatogr – Modern TLC 1: 182.

5. Sun YQ, et al. (1990) Thin layer chromatography scanning and its application in pharmaceutical analysis. Beijing. People's Medical Publishing House, Beijing, China.

6. Nyiredy S, Dallenbach-Toelke K, Sticher O (1988) The 'PRISMA' Optimization System in Planar Chromatography. J Planar Chromatogr – Modern TLC 1: 336.

7. Peishan X, Yuzhen Y (1987) HPTLC fingerprint identification of commercial ginseng drugs - reinvestigation of HPTLC of ginsenosides. Journal of High Resolution Chromatography and Chromatography Communication 10: 607-613.

8. Peishan X, Yuzhen Y, Haoquan Q, Qiaoling L (2001) Fluorophotometric thin-layer chromatography of ginkgo terpenes by postchromatographic thermochemical derivatization and quality survey of commercial ginkgo products. J AOAC Int 84: 1232-1241.

9. Fenimore DC, Davis CM (1981) High performance thin layer chromatography. Anal Chem 53: 252A–266A.

10. Peishan X, Yuzhen Y (1992) Optimization of the TLC of Proberberine Alkaloids and Fingerprint evaluation of the Coptis Rhizome [J]. J Planar Chromatogr - Modern TLC 5: 302 – 307.

11. Shrikumar S, Maheswari MU, Suganthi A, Ravi TK (2006) The growth of pharmaceutical industry is based on continuing success in producing new products whether they are used as therapeutic or prophylactic agents. The role of R&D is pivotal in this endeavor. Pharmainfo net.

12. Schibli A, Reich E (2005) Modern TLC: A Key Technique for Identification and Quality Control of Botanicals and Dietary Supplements. J Planar Chromatogr - Modern TLC 18: 34 -38.

13. Upton RT (2010) Use of high-performance thin layer chromatography by the American Herbal Pharmacopoeia. J AOAC Int 93: 1349-1354.

14. Reich E (2000) Chromatography: Thin-Layer (Planar). Historical Development, Reference Module in chemistry, molecular Science and chemical Engineering. Encyclopedia of Separation Science 834-839.

15. Kunle OF, Egharevba HO, Ahmadu PO (2012) Standardization of herbal medicines - A review, International Journal of Biodiversity and Conservation.

16. Peishan X, Zhong-zhi Q (2009) TLC Atlas of Chinese Crude Drugs in Pharmacopoeia of the People's Republic of China. Chinese Pharmacopoeia Commission. People's Medical Publishing House.

17. Xie PS, Leung AY (2009) Understanding the traditional aspect of Chinese medicine in order to achieve meaningful quality control of Chinese materia medica. J Chromatogr A 1216: 1933-1940.

18. Renger S (2000) Thin-Layer (Planar) Chromatography. Encyclopedia of Separation Science.

19. Wall PE (2000) Chromatography:Thin-Layer (Palnar) Spray Reagents. Reference Module in Chemistry, Molecular Science and Chemical Engineering. Encyclopedia of Seperation Science.

20. Wall PE (2000) Chromatography: Thin-Layer (Planar) |Densitometry and image Analysis, Reference Module in Chemistry, Molecular Science and Chemical Engineering, Encyclopedia of Seperation Science.

21. Biringanine G, Chiarelli MT, Faes M, Duez P (2006) A validation protocol for the HPTLC standardization of herbal products: application to the determination of acteoside in leaves of Plantago palmata Hook. f.s. Talanta 69: 418-424.

22. Rajani M, Kanaki NS (2008) Phytochemical standardization of herbal drugs and polyherbal formulations. Bioactive Molecules and Medicinal Plants. Springer, India.

23. Jain A, Lodhi S, Singhai AK (2009) Simultaneous estimation of quercetin and rutin in Tephrosia purpurea Pers by high performance thin layer chromatography. Asian J Tradit Med 4: 104-109.

24. Rajkumar T, Sinha BN (2010) Chromatographic fingerprint analysis of budmunchiamines in Albizia amara by HPTLC technique. Int J Res in Pharma Sci 3: 313.

25. Butz S, Stan HJ (1995) Screening of 265 pesticides in water by Thin-Lyer Chromatography with automated Multiple Development. Anal Chem 67: 620-630.

26. Ram M, Abdin MZ, Khan MA, Jha P (2011) HPTLC Fingerprint Analysis: A Quality Control for Authentication of Herbal Phytochemicals. High-Performance Thin-Layer Chromatography (HPTLC) 201: 105-116.

27. Milojkovic-Opsenica D, Ristivojevic P, Andric F, Trifkovic J (2014) Planar chromatographic systems in pattern recognition and fingerprint analysis. J Chemometrics.

28. Peishan X, Yuzhen Y, Haoquan Q, Qiaoling L (2001) Fluorophotometric Thin-Layer Chromatography of Ginkgo Terpenes by Postchromatographic Thermochemical Derivatization and Quality Survey of Commercial Ginkgo Products. J AOAC Int 84: 1232-1241.

29. Peishan X, Zhongzhi Q (2009) TLC Atlas of Chinese Crude Drugs in Pharmacopoeia of People's Republic of China. (English Edition), People's Medical Publishing House, Beijing, China.

30. Peishan X, Yuzhen Y (1992) Optimization of the TLC of Proberberine Alkaloids and Fingerprint evaluation of the Coptis Rhizome [J]. Journal of Planar Chromatography - Modern TLC 5: 302-307.

31. Tian RT, Xie PS, Liu HP (2009) Evaluation of traditional Chinese herbal medicine: Chaihu (Bupleuri Radix) by both high-performance liquid chromatographic and high-performance thin-layer chromatographic fingerprint and chemometric analysis. J Chromatogr A 1216: 2150-2155.

32. Liu HP, Xie PS, Tian RT (2008) HPTLC fingerprint analysis of bupleurum spp. (*Chai-Hu*) (Chinese). Trad Chin Drug Res & Clin Pharmacol 19: 39-42.

33. Xie PS, Guo LG, Zhao Y, Bensky D, Stoeger E (2014) Searching the clue of the relationship between the alteration of bioactive ingredients and the herbal 'property' transformation from raw Rehmanniae radix (Sheng-Di-Huang) to steam-heating-processed Rehmanniae radix (Shu-Di-Huang) by chromatographic fingerprint analysis. Chinese Medicine 5: 47-60.

34. Jork H, Funk W, Fischer W, Wimmer H (1990) Thin Layer Chromaatography. Physical and Chemical Detection Methods. VCH, Weinheim.

35. Xie PS, Ma SC, Tu PF, Wang ZT, Stoeger E, et al. (2013) The Prospect of Application of Extractive Reference Substance of Chinese Herbal medicines. Chinese Medicines 4: 125-136.

29

Size-Separation of Silver Nanoparticles using Sucrose Gradient Centrifugation

Anil K Suresh[1]*, Dale A Pelletier[2], Ji-Won Moon[3], Tommy J Phelps[3] and Mitchel J Doktycz[2,4]

[1]*Department of Biotechnology, School of Life Sciences, Pondicherry University, Puducherry, India*
[2]*Biosciences Division, Oak Ridge National Laboratory, Oak Ridge, Tennessee, USA*
[3]*Environmental Sciences Division, Oak Ridge National Laboratory, Oak Ridge, Tennessee, USA*
[4]*Center for Nanophase Materials Science, Oak Ridge National Laboratory, Oak Ridge, Tennessee, USA*

Abstract

Size and shape distributions of nanoparticles can drastically contribute to the overall properties of nanoparticles, thereby influencing their interaction with different chemotherapeutic molecules, biological organisms and or materials and cell types. Therefore, to exploit the proper use of nanoparticles for various biomedical and biosensor applications, it is important to obtain well-separated monodispersed nanoparticles. However, gaining precise control over the morphological characteristics of nanoparticles during their synthesis is often a challenging task. Consequently, post-synthesis separation of nanoparticles is necessary. In the present study, demonstration on the successful one-pot post-synthesis separation of anisotropic silver nanoparticles to near monodispersities using sucrose density gradient sedimentation. The separation of the nanoparticles was evidenced based on optical confirmation, and spectrophotometric and transmission electron microscopy measurements. Our results clearly demonstrate the facile separation of anisotropic silver nanoparticles using sucrose density gradient sedimentation and can enable the use of nanoparticles for various labeling, detection and biomedical applications.

Keywords: Anisotropic; Density gradient; Fractionation; Sucrose; Separation; Nanoparticles

Introduction

Nanoparticles (NPs) possess unique physico-chemical characteristics, which are dependent on their size and shape distributions [1-5]. Consequently, selection of morphological and structural properties of NPs is critical for their applications in drug and gene delivery, sensors, solar systems, storage and photovoltaic devices etc. [6-10]. Therefore, obtaining monodispersed NPs, in order to better define and exploit their characteristic properties is imperative [11,12]. Even though tremendous efforts are directed towards optimizing synthesis strategies, typical, as-synthesized suspensions still tend to include polydispersed sizes and shapes [5,13]. With an aim towards achieving monodispersed nanoparticles, significant efforts are focused on the post-synthesis separation of nanoparticles [5,13,14]. Various separation techniques including chromatography, magnetic separation, electrophoresis, selective precipitation, membrane filtration, solvent or solution-based extractions and density gradient centrifugation have been used for isolating nanoparticles [5,13-16]. Density-based centrifugation is considered as a particularly convenient method as it is economic, non-laborious, and does not involve the use of liquid-solid phase interactions and/or hazardous chemical reactions. This technique makes use of the sedimentation coefficients of nanoparticles in the surrounding medium, which can vary with nanoparticles form and mass [5], and has been successfully implemented to separate a wide range of nanoparticle using density gradient centrifugation rate [5,13]. For example, the separation of various sizes of FeCo@C nanocomposites to defined size distributions was demonstrated using iodixanol gradient solution and by varying step-gradient density parameters and centrifugation duration [14]. Likewise, dimers and trimers of gold nanoparticles were isolated using high-density cesium chloride ($CsCl_2$) suspensions [5,10]. Similarly, rapid separation of metal and quantum dot (CdSe) nanoparticles was demonstrated using non-hydroxylic organic as a density gradient [5].

Silver nanoparticles are widely used as antibacterial agents against diverse microorganisms due to their potent anti-microbial properties [17-20]. Additionally, they have been exploited in biomedical applications for treating burns and wounds, in dental materials, as coatings for stainless steel and textiles, for water treatment, and sunscreen lotions [20]. AgNPs can possess low cytotoxicity, high thermal stability and low volatility [20]. Therefore, uniform monodispersed silver nanoparticles are highly desirable, especially for meeting the needs of these various applications [11,20]. Facile separation of mixed shapes (rods, spheres, triangles pyramids, stars and pentagons) of gold nanoparticles, synthesized using fungus, to discrete shapes using sucrose density gradient using a tabletop centrifuge was demonstrated [7]. Using similar technique, the separation of gold and silver nanoparticles synthesized using a plant extract was also demonstrated recently [21].

The present investigation demonstrates the separation of anisotropic AgNPs using a simple one-step sucrose density gradient sedimentation. This separation technique is commonly used for isolating cellular organelles and viruses by high-speed ultracentrifugation. Our results permitted the separation of highly dense silver nanoparticles at low speed (4000 g) using table top-centrifugation. The sucrose density gradient method has been previously used to separate polymeric nanoparticles [7]. The ease and facile nature of the adopted methodology offers new insights for sustainable separation and can be extended to separate other forms of biologically relevant metallic nanoparticles.

---

*Corresponding author: Anil K Suresh, Department of Biotechnology, School of Life Sciences, Pondicherry University, Kalapet, Puducherry 605014, India
E-mail: anil.dbt@pondiuni.edu.in

Materials and Methods

The bacterium *Pseudomonas* sp. strain GM102 used to produce anisotropic AgNPs was isolated from surface sterilized root tissue collected from native *Populus deltoids* [21]. All other chemicals and reagents were from standard commercial sources and of the highest quality available.

Preparation of density gradient sucrose suspension

A discontinuous gradient of sucrose was prepared by successive layering of dilute sucrose solution upon one another as described earlier with slight modifications [22]. Briefly, 5 mL each of 2.5%, 5%, 7.5%, 10%, 12.5%, 15%, 17.5% and 20% sucrose were layered on top of each other in a 40 mL polystyrene centrifuge tube. The tube was marked in increments of single percentage points within each region.

Synthesis and fractionation of anisotropic silver nanoparticles

For the synthesis of anisotropic AgNPs, a single *Pseudomonas* sp. bacterial colony from a 12 h cultured R2A agar Petri dish served as inoculum for 100 mL of R2A broth in a 500 mL Erlenmeyer flask, followed by incubation at 25°C on a shaker (200 rpm) for 24 h. The bacteria were collected by centrifugation (5000 × g, 25°C, 30 min) and washed with sterile distilled water under sterile conditions. In a 500 mL Erlenmeyer flask, ~3–5 g wet bacterial biomass was suspended in 100 mL of 1 mM $AgNO_3$ solution and incubated at 25°C under shaking (200 rpm). Synthesis of AgNPs was monitored by UV-vis spectra (200-700 nm). After completion of the reaction (48 h), the reaction mixture was centrifuged (5000 × g, 30 min) to remove the bacteria, filtered using a sterile 0.1 μm syringe filter and the AgNPs were collected by performing ultra-centrifugation (100,000 × g, 1 h). After washing twice with Milli Q water the anisotropic AgNPs were used for further characterization and separation.

Separation was accomplished by forming a discontinuous gradient of sucrose by successive layering of dilute sucrose solution upon one another as described above. Then, 5 mL of the above-prepared anisotropic AgNPs suspension was loaded on top of the sucrose gradient and centrifuged at 4000 g for 1 h using a tabletop centrifuge (Spinchtron™ R, Beckman Coulter, CA). Fractions of 2 mL were carefully collected using a micropipettor and further purified by dialysis using a molecular weight cut-off of 1000 Da (Spectrum Laboratory Inc.) against Milli Q water. These samples were concentrated to half of their original volumes using a Speed Vac SC 100 (Savant Instruments Inc., NY) and used for further analysis.

Physical Characterization of Nanoparticles

UV-vis absorbance was recorded on a CARY 100 Bio spectrophotometer (Varian Instruments, CA) operated at a resolution of 1 nm. Fourier transform infrared (FTIR) analysis of the samples deposited on a ZnSe window was performed on a Nicolet Magna-IR 760 spectrophotometer at a resolution of 4 cm⁻¹. X-ray diffraction (XRD) of the dried AgNPs powder was performed on a Discover D8 X-ray diffractometer with a Xe/Ar gas–filled Hi-Star area detector and an XYZ platform, operated at 40 kV and at a current of 40 mA. Transmission electron microscopy (TEM) measurements for the NPs samples prepared on carbon-coated copper grids were performed on a Hitachi HD-2000 STEM operated at an accelerating voltage of 200 kV. For Atomic Force Microscopy (AFM), nanoparticle samples were imaged in either contact mode or intermittent contact mode with a PicoPlus AFM (Aligent Technologies, Tempe, AZ) using a 100 μm scanning head at 128–512 pixels per line scan and a scan speed of 0.5 line/s. The cantilevers used were Veeco silicon nitride probes (MLCT-AUHW, Veeco, Santa Barbara, CA).

Results and Discussion

Anisotropic AgNPs with a wide range of size and shape distributions were synthesized using a previously reported bacterial-based biosynthesis method [20,23]. This technique results in a mixture of diverse particle sizes and shape distributions and illustrates the diverse NP mixtures that can result from biological and chemical based synthesis methods [11,20-24]. Therefore this material was used to illustrate the size/shape separations that can be accomplished using simple sucrose density gradient sedimentation. Figure 1 shows transmission electron microscopy and atomic force microscopy measurements of the as-prepared parent AgNP suspension; the diversity of AgNP sizes and shapes can be seen in TEM images of the particles taken from different areas of the grid and at increasing magnifications. These images reveal a heterogeneous distribution of AgNPs sizes and shapes. The particles included polydispersed spherical (55%), polygonal and hexagonal (24%), parallelograms (8%) and triangular (13%) shaped nanoparticles (Figure 1 and Supplementary Figure 2). A particle size histogram plot, obtained by counting ~100 particles from the TEM images, showed that AgNPs varied in size and ranged from ~2 to 70 nm (Figure 1B and Supplementary Figure 1). The presence of polydispersed AgNPs was further confirmed based on AFM imaging, which revealed well-separated heterogeneous nanoparticles with a particle height ranging from ~2 to 70 nm (Figure 1C and 1D), which correlates well to the size and shape distributions observed using TEM (Figure 1A).

The anisotropic nature of the AgNPs was also evident from UV-vis spectroscopy measurements where a broad surface plasmon resonance (SPR) peak (315-600 nm) was observed, confirming the polydisperse nature of the AgNPs. Also, based on the intensity of the transverse SPR peak, it can be inferred that the majority of the particles are spherical (Figure 2A).

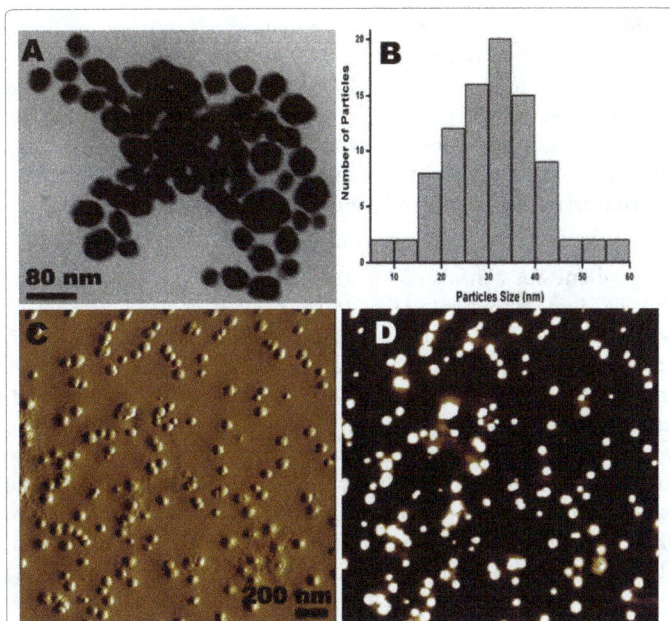

Figure 1: Transmission electron microscopy (A), and atomic force deflective (C) and topographical (D) images of the as-prepared parent silver nanoparticles suspension showing the anisotropic size and shape distributions of nanoparticles. (B) Surface height measurements obtained from the TEM image by counting ~100 particles.

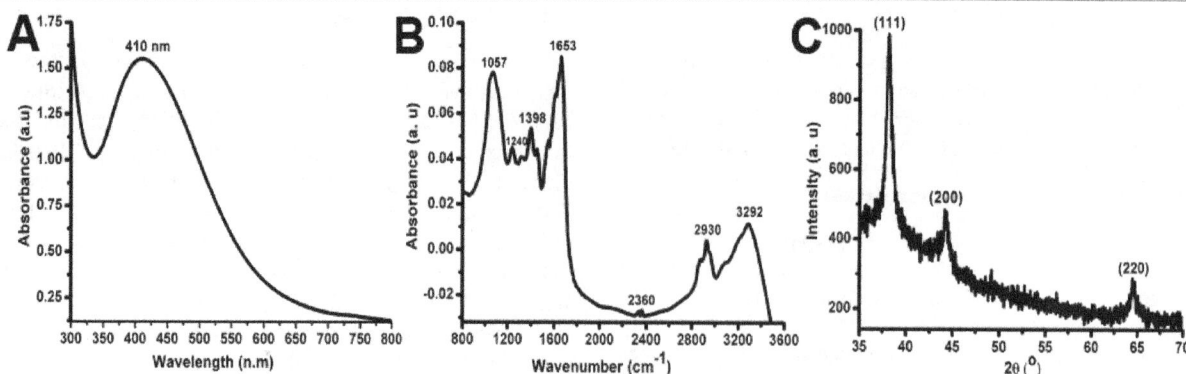

Figure 2: UV-Vis, FTIR and XRD measurements of the as-prepared parent silver nanoparticle suspension. (A) UV-Vis spectrum of the as-prepared parent silver nanoparticles showing a broad peak indicating the presence of anisotropic silver nanoparticles. (B) Fourier transform spectra of the as-prepared silver nanoparticles, significant vibration bands are labeled. (C) XRD analysis of the as-prepared silver nanoparticle powder.

To understand the stabilizing agent that might be associated with the AgNPs [20], and which might influence the separation of the particles, FTIR spectroscopy was performed. FTIR spectra revealed bands at 1057, 1240, 1398, 1653, 2360 and 2930 along with an intense, broad band at 3292 cm^{-1} (Figure 2B). The bands at 1057 and 1653 cm^{-1} can be attributed to the–N-H and carbonyl (-C-O-C- or –C-O-) stretch vibrations in the primary and secondary amide linkages. A peak for tertiary amide was also observed at 1398 cm^{-1}. Collectively these peaks are consistent with the presence of protein or peptide on the NP surfaces those likely acts as a capping molecule [11,25].

The crystalline nature of the AgNPs was evaluated by performing XRD analysis. Intense peaks at 111, 200 and 220 are observed, confirming the crystalline nature of the AgNPs (Figure 2C). These peaks are due to the line broadening of Bragg reflections in the 2θ range of 35°–70° at 38.1, 44.8 and 64.9 respectively based on the face-centered cubic structure of silver and agree with reported values [20].

Size fractionation of the AgNPs could be achieved by creating a discontinuous density gradient of 2.5% to 20% sucrose by layering successively dilute sucrose solutions upon one another, and centrifugation at 4000 g for 1 h (Figure 3). It was observed that the nanoparticles, initially brown and at the top of the sucrose gradient, form a gradient across the length of the centrifuge tube upon centrifugation (Figure 3B), indicating a density-based separation.

The purity and separation of individual fractions collected from the gradient was confirmed by UV–vis spectroscopy (200-900 nm) and TEM measurements. The fractions that showed prominent peak shifts; hence forth designated as fractions a, b, c, d, e and f, occurring at density gradient points of 2.5%, 4.8%, 5.3%, 7.1%, 10.4% and 16.2% respectively, were subjected to further characterization. As seen in the UV-vis spectra of these fractions, blue shift for the fractions a and b, and red shift for the fractions d, e and f in the surface plasmon peak was observed when compared to that of the as-prepared parent suspension (Figures 2A and 4). The absorbance of the isolated sample fractions displayed peaks at 403 nm, 405 nm, 410 nm, 416 nm, 420 nm and 425 nm, respectively for the fractions a, b, c, d, e and f (Figure 4). The differences in the observed SPR peaks show differences in NP size distribution [26]. The size-dependent absorbance of light by metal nanoparticles is a well-known phenomenon that occurs due to the coherent oscillations of the conductance band electrons generated by interaction with the electromagnetic field [27].

Transmission electron microscopy confirmed distinct morphological characteristics for the various fractions. Fraction a, concentrated in the 2.5% gradient, consisted of the smallest spherical shaped particles of 4.5 ± 0.5 nm (Figure 5A); fractions b and c, separated in the 4.8% and 5.3% gradient points, respectively, included monodispersed spheres of 15 ± 3 nm (Figure 5B and 5C); fraction d, contained spheres and triangles of 28 ± 4 nm and eluted at the 7.1% gradient point (Figure 5D); fraction e, included pentagonal and hexagonal shaped particles of 45 ± 5 nm and partitioned at the 10.4% gradient point (Figure 5E); whereas the largest spheres and largest triangles 60 ± 10 nm and above, were collected in fraction f, at the 17.3% gradient point (Figure 5F).

Overall, our results demonstrate the utility of density gradient techniques to separate mixtures of different size and shape distributions of nanoparticles to near monodispersities, illustrating the separation of AgNPs using sucrose density gradient sedimentation. The separation of gold nanoparticles has been previously reported [7]. However, in that study, the authors used a gradient of sucrose with higher percentages 30-70%, compared to the 2-20% gradient used here. These differences may be the result of differences in the average sizes of the nanoparticles being separated as well as the core material of the nanoparticle. Nanoparticle density may depend on nanoparticle type, and consequently require different percentages of sucrose gradients for their separation [28]. Alternatively, hydrodynamic characteristics of the nanoparticle may contribute to their separation in density gradients. Ongoing experiments to further characterize the bases for nanoparticle separation in density gradients and assess if this simple technique can be extended to separate, a mixture of diverse types of nanoparticles is under investigation.

Conclusions

In summary, sucrose density gradient have been successfully used as a separation method to fractionate anisotropic AgNPs containing particles of different sedimentation rates. Compared to the existing conventional methods our described sucrose density based separation methodology is scalable, economic, avoids the use of corrosive and toxic organic gradient solutions, and can be achieved with commonly accessible laboratory equipment, a tabletop centrifuge. The ease and facile nature of the adopted methodology offers new insights for sustainable separations, and can be extended to other types of metallic nanoparticles.

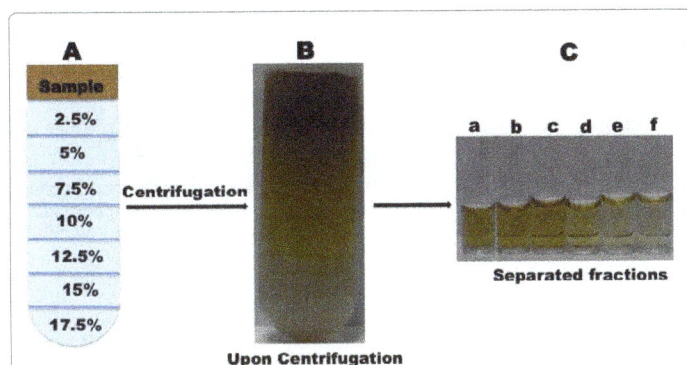

Figure 3: Optical images showing the sucrose density gradient separation of anisotropic silver nanoparticles. Images of the as-prepared parent silver nanoparticle mixture (left); anisotropic silver nanoparticles in sucrose gradient after fractionation (center); and the color images (right) of the individual fractions upon successful size fractionation.

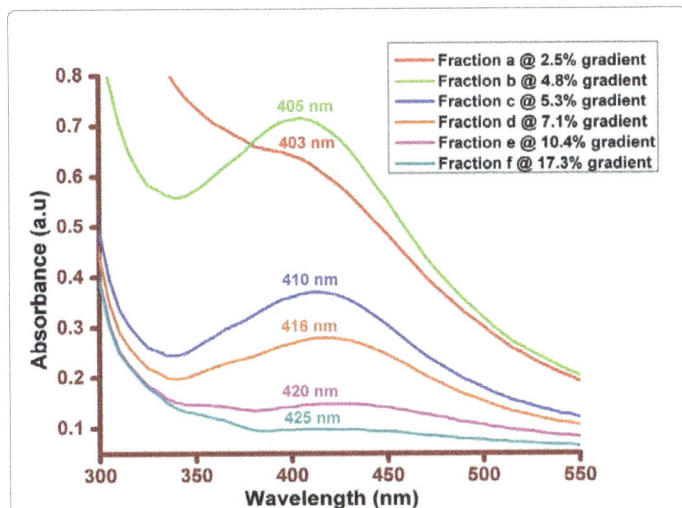

Figure 4: UV-Vis spectroscopy measurements of the various pooled fractions a, b, c, d, e and f sedimented at sucrose gradients of 2.5%, 4.8%, 5.3%, 7.1%, 10.4% and 17.3% respectively. The individual samples and their surface plasmon peak are labeled.

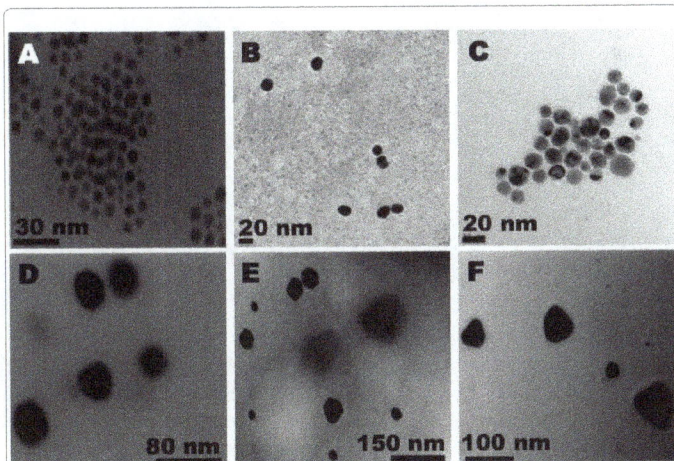

Figure 5: Transmission electron microscopy images of the various pooled fractions; A, B, C, D, E and F upon sedimentation using sucrose density gradient at 2.5%, 4.8%, 5.3%, 7.1%, 10.4% and 17.3% respectively, following separation of the anisotropic silver nanoparticles.

Acknowledgments

A. K. S. thanks the Department of Biotechnology, Govt. of India, and New Delhi for financial support through the Ramalingaswami Fellowship. The authors acknowledge support from the Office of Biological and Environmental Research, U. S. Department of Energy (DOE). Oak Ridge National Laboratory is managed by UT-Battelle, LLC, for the U.S. DOE under contract DE-AC05-00OR22725.

References

1. Suresh AK, Pelletier DA, Doktycz MJ (2013) Relating nanomaterial properties and microbial toxicity. Nanoscale 5: 463-474.

2. Huang X, Jain PK, El-Sayed IH, El-Sayed MA (2008) Plasmonic photothermal therapy (PPTT) using gold nanoparticles. Lasers Med Sci 23: 217-228.

3. Henglein A (1993) physicochemical properties of small metal particles in solution - microelectrode reactions, chemisorption, composite metal particles, and the atom-to-metal transition. J Phys Chem 97: 5457-5471.

4. Kamat PV (2002) Photophysical, photochemical and photocatalytic aspects of metal nanoparticles. J Phys Chem B 106: 7729-7744.

5. Xiong B, Cheng J, Qiao Y, Zhou R, He Y, et al. (2011) Separation of nanorods by density gradient centrifugation. J Chromatogr A 1218: 3823-3829.

6. Peer D, Karp JM, Hong S, Farokhzad OC, Margalit R, et al. (2007) Nanocarriers as an emerging platform for cancer therapy. Nat Nanotechnol 2: 751-760.

7. Kumar SA, Peter YA, Nadeau JL (2008) Facile biosynthesis, separation and conjugation of gold nanoparticles to doxorubicin. Nanotechnology 19: 495101.

8. Parak WJ, Gerion D, Pellegrino T, Zanchet D, Micheel C, et al. (2003) Biological applications of colloidal nanocrystals. Nanotechnology 14: R15.

9. Latorre M, Rinaldi C (2009) Applications of magnetic nanoparticles in medicine: magnetic fluid hyperthermia. P R Health Sci J 28: 227-238.

10. Kowalczyk B, Lagzi I, Grzybowski BA (2011) Nanoseparations: Strategies for size and/or shape-selective purification of nanoparticles. Current Opinion in Colloid & Interface Science 16: 135-148.

11. Suresh AK, Doktycz MJ, Wang W, Moon J-W, Gu B, et al. (2011) Monodispersed biocompatible silver sulfide nanoparticles: Facile extracellular biosynthesis using the gamma-proteobacterium, *Shewanella oneidensis*. Acta Biomaterialia 7: 4253-4258.

12. Sun S, Murray CB, Weller D, Folks L, Moser A (2000) Monodisperse FePt nanoparticles and ferromagnetic FePt nanocrystal superlattices Science 287: 1989-1992.

13. Sharma V, Park K, Srinivasarao M (2009) Shape separation of gold nanorods using centrifugation. Proc Natl Acad Sci U S A 106: 4981-4985.

14. Sun X, Tabakman SM, Seo WS, Zhang L, Zhang G, et al. (2009) Separation of nanoparticles in a density gradient: FeCo@C and gold nanocrystals. Angew Chem Int Ed Engl 48: 939-942.

15. Wei GT, Liu FK, Wang CR (1999) Shape separation of nanometer gold particles by size-exclusion chromatography. Anal Chem 71: 2085-2091.

16. Hanauer M, Pierrat S, Zins I, Lotz A, Sönnichsen C (2007) Separation of nanoparticles by gel electrophoresis according to size and shape. Nano Lett 7: 2881-2885.

17. Morones JR, Elechiguerra JL, Camacho A, Holt K, Kouri JB, et al. (2005) The bactericidal effect of silver nanoparticles. Nanotechnology 16: 2346-2353.

18. Kim JS, Kuk E, Yu KN, Kim JH, Park SJ, et al. (2007) Antimicrobial effects of silver nanoparticles. Nanomedicine 3: 95-101.

19. Duran N, Marcato PD, De Souza GIH, Alves OL, Esposito E (2007) Antibacterial effect of silver nanoparticles produced by fungal process on textile fabrics and their effluent treatment. Journal of Biomedical Nanotechnology 3: 203-208.

20. Suresh AK, Pelletier DA, Wang W, Moon J-W, Gu B, (2010) et al. Silver Nanocrystallites: Biofabrication using Shewanella oneidensis, and an Evaluation of Their Comparative Toxicity on Gram-negative and Gram-positive Bacteria. Environ Sci Technol 44: 5210-5215.

21. Brown SD, Utturkar SM, Klingeman DM, Johnson CM, Martin SL, et al. (2012) Twenty-one genome sequences from Pseudomonas species and 19 genome sequences from diverse bacteria isolated from the rhizosphere and endosphere of Populus deltoides. J Bacteriol 194: 5991-5993.

22. Kumar SA, Peter YA, Nadeau JL (2008) Facile biosynthesis, separation and conjugation of gold nanoparticles to doxorubicin. Nanotechnology 19: 495101.

23. Lengke MF, Fleet ME, Southam G (2007) Biosynthesis of silver nanoparticles by filamentous cyanobacteria from a silver(I) nitrate complex. Langmuir 23: 2694-2699.

24. Mohanpuria P, Rana NK, Yadav SK (2008) Biosynthesis of nanoparticles: technological concepts and future applications. Journal of Nanoparticle Research 10: 507-517.

25. Suresh AK, Pelletier DA, Wang W, Broich ML, Moon JW, et al. (2011) Biofabrication of discrete spherical gold nanoparticles using the metal-reducing bacterium Shewanella oneidensis. Acta Biomater 7: 2148-2152.

26. El-Brolossy TA, Abdallah T, Mohamed MB, Abdallah S, Easawi K, et al. (2008) Shape and size dependence of the surface plasmon resonance of gold nanoparticles studied by Photoacoustic technique. European Physical Journal-Special Topics 153: 361-364.

27. El-Sayed MA (2001) Some interesting properties of metals confined in time and nanometer space of different shapes. Acc Chem Res 34: 257-264.

28. Lee SO, Salunke BK, Kim BS (2014) Sucrose density gradient centrifugation separation of gold and silver nanoparticles synthesized using Magnolia kobus plant leaf extracts. Biotechnology and Bioprocess Engineering 19: 169-174.

30

Quantitative Analysis of 25-OH Vitamin D using Supported Liquid Extraction and Liquid Chromatography - Mass Spectrometry

Eduard Rogatsky*, Shaynah Browne, Min Cai, Harsha Jayatillake and Daniel Stein

Biomarker Analytical Resource Core, Einstein-Montefiore Institute for Clinical and Translational Research. Albert Einstein College of Medicine, USA

Abstract

We report a low sample volume LC/MS method for 25OH Vitamin D analysis. The method requires only 40 µl serum, is fully automatable, fast and sensitive. The method was successfully implemented for clinical analysis of infants for vitamin D deficiency. The 6 min method has a sensitivity range from 2 to 100 ng/ml. A 3 min method was validated from 5 to 100 ng/ml range. Fast analysis, low sample volume and high sensitivity were achieved by combination of supported liquid extraction sample preparation, UHPLC chromatography and fused-core chromatographic column for separation, and highly sensitive mass spectrometry for detection. We emphasize the importance of sample preparation quality for rugged LC/MS analysis. Using SLE-based sample preparation we successfully used only 40 µl of serum while achieving a LLOQ of 2 ng/ml. We found from assayed samples, that 3-12 month old infants were not Vitamin D deficient, compared to adults. The average level of 25(OH)D3 in infants was 57.3 compared to 38.0 ng/ml for adults in a Bronx, NY patient population.

Keywords: Vitamin D3; Mass spectrometry; Liquid Chromatography

Introduction

Vitamin D3 (cholecalciferol) and vitamin D2 (ergocalciferol) are the most abundant forms of vitamin D in the human body. Vitamin D is synthesized in the skin from 7-dehydrocholesterol in response to UV light (sunlight). 25OH Vitamin D2 and 25-OH vitamin D3 are derived from 25-hydroxylation of ergocalciferol and cholecalciferol, respectively, in the liver. They are precursors of active $1,25(OH)_2$ vitamin D2 and D3. The latter 1,25 dihydroxy vitamin D metabolites have a short half life, exist at low abundance, and their levels must be interpreted in the context of precursor abundance and renal function (responsible for 1-hydroxylation). 25-OH vitamin D2 and D3 levels are felt to represent vitamin D stores in the human body and are used to diagnose vitamin D deficiency or toxicity and are tightly linked endocrinologically to inadequate bone mineralization and elevations in PTH with deficiency and hypercalcemia with toxicity. Vitamin D has long been recognized as essential for healthy bone and mineral metabolism, and its replacement in vitamin D deficiency has proven to cure rickets. Vitamin D is now recognized as a prohormone, which has multiple roles in maintaining optimum health [1-3]. Since Vitamin D deficiency is a widespread clinical problem and has been associated with many adverse health outcomes, analysis of Vitamin D2 (ergocalciferol) and D3 (cholecalciferol) and their major metabolites 25(OH)D2 and 25(OH)D3 has become a high priority topic in clinical analysis.

Traditional immunoassay-based analysis of 25-OH Vitamin D requires a large sample volume, typically 1-1.5 ml of serum. Such volume is easily available for analysis of adults. In contrast, for infants and newborns, sample volume is limited; as a result clinical analysis targets only critical analytes, while indicative, but non-essential analytes are routinely omitted, due to very limited sample availability. In addition, the Vitamin D immunoassays, such as RIA (DiaSorin) and automated chemiluminescent immunoassays, for example Abbott Diagnostics (Architect), DiaSorin (LIAISON), IDS (ISYS), Roche Diagnostics (E170), and Siemens (Centaur) are only able to measure total Vitamin D, and do not distinguish between the D2 and D3 forms as well as their inactive epimers [4]. Another disadvantage of immunoassays is higher assay variability and lack of specificity compared to LC/MS based assays. Typical LC/MS assays however consume 0.1-0.2 ml of serum, which is high for pediatric specimen analysis [5].

We successfully overcame this constraint by a combination of a highly sensitive triple quad mass spectrometer 6490 (Agilent Technologies) for detection and quantitation, use of a fused-core chromatographic column (Supelco) for a fast, 6 min separation, and supported liquid-liquid extraction (Biotage) as a simple and automatable sample preparation step. This combination allowed the use of only 40 microliter of serum per assay, while maintaining a LLOQ of 2 ng/ml. In a subsequent development step, the method was shortened to a 3 min run time with a validated LLOQ of 5 ng/ml [6].

Material and Methods

Chemicals and reagents

25-OH D (25-OH D2 Cat. # S4176UNL-0.1 and 25-OH D3 Cat. # S4163UNL-0.1) certified stock solutions were purchased from IsoSciences (PA, USA). Human serum Tri-Level Vitamin D Plus controls were purchased from Utak (CA, USA). 25-OH D2 (26,26,26,27,27,27-d6) (IS1) Cat. D5337 Lot 303 was obtained from Medical Isotopes Inc. and 25-OH D3(26,26,26,27,27,27-d6) (IS2) Cat. I1-11576D DLM-7708-0 was obtained from Cambridge Isotopes Inc. Optima LC/MS Formic acid was obtained from Fisher Scientific; HPLC grade Methanol, 2-propanol and Heptane were obtained from Fisher Scientific; Millipore Q deionized water; Stripped human serum cat. # MSG 1000 was purchased from Golden West Biologicals, Inc. (CA, USA). ISOLUTE SLE+200 mg Supported Liquid Extraction 96 Fixed Well Plates were obtained from Biotage. 0.5 mL 96 well plates were obtained from Agilent. The collection 2 mL well/Con B plates and

---

*Corresponding author: Eduard Rogatsky, Biomarker Analytical Resource Core, Einstein-Montefiore Institute for Clinical and Translational Research, Albert Einstein College of Medicine, USA
E-mail: eduard.rogatsky@einstein.yu.edu

piercible cover mats were obtained from Microliter Analytical supplies, now part of Wheaton (Millville, NJ).

Stock solutions, calibration standard, Internal standard and controls

An internal standard solution containing two standards (IS1and IS2) was prepared at the concentration of 41.6 ng/mL 25-OH D2 (26,26,26,27,27,27-d6), 80 ng/mL 25-OH D3 (26,26,26,27,27,27-d6) in 50% 2-propanol.

Procedure

- 5.0 g Hydroxy Propyl β-cyclodextrin is dissolved in 121.5 g water to get 125 mL 4% HPB solution.

- 25-OH D2 (26,26,26-,27,27,27-d6) 1.3 mg is added to 4% HPB water solution to get 10.4 µg/mL 4% HPB water solution, aliquoted 0.5 mL per 5 mL Vials, then dried by N2 and stored at -40°C freezer.

- 25-OH D3 (26,26,26,27,27,27-d6) 5.0 mg is added to 4% HPB water solution to get 40 µg/mL 4% HPB water solution, aliquoted 0.5 mL per 5 mL Vials, then dried by N2 and stored at -40°C freezer.

- The 25-OH D2 (26,26,26,27,27,27-d6) of 2 Vials and 25-OH D3 (26,26,26,27,27,27-d6) of 1 Vial are dissolved in 50% 2-propanol, transferred into 250 mL volumetric flask, then wash the vials 3 times by 50% 2-propanol, at last, add 50% 2-propanol to the line of the flask.

- The calibrators for calibration curves were prepared from the Isosciences 25-OH D2 (101.75 µg/mL) and 25-OH D3 (104.5 µg/mL) certified solutions by appropriate dilution in stripped serum (MSG 1000) at 7 different concentrations above the zero calibrator (2 ng-104 ng/mL (25-OH D3) and 2 ng-102 ng/mL (25-OH D2)), procedures as followed.

- Stock solution 101.75 ng/mL 25-OH D2 and 104.5 ng/mL preparation First, add MSG 2000 strip serum (less than 90 mL, received 04/2012) into 100 mL volumetric flask, then add 100 µL 101.75 ug/mL 25-OH D2, 104.5 ug/mL 25-OH D3 (200 uL fisher scientific pipet), at the end, add strip serum to the line. Aliquot 2 mL (1000 uL fisher scientific pipet) of solution into 3 mL brown plastic vials (50 tubes total).

- Use one of the 2 mL stock solution to get the calibrators (2 ng/mL, 5 ng/mL, 10 ng/mL, 20 ng/mL,40 ng/mL, 60 ng/mL)

- 60 ng/mL- 600 uL (100 ng/mL) + 400 uL strip serum=1 mL into 1.5 mL glass vial

- 40 ng/mL- 400 uL (100 ng/mL) +600 uL strip serum=1 mL into 1.5 mL glass vial

- 20 ng/mL 200 uL (100 ng/mL) +800 uL strip serum=1 mL into 1.5 mL glass vial

- 10 ng/mL 100 uL (100 ng/mL) + 900 uL strip serum=1 mL into 1.5 mL glass vial

- 5 ng/mL 50 uL (100 ng/mL)+ 950 uL strip serum=1 mL into 1.5 mL glass vial

- 2 ng/mL 100 uL (20 ng/mL) + 900 uL strip serum=1 mL into 1.5 mL glass vial

- Store the stock solution and calibrators in freezer (-40°C)

- Tri-Level Vitamin D Plus controls (Low 13 ng/mL 25-OH D2, 14 ng/mL 25-OH D2, 25-OH D3, Lot # 7038, ref # 10060; Level 1 32 ng/mL 25-OH D2, 25-OH D3, Lot # 7039, ref # 10061; Level 2 73 ng/mL 25-OH D2, 75 ng/mL 25-OH D3 Lot # 7040, ref # 10062, Utak, CA, US) were reconstituted as per company's recommendations. Procedure as followed: Remove cap from each vial to be used.

- Reconstitute the dried control material by adding 5.0 mL of deionized water, using a 5 mL volumetric pipette. Replace cap and let sit 10-15 minutes

- Swirl gently 3-4 minutes to ensure a homogeneous mixture

- After each use, promptly return the reconstituted control material to -40°C freezer

Patients sample (serum)

All human samples were collected through protocols approved by the Albert Einstein College of Medicine Committee on Clinical Investigations (IRB). Blood was collected and processed following routine clinical laboratory procedure. The collected serum was stored at -20°C until assayed.

Sample preparation

The SLE method was adapted. Briefly, 40 µL of patient sample calibrator or control was transferred in a 0.5 mL 96 well plate (Agilent). After dispensing of 40 µL of internal standards IS1/IS2, the mixture was shaken for 3 min manually. 80 µL 25% 2-propanol was added and the mixture was shaken for 3 min manually again. Then, the mixture was transferred into SLE+ 200mg 96 fixed well plate (Biotage) with 96 wells collection plate (Microliter Analytical supplies, 2 mL well/Con B) in collection position. After 5 minutes of waiting, 900 uL of heptane was added to the each well, and heptane by gravity flow was passed through the SLE sorbent. After 10 minutes, the collection plate was placed into a Speedvac concentrator to evaporate solvent at room temperature for 50 minutes followed by reconstitution in 80 µL 50% methanol. The collection plate was covered with a piercible cover and agitated at 200 RPM for 10 minutes prior to LC/MS analysis.

Chromatography

LC–MS/MS configurations

A Agilent LC system (Alpharetta, GA, USA) was used for HPLC chromatography. The system includes a 1290 infinity autosampler G226A, 1290 Flex Cube G4227, 1290 TCC G1316C, and 1290 Bin pump G4220A. For MS/MS analysis, an Agilent 6490 Triple Quadrupole LC/MS equipped with Jet Stream source. The instrument was operated in positive ion electrospray mode. The source parameters were: gas temperature 200°C, gas flow 11 L/min, nebulizer 30 psi, sheath gas temperature 250°C, sheath gas flow 11 L/min, capillary voltage 3000 V, Nozzle voltage 2000V. Agilent MassHunter software was used for data analysis and operation of the LC/MS system. Table 1 summarizes

Analyte	Q1 mass (Da) precursor	Q3 product	Dwell Time, ms.	Collision Energy, V
25-OH d6D2	419.3	355.3	50	7
25-OH D2	413.3	355.3	50	7
25-OH d6D3	407.3	371.3	50	9
25-OH D3	401.3	365.3	50	8

Table 1: Quantification MRM precursor/product ion transitions selected for measurement of 25-OH D.

Time	Mobile Phase B
0	50%
2.60	85%
3.80	100%
4.80	100%
5.00	50%
6.00	50%

Table 2: LC/MS 6 min method time table.

Time	Mobile Phase B
0	50%
2.0	85%
2.70	100%
2.80	100%
3.00	50%

Table 3: LC/MS 3 min method time table.

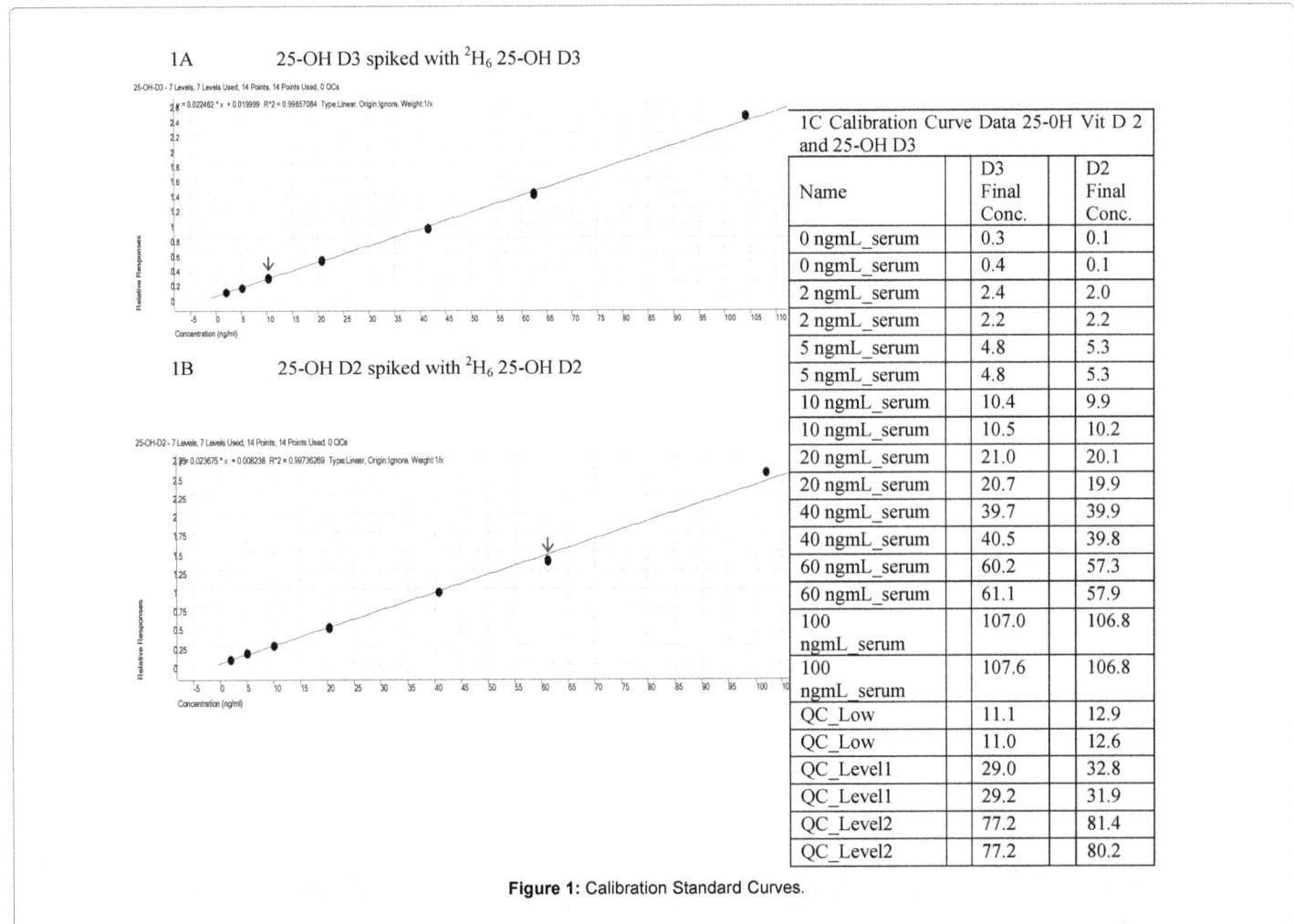

1A   25-OH D3 spiked with $^2H_6$ 25-OH D3

25-OH-D3 - 7 Levels, 7 Levels Used, 14 Points, 14 Points Used, 0 QCs

y = 0.022482 * x + 0.019999 R^2 = 0.99857084 Type:Linear, Origin:Ignore, Weight:1/x

1B   25-OH D2 spiked with $^2H_6$ 25-OH D2

25-OH-D2 - 7 Levels, 7 Levels Used, 14 Points, 14 Points Used, 0 QCs

y = 0.023675 * x + 0.008238 R^2 = 0.99736269 Type:Linear, Origin:Ignore, Weight:1/x

1C Calibration Curve Data 25-0H Vit D 2 and 25-OH D3

Name	D3 Final Conc.	D2 Final Conc.
0 ngmL_serum	0.3	0.1
0 ngmL_serum	0.4	0.1
2 ngmL_serum	2.4	2.0
2 ngmL_serum	2.2	2.2
5 ngmL_serum	4.8	5.3
5 ngmL_serum	4.8	5.3
10 ngmL_serum	10.4	9.9
10 ngmL_serum	10.5	10.2
20 ngmL_serum	21.0	20.1
20 ngmL_serum	20.7	19.9
40 ngmL_serum	39.7	39.9
40 ngmL_serum	40.5	39.8
60 ngmL_serum	60.2	57.3
60 ngmL_serum	61.1	57.9
100 ngmL_serum	107.0	106.8
100 ngmL_serum	107.6	106.8
QC_Low	11.1	12.9
QC_Low	11.0	12.6
QC_Level1	29.0	32.8
QC_Level1	29.2	31.9
QC_Level2	77.2	81.4
QC_Level2	77.2	80.2

Figure 1: Calibration Standard Curves.

the selected MRM transitions and other optimized mass spectrometry parameters.

Chromatographic conditions

A Supelco C18 Ascentis Express column 50 × 2.1 mm (Cat# 53822-U, Sigma-Aldrich Inc., PA, USA), 2.7 μm was used. Mobile phase A contained methanol: 2-propanol: water (40:10:50) with 0.4% formic acid with 0.1% formic acid. Mobile phase B contained methanol: 2-propanol (90:10) with 0.1% formic acid. Chromatographic separation

was achieved at 40°C, gradient flow rate was 0.5 mL/min; injection volume was 10 μL for both 3 and 6 min methods (Tables 2 and 3).

Results

Assay linearity was estimated using a curve with 7 non-zero calibrators in duplicates and a best fit curve was established using linear or quadratic weighting 1/x mode, respecting a least squares linear regression with an r > 0.98. Figure 1A and 1B LLOQ was determined using 2 ng/mL 25-OH D2, 25-OH D3 calibrator prepared in stripped

Serum Repeatability	QC1 25OH Vit. D3		QC2 25OH Vit. D3		QC3 25OH Vit. D3		QC1 25OH Vit. D2		QC2 25OH Vit. D2		QC3 25OH Vit. D2	
number of determinations	75		75		75		75		75		75	
Mean (ng/ml)	13.0		36.0		81.6		13.1		41.1		92.6	
SD (ng/ml)	0.32		0.77		2.41		0.51		0.96		4.79	
CV (%)	2.4%		2.1%		3.0%		3.9%		2.3%		5.2%	

Table 4: Assay imprecision for SLE-based LC/MS assay.

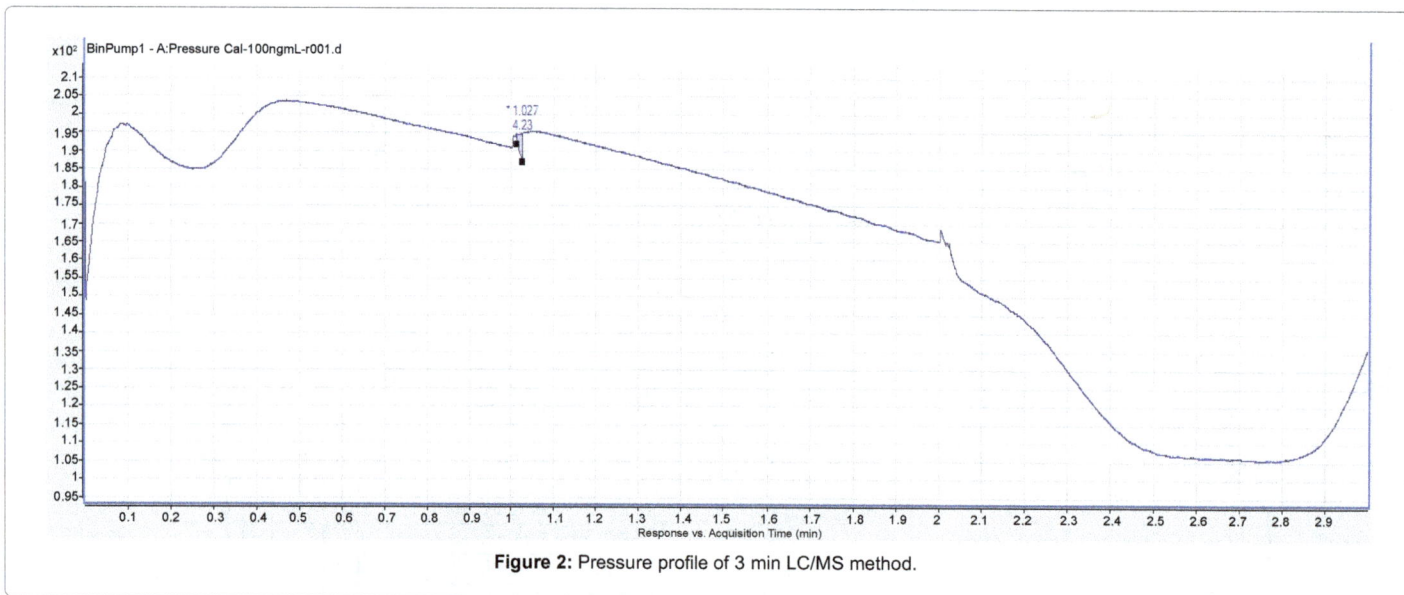

Figure 2: Pressure profile of 3 min LC/MS method.

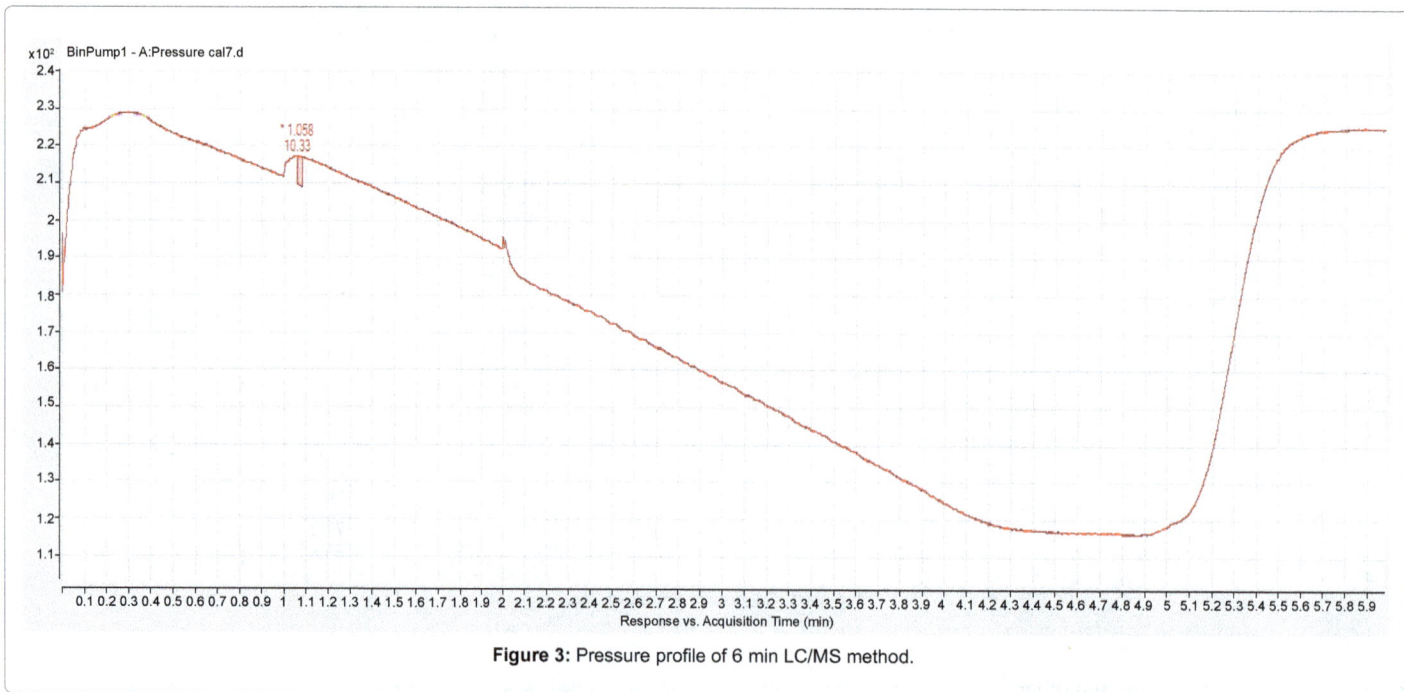

Figure 3: Pressure profile of 6 min LC/MS method.

serum and established. 2 ng/mL 25-OH D2, 25-OH D3 were accepted as the LLOQ because the following conditions are met: 1. The analyte response at the LLOQ is at least 5 times the response compared to blank response. 2. Analyte peak (response) is identifiable and reproducible with a precision of 20% and accuracy of 80%-120%. Accuracy was calculated from calibration curves and found that error is around 7% for ULOQ (Figure 1C).

Assay imprecision was evaluated using quality control samples (low, medium, high) with defined calibrators in a stripped serum matrix or defined calibrators in stripped serum matrix. Samples were measured in duplicate 15 times over 5 separate days (75 total measurements) to determine the repeatability and reproducibility of the assay (i.e. within assay and between assay variability), see Table 4. Analytes recovery during SLE was found around 70% including analyte decomposition

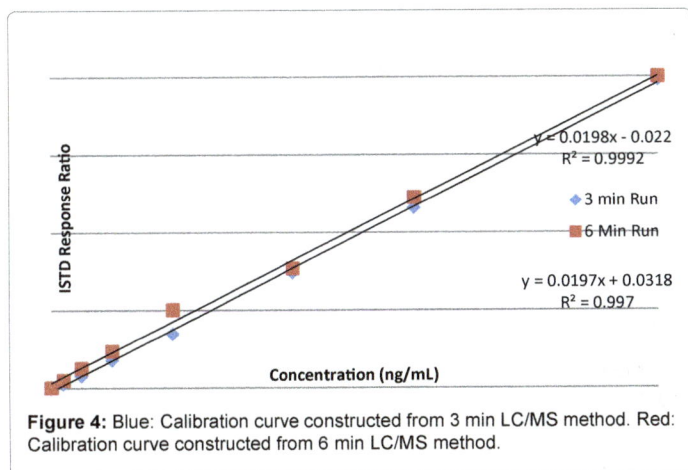

Figure 4: Blue: Calibration curve constructed from 3 min LC/MS method. Red: Calibration curve constructed from 6 min LC/MS method.

during drying.

Discussion

Typically, lowering of analysis cost is achieved by simplification of sample preparation. Paradoxically, we found that plasma protein precipitation in the long run is not cost efficient, since this approach reduces column lifetime and requires much more frequent instrument maintenance. We were able to lower sample volume to 40 ul and implement supported liquid-liquid extraction (Biotage) as a very simple sample preparation step. Unusually low sample volume was compensated by high sensitivity of mass spectrometry detection using a 6490 triple quad mass spectrometer from Agilent Technologies. Currently, it is one of the most sensitive instruments in the market. Fast gradients were accomplished by using ultra low dead volume 1290 UHPLC liquid chromatography and sharper chromatographic peaks were achieved by using fused core chromatographic column at higher flow rates. As a result of the combination of these method and instrumental features, we were able to maintain a LLOQ of 5 ng/ml for a 3 min run time and 2 ng/ml for a 6 min run.

We observed that reduction of the LC run time from 6 (Figure 2) to 3 min (Figure 3) leads to incomplete column equilibration. In general, incomplete column equilibration can cause instability of retention time for early eluting analytes. However, Vitamin D is a hydrophobic analyte and its retention time stability was not adversely impacted. In addition, we found that shortening the chromatographic separation did not impact assay linearity and selectivity, since the analyte to internal standard ratio was not affected (Figure 4).

As seen in Figure 4, the analyte (25OHD3) and its stable isotope labeled internal standard ratio is very similar for the 3 and 6 minute chromatographic methods, within the full calibration curve range. This overlap of calibration curves is evidence, that the quality of assay performance was maintained despite shortening the duration of the chromatographic separation by 50%.

Conclusions

Using SLE-based sample preparation we successfully used only 40 ul of serum while maintaining a robust and clinically relevant LLOQ of 5 ng/ml for 3 min method and 2 ng/ml for 5 min method. We found that reduction of LC separation leads to incomplete pressure equilibration and only partial mobile phase equilibration of the chromatographic column. Despite this fact however, assay performance was maintained.

We found that in this cohort, 3-12 month old infants were not Vitamin D deficient compared to adults. The infant's average 25(OH) D3 was 57.3 ng/ml compared to the 38.0 ng/ml for adults living in the Bronx, NY.

References

1. Holick MF (2009) Vitamin D status: measurement, interpretation and clinical application. Ann Epidemiol 19: 73-78.

2. DeLuca HF (2004) Overview of general physiologic features and functions of vitamin D. Am J Clin Nutr 80: 1689S–1696S.

3. Zhang R, Naughton DP (2010) Vitamin D in health and disease: current perspectives. Nutr J 9: 65.

4. Farrell CJ, Martin S, McWhinney B, Straub I, Williams P et al. (2012) State-of-the-Art Vitamin D Assays:A Comparison of Automated Immunoassays with Liquid Chromatography–Tandem Mass Spectrometry Methods. Clinical Chemistry 58: 531-542.

5. Yazdanpanah M, Bailey D, Walsh W, Wan B, Adeli K (2013) Analytical measurement of serum 25-OH-vitamin D3, 25-OH-vitamin D2 and their C3-epimers by LC–MS/MS in infant and pediatric specimens. Clin Chem 46: 1264-1271.

6. Rogatsky E, Stein DT. Low Cost, Low Sample Volume, Rapid Vitamin D Assay. Poster 33b at MSACL conference. San Diego, CA 1-4 March 2014

31

Updated Review on Micellar Electro kinetic Chromatography

Alagar Raja M[1]*, Bhargav KS[1], Banji D[1], Selva Kumar D[2]

[1]*Department of Pharmaceutical Analysis & Quality Assurance, Nalanda College of Pharmacy, Nalgonda, India*
[2]*School of Pharmacy, Taylors University, Subangjaya, Malaysia*

Abstract

In this review article the basic principle of separation process during Micellar electro kinetic chromatography (MEKC) are described. The separation mechanism in MEKC is based on differences in equilibrium between an aqueous Phase and micellar Phase. These Phases are moving different velocities, due to a combination of electrophoresis & electro osmosis. It is a useful branch of capillary electrophoresis (CE) that utilizes surfactant above critical micelle concentration (CMC) as pseudo-stationary phase. MEKC can be employed to separate both charged and neutral molecules, individually or simultaneously, including chiral compounds. MEKC benefits from high peak efficiency due to electro-osmotic flow (EOF) in the separation capillary, compounded with large variety of synthetic surfactants, organic modifiers, temperature and variable separation voltage has made MEKC the method of choice for separation scientists. In this review, we present the introduction of CE, fundamentals of surfactant chemistry as it relates to MEKC, separation principles in MEKC including equations involved in calculating separation parameters (capacity factor, resolution etc.).

Keywords: MEKC; Electrophoresis; Electro osmosis; Electro osmotic flow; Capacity factor

Introduction

According to International Union of Pure and Applied chemistry (IUPAC), chromatography can be defined as: "A Physical method of separation in which the components to be separated are distributed between 2 phases, one of which is stationary (stationary phase) while the other (the mobile phase) moves in a definite direction". The mobile phase flow can be controlled by gravity (e.g., column chromatography), by applying pressure (e.g., high pressure liquid chromatography), and by electricity (e.g., electrophoresis).Capillary Electrophoresis (CE) is an electro-driven separation techniques, it calls for low reagent consumption, high efficiency and selectivity with reasonably short analysis time. In CE, the capillary is filled with suitable buffer and after injecting analytes from anode side (under normal polarity conditions), high voltage is applied at its both ends. The analytes (positively and negatively charged) will move with different velocity and can be separated based on their electrophoretic mobility. However, in case of neutral molecules, since they do not bear any charge, move with the solvent front and elute as a single band and thus, cannot be separated. To solve this problem, charged surfactants above Critical Micelle Concentration (CMC) are added in the CE running buffer, which allows separation of uncharged molecules along with charged ones. Electrokinetic chromatography (EKC) is a family of electrophoresis techniques named after electrokinetic phenomena, which include electro osmosis, electrophoresis, and chromatography. Miceller electrokinetic chromatography (MEKC) is a mode of EKC in which surfactants (micelles) are added to the buffer solution. Surfactants are molecules which exhibit both hydrophobic and hydrophilic character. They have polar "head" groups that can be cationic, anionic, neutral, or zwitterion and they have nonpolar, hydrocarbon tails. The formation of micelles or "micellization" is a direct consequence of the "hydrophobic effect." The surfactant molecules can self-aggregate if the surfactant concentration exceeds a certain critical micelle concentration (CMC). The hydrocarbontails will then be oriented toward the center of the aggregated molecules, whereas the polar head groups point outward. Micellar solutions may solubilize hydrophobic compounds which otherwise would be insoluble in water. The front cover picture shows an aggregated SDS molecule. In the center of the aggregate, p-fluorotoluene is situated depicting the partitioning of a neutral, hydrophobic solute

into the micelle. Every surfactant has a characteristic CMC and aggregation number, *i.e.*, the number of surfactant molecules making up a micelle (typically in the range of 50-100). The size of the micelles is in the range of 3 to 6 nm in diameter; therefore, micellar solutions exhibit properties of homogeneous solutions. Micellar solutions have been employed in a variety of separation and spectroscopic techniques. In 1980, Armstrong and Henry pioneered the use of micellar solutions as mobile phase's forreversed-phased liquid chromatography (RPLC).

Separation Principle

MEKC is based on the addition to the buffer solution of a micellar "pseudostationary" phase, which interacts with the analytes according to partitioning mechanisms, just like in a chromatographic method. The "pseudostationary" phase is composed of a surfactant added to the buffer solution in a concentration above its critical micellar concentration (CMC). In this system, EOF acts like a chromatographic "mobile phase". From a "chromatographic point of view", the EOF's "plug-like" flow profile is almost ideal as it minimizes band broadening, which can occur during the separation process. The most commonly used surfactant sodium dodecyl sulfate (SDS), an anionic surfactant. The anionic SDS micelles are electrostatically attracted towards the anode. The EOF transports the bulk solution towards the negative electrode due to the negative charge on the internal surface of the silica capillaries. But the EOF is usually stronger than the electrophoretic migration of the micelles and therefore the micelles will migrate also toward the negative electrode with a retarded velocity (Figure 1).

When a neutral analyte is injected into the micellar solution, a fraction is incorporated into the micelle, while the remaining fraction

---

*Corresponding author: Alagar Raja M, HOD, Department of Pharmaceutical Analysis & Quality Assurance, Nalanda College of Pharmacy, India, E-mail: rajampharman_1982@rediffmail.com

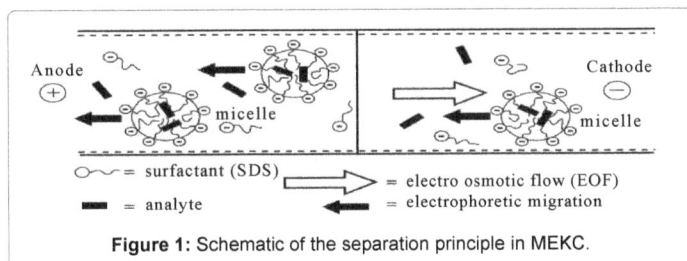

Figure 1: Schematic of the separation principle in MEKC.

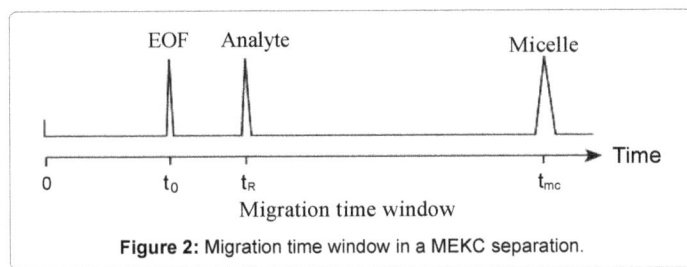

Figure 2: Migration time window in a MEKC separation.

Name of Surfactant	CMC^a/10^{-3} M	n	Kp
Sodium dodecyl sulfate (SDS)	8.1	6.5	16
Sodium tetradecyl sulfate (STS)	2.1 (50°C)	138^b	32
Sodium decanesulfonate	40	40	-
Sodium dodecanesulfate	7.2	54	37.5
Sodium N-lauroyl-N-methyltaurate (LMT)	8.7	-	<0
Sodium polyoxyethylene dodecyl ether sulfate	2.8	66	-
Sodium N-dodecanoyl-L-valinate (SDVal)	5.7 (40°C)	-	-
Sodium cholate	13-15	2-4	-
Sodium deoxycholate	4-6	4-10	-
Sodium taurocholate	10-15	5	-
Sodium taurodeoxycholate	2-6	-	-
Potassium perfluoroheptanoate	28	-	25.6
Dodecyltrimethylammonium chloride (DTAC)	16 (30°C)	-	-
Dodecyltrimethylammonium bromide (DTAB)	15	56	-
Tetradecyltrimethylammoniumbromide (TTAB)	3.5	75	-
Cetyltrimethylammonium bromide (CTAB)	0.92	61	-

^{a} 25°C
^{b} In 0.10M NaCl

Table 1: Critical Micelle Concentration, Aggregation Number (n), and Kraft point (Kp) of Selected Ionic Surfactants.

of the analyte migrates with the electro-osmotic velocity. Consequently, micelles decrease selectively the migration of neutral solutes they interact with (by partitioning mechanism), which otherwise would migrate with the same velocity as the EOF.

The separation depends on the individual partitioning equilibrium of the different analytes between the micellar and the aqueous phase. The greater percentage of analyte is distributed into the micelle, the slower it will migrate. Therefore, analytes that have greater affinity for the micelles exhibit slower migration velocities compared with analytes that are mostly distributed in the bulk solution with SDS micelles, the general migration order will be exactly the opposite as in ECZ: anions, neutral analytes and cations. Anions will remain mostly in the bulk solution due to electrostatic repulsions from the micelle; neutral molecules will be separated exclusively due to their hydrophobicity; while cations will migrate last due to the strong electrostatic attraction [1,2]. This generalization regarding the migration order can be sometimes useful, but strong hydrophobic interaction between analytes and micelles can overcome repulsions and attractions. Likewise, the own electrophoretic mobilities of the analytes can also modify the

migration order. Analytes which are highly retained by the micelle will have longer migration times, close to the EOF (t_0). Very hydrophobic compounds may be totally included into the micelle and will migrate with the micelles velocity (t_{mc}). Methanol is not retained by the micelles and migrates with t_0 being used as marker for the EOF, while a dye Sudan III is totally included into the micelle and can be used as a micellar marker. The period between the migration time of the bulk solution and the migration time of the micelle is often referred to in the literature as migration time window (Figure 2).

A relatively recent development in MEKC has been to perform separations in the absence of EOF. This may be achieved using coated capillaries or at low pH values. This could be especially useful in the separation of acidic analytes, which would ionize at high pH values and would not interact with the negatively charged SDS micelle. Cationic surfactants can be used in MEKC to reverse the charge on the capillary wall, by absorption on the capillary wall surface through a mechanism involving electrostatic attraction between the positively charged ammonium moieties and the negatively charged Si-O-groups; when a reversal of the EOF takes place.

Composition of the Micellar Solution

Ionic surfactants are essential for MEKC. Numerous ionic surfactants are commercially available. The surfactants suitable for MEKC should meet the following criteria

1. The surfactants must have enough solubility in the buffer solution to form micelles.

2. The micellar solution must be homogeneous and UV transparent.

3. The micellar solution must have a low viscosity.

Table 1 lists the CMC, aggregation number, and Kraft point of some selected ionic surfactants available for MEKC. The Kraft point is the temperature above which the solubility of the surfactant increases steeply due to the formation of micelles. In order to obtain a micellar solution, the concentration of the surfactant must be higher than its CMC. The surfactants enough solubility to form micelles only at temperatures above the Kraft point as mentioned above. The counter ion of the ionic surfactant doesaffect the Kraft point. For example, the Kraft point of sodium dodecyl sulfate (SDS) is 16°C but potassium dodecyl sulfate has a Kraft point of approximately 35°C. Therefore, if SDS is dissolved in a buffer containing potassium ions, the solubility of SDS will be less than its CMC at ambient temperature because of the exchange reaction of the counter ions. The actual CMC in the buffer solution is usually lower than the value listed in Table 1, as these values were obtained with pure water as solvent. High concentrations of surfactant (>200 mM) result in relatively high viscosities (and high currents) and should therefore be avoided in MEKC. It is recommended that the concentration of the buffer salt should not be lower than 10 mM. A higher concentration of buffer relative to that of surfactant is preferred to keep the pH constant during the run. It should be remembered that, in electrophoresis, electrolysis occurs at the electrodes, resulting in a reduction of the pH of the anodic solution. This solution may enter the capillary during the run when an anionic micelle (e.g., SDS) is employed. The cathodic solution becomes alkaline during electrophoresis and may enter the capillary when a cationic surfactant (e.g., CTAB) is employed. In this case the EOF is reversed.

Surface-Active Agents (Surfactants)

The surfactants are amphiphilic in nature and are miscible with both polar and apolar substances. A typical amphiphilic molecule itself

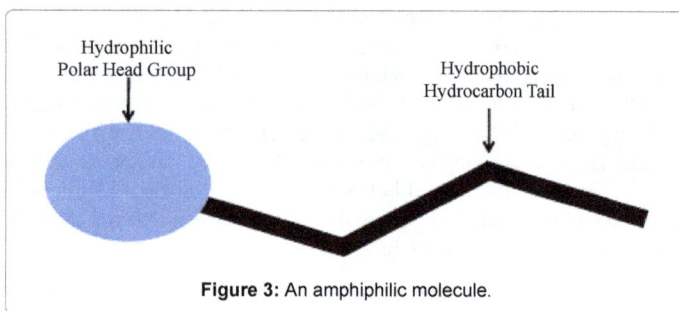

Figure 3: An amphiphilic molecule.

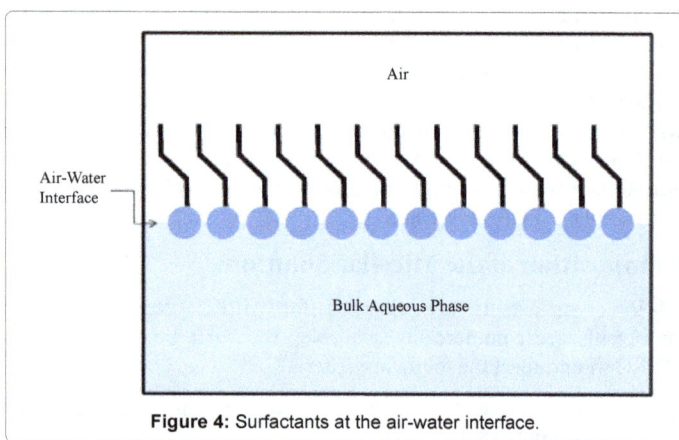

Figure 4: Surfactants at the air-water interface.

consists of polar (hydrophilic) group (e.g., alcohol, ether, carboxylate, sulfate, sulfonate, phosphate, amine, ammonium etc) and apolar (hydrophobic) group (e.g., usually a long hydrocarbon chain) as represented in Figure 3.

The hydrophilic portion exhibits a strong affinity for water, while the hydrophobic part tends to accumulate together (hydrophobic effect) due to mutual antipathy for water. Because amphiphiles conation both water loving and repelling groups, they often tend to migrate at the interface of an aqueous solution, such that hydrophilic part is in water and hydrophobic part away from water (in the air) as represented in Figure 4.

Due to accumulation at the air-water interface, the surface tension of water drops and these molecules are accordingly dubbed, surface active agents. There are many substances, such as medium- or long-chain alcohols that are surface active (e.g., *n*-hexanol, dodecanol) but they are not considered as surface-active amphiphiles (surfactants). Specifically, surfactants are distinguished by self-assembly structures (micelles, vesicles) in bulk phases [3-9] and ability to form oriented monolayers at the interface. Surfactants are also responsible for the fundamental physical effects, such as, wetting, dispersion or deflocculation and emulsification. Alternatively, surfactants interfere with the ability of the molecules of a substance to interact with one another (especially at the interface) and thereby, lower the surface tension of the substance.

Classification of Surfactants

Surfactants are characterized based on the charge present in the hydrophilic portion of the molecule (after dissociation in aqueous solution). There are four categories of surfactant

(1) Anionic, (2) Cationic, (3) Nonionic and (4) Zwitterionic.

Anionic surfactants, when dissolve in water, dissociate into hydrocarbon chain bearing anion (e.g., $-COO^-$, $-SO_3^-$, $-PO_4^{-3}$, $-SO_4^{-2}$), and a counter cation (e.g., Na^+, K^+) and are the most commonly used

type of surfactants. Cationic surfactants on the other hand, when dissolved in water, dissociate into hydrocarbon chain bearing cationic head group [e.g., $(R)_4N^+$, $(R)_4P^+$] and a counter anion (e.g., Cl^-, Br^-). A very large proportion of this class corresponds to fatty amine salts and quaternary ammoniums, with one or several long chain of the alkyl type, often coming from natural fatty acids. The quaternary ammonium group containing surfactants are well known for displaying emulsifying properties, antimicrobial activity, anti-corrosive effects and are used in cosmetic formulations and as phase transfer catalyst in organic synthesis. Zwitterionic surfactants contain both anionic and cationic portion within the surfactant backbone and are also known as amphoteric surfactants. Some zwitterionic surfactants stay zwitterionic at all pH, while few are cationic at low pH and anionic at high pH. They are generally quite expensive as they are not very easy to make and thus are used in special circumstances, for instance in cosmetics, due to high biological compatibility and low toxicity. Nonionic surfactants, as name indicates, are devoid of charges. The hydrophilic group usually is alcohol, phenol, ether, ester or amide. Large proportions of these nonionic surfactants are hydrophilic by the presence of a polyethylene glycol chain and are called polyethoxylatednonionics. Sugar-derived nonionic surfactants are also in use as they exhibit very low toxicity and good have excellent biodegradability.

Buffer Additives

Since MEKC is often applied in the separation of analytes with very similar hydrophobicities and chemical characteristics, sometimes is useful to extend the concept of using a "mobile phase" and a "pseudostationary phase" to the use of buffer additives such as organic modifiers and cyclodextrines. Organic solvents (methanol, acetonitrile) are used in CZE in order to increase solubility of the analytes, but their role in MEKC is more complex and profound. Organic solvents reduce EOF, consequently increase the migration times and migration time window of the analytes. Also, organic additives reduce the hydrophobic interactions between the micelle and the analyte and can be useful in the separation of analytes which otherwise are almost completely incorporated in micelles. The addition of organic solvents will increase the migration velocity of these hydrophobic analytes, by reducing the partition coefficient between the micelle and the bulk solution. However high concentration of organic solvents may break down the micellar structure, consequently concentrations above 25-30% should be avoided. Cyclodextrines (CD) are cyclic oligosaccharides with truncated cylindrical molecular shapes, having an external hydrophilic surface and an internal hydrophobic cavity, in which they can include other compounds by hydrophobic interactions. The inclusion mechanism is sterically selective, because analytes must fit the size of the cavity, the diameter of which depends on the number of glucose units in the CD structure. There is a wide range of both natural and derivatised CD commercially available. The native CD, α-, β-, and γ-CD possess different numbers of glucose sub-units, six, seven and eight respectively. These surface hydroxyl groups can be chemically replaced with groups such as hydroxypropy and dimethyl groups. Ionic chargeable CD offers the possibility of separation of neutral drug enantiomers or enhanced separation of ionic drugs. Several CE specific derivatised CDs have been produced with amino, sulfate or carboxylic groups. Because of the chirality of the hydroxyls in the glucose molecules that form the rim of the CD cavity, the inclusion complex formation will be chirally selective. If the enantiomers of a compound have different binding constants, then chiral separation is possible by adding the proper CD in the buffer electrolyte. CDs are neutral from electrophoretic point of view, and are not incorporated in micelles, because of the hydrophilic nature of the outside surface of the molecules. Therefore, an analyte included in

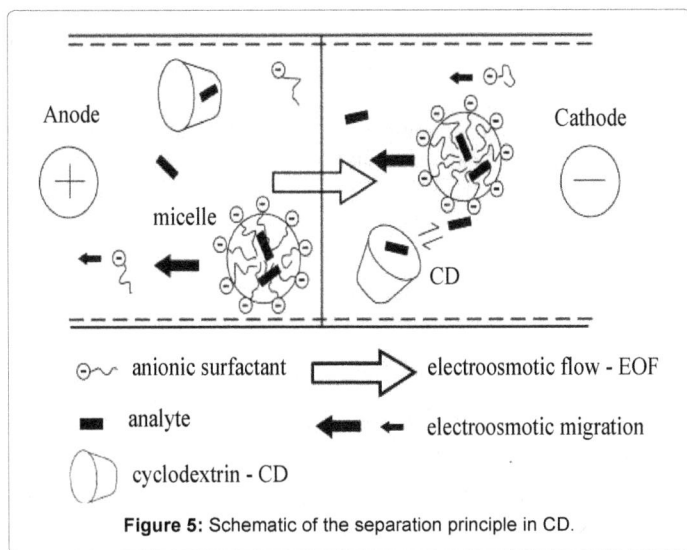

Figure 5: Schematic of the separation principle in CD.

the CD will migrate with the same velocity as the EOF. The addition of cyclodextrines reduces the apparent distribution coefficient of the analytes between the two phases. Hydrophobic analytes can become incorporated into either the CD cavity or the micelle. Effectively the addition of the CD establishes two "pseudo stationary" phases in the electrolyte, which can reduce analysis times and offer the possibility of improved separation. CDs have advantages over organic solvents, as they are UV transparent and non-volatile. The schematic principle in cyclodextrin modified micellarelectrokinetic chromatography is presented (CD-MEKC) (Figure 5).

Marker of the Electro-osmotic Flow and the Micelle

In order to calculate the capacity factor, it is necessary to know the migration time of the bulk solution, t_0, the migration time of the micelle, t_{mc}, as well as the migration time of the analyte, t_R. Since the whole capillary is filled with micellar solution, the markers of the bulk solution and the micelle are required to obtain t_0 and t_{mc}. Strictly speaking, no ideal marker is available for MEKC. The marker for the bulk solution must be electrically neutral as well as totally excluded from the micelle. Mesityl oxide, often used in CZE to measure t_0, is not an appropriate choice in MEKC, because it is partially incorporated into the micelle. Methanol often serves to measure t_0, because its distribution coefficient is almost negligible. Furthermore, it can be detected by UV absorption due to a change in refractive index as the methanol peak passes through the detection zone. The marker for the micelle must be totally incorporated into the micelle. Sudan III or IV are often used to measure t_{mc}. Both solutes are not soluble in water and can be dissolved in methanol or in the micellar solution. However, because of the poor solubility in water, it is not always possible to observe the peaks of Sudan III or IV in the electropherograms. As an alternative, compounds that are insoluble in water but soluble in the micellar solution can be employed to measure t_{mc}. Timepidium bromide or quinine hydrochloride are good markers for anionic SDS micellar systems [10,11].

Theoretical Aspects

In MECC we can define the capacity factor (k) similarly as in chromatography

$$k = n_{mc} / n_{aq}$$

where n_{mc} and n_{aq} are the amount of analyte incorporated into the micelle and in the aqueous respectively. It can be calculated from the migration time of the analyte (t_R), of the EOF (t_0) and of the micelle (t_{mc})

$$k = t_R - t_0 / t_0 (1 - t_{R/} t_{mc})$$

When k = 0, the migration time of the analyte is equal to t_0, which means that the analyte does not interact with the micelle; and when k is infinity, the migration time of the analyte is equal to t_{mc}, which means that the analyte is totally incorporated into the micelle.

The capacity factor is a fundamental term in chromatography while the electrophoretic mobility is characteristic to the electrophoretic process. In ECZ the migration velocity (v) of the analyte is expressed as

$$v = (\mu_{eo} + \mu_a) E$$

Where μ_{eo} and μ_a are the electrophoretic mobilities of the EOF and analyte respectively and E is the electric field strength. We can apply this equation to MECC by defining the effective electrophoretic mobility of a neutral analyte (μ_{na}) as

$$\mu_{na} = \mu_{mc} k/1+k$$

Where μ_{mc} is the electrophoretic mobility of the micelle and k/1+k represents the fraction of analyte incorporated into the micelle. Thus, the velocity of a neutral analyte in MECC is given as

$$v = (\mu_{eo} + \mu_{na}) E$$

The capacity factor provides quantitative information about the analyte distribution between the two phases, while the electrophoretic mobility only gives qualitative information about it.

The resolution equation in MECC can be given by the following equation

$$R_S = \frac{\sqrt{N}}{4} \left(\frac{\alpha - 1}{\alpha} \right) \left(\frac{k_2}{1 + k_2} \right) \left(\frac{1 - (t_0 / t_{mc})}{1 + (t_0 / t_{mc}) k_1} \right)^4$$

Where N is the theoretical plate number, α the separation factor between the two analytes and k_1 and k_2 their capacity factor. The separation factor (α) is determined by the micellar solubilization process and is influenced by the chemical nature of both the micellar phase and the surrounding aqueous phase. Various surfactant systems can be used as well as mixed micelles, possessing different solubilization characteristics, in order to control migration behavior of the analytes and optimize selectivity.

Enantiomeric Separations

In the pharmaceutical industry, the determination of the optical purity or separation and determination of enantiomers is becoming increasingly important. The non-active enantiomer in a drug formulation may be considered as an impurity. High resolution separation methods are required to achieve chiral separations. The approaches used in CE, discussed next, are relatively simple compared to HPLC methods in which expensive, chiral stationary phases are frequently used. In CE only minute amounts of chiral selectors are required to determine enantiomeric purity. Two approaches can be used to perform enantiomeric separations in MEKC

1. Use of chiral surfactants

2. Use of chiral additives

Chiral surfactants

Bile salts are widely available commercially and have shown to be

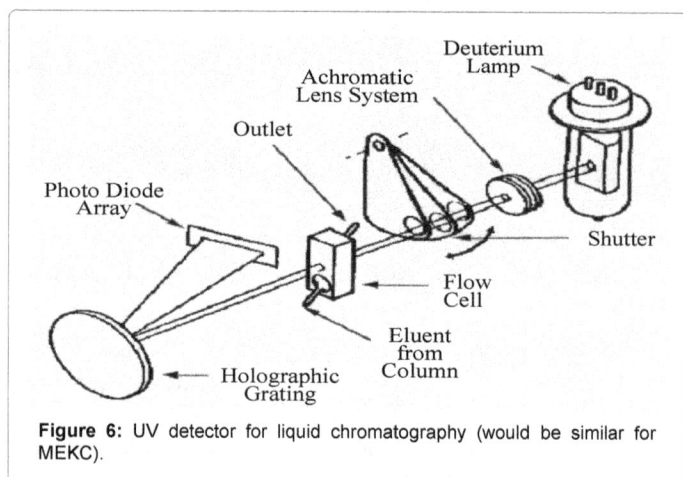

Figure 6: UV detector for liquid chromatography (would be similar for MEKC).

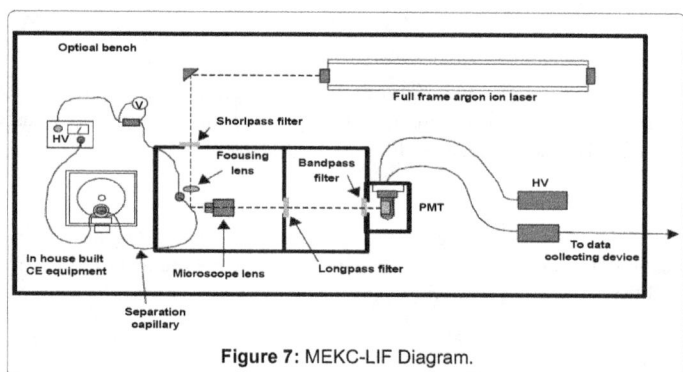

Figure 7: MEKC-LIF Diagram.

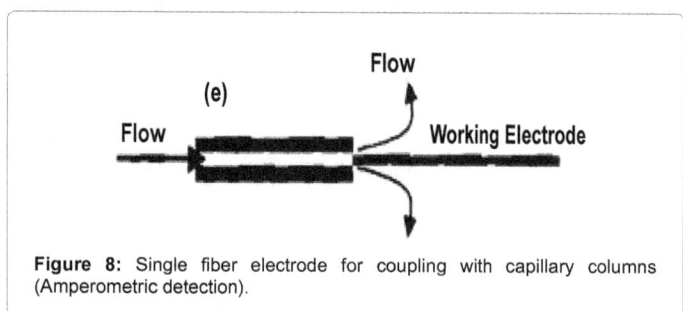

Figure 8: Single fiber electrode for coupling with capillary columns (Amperometric detection).

useful chiral surfactants. Sodium cholate or sodium deoxycholate can be used under neutral or alkaline conditions to ionize the carboxyl group of the surfactant. Taurine conjugates of bile salts can also are used in acidic conditions because taurine has a sulfonic acid group. Amino acid-derived surfactants (*e.g.*, sodium N-dodecanoyl-Lvalinate (SDVal) are another group of chiral surfactants that are commercially available. They also must be used under neutral or alkaline conditions. In order to use these surfactants under acidic conditions, SDS can be added to form mixed micelles with appreciable electrophoretic mobilities. The addition of a small amount of methanol and/or a relatively high concentration of urea often improves resolution while sharpening peak profiles.

Chiral Additives

The second, more popular, method of enantiomeric separation by MEKC is to add CD to the micellar solution. The SDS micelle may be conveniently used for this approach. Various CDs or CD derivatives may be tried. The concentrations of SDS and CD should be optimized to yield optimal capacity factors. Cicletanine is a member of a new

class of antihypertensive drugs, the fusopyridines. Chiral selectivity was obtained by using a buffer consisting of 100 mM borate, 50 mM SDS, pH 8.6 and 50 mM b-CD. The addition of methanol and/or urea often enhances solubility and improves resolution. When g-CD is employed, a second chiral component, such as *d*-camphor-10-sulfonate or *l*-menthoxyacetic acid, may enhance resolution.

Detectors for MEKC

Photometric detection (LOD is over 10^{-3} M) (Figure 6).

Laser induced fluorescence (LIF) detection (LOD is below 10^{-9} M) (Figure 7).

Electrochemical Detection (Figure 8).

MEKC Applications in the Analysis of Pharmaceutical Substances

In principle MEKC is used for the analysis of neutral compounds, or when analyzing mixtures of neutral and charged solutes. But MEKC conditions are also employed when selectivity requirements for a separation exceed the simple mobility differences obtainable in CZE.

MEKC can be especially useful for the determination of drugs in samples having a high protein content (clinical samples, biofluids) reducing the disadvantageous matrix effects caused by organic materials, while CZE through its simplicity and operation stability could be advantageous for pharmaceutical determinations.

MEKC can be usually applied in simultaneous separation from complex mixtures of pharmaceutical substances with very similar structural and physico-chemical characteristics. Many reports have been published detailing the use of MEKC for pharmaceutical applications; Table 2 presents briefly selected pharmaceutical applications and the description of the electrophoretic conditions.

Another application of MEKC is the chiral separation of optically active pharmaceutical substances. Enantiomer separation by MEKC involves the addition of a chiral agent such as chiral surfactants, crown ethers, or CDs to the background electrolyte with chiral/achiral micelles. Chiral MEKC with chiral surfactants is an important separation mode for chiral compounds, with chiral surfactants including also naturally occurring compounds such as bile salts, amino acids or glucose. Chiral separation in MEKC is affected by the affinity of the enantiomers toward the micelles, and the concentration of the micellar phase, which depends on the aggregation properties of the chiral surfactants. MEKC can be used for the separation of structural related impurities from the main active drug, and has been proven an alternative to HPLC for quantitation of compounds and the determination of drug-related impurities. The structurally related impurities of a drug will possess similar structural and physico-chemical characteristics to the main component, which makes their separation and determination a challenging task. The high separation efficiencies possible for CE often allow a small degree of selectivity to provide an acceptable resolution. The separation and determination of drug-related impurities using CE has been extensively studied, and the method performance and validation data obtained clearly shows that CE methods are successful applications in this area.

Conclusion

MEKC is relatively new technique in chromatographic separation and has wider applications in pharmaceutical drug developmental and drugs ranging from combinatorial chemistry, chiral (Enantiomers), separation and purification, clinical fluid analysis. This technique has

Pharmaceutical class	Substances	Electrophoretic conditions
Penicillins	Amoxicillin, Ampicillin, Benzylpenicillin, Phenoxymethypenicillin, Oxacillin, Cloxacilin	40 mM sodium tetraborate +100 mM SDS, pH 9.3 Voltage: +10 kV, temperature: 20°C, UV detection 210 nm
Cephalosporins	Cefazoline, Cefuroxime, Ceftriaxone, Cefoperazone, Ceftazidime	20 mM sodium tetraborate+15 mM disodium hydrogenophosphate+50 mM SDS pH 6.5 Voltage: +18 kV temperature: 20°C, UV detection 214 nm
Macrolides	Erythromycin, Tylosin and related substances	80 mM sodium phosphate+20 mM sodium cholate+7 mM Acetyltrimethyl ammonium bromide, pH 7.5 Voltage: +15 kV, temperature: 25°C, UV detection 280 nm
Aminoglycosides	Gentamicin, Sisomicin, Netilmicin, Kanamycin, Amikacin, Tobramycin	100 mM sodium tetraborate + 20 mM sodium deoxycholate + 15 mM beta-cyclodextrin, pH – 10 voltage: + 20kV, temperature: 25°C
Tetracyclines	Tetracycline, Oxytetracicline, Democlocycline, Chlortetracycline, Doxycicline, Minocycline	15 mM ammonium acetate + 20 mM SDS, pH 6.5 voltage: + 15kV, temperature: 25°C
Sulfonamides	Sulfamethazine, Sulfamerazine, Sulfathiazole, Sulfachloropyridazine, Sulfamethoxazole, Sulfacarbamide, Sulfaguanidide	13.32 mM disodium hydrogen phosphate, 6.67 mM potassium dihydrogen phosphate + 40 mM SDS, pH 7.5 voltage: + 21 kV, temperature: 25°C fluorescence detection
Fluoroquinolones	Norfloxacin, Ciprofloxacin, Ofloxacin, Enrofloxacin, Danofloxacin	25 mM sodium carbonate + 100 mM SDS, pH 9.2 voltage: + 20kV, temperature: 30°C, UV detection 280 nm
Antifungal azoles	Fluconazole, Voriconazole, Itraconazole, Posaconazole	25 mM phosphoric acid+100 mM SDS+13% acetonitrile+13% tetrahydrofuran, pH 2.2
Barbiturates	Phenobarbital, Amobarbital, Pentobarbital, Secobarbital, Butabarbital	10 mM sodium tetraborate+10 mM disodium hydrogenophosphate+100 mM SDS+15% acetonitrile. pH 8.5 Voltage: +20 kV, UV detection 214 nm
Benzodiazepines	Flunitrazepam, Diazepam, Midazolam, Clonazepam, Bromazepam, Temazepam, Oxazepam, Lorazepam	25 mM phosphate/borate+75 mM SDS, pH 9.3
Phenotiazines	Promethazino, Ethopropazine, Trimeprazine, Methoprimeprazine, Thioridazine	80 mM citric acid+10 mM tetradecyltrimethyl ammonium bromide+7 mM β-CD (9 mM HP β-CD), pH 3.5 voltage: +20 kV, temperature: 25°C, UV detection 254 nm enantiomer separation
Tricyclic antidepressants	Imipramine, Amitriptyline, Desipramine, Nortriptyline, Doxepin, Trimipramine	37.5 mM phosphate+25 mM dodecyltrimethylammonium bromide+2 M urea, pH 8 Voltage: +25 kV
Benzodiazepines	Alprazolam, Bromazepam, Chlordiazepoxide, Diazepam, Flunitrazepam, Medazepam, Oxazepam, Nitrazepam	25 mM sodium tetraborate+50 mM SDS+12% methanol, pH 9.3 Voltage: +25 kV temperature: 20°C, UV detection
Xanthines	Caffeine, Theobromine, Theophylline, Pentoxifylline	20 mM sodium tetraborate+100 mM SDS, pH 9.3 Voltage: +30 kV, temperature: 25°C, UV detection 274 nm

Table 2: Applications of MEKC in the analysis of different pharmaceutical substances.

advantages over others as small sample and solvent requirements with high resolutions. The micelle formation by surfactants further advances the technique as capillary electrophoresis.

References

1. Rizvi SAA (2011) Fundamentals of micellar electro kinetic chromatography. European Journal of Chemistry 2: 276-281.

2. Terabe S, Otsuka K (1995) Beckman Miscellar electro kinetic chromatography. Himeji Institute of Technology. Kamigori, Hyogo, Japan.

3. Andriy Gryshchenko (2010) Micellar electrokinetic chromatography theory and application. Memorial University Chemistry Department.

4. Gabriel Hancul (2013) Principles of Micellar Electrokinetic Capillary Chromatography Applied in Pharmaceutical Analysis. Advanced Pharmaceutical Bulletin 3: 1-8.

5. Erim FB, Xu X, Kraak JC (1995) Application of micellar electrokinetic chromatography and indirect UV detection for the analysis of fatty acids. Journal of Chromatography A 694: 471-479.

6. Terabe S (2009) Capillary Separation: Micellar electrokinetic chromatography. Annual Review of Analytical Chemistry 2: 99-120.

7. Silva M (2013) Micellar electrokinetic chromatography: A review of methodological and instrumental innovations focusing on practical aspects. Electrophoresis 34: 141-158.

8. Deeb SE, Iriban MA, Gust R (2011) MEKC as a powerful growing analytical technique. Electrophoresis 32: 166-183.

9. Loco VJ, Caroli S (2007) Method validation for food analysis: concepts and use of statistical techniques. The determination of chemical elements in food: Application for atomic and mass spectrometry. Wiley & Sons Inc. 138.

10. Hommels G, Waterval JCM, Bestebreurtje P, Versluis C, Underberg WJM (2001) Capillary electrophoretic bioanalysis of therapeutically active peptides with UV and mass spectrometric detection after on-capillary pre-concentration. Electrophoresis 22: 2709-2716.

11. Kahle KA, Foley JP (2007) Review of aqueous chiral electro kinetic chromatography (EKC) with an emphasis on chiral micro emulsion EKC. Electrophoresis 28: 2503-2526.

32

Sensitive Analysis of Off-flavor Compounds, Geosmin and 2-Methylisoborneol, in Water and Farmed Sturgeon by using Stir Bar Sorptive Extraction Coupled with Thermal Desorption and Gas Chromatography-Mass Spectrometry

Ruan ED*, Aalhus J and Juarez M

Lacombe Research Centre, Agriculture and Agri-Food Canada, 6000 C&E Trail, Lacombe, AB T4L 1W1, Canada

Abstract

The semi-volatile cyclic alcohols 2-methylisoborneol (2-MIB) and geosmin (GSM) impart muddy or musty flavors to water and fish products. A rapid quantitative analytical technique has been developed based on stir bar sorptive extraction (SBSE), coupled with thermal desorption and gas chromatography-mass spectrometry (TD-GC-MS). SBSE is used to extract and concentrate the off-flavor compounds of GSM and 2-MIB in coating material polydimethylsiloxane (PDMS). The analytes are thermally released from SBSE and analyzed by GC-MS. Limits of detection (LOD) and limits of quantification (LOQ) of GSM and 2-MIB are 0.3 ng/L and 1 ng/L, based on the main fragment ions of m/z 112 and m/z 95, respectively. This methodology allows quantitative analysis of 2-MIB and GSM in both water and fish tissue successfully.

Keywords: Off-flavor; Geosmin; 2-Methylisoborneol; Stir Bar Sorptive Extraction (SBSE); Gas Chromatography-Mass spectrometry (GC-MS)

Introduction

Muddy-earthy-musty-type flavours, such as Geosmin (1a,10h-dimethyl-9a-decalol, GSM) and 2-methylisoborneol (1-R-exo-1,2,7,7-tetramethylbicylo-[2,2,1]-heptan-2-ol, 2-MIB), are semi-volatile off-flavor compounds with similar chemical structures that are produced by certain species of actinomycetes, fungi and blue-green algae [1,2]. These off-flavor compounds have been shown to be the main cause of earthy-musty odorants in water from aquaculture facilities, and the occurrence of such flavours has also been reported for a diverse range of freshwater aquaculture species. 2-MIB and GSM tend to bio-accumulate within fish flesh dependent on the concentration of the compound in the water supply, water temperature, fat content and mass of fish, and other abiotic and biotic factors. Although off-flavor compounds are harmless to human health, fish presenting with these flavour characteristics are often referred to as being 'off-flavour' or 'tainted' and are commonly considered to be spoiled or of low quality [3]. Placing tainted fish in the marketplace typically lowers consumer confidence in the cultured product and ultimately results in significantly lower commercial returns. Uptake of these tainting compounds is primarily via the gills and accumulation in the flesh is influenced by the concentration of the compounds in the holding water, water temperature, and the physiology and lipid content of the fish [2,4]. These compounds are known to be particularly problematic in aquaculture systems due to persistent and elevated nutrient loading.

Recently, this issue has been highlighted as the primary cause of an escalation in negative consumer perceptions of aquaculture fish and are ultimately eroding the market value of end products. This lack of information is constraining efforts to understand the mechanisms of off-flavour tainting and the implementation of practices aimed at regaining consumer confidence in the quality of farmed fish. As such, it is essential for fish farm managers to have access to reliable measurement techniques for 2-MIB and GSM in water and in fish, which would allow the optimization of off-flavor depuration protocols and ultimately enhance product quality. The threshold concentration of off-flavor compound odor in water has been reported to be around 30 ng/L for 2-MIB and 10 ng/L for GSM [5-7]. Traditionally, the analysis of off-flavor compounds, as GSM and 2-MIB, in water and fish tissue requires several steps for sample preparation and concentration of the target compounds, like liquid/liquid extraction (LLE), solid phase extraction (SPE) or distillation techniques [6,8]. The approach of solid phase microextraction (SPME) is simple and fast, but it is limited by capacity of sorptive fibers [5,9]. A novel approach using stir bar sorptive extraction (SBSE), coated with polydimethylsiloxane (PDMS), recently has been wide used, as a solvent-less sample preparation method for the extraction and enrichment of organic compounds from different matrices [10,11]. After the extraction, the stir bar is removed from sample and transferred to a thermal desorption instrument where the analytes are thermally released and delivered to the GC column.

In this study, SBSE was used for the sample enrichment of 2-MIB and GSM in water and fish samples, followed by TD-GC-MS analysis. The aim of this study is to develop a sensitive, selective, and simple method determining and monitoring the level of 2-MIB and GSM at low in water and fish.

Materials and Methods

Materials

Standard solution (100 µg/mL in methanol) of 2-MIB and GSM were purchased from Sigma–Aldrich (St. Louis, MO, USA). The different concentration of standards was diluted to prepare the working

---

*Corresponding author: Eric Ruan, Lacombe Research Centre, Agriculture and Agri-Food Canada, 6000 C&E Trail, Lacombe, AB T4L 1W1, Canada
E-mail: Eric.Ruan@agr.gc.ca

standard solutions. All other chemicals were of analytical grade. The commercial stir bars [Twister™] were incorporated in a glass jacket and coated with PDMS (length: 10 mm, thickness: 0.5 mm), as well as the 10 mL vials and related equipment were obtained from Gerstel (Linthicum, MD, USA).

Samples of water and fish (sturgeon) were provided by a local fish farm. Water was collected with fish samples. Skin-off fillets were collected, vacuum sealed in individual plastic bags, and immediately frozen. Fish fillet samples arrived at the laboratory frozen in separate labelled vacuum sealed plastic bags. The fillets were thawed in a cold room (4°C) for 2–3 h before being cut into small pieces for grounding using a Mini-Prep Chopper/Grinder (Cuisinart®, Canada). Ground fish tissue (≤ 1 ± 0.05 g) was put into 10 mL amber vials with 9 mL saturated NaCl solution.

Methods

For sample extraction using SBSE, one stir bar was used for 2 h at 1000 rpm in each screw-capped vial. After extraction, the stir bar was removed with a forceps, rinsed twice with Millipore water, and dried with a lint-free tissue. Two stir bars were placed in a glass thermal desorption tube for TD–GC–MS analysis. The desorption tube was then placed in TDU, where the stir bars were thermally desorbed by programming the TDU from 40°C (held for 0.5 min) to 280°C (held for 3 min) at 240°C/min. Transfer temperature was fixed at 275°C. The desorbed compounds were Cryo focussed in the CIS 4 with a glass wool notched liner at −120 °C. After desorption, the CIS 4 was programmed from 280°C (held for 3 min) at 12°C/s to inject the trapped compounds onto the analytical column. Injection was performed in the programmable temperature vaporization (PTV) solvent vent mode, and purge flow to split vent was 36 mL/min at 1 min.

The separations were carried out on a HP-5ms fused-silica capillary column [30 m (length)×250 μm (I.D.)×0.25 μm (film thickness); Agilent Technologies, Mississauga, ON, Canada]. The oven temperature was programmed from 50°C (held for 1 min) to 150°C (held for 1 min) at 15°C/min, then to 280°C (held for 0.8 min) at 25°C/min, and total run time was 15 min. Helium was used as the carrier gas at a flow rate of 1.2 mL/min. The mass spectrometer (5975C, MSD, Agilent Technologies) was operated in the full scan mode. The used stir bars were cleaned by soaking in Milli-Q purified water for 1 h and then a mixture of methylene chloride–methanol (1:1) for 2 h. The stir bars were dried on a clean surface at room temperature for 2–4 h and reconditioned using a Tube Conditioner (TC2, Gerstel, MD, USA) at 280°C for 2 h in a flow of N_2.

Results and Discussion

SBSE is by nature an equilibrium technique and the extraction of compounds from the matrixes into the PDMS phase is controlled by the partitioning coefficients with the octanol–water distribution ($K_{o/w}$) [11]. Both the distribution coefficient ratio and phase ratio control the extraction recovery in SBSE process. High hydrophobicity of compound has high $K_{o/w}$ values with high extraction efficiency. The $\log K_{o/w}$ values of 2-MIB and GSM were 3.31 and 3.57 respectively and theoretical recoveries of 2-MIB and GSM for SBSE using smaller sample volumes are higher. Water samples were extracted by SBSE directly and fish samples (fillets) were well grounded. As shown in Figure 1, water sample (1 mL) or fish sample (1 ± 0.05 g) with 9 mL saturated NaCl solution was extracted by one stir bar with a stirring rate was 1000 rpm in 10 mL vial. In process of SBSE of GSM and 2-MIB over time, the equilibrium of extraction reached after 2 hours in our previous study [12-17]. So, a SBSE time of 2 hours was chosen for all experiments, without further tests.

The typical MS spectra and ion extraction chromatograms of 2-MIB and GSM were shown in Figure 2, obtained by 2 hours SBSE of water sample with fortified at 100 ng/L standards followed by TD–GC–MS analysis. The linearity was initially examined over a range of 0.3–100 ng/L (n=3) for 2-MIB and GSM standards, with correlation coefficients (r²>0.99), as shown in Figure 3. Limit of detections (LODs) were calculated as three times of the signal-to-noise (S/N) of blank sample and limit of quantifications (LOQs) were determined as ten times of S/N of blank. The values of LODs (n=3) and LOQs (n=3) of 2-MIB and GSM were ~0.3 ng/L and ~1 ng/L based on the main fragment ions of m/z 112 and m/z 95, respectively.

Fish fillets and water samples were analyzed using the developed method to obtain an average value for monitoring and evaluating the process of depuration. The recovery of water samples were shown in Table 1 and results of GSM and MIB in water and fish sampels were displayed in Table 2. The described SBSE–TD–GC–MS method was developed for accurate analyses of the 2-MIB and GSM off-flavor compounds in fish and water.

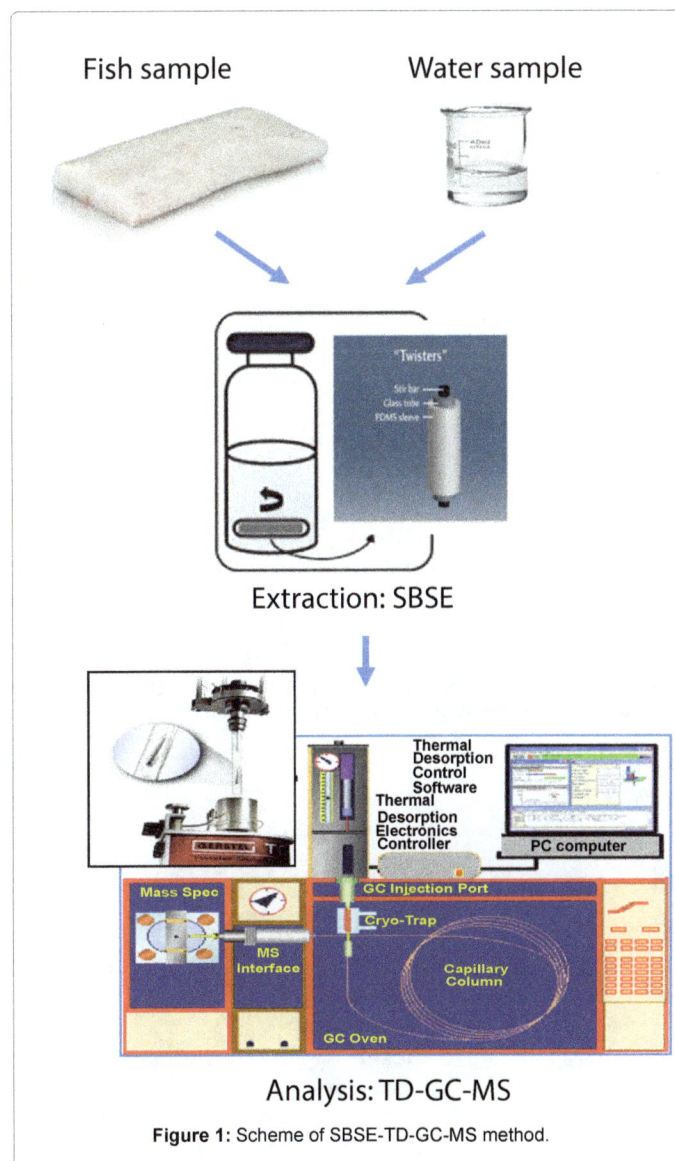

Figure 1: Scheme of SBSE-TD-GC-MS method.

	Recovery (%)	
	Spike-in 100 ng/L	Spike-in 30 ng/L
MIB	98.2 ± 5.0	93.7 ± 1.8
GSM	93.7 ± 1.0	94.0 ± 4.0

Table 1: Recovery (%) of 2-MIB and GSM in water samples using SBSE-TD-GC-MS method.

	2-MIB	GSM
Rt (min)	8.97	12.19
Water (ng/L)	16.2 ± 1.5	10.9 ± 1.1
Fish (ng/Kg)	474.6 ± 35.4	339.6 ± 32.5

Amount of off-flavor: mean (± S.E.) level of triplicates.

Table 2: Results of 2-MIB and GSM (ng/kg) determined using SBSE-TD-GC-MS method.

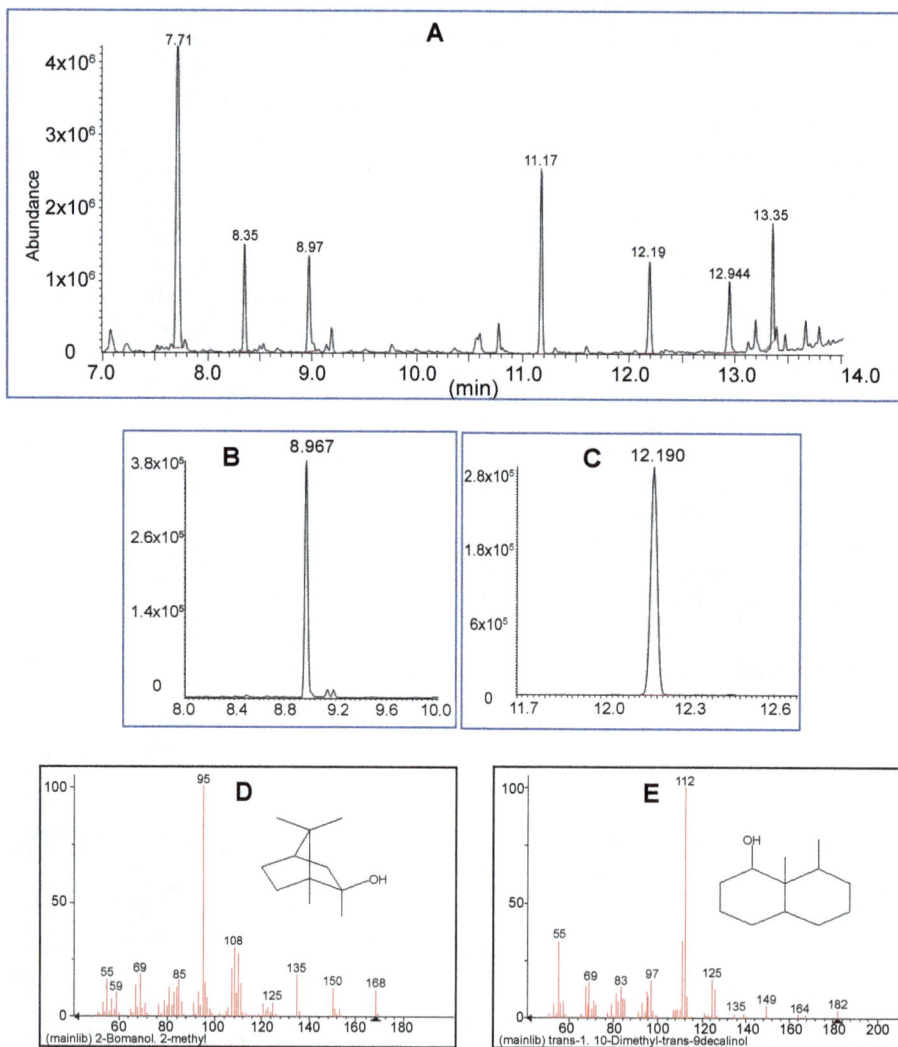

Figure 2: Typical GC-MS chromatography and spectra of GSM and 2-MIB. (A) Total ion chromatography (TIC) of water sample with 2-MIB and GSM spiked, extraction ion chromatograms (XIC) of m/z 95 for 2-MIB (B) and m/z 112 for GSM (C), spectra of 2-MIB (D) and GSM (E).

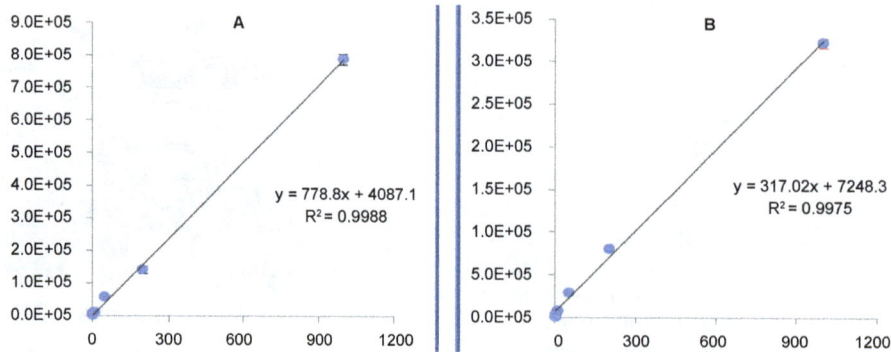

Figure 3: Standard curves of 2-MIB (A) and GSM (B).

Conclusion

The rapid, simple and solvent-less method of SBSE–TD–GC–MS can be applied to detect off-flavor compounds at very low concentration, by using significantly smaller amounts of water and fish samples, without requirement of any pre-concentration, like microwave distillation, or liquid extraction. The method can be reliably utilized by the farmed fish industry for studying accumulation and dissipation of 2-MIB and GSM in fish tissues associated with altering management practices.

Acknowledgements

Dr. E. Ruan acknowledges the receipt of a Natural Sciences and Engineering Research Council (NSERC) fellowship funded through Agriculture Agri-Food Canada (AAFC) A-base program.

References

1. Korth W, Ellis JBowmer K (1992) The stability of geosmin and MIB and their deuterated analogues in surface waters and organic solvents. Water Science and Technology 25: 115-122.

2. Auffret M, Yergeau É, Pilote A, Proulx É, Proulx D, et al. (2013) Impact of water quality on the bacterial populations and off-flavours in recirculating aquaculture systems. FEMS Microbiol Ecol 84: 235-247.

3. Kankaanpää HT, Holliday J, Schröder H, Goddard TJ, von Fister R, et al. (2005) Cyanobacteria and prawn farming in northern New South Wales, Australia--a case study on cyanobacteria diversity and hepatotoxin bioaccumulation. Toxicol Appl Pharmacol 203: 243-256.

4. Tucker CS (2000) Off-Flavor Problems in Aquaculture. Reviews in Fisheries Science 8: 45-88.

5. Beveridge M, Robertson RF, Jauncey K, Lawton LA (2005) Depuration rates and the sensory threshold concentration of geosmin responsible for earthy-musty taint in rainbow trout, Onchorhynchus mykiss. Aquaculture 245: 89-99.

6. Zhang HG, Ma SS, Li QF, Fu XJ, Zhang Y, et al. (2011) [Analysis of the changes of microbial community structure on bio-carrier of recirculating aquaculture systems (RAS)]. Huan Jing Ke Xue 32: 231-239.

7. Lin TF, Liu CL, Yang FC, Hung HW (2003) Effect of residual chlorine on the analysis of geosmin, 2-MIB and MTBE in drinking water using the SPME technique. Water Res 37: 21-26.

8. Hurlburt B, Lloyd SW, Grimm CC (2009) Comparison of analytical techniques for detection of geosmin and 2-methylisoborneol in aqueous samples. J Chromatogr Sci 47: 670-673.

9. Morales-Valle H, Silva LC, Paterson RR, Oliveira JM, Venâncio A, et al. (2010) Microextraction and Gas Chromatography/Mass Spectrometry for improved analysis of geosmin and other fungal "off" volatiles in grape juice. J Microbiol Methods 83: 48-52.

10. Ma X, Gao N, Chen B, Li Q, zhang QGu G (2007) Detection of geosmin and 2-methylisoborneol by liquid-liquid extraction-gas chromatograph mass spectrum (LLE-GCMS) and solid phase extraction-gas chromatograph mass spectrum (SPE-GCMS). Front Environment Science Engineering in China 1: 286-291.

11. Pena-Pereira F, Lavilla I, Bendicho C (2010) Liquid-phase microextraction approaches combined with atomic detection: a critical review. Anal Chim Acta 669: 1-16.

12. Jelen HH, Majcher M, Zawirska WR, Wiewiorowska M, Wasowicz E (2003) Determination of geosmin, 2-methylisoborneol, and a musty-earthy odor in wheat grain by SPME-GC-MS, profiling volatiles, and sensory analysis. J Agric Food Chem 51: 7079-7085.

13. Ochiai N (2005) Application of Stir Bar Sorptive Extraction (SBSE) Coupled to Thermal Desorption GC-MS for Determination of Ultra-Trace Level Compounds in Aqueous Samples. AppNote 5: 1-3.

14. David F, Sandra P (2007) Stir bar sorptive extraction for trace analysis. J Chromatogr A 1152: 54-69.

15. Prieto A, Basauri O, Rodil R, Usobiaga A, Fernandez LA, et al. (2010). Stir-bar sorptive extraction: A view on method optimisation, novel applications, limitations and potential solutions. J Chromatogr A 1217: 2642-2666.

16. Ruan ED, Aalhus JL, Summerfelt ST, Davidson J, Swift BJuarez M (2013) Determination of off-flavor compounds, 2-methylisoborneol and geosmin, in salmon fillets using stir bar sorptive extraction-thermal desorption coupled with gas chromatography-mass spectrometry. J Chromatogr A 1321: 133-136.

17. Davidson J, Schrader K, Swift B, Juarez M, Ruan E, et al. (2014) The effects of water aeration media and hydrogen peroxide disinfection of depuration systems on off-flavor removal from Atlantic salmon, Salmo salar cultured in recirculating aquaculture systems. Aquacultural Engineering.

33

Volatile Elements of Coconut Toddy (*Cocos Nucifera*) by Gas Chromatography – Mass Spectrometry

Karthikeyan R*, Suresh Kumar K, Singaravadivel K and Alagusundaram K

Department of Food Microbiology, Indian Institute of Crop Processing Technology, Ministry of Food Processing Industries, Govt. of India, Thanjavur (TN), India

Abstract

Volatile components of coconut toddy were analyzed using GC Clarus 500 Perkin Elmer (Germany)-MS and identified by matching the mass spectra obtained with those present in the "NIST 2005" Library. Totally Thirty one volatile components were identified and were different quantities amounts of volatile components. The highest peak of volatile components in the fresh toddy observed was Lupeol and Squalane.

Keywords: Volatile components; Coconut toddy; *Cocos nucifera*

Introduction

Toddy is one of the traditional, social and local drink extracted from either a coconut tree or a Palm tree Spadix. In the states of Tamil Nadu and Kerala, it is known as "Kallu", Indonesia known as nira, Sri Lanka as toddy, Thailand as maprau, North Africa as lagbi. The fermented beverages are those which are produced from fruit juice or plant sap by natural fermentation. In India, wild dates (Phoenix sylvestris), coconut palm (*Cocos nucifera*), palmyara (*Borassus flabellifer*) etc. are frequently used for this purpose. Toddy contains 4-5% of alcohol [1-3].

The methods of tapping the coconut palm are very ancient. When the palm has reached the normal bearing stage, every leaf axis produces a spadix or inflorescence. This unopened flower (spathe) is prepared by slightly brushing it with gentle taps of small mallet. To prevent it from opening, the spathe is slightly bound round with fibre. When the spathe is nearly ready to produce toddy, which is after about three weeks, about two to three inches is cut from the end. During the preparation the spathe is gradually bend over, so that by the time toddy flows, a receptacle can be placed at the end for its collection. The flow of juice increases gradually and the pot should be changed twice daily, at the same time shaving a thin slice from the end of the spathe, tapping slightly with the mallete and smearing on a mixture of bruised leaves for stimulating the flow of toddy from the spathe [1]. By the time collection of the sap is over, fermentation sets-in immediately and the sap is converted into fresh toddy or sweet toddy. This product is a milky white and effervescent.

Fresh toddy is a sweet, oyster white and translucent fluid and it is considered to be as pure as mothers' milk and traditionally believed that it has many medicinal properties and a refreshing health drink. If it is kept undisturbed, fermentation continues, which results in the production of acetic acid. A characteristic smell in the fresh toddy changed to fermented acid smell. To collect systematic data available on the volatile chemical and biological composition of toddy this study was conduct.

Materials and Methods

Materials

Toddy was obtained from the coconut plantation in Thanjavur, Tamil Nadu, India. The samples were taken at random after morning and afternoon tapping.

Methods

50 ml of toddy sample was taken into the separating funnel and 50 ml of organic solvent was added to same. The flask was stoppered tightly and shaken well. The gas produced inside the separating funnel was released at successive intervals by opening the knob. The shaking was done until there was less pressure inside the flask. The flask was kept undisturbed for 10 minutes in a stand. Separation of two layers inside the flask was observed. The aqueous layer was drained out. To the organic layer 25 ml of the organic solvent was added. Then shaking and draining was done as such in the first step. The organic layer was concentrated to about 1–2 ml by giving nitrogen flushing to the sample. Then it was filtered through a Whatmann filter No. 4 paper containing sodium sulphate to get clear solution. Then it was injected into the GC-MS for analysis [3].

The GC-MS analyses of the extracts were performed using a set of GC-MS, GC Clarus 500 Perkin Elmer (Germany) with Elite-5MS (5% Diphenyl/95% Dimethyl poly siloxane), 30×0.25 mm×0.25 µm df column of 110°C -2 min hold at temperatures. The injector temperature was 250°C with mode split 10:1. Helium was used as a carrier gas at a pressure of 12 psi and the ion source working in electron impact(EI) mode at 70 eV was held about 200°C [3,4].

The identification was done by matching the mass spectra obtained with those present in the "NIST 2005" Library. The internal standard used for quantification of volatile using GC-MS was 1, 4–dichlorobenzene [5,6].

A method was standardized using different organic solvents viz., Chloroform Dichloromethane, Diethyl ether, chloroform, for the extraction of phytochemical compounds present in the Toddy sample. The extract from diethyl ether gave few expected compounds like Ethyl hydrogen succinate, Hydroquinone, Phenyl ethyl Alcohol, 2,4,6,8-Tetraazabicyclo [3.3.0] octan-3-one, 7-nitroimino, Dibutyl phthalate, Oleic acid, Sqyalene and n-Hexadecanoic. The extract

*Corresponding author: Karthikeyan R, Department of Food Microbiology, Indian Institute of Crop Processing Technology, Ministry of Food Processing Industries, Govt. of India, Thanjavur-613 005 (TN), India, E-mail: krbiotech@gmail.com

S.No	RT	Name of the compound	Molecular Formula	MW	Peak Area %
1	6.05	Ethyl hydrogen succinate	$C_6H_{10}O_4$	146	32.44
2	9.19	2-Buten-1-ol, propanoate	$C_7H_{12}O_2$	128	0.37
3	11.63	Dichloroacetic acid, 2,2-dimethylpropyl ester	$C_7H_{12}Cl_2O_2$	198	0.46
4	14.25	3,4-Hexanediol, 2,5-dimethyl	$C_8H_{18}O_2$	146	0.46
5	15.08	1,2-Benzenedicarboxylic acid, diheptyl ester	$C_{22}H_{34}O_4$	362	0.93
6	16.43	1,2-Benzenedicarboxylic acid, butyl octyl ester	$C_{20}H_{30}O_4$	334	1.85
7	16.57	Nonanoic acid	$C_9H_{18}O_2$	158	4.63
8	17.16	3-Pentanol, 2-chloro-4-methyl-, (R*,S*)-(ñ)-	$C_6H_{13}ClO$	136	1.39
9	18.58	3-Pentanol, 2,3-dimethyl-	$C_7H_{16}O$	116	1.85
10	19.26	Pentanoic acid, 10-undecenyl ester	$C_{16}H_{30}O_2$	254	18.54
11	25.19	Di-n-octyl phthalate	$C_{24}H_{38}O_4$	390	23.17
12	29.58	1-Monolinoleoylglycerol trimethylsilyl ether	$C_{27}H_{54}O_4Si_2$	498	13.90

Table 1: Compounds identified by GC-MS in Dichloromethane extract.

S.No	RT	Name of the compound	Molecular Formula	MW	Peak Area %
1	6.07	Aminocyanoacetic acid	$C_3H_4N_2O_2$	100	0.41
2	10.60	Butanoic acid, 2-oxo-, methyl ester	$C_5H_8O_3$	116	0.00
3	14.11	à-D-Mannopyranoside, methyl 3,6-anhydro-	$C_7H_{12}O_5$	176	0.00
4	16.56	Propanedioic acid, propyl-	$C_6H_{10}O_4$	146	0.51
5	19.24	2,3-Epoxyhexanol	$C_6H_{12}O_2$	116	1.02
6	25.20	1,2-Benzenedicarboxylic acid, diisooctyl ester	$C_{24}H_{38}O_4$	390	3.05
7	29.63	Squalene	$C_{30}H_{50}$	410	41.16
8	33.91	Lupeol	$C_{30}H_{50}O$	426	53.86

Table 2: Compounds identified by GC-MS in Chloroform extract.

S.No	RT	Name of the compound	Molecular Formula	MW	Peak Area %
1	2.74	2,4,6,8-Tetraazabicyclo[3.3.0] octan-3-one, 7-nitroimino	$C_4H_6N_6O_3$	186	16.31
2	3.59	Hexanoic acid	$C_6H_{12}O_2$	116	2.95
3	5.38	Phenylethyl Alcohol	$C_8H_{10}O$	122	12.67
4	6.12	Ethyl hydrogen succinate	$C_6H_{10}O_4$	146	33.61
5	7.69	Hydroquinone	$C_6H_6O_2$	110	34.35
6	8.73	n-Decanoic acid	$C_{10}H_{20}O_2$	172	0.35
7	16.42	Dibutyl phthalate	$C_{16}H_{22}O_4$	278	0.43
8	16.56	n-Hexadecanoic acid	$C_{16}H_{32}O_2$	256	1.20
9	19.27	Oleic Acid	$C_{18}H_{34}O_2$	282	4.53
10	23.91	Squalene	$C_{30}H_{50}$	410	1.31
11	25.22	1,2-Benzenedicarboxylic acid, mono(2-ethylhexyl) ester	$C_{16}H_{22}O_4$	278	0.59

Table 3: Compounds identified by GC-MS in Diethyl ether extract.

from chloroform revealed the presence of the following compounds like Aminocyanoacetic acid, Lupeol, Squalene. Dichloromethane extract gave few compounds like Ethyl hydrogen succinate, Di-n-octyl phthalate and Non-anoic acid etc. When the solvents like Acetone and Acetonitrile were used for standardization, both the solvents were miscible with the toddy samples taken for analysis (Table 1). Since the density of the solvents and Toddy samples were found to be close to each other [7-10].

Results

The major volatile components identified in the fresh toddy were Lupeol and Squalane (Table 2). Propanedioic acid was found in coconut toddy whilst Di-n-octyl phthalate was found in smaller amounts. In Phenyl ethyl alcohol (Figure 1) and 2,4,6,8-Tetraazabicyclo (3.3.0) octan-3-one, 7-nitrimino was volatile components that were found only in fresh coconut toddy the results are shown in Table 3. Thirty one volatile components were identified in coconut toddy from, Thanjavur, Tamil Nadu, India.

Discussion

In this study, only 12 volatile components were identified, whilst Apriyanton et al. [6] reported that around 70 volatile components were identified in coconut toddy.

Toddy of coconut contains a small amount of protein, fat, minerals and vitamins as well as sugar components. These could possibly interact during processing and form some of the volatile components and non-enzymatic browning intermediates as well as Maillard products. (2,5,6,8,9), found glutamic acid, threonine, aspartic acid and serine as major amino acids in fresh sap, whilst proline, methionine, triptophane, and histidine were 4.00 mg/100 g fresh sap [10] reported that toddy of coconut contain sucrose, ash, protein, vitamin C and acids, such as succinic acid, and citric acid (Figure 2).

Conclusion

The volatile components identified in coconut toddy were Lupeol

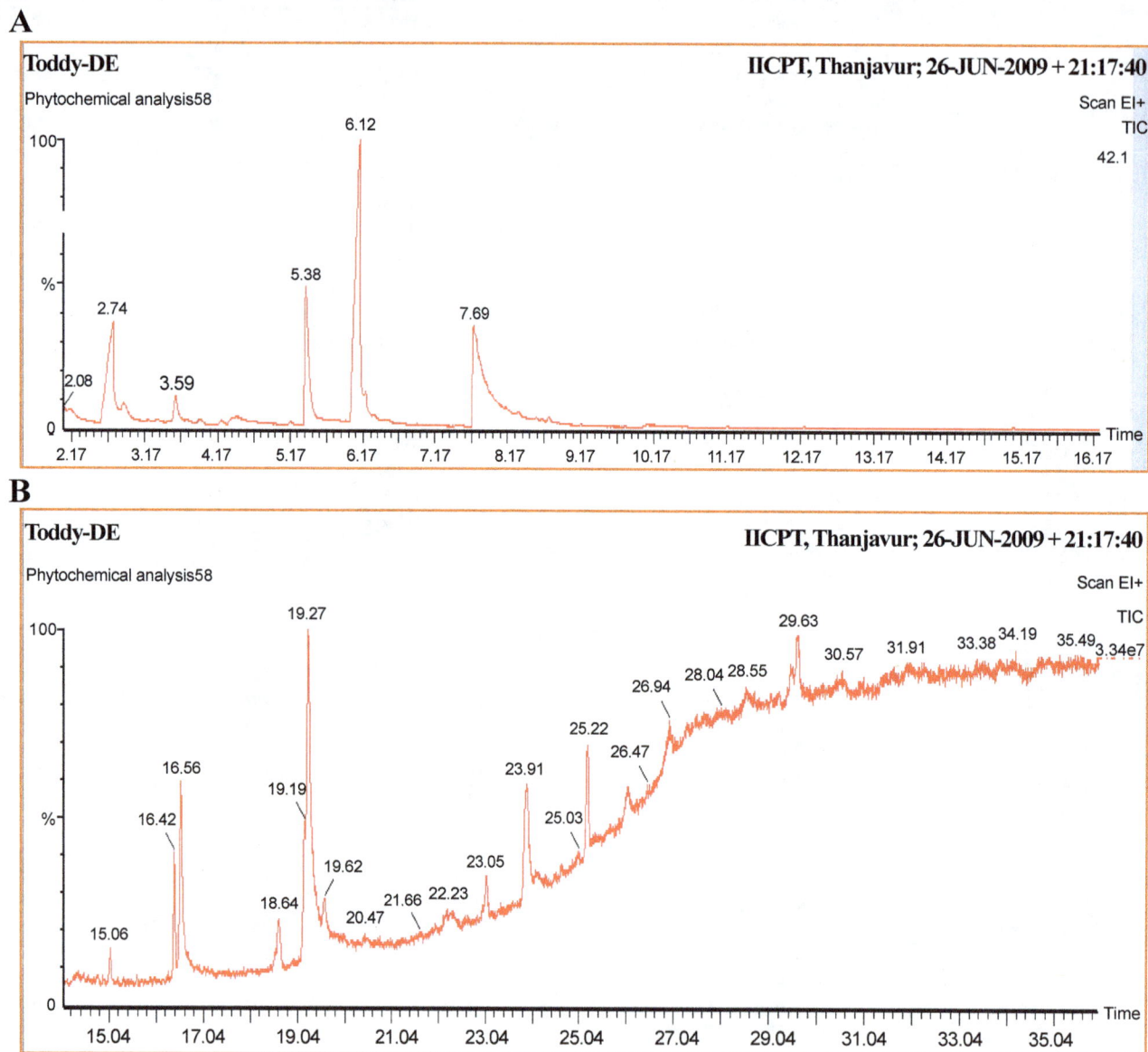

Figure 1: Using Diethyl ether extract.

(Figure 3) and Squalane. The results are shown in Table 2. There were 31 volatile components identified in coconut toddy from this constituency.

Acknowledgement

The authors are grateful to National Agricultural Innovation Project and Indian Council of Agricultural Research for the funds provided. Also thanks are due to K. Alagusundaram, Director, Indian Institute of Crop Processing Technology, Thanjavur for providing all the facilities, encouragement and support used to carry out the work. Also thanks are due to Dr.K.Singaravadivel, Principal Scientist, Department of Food microbiology, Indian Institute of Crop Processing Technology, Thanjavur for useful discussions.

References

1. Child R (1972) Toddy products. In Coconuts. (2nd edition) 296-306. Ed. By R. Child., Logman, London.

2. Hortwitz W (1970) Beverages Distilled liquors In AOAC, Ed By Hortwitz, Wasington, USA 144-147.

3. Shamala TR, Srikantiah KR (1988) Microbial and biochemical studies on traditional indian palm wine fermentation. Food Microbiology5: 157-162.

4. Alli L, Bourque J, Metussin R, Liang R, Yaylalayan (1990) Identification of pyrazines in mapple syrup. Journal of Agriculture Food Chemistry 38: 1242-1244.

5. Akochi EK, Alli I, Kernasha S (1997) Characterization of the pyrazines formed during the processing of mapple syrup. J Agric Food Chem 45: 3368-3373.

6. Jatmika A, Mahlil-Hamzah A, Siahaan D (1990) Alternative processed coconut sap product) Bulletin Kelapa Manggar 3:37.

7. https://getinfo.de/app/Analysis-of-Volatiles-of-Kecap-Manis-a-Typical/id/BLCP%3ACN018073215

8. Hori K, Somoda J, Suryoseputro S, Purboyo RBRA, Purnomo H (2001)

A

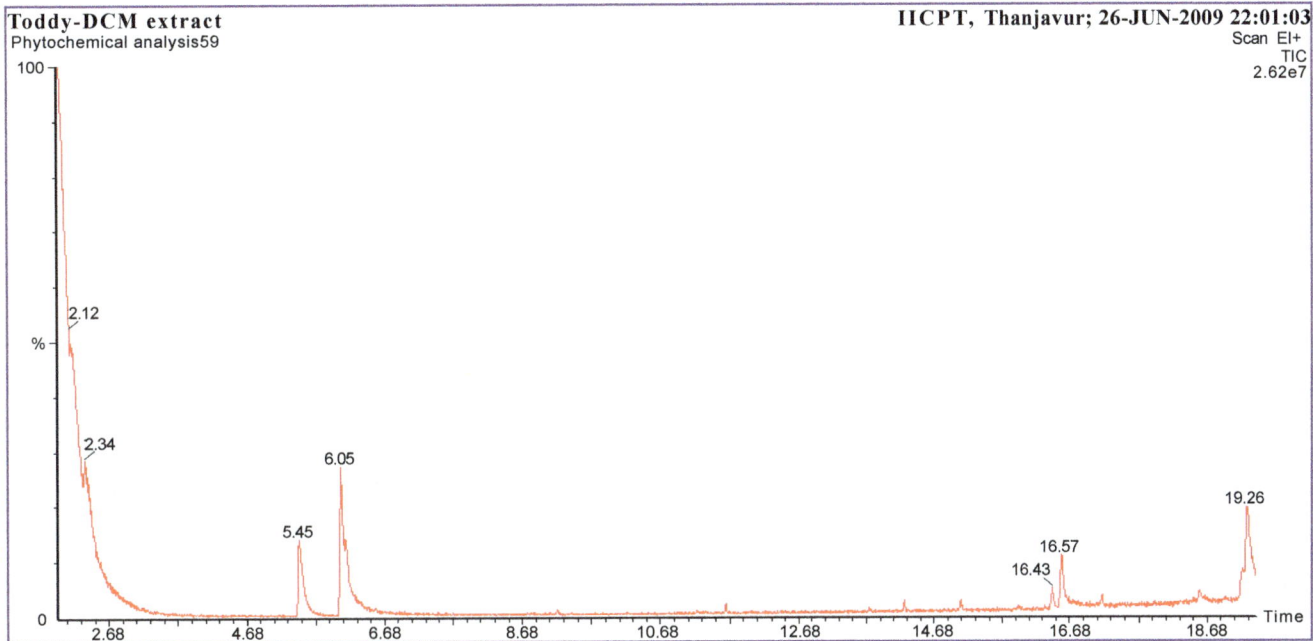

B

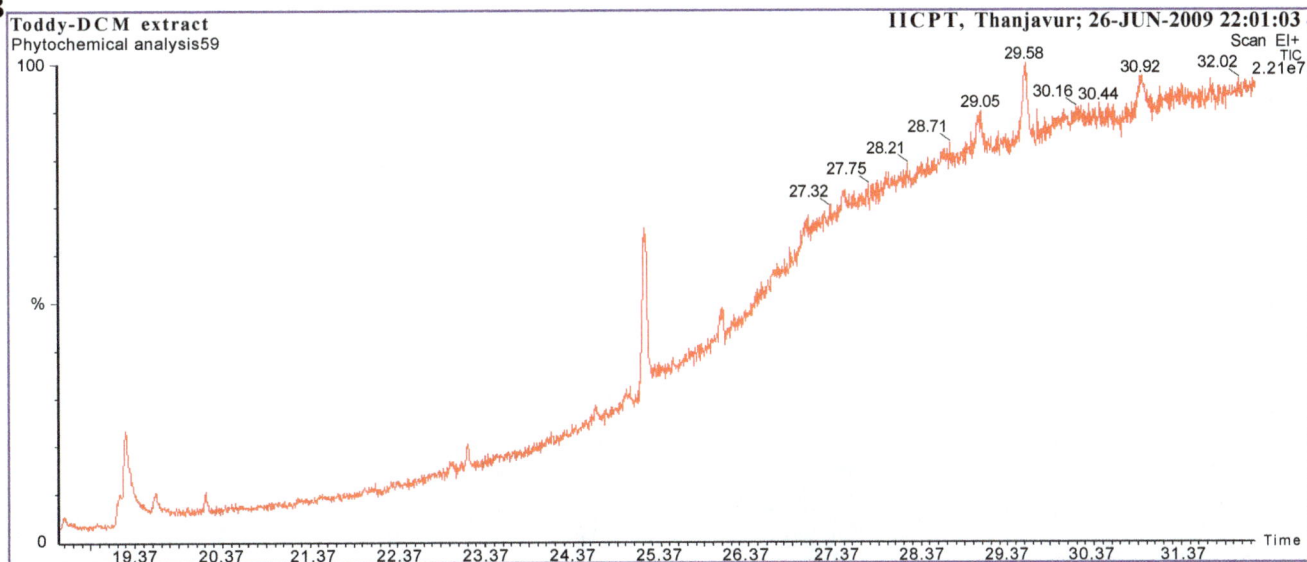

Figure 2: Using Dichloromethane extract.

A

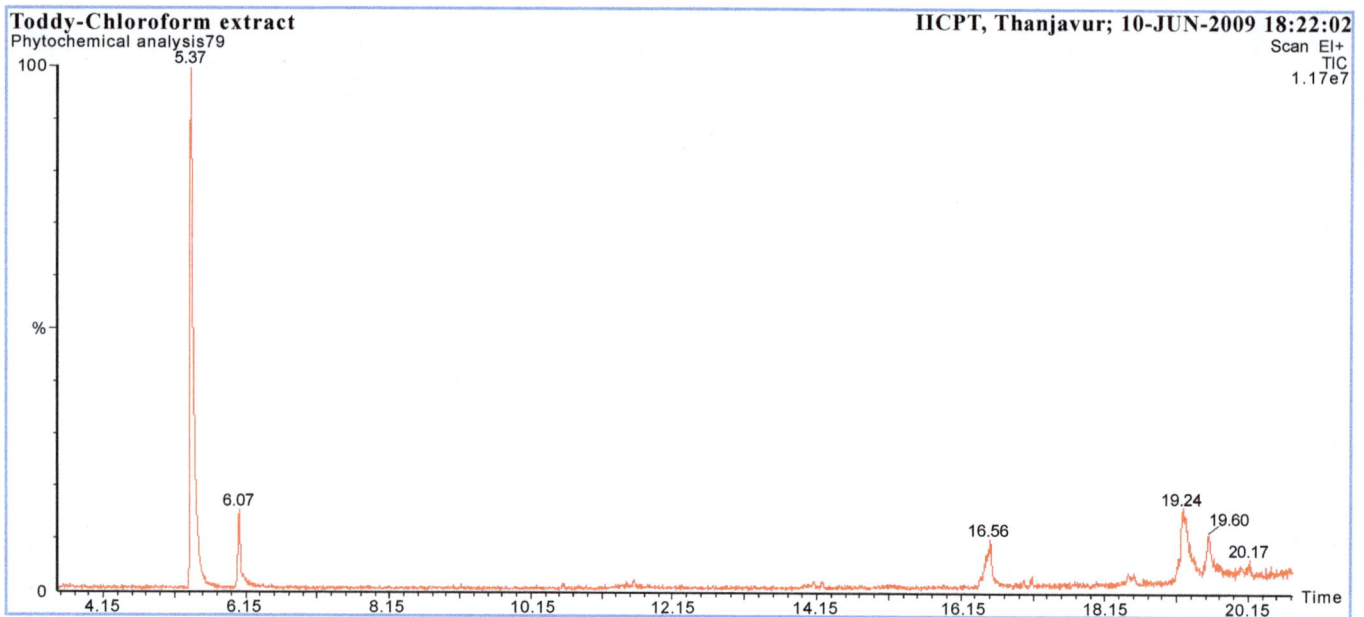

B

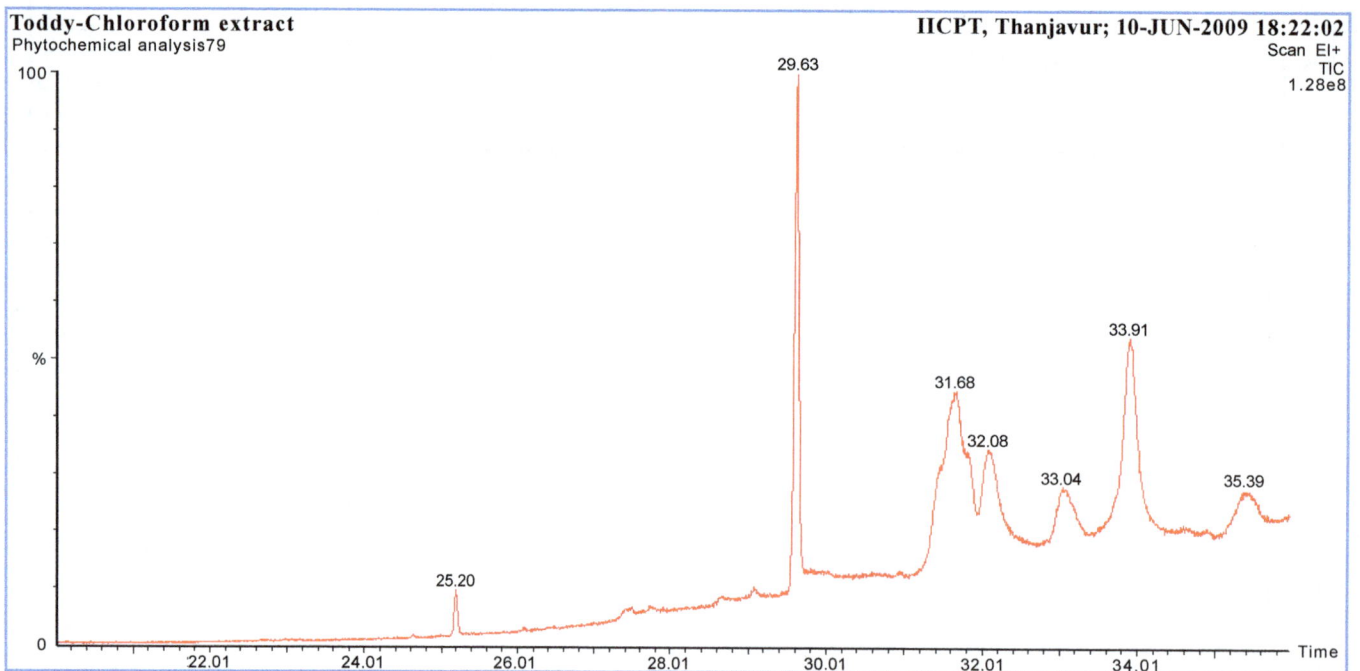

Figure 3: Using Chloroform extract.

Utilization of/and preference for palm sugar by Indonesian and Japanese panelists. Journal of Asian Regional Association for Home Economics8: 180-185.

9. Hori K, Suryoseputro S, Purnomo H, Foe K, Hashimura F (2001). Indigenous technology of coconut sugar production in the village of Genteng, Banyuwangi (East Jawa) and Dawan, Klungkung (Bali) and the knowledge about palm sugar

by Japanese young people. Bulletin of Fukuoka University of Education 50: 109-118.

10. Itoh T, Matsuyama A, Widjaya CH, Nasution MZ, Kumendong J (1982) Compositional of nira palm juice of high sugar content from palm tree. Proceedings of IPB-JICA International Symposium on Agricultural productand processing technology.

34

Evaluation of Population and Individual Variances of Urinary Phthalate Metabolites in terms of Epidemiological Studies

Ralph Feltens[1], Stefan Roeder[2], Wolfgang Otto[3], Michael Borte[4], Irina Lehmann[2], Martin von Bergen[1,3,5] and Dirk K. Wissenbach[1]*

[1]Department of Metabolomics, Helmholtz Centre for Environmental Research-UFZ, Leipzig, Germany
[2]Department of Environmental Immunology, Helmholtz Centre for Environmental Research-UFZ, Leipzig, Germany
[3]Department of Proteomics, Helmholtz Centre for Environmental Research-UFZ, Leipzig, Germany
[4]Children's Hospital, Municipal Hospital St. Georg Leipzig, Affiliated to the University of Leipzig, Germany
[5]Department of Biotechnology, Chemistry and Environmental Engineering Aalborg University, Aalborg, Denmark

Abstract

In the context of an epidemiological study, urinary concentrations of nine phthalic diester metabolites (monoethyl-, mono-(3-carboxypropyl)-, mono-n-butyl-, monoisobutyl-, monobenzyl-, mono-(2-ethylhexyl)-, mono-(5-hydroxy-2-ethylhexyl)-, mono-(5-oxo-2-ethylhexyl)- and mono-(5-carboxy-2-ethylpentyl)-phthalate) were quantified via LC-MS/MS. As in the majority of epidemiological studies only single spot samples were available for urine analysis, the implicit assumption in this case is, that exposure data obtained from single spot samples are representative for a longer exposure period. To validate the relevance of single spot analyses we quantified the respective intra-individual variances of urine samples collected from ten volunteers once daily over a period of 30 days. Using the values for the daily variances, approximate values for the underlying population variances in the cohort samples representing the differences between the average individual metabolite levels were calculated. For most of the volunteers, daily metabolites variations were lower, than the variations observed in the epidemiological setup. The results showed that by accounting for the contribution of daily variance, the standard deviations of the log-transformed phthalate values of the cohort samples are reduced (14% to 28%) but still larger (3% to 66%) than daily standard deviation values, with the exception of MCPrP concentrations.

Keywords: Endocrine disruptors; Phthalates; Metabolites; Epidemiology; Variance; Individual

Introduction

Phthalic diesters (phthalates) are a group of organic compounds that are produced at a scale of millions of tons per year. They have been in use for several decades now as plasticizers and additives in a large variety of consumer products. Since they are not covalently bound to other compounds within their respective formulations, significant amounts of these chemicals have accumulated in the environment and exposure is ubiquitous. For di-(2-ethylhexyl)-phthalate, diet is thought to be the main source of exposure, but non-dietary pathways can also be substantial, as has been shown for other phthalates [1,2].

Over the past 15 years, animal studies have shown the potential of several phthalates to act as endocrine disruptors in an estrogenic or anti-androgenic fashion, raising concerns over their health impacts [3-6]. However, compared with the large body of experimental evidence suggesting reproductive or developmental toxicity of phthalates, human data have been rather limited [7-9]. More recently, phthalates have been proposed to act as obesogens, endocrine disruptors which induce or contribute to adipogenesis and the development of obesity [10,11]. In this context, cell culture and animal studies have established the ability of mono-(2-ethylhexyl)-phthalate to selectively bind to peroxisome proliferator-activated receptors (PPARs) which belong to the nuclear receptor superfamily of ligand-activated transcription factors and to modulate their functions [12,13]. PPARα and PPARγ are expressed in adipocytes and play essential roles in the regulation of cellular differentiation, development, and metabolism.

Quantification of phthalate metabolites in urine is a good measure of exposure since urinary enzymatic activity is negligible and metabolites thus generally reflect internal exposure rather than contaminants introduced during sample collection and processing [14]. However, hydrolytic genesis of simple monoesters from the ubiquitous diesters seems to occur to a limited extent also via non-biogenic processes, and oxidatively transformed monoesters, if actually produced in significant amounts, represent the more reliable metabolites [15]. There have been several studies showing that phthalic diesters are metabolized and excreted fairly quickly with biologic half-lives of about ten hours and urinary phthalate levels have been shown to vary, probably in synchrony with the intermittent incorporation of their parent compounds during food uptake or through the use of personal care products [1,16-18].

Between 2006 and 2008, pregnant woman were recruited for the German Lifestyle and environmental factors and their Influence on Newborns Allergy risk (LINA) cohort study with the aim to investigate environmental influences during pregnancy on health risks of the child, in particular asthma and allergic diseases [19]. Since urinary phthalate metabolites are among the exposure parameters to be quantified, the question arose to what extent values obtained from individual spot samples at the 36th week of pregnancy constitutes a representative measure for exposure during the whole pregnancy. Several studies focused on the variability of phthalate and metabolites in human urine even for gestation periods [20-26]. However still less is known about the common approach to use a single spot urine analysis for correlation to health outcomes in the context of epidemiological studies. In order to obtain more information on this, phthalate metabolite levels in urine samples from more than 600 women in the 36th week of pregnancy

*Corresponding author: Dirk K. Wissenbach, Department of Metabolomics, Helmholtz Centre for Environmental Research-UFZ, Leipzig, Germany
E-mail: dirk.wissenbach@ufz.de

were quantified and compared to day-to-day variability observed in the urine samples of ten volunteers over a period of 30 days. Aim of this study was I. to describe the interday variability and II. To evaluate the possibility to use single spot analysis of those phthalate metabolites for epidemiological studies.

Materials and Methods

Materials and reagents

N-(2-acetamido)-iminodiacetic acid (ADA) and formic acid (analytical grade / mass spec grade) were obtained from Fluka (Taufkirchen, Germany), acetic acid from Biosolve (Valkenswaard, The Netherlands). Acetonitrile and methanol (picograde) were obtained from LGC Promochem (Wesel, Gemany), methanol (LC-MS grade) from Merck (Darmstadt, Germany). Ultrapure water was produced by a MiliporeQ (MerckMillipore, Darmstadt, Germany) water purification system. Isolute C18 SPE 200 mg/3 ml columns were obtained from Biotage (Düsseldorf, Germany), autosampler vials from Wicom (Heppenheim, Germany) and deglucuronidase / arylsulfatase (*helix pomatiae*) from Roche (Mannheim, Germany). $^{13}C_4$-labeled and unlabeled standards for monoethyl-, mono-(3-carboxypropyl)-, mono-n-butyl-, monobenzyl-, mono-(2-ethylhexyl)-, mono-(5-hydroxy-2-ethylhexyl)-, mono-(5-oxo-2-ethylhexyl)- and mono-(5-carboxy-2-ethylpentyl)-phthalate (MEP, MCPrP, MnBP, MBzP, MEHP, MEHHP, MEOHP and MECPP, respectively), $^{13}C_4$-labeled methylumbelliferone (MeUmb) and unlabeled mono-isobutylphthalate (MiBP) were purchased from LGC (Wesel, Germany). Unlabeled MeUmb was obtained from Aldrich (Steinheim, Germany), methylumbelliferone glucuronide (MeUmb-gluc) from Sigma-Aldrich (Seelze, Germany). Lyophilised ClinChek® control urine samples (level I and II) were purchased from Recipe Chemicals+Instruments (Munich, Germany).

Description of the study population

Six-hundred and twenty-two pregnant woman were recruited within the prospective mother-child-study LINA (Lifestyle and environmental factors and their Influence on Newborns Allergy risk) from March 2006 until December 2008 in Leipzig, Germany [27,28]. Participation in this study was voluntary and informed consent was given. This study was approved by the Ethics Committee of the University of Leipzig, Germany (file reference 046-2006, 160-2008). All participants agreed to the usage of the results in the context of the LINA study.

Urine samples and sample preparation

Within the LINA study maternal urine samples were collected during the 36th week of pregnancy. 300 additional longitudinal urine samples were collected from ten volunteers (25 to 49 years old; four male, six female) who gave informed consent. Samples and sample aliquots were stored separately at -80°C until further analysis.

Phthalate quantification was carried out for 610 samples of the LINA cohort and all 300 samples of the longitudinal study. For this, 100 µl aliquots were buffered with 150 mM Na·ADA, pH 6.6, spiked with 25 µl 50% methanol containing 2 ng each of $^{13}C_4$-labelled MEP, MCPrP, MnBP, MBzP, MEHP, MEHHP, MEOHP, MECPP, MeUmb and unlabeled MeUmb-gluc, and incubated for 5 hr at 37°C with 3.5 µl deglucuronidase/arylsulfatase in a total volume of 210 µl to achieve deglucuronation of phthalic monoesters. Reactions were stopped by addition of 510 µl 700 mM formic acid (FA) with 47 mM acetic acid. Purification of samples was performed via solid phase extraction (SPE) on Isolute C18 columns via gravity flow. After equilibration of the columns using 3 ml 100% methanol, followed by 3 ml of 2% FA in 5% methanol, samples were applied, followed by two washing steps of 3

ml 2% FA in 10% methanol. Analytes were eluted in 1.5 ml of 2% FA in 90% methanol, vacuum-dried at 45°C, resuspended in 100 µl 35% acetonitrile and transferred to insert-containing auto sampler vials, which were crimp-sealed.

With the preparation of each batch of about 40 samples, aliquots of three quality control urine samples (two commercially obtained, one self-prepared by spiking of unlabeled standards into an arbitrarily chosen urine sample at 20 ng/ml) were included in order to ensure correctness of the results obtained for each batch.

LC-MS/MS

10 µl aliquots were separated on a Dionex™ UltiMate™ 3000 UPLC System (Thermo Scientific, MA, USA) using a sigmoidal 21 min gradient from 5% to 100% methanol (eluent B) with 0.01% FA (Supplementary Table 1). Separation was achieved via reversed phase chromatography on an Acquity UPLC BEH C18, 1.7 µm, 2.1 mm × 100 mm column (Waters, Eschborn, Germany) at 50°C and 200 µl/min. Detection and quantification of analytes after elution and post-column infusion (50 µl/min in 0.2% NH$_3$) was achieved on a Q-Trap 5500 triple quadrupole mass spectrometer (AB Sciex, Framingham, MA, USA) with electrospray ionisation at 350°C and 4500 V in negative mode via previously established, scheduled SRMs (Supplementary Table 2). Between samples, the column was washed for 5.5 min with 100% methanol, 0.01% FA and an injection of 100 µl isopropanol/acetonitrile/acetone (45:45:10), followed by re-equilibration in 5% methanol, 0.01% FA. Absolute concentrations of phthalate analytes and the deglucuronation standard MeUmb were calculated with respect to the known, spiked-in concentrations of the isotopically labelled standards and previously obtained calibration curves, using dedicated software (Analyst, AB Sciex). Values for samples with concentrations below LOQ (23 and 43 instances for MCPrP and MEP, respectively) were set to LOQ/2 for data treatment. An overview of the covered analytes, the used internal standards, and calibration range including LOQ data is given in Table 1.

Finally, all metabolite concentrations were normalized against urinary creatinine concentrations. All samples were analyzed once with the described LC-MS/MS procedure.

Data analysis

Data analyses including numerical simulations was done using Excel, Past (version 3.0.1) or R (version 3.1.1) using RStudio (version 0.98.977) with packages xlsx, reshape2 and ggplot2. Graphics were prepared using Statistica (version 12), Excel or R. Before pooling the values of the ten volunteers from the longitudinal study into a single dataset, for each metabolite the concentration values of an individual were normalized by multiplication with k (k=median$_{total}$/median$_{individual}$).

Analyte	Internal Standard	Calibration Range [µg/l]	LOQ [µg/l]
MeUmb	13C4-MeUmb	10-100	10
MEP	13C4-MEP	0.315-1000	0.315
MiBP	13C4-MnBP	3.15-1000	3.15
MnBP	13C4-MnBP	3.15-1000	3.15
MCPrP	13C4-MCPrP	0.315-1000	0.315
MBzP	13C4-MBzP	0.315-1000	0.315
MEHP	13C4-MEHP	0.315-315	0.315
MEHHP	13C4-MEHHP	1.0-1000	1.0
MEOHP	13C4-MEOHP	0.315-315	0.315
MECPP	13C4-MECPP	0.315-315	0.315
BPA	13C12-BPA	3.15-1000	3.15

Table 1: Overview covered analytes, used internal standard, calibration range [µg/l] and limit of quantification (LOQ) [µg/l].

	measure	MEP	MiBP	MnBP	MCPrP	MBzP	MEHP	MEHHP	MEOHP	MECPP
Cohort	Median	3.88	4.18	4.66	0.37	1.90	2.05	2.60	2.27	2.49
	μ_{cohort}	3.91	4.21	4.64	0.38	1.89	2.09	2.60	2.27	2.49
	Std. Dev.	1.12	0.65	0.65	0.77	0.84	0.72	0.82	0.80	0.71
	sd_{cohort}	1.10	0.54	0.57	0.52	0.81	0.64	0.74	0.71	0.62
Longitudinal study	Median	3.20	3.25	3.89	0.18	1.63	1.36	2.03	1.43	1.71
	μ_{TS}	3.22	3.24	3.91	0.21	1.64	1.36	2.06	1.46	1.73
	Std. Dev.	0.88	0.34	0.36	0.68	0.73	0.64	0.74	0.72	0.67
	sd_{TS}	0.61	0.28	0.31	0.42	0.46	0.42	0.48	0.48	0.43
Cohort$_{Population}$	μ_{Pop}	4.09	4.25	4.69	0.47	2.00	2.18	2.72	2.39	2.58
	sd_{Pop}	0.91	0.46	0.47	0.31	0.66	0.48	0.56	0.53	0.45

Table 2: Medians, means (μ) and standard deviations (sd) of phthalate metabolite concentration (ln-values) distributions. Means and standard deviations designated as μ_{cohort}, μ_{TS}, sd_{cohort} and sd_{TS} were obtained via iterative nonlinear least-squares (nls) regression. Means and standard deviations for the cohort (μ_{Pop} and sd_{Pop}) accounting for the daily variation (representing population variance only) were calculated from μ and sd-values obtained of the cohort (representing population plus daily variance) and the sd-value of the longitudinal study cohort (sd_{TS}, representing daily variance only) as described in Materials and Methods.

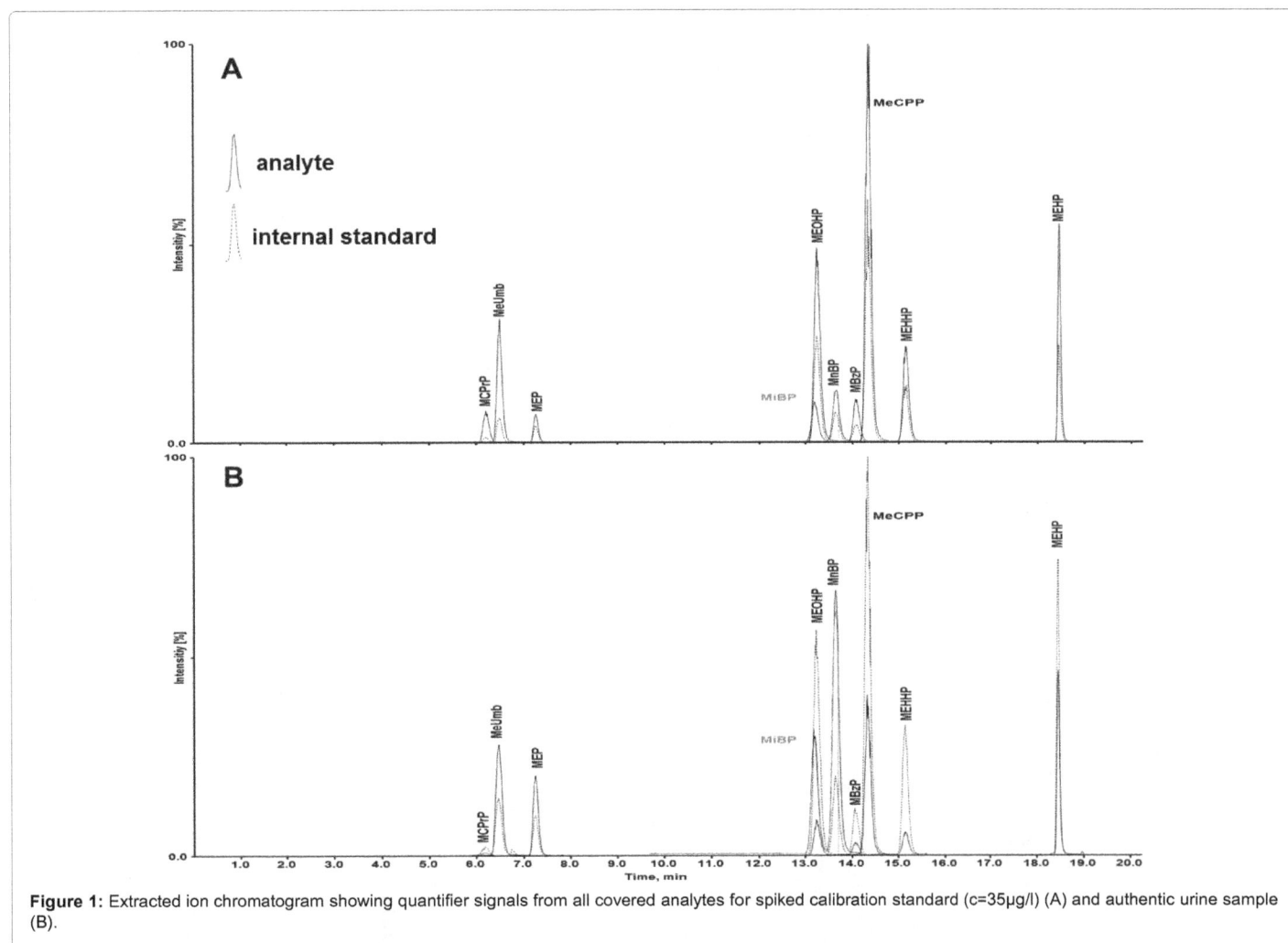

Figure 1: Extracted ion chromatogram showing quantifier signals from all covered analytes for spiked calibration standard (c=35µg/l) (A) and authentic urine sample (B).

Concentration values of zero were excluded from analyses based on logarithmically transformed data. Means and variances for the empirical cumulative distributions of the log-transformed concentration values were calculated using the iterative nonlinear least-squares (nls) curve fitting function for nonlinear models in R. From the obtained values for each metabolite, the mean and standard deviation (μ_{Pop} and sd_{Pop}) of the "true" distribution (i.e., a distribution being based on average individual values) underlying the observed LINA cohort distribution was calculated as $sd_{pop} = \sqrt{sd_{cohort}^2 - sd_{TS}^2}$

(Formula 1) and $\mu_{pop} = \mu_{cohort} + \frac{1}{2}sd_{TS}^2$ (Formula 2), with sd_{cohort} and μ_{cohort} being the standard deviation and mean from the cohort study and sd_{TS} the standard deviation of the longitudinal study.

Results

Good chromatographic separation was obtained for all the metabolites as shown in Figure 1(A) for a methanolic calibration sample as well in an authentic urine matrix as given in Figure 1(B). In addition low LOQ and good linearity were achieved as given in Table 1. High selectivity of the MS method was observed by the comparison of a methanolic standard (Figure 1(A)) and an authentic urine sample (Figure 1(B)).

In order to decide whether urinary creatinine concentrations should be used to normalize urinary phthalate metabolite levels, Spearman correlation analyses were carried out for datasets of the cohort and the longitudinal study. For both datasets strong and highly significant correlations (0.30 to 0.67; $p_{uncorr.} < 1.6 \times 10^{-7}$) were found between creatinine and phthalate concentrations. However, when phthalate concentrations were normalized against their respective creatinine concentrations, correlations were drastically reduced for all metabolites (Supplementary Figure 1). For most metabolites, remaining correlations were weak and statistically not significant ($p_{uncorr.} < 0.05$). For metabolites MiBP, MnBP, MEHP and, in the case of the longitudinal study, MCPrP, however, normalizing against creatinine led to "overcompensation", resulting in statistically significant negative correlations. According to these results, all analyses of the datasets were based on creatinine-normalized concentrations.

As has been reported for other cohort studies, concentrations of phthalate metabolites in urine samples collected for both studies showed considerable variation with an approximately log-normal distribution pattern (Figure 2, Supplementary Figure 1) [29-32].

Similarly, values documenting daily fluctuations in the urinary metabolite concentrations of individuals showed log-norm distributions as well. Trends in the rise and fall of concentrations over time courses of 2 days or more were not evident upon visual inspection of the respective plots (Supplementary Figure 2). Spearman correlation analyses of concentrations on consecutive days, however, yielded coefficients that in some cases (14 out of 90) were significant with $p_{uncorr.} < 0.05$. Calculating Fisher-weighted means of Spearman correlation coefficients from all ten individuals for a given metabolite showed weak (0.14 to 0.24) but significant ($p_{uncorr.} \leq 0.018$) correlations for all metabolites except for MEHP and MiBP ($p_{uncorr.} > 0.05$).

Median metabolite concentrations obtained in the longitudinal study were generally lower than the respective median concentrations from the cohort samples (Figure 2). Specifically, median creatinine-normalized concentration for MEP was 48.6 µg/g for cohort samples

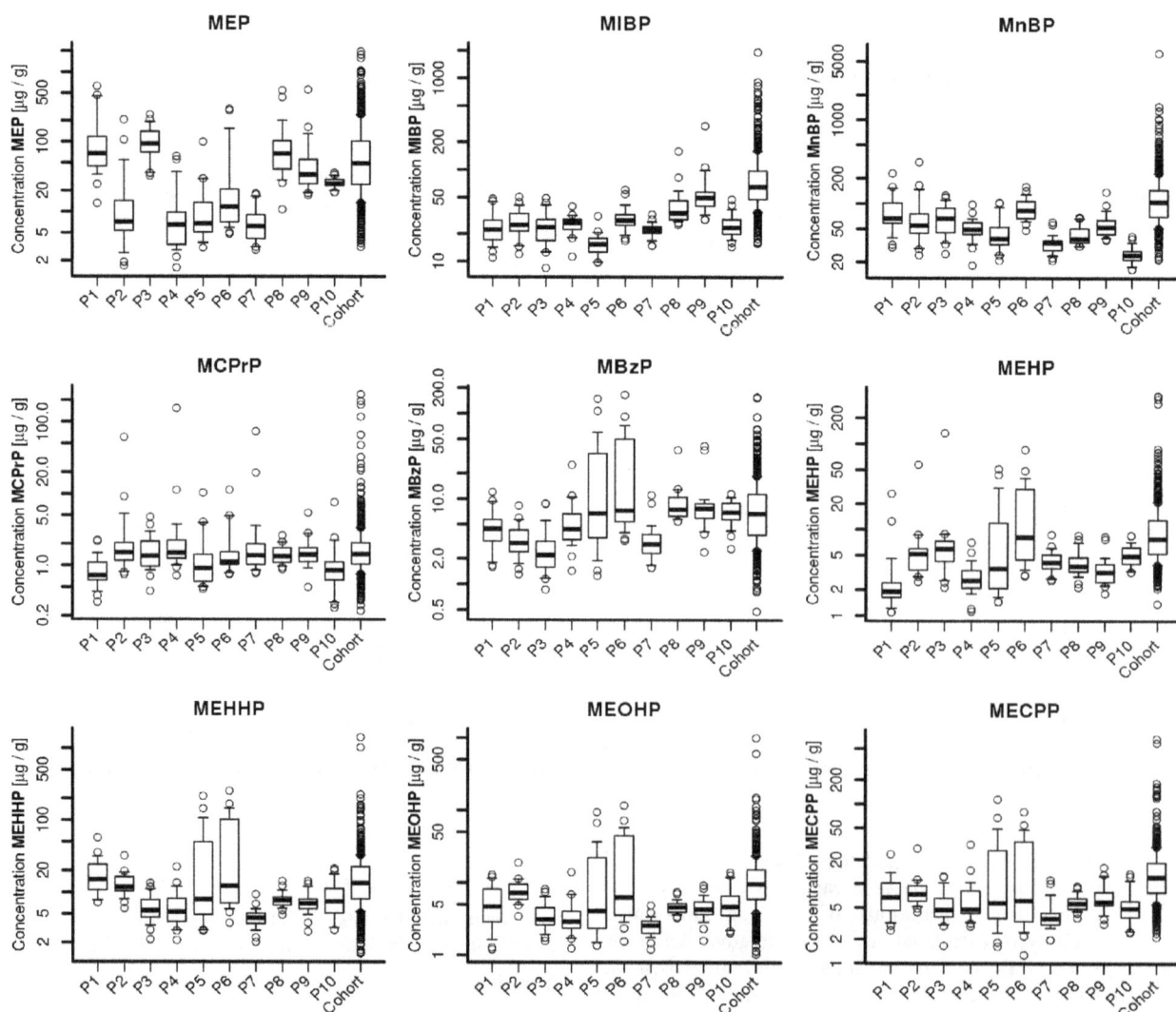

Figure 2: Boxplot showing the 10-90 percentile ranges for phthalate concentrations (log values) obtained the longitudinal study (ten individuals, n=30 each) and the cohort study (n=610).

while averaged median creatinine-normalized concentration was 32.6 µg/g for the longitudinal study. For MiBP 65.6 µg/g and 27.0 µg/g, for MnBP 105.4 µg/g and 50.7 µg/g, for MCPrP 1.5 µg/g and 1.2 µg/g, for MBZP 6.7 µg/g and 5.3 µg/g, for MEHP 7.8 µg/g and 4.3 µg/g, for MEHHP 13.4 µg/g and 8.4 µg/g, for MEOHP 9.7 µg/g and 4.4 µg/g, and for MECPP 12.1 µg/g and 5.5 µg/g were determined as creatinine-normalized concentrations for cohort samples and as average creatinine-normalized concentrations for the longitudinal study, respectively. For most of the volunteers, daily metabolites variations were lower, than the variations observed in the epidemiological setup.

In order to obtain an estimate to what extent the variability of metabolite concentrations seen in the cohort can be accounted for by day-to-day variance and which proportion is due to actual differences in (average) exposure of the individual participants (i.e., population variance), the standard deviation values of the respective log-scaled metabolite concentrations were determined. When the cumulative distribution plots of the data points were overlaid with the respective cumulative distribution functions (CDFs) using means and standard deviations calculated from log-scaled values, it was found that the resulting CDF curves did not accurately match the course of the data points (not shown). Using median values instead of means led to only slight improvements (Figure 3). The reason for this was a number of exceedingly high concentration values (outliers), which led to

standard deviation values not adequately capturing the spread of the majority of data points. In order to obtain a better representation of the standard deviations without removing or arbitrarily rescaling outliers, a nonlinear least squares regression algorithm was used, which led to more accurate results (Figure 3) with standard deviations that were smaller on average by 13% for the cohort and 31% for the longitudinal cohort study (Table 2).

For each metabolite, the standard deviations obtained from the log-transformed concentration values of the cohort (representing population plus daily variance) and the longitudinal study cohort (representing daily variance only) were used to calculate standard deviation values representative for the average individual phthalate concentrations (i.e., population variance, based on log-transformed values) within the cohort (Formula 1). The results showed that by accounting for the contribution of daily variance, the standard deviations of the log-transformed phthalate values of the cohort samples are reduced (14% to 28%) but still larger (3% to 66%) than daily standard deviation values, with the exception of MCPrP concentrations (standard deviation reduced by 40%; day-to-day standard deviation 33% larger than the standard deviation of the cohort population; Table 2). Using formula 2 and the standard deviations observed in the longitudinal study, the means observed in the cohort increased by an average of 5.8% (untransformed concentrations; range 0.9%

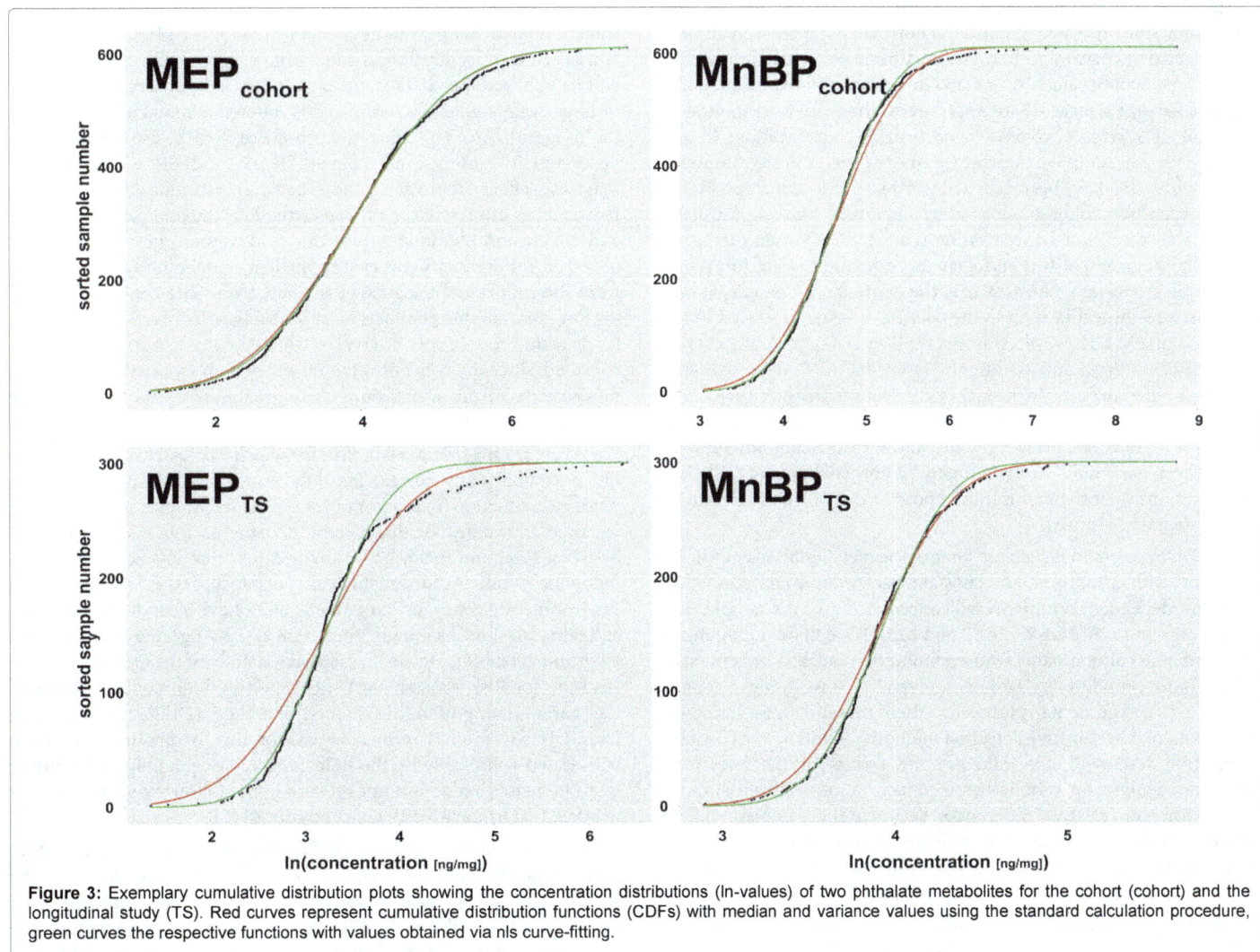

Figure 3: Exemplary cumulative distribution plots showing the concentration distributions (ln-values) of two phthalate metabolites for the cohort (cohort) and the longitudinal study (TS). Red curves represent cumulative distribution functions (CDFs) with median and variance values using the standard calculation procedure, green curves the respective functions with values obtained via nls curve-fitting.

to 5.6%, except for MCPrP with an increase of 23%), yielding means representing average concentration values in the absence of daily fluctuations (Table 2).

Discussion

One major problem by the analysis of phthalate is the ubiquitous presence of such compounds in the environment including laboratories. This background may be reduced e.g., by the usage of ultra-pure solvents and glass ware. However, high variations for phthalate background levels can never be excluded. This is especially true for this study, as the samples from the epidemiological study were obtained by different hospitals. By analysis of phthalate metabolites this problem can be overcome. This applies in particular for analysis of phase II metabolites. However, as phase I and phase II metabolites can be detected simultaneously in human urine, analytical techniques can either detect phase I and II metabolites independently or as a sum parameter of phase I and II metabolism after deconjugation. While the first option provides detailed information about the phase I and II ratio, the second one leads to lower detection limits and thus to higher detection rates. In this study the sum of phase I and II metabolites were detected as an enzymatic deconjugation step was carried out during samples preparation.

For urinary phthalate metabolite concentrations in the literature volume-based and creatinine normalized concentrations are given [2,32-39]. A correlation analysis of samples from our cohort and longitudinal study revealed that without normalizing against creatinine, there is a strong correlation between creatinine concentration and all phthalate metabolite values. As metabolite concentrations are generally used as a surrogate measure for exposure to the parent compound in epidemiological studies, volume-based urinary concentrations ideally would have to be multiplied by urine volume to provide the total amount of metabolites that have been excreted within a given time interval (the time between two voidings). However, since total volume is usually not available, creatinine concentration is a commonly used surrogate. Under the premise that this endogenous compound is produced and excreted at a constant rate which is the same for all people, it will behave in a reciprocal fashion to the volume of urine produced in the same time interval. However, this assumption is not generally correct as described by Wyss and Kaddurah-Daouk [40]. We have also seen individual differences in the results of correlation analyses performed for each of the ten participants in the longitudinal study, implying that this analyte is not perfectly suited in all cases (data not shown). Nevertheless, our results strongly support a normalization of phthalate concentrations against creatinine as opposed to no normalization at all (Supplementary Figure 3).

Negative correlations seen for some of the metabolites may indicate a problem with a more or less constant background contamination that is not dependent on the overall concentration of the sample. The parent compounds of MiBP, MnBP and MEHP are known ubiquitous environmental contaminants whose monoesters can also be generated under abiotic conditions [16,41,42]. Even if constituting a minor fraction of environmental phthalates, their presence may influence results obtained by sensitive detection methods. Addition of a constant background level will not influence the results of the Spearman correlation analysis on volume-based concentrations, since it is rank-based. However, relative differences between the samples will be reduced. When relatively similar values are divided by more dissimilar creatinine concentrations, creatinine values become the dominant determinant for the ranking, resulting in a negative correlation.

The fact that urinary concentrations taken 24 hr apart were only slightly correlated shows that metabolism and elimination of phthalates is fast and that even peak metabolite levels observed on one day will almost have returned to background levels on the following day. Such a result was to be expected, as experiments with human volunteers ingesting a dose of isotopically labeled phthalates with shorter (di-n-butyl phthalate, DnBP and di-(isobutyl)-phthalate, DiBP) or longer side chains (DEHP and di-(isononyl)-phthalate, DiNP) have shown these compounds to be quickly resorbed, metabolized and eliminated, with urinary peak concentrations occurring after 2-4 hr and half-lives of elimination - dependent on the type of metabolite - in the range of 2.6 to 4.2 hr or 2 to 5 hr, respectively, during the first 8 to 24 hr [43-45].

By comparing the obtained phthalate metabolite concentrations for cohort samples with other values from the literature (Table 3), it must be kept in mind, that those studies were either performed in different countries/continents, at different time points or with non-pregnant females or females and males. The study described by Zeman et al. represents in the authors opinion the most similar setup in comparison to our mother-child cohort [35]. Creatinine normalized phthalate concentrations were similar for MEP (48.6 µg/g and 34.3 µg/g) and MiBP (65.6 µg/g and 68.7 µg/g) by comparison of LiNA cohort values with those of Zeman et al. [35]. In comparison a ~ 2 fold higher median MnBP concentration was found in the cohort sample (105.4 µg/g vs. 45.5 µg/g) [35]. This difference may be explained by the different countries in which the studies were conducted, since an even higher median MnBP concentration had been observed in the German study described by Koch et al. in 2002 (157 µg/g) [37]. The same is true for median MEHP but interestingly not for the MEHHP, MEOHP, and MECPP concentrations. Concerning the metabolite ratio of DEHP and its oxidized metabolites, different ratios were observed in contrast to large scale studies [30,46,47]. This difference may be explained by the usage of *helix pomatiae* glucuronidase, which also provides low lipase activity and thus may convert DEHP to MEHP artificially. This may also affects the other monoester concentrations in this study. However, as this is a observational error, data and comparisons, results and conclusion obtained within this study should be still valid. The observed differences between the median metabolite concentrations from the cohort and the average median metabolite concentrations of the longitudinal study cohorts most probably reflect declining exposure levels within the six year interval in which the two sets of samples were collected, since the concentration ratios are close to the values that can be calculated from estimations of approximately exponential decay rates documented for the years 1988-2003 [39].

Apart from MEP and MCPrP, for which Wittassek and coworkers did not supply any values, the only exception is MiBP, where the observed reduction by a factor of 2.4 is not forecast by a corresponding trend, as the values for this metabolite, after an initial slight increase between 1988 and 1996, had remained more or less constant in the following 7 years. According to data available from the German Federal Environment Agency, urinary MiBP values have been declining, albeit at a fairly low and seemingly linear rate of ~ 1.7 µg/l per year, between 1998 and 2008 [48]. However, following a study of the German Federal Institute for Risk Assessment (BfR) reporting high concentrations of the parent compound di-(isobutyl)phthalate (DiBP) in packaging materials for food, commercial use of this compound had been withdrawn voluntarily by the industry in 2009, possibly contributing to an increased reduction in exposure in the following years and thus to decreased urinary MiBP concentrations [49].

Another aspect accounting for the metabolite concentration differences observed between the cohort samples and those of the longitudinal study may be sampling bias. The longitudinal study consisted of people with a higher education level, whereas the participants of the cohort study represented a cross section of different

Study	cohort study	32 instead of 2	33 instead of		34 instead of 4	35 instead of 1	31 instead of 5	37 instead of 6
N	610	108	83	74	100	279/ 139	1310	85
Sex	f (p)	f	f	m	f (p)	f (p)	f	f/m
Time	2006-7	2007	2012		2001	2007	2007-8	2002
Country	GER	MEX	ITA		TWN	FRA	USA	GER
Value	Med.	GM	GM		Med.	Med.	GM	Med.
Conc.	µg/g	µg/g	µg/g		µg/g	µg/g	µg/g	µg/g
Phthalate Metabolite								
MEP	48.6[3.1-1944]	83.2	77.1	56.4	/	34.3	106	73.3
MiBP	65.6 [15.9-1928]	8.4	/	/	15.2	68.7	8.18	/
MnBP	105.4 [21.4-6294]	72.4	20.3	17.6	87.5	45.5	23.1	157
MCPrP	1.45 [0.24-241.30]	3.91	/	/	/	/	/	/
MBzP	6.70 [0.45-124]	4.37	14.7	16.4	2.07	13.6	8.07	17.2
MEHP	7.8 [1.4-360.4]	5.2	3.4	2.8 10.8	16.4	17.9	3.02	9.2
MEHHP	13.41 [1.42-1416]	45.84	12.7/	/	32.2	44.4	25.2	40.2
MEOHP	9.7[1.0-1007]	31.8	/	/	29.5	32.9	14.2	30.4
MECPP	12.1 [2.1-661]	71.9	/	/	44.7	58.1	38.7	/

Table 3: Comparison of median (Med.) or geometric mean (GM) of phthalate concentrations from the cohort study with respective data from other publications. Values indicating concentration ranges are shown in brackets; values shown in gray (study [1]) refer to a sub-sample of the cohort. N: number of samples in cohort; Sex: female (f) or male (m) study participants, (p): pregnant; Time: year of sampling; Conc.: concentrations referring to creatinine mass (µg/g). Metabolites are grouped with respect to their different parent compounds (DEP, DnBP, DiBP, BBzP and DEHP).

educational backgrounds. However, to our knowledge, there has been no study to date showing phthalate exposure to be correlated with socio-economic factors. Furthermore, apart from the fact that pregnant women (cohort study) were compared to a population of mixed genders (longitudinal study), it has also to be kept in mind that the number of individuals in the latter sample set is only ten and thus might not be sufficiently representative for the general population.

Comparing location and extent of the 10-90 percentile ranges of phthalate metabolite concentrations in the longitudinal study, differences between individual participants are apparent (Figure 2). However, no influence of gender on phthalate metabolite concentrations was found via unpaired two-tailed student t-test (data not shown). For MEP, the strong differences in concentration medians with respect to interday variability suggests a meaningful stratification of participants into groups of low exposed (p 2, 4, 5, 6 and 7) and high exposed persons (p 1, 3, 8, 9 and 10). With regard to daily variations, also differences between the ten volunteers were observed. Comparing minimum and maximum spread values, p 6 showed an ~25 fold higher spread of MEP concentration values than p 10 (based on 10-90% percentiles after normalization to their respective medians). A factor of~ 12 was observed for MBzP concentrations by comparing p 6 and p 9. For the DEHP metabolites MEHP, MEHHP, MEOHP and MECPP, p 5 showed daily variations larger by ~ 10, ~21, ~18 and ~12 times in comparison to p 9 or p 8.

In order to get a measure for the day-to-day variance and to correct for variance between different individuals, concentration values of the ten participants of the longitudinal study were normalized by a correction factor k (k=median$_{total}$/median$_{individual}$), thereby adjusting individual medians to the total median. As this scaling results in unchanged relative standard deviations or, with respect to using logarithmic values for data analysis, unchanged absolute standard

deviations for each individual, the resulting values could be collated into a single dataset representative for the daily variance of a single person.

The distributions of urinary metabolite concentrations observed in the cohort study do not solely represent the different average exposure levels of the study participants (i.e., population heterogeneity), but time-dependent variation as well, since short-term fluctuations in exposure have a strong influence on the results obtained from spot samples. On the assumption that average individual phthalate exposure levels are log-normally distributed within a population (e.g., the levels of several phthalate diesters have been shown to be log-normally distributed in house dust samples) and that daily variations in the metabolite concentrations follow a log-normal distribution as well (as suggested by the data from our longitudinal study), it can be shown that population (sd^2_{pop}) and day-to-day variances (sd^2_{TS}) of the log-transformed concentrations behave in an additive manner ($sd^2_{pop} + sd^2_{TS} = sd^2_{cohort}$) [50]. Using log-transformed concentration values, it is thus possible to use the standard deviations from the cohort and longitudinal study to obtain the standard deviation for the log-transformed concentrations of a given metabolite for the population (Formula 1; Table 2).

With respect to log-transformed concentration values, knowing day-to-day variance (sd^2_{TS}) and the mean observed for the cohort study (μ_{cohort}), the mean value of the respective underlying population (μ_{Pop}) can be derived as well. Whereas the expected value E_{Pop} for a given metabolite concentration in the population is (approximately) represented by the mean of the observed concentration values in the LINA cohort study (E_{cohort}), the same is not true for the expected values of the log-transformed values (i.e., μ_{Pop}, μ_{cohort}), since for log-normal distributions the respective relations are defined by in

$(E_{pop}) = \mu_{pop} + \frac{1}{2}sd^2_{pop}$ and in $(E_{cohort}) = \mu_{cohort} + \frac{1}{2}sd^2_{cohort}$. It follows that $\mu_{pop} = \mu_{cohort} + \frac{1}{2}sd^2_{cohort} - \frac{1}{2}sd^2_{pop}$ and, with formula 1 rearranged to $\frac{1}{2}sd^2_{cohort} = \frac{1}{2}sd^2_{pop} + \frac{1}{2}sd^2_{TS}$, it follows that $\mu_{pop} = \mu_{cohort} + \frac{1}{2}sd^2_{TS}$ (Formula 2; Table 2). Comparing the values obtained for standard deviations of the log-transformed concentration values (Table 2), it is evident that after accounting for the contribution of daily variance, the standard deviations representing the spread of average concentrations (and thus exposure) within the population (sd_{Pop}) are of a similar magnitude for MCPrP, MEHP, MEHHP; MEOHP and MECPP, as the standard deviations representing daily variance (sd_{TS}). For MEP, MiBP, MnBP and MBzP, lower standard deviations were observed for the longitudinal study in comparison to the cohort (approx. -30 to -40% based on log values). Together with the observation that urinary concentrations of samples taken only 24 hr apart show only marginal correlation, these results call into question the suitability of phthalate metabolite concentration values obtained from singular samples to represent average individual exposures. In a more general vein, if exposure to a given compound is subjected to strong daily variations and the respective compound does not bioaccumulate, but instead is excreted within a relatively short time frame, concentration values obtained from single samples may be of limited or even negligible value in the context of epidemiological studies, as the presence of actually existing correlations with other parameters will be obscured and not be detected by statistical evaluation, leading to false negative results [16,51-53]. This is true for all investigated phthalate metabolites except of MEP, MiBP and MnBP, as intra individual fluctuations were smaller in the longitudinal in contrast to the cohort study. However, in epidemiological studies with sufficient statistical power, correlations to certain outcomes may be detected even with analytes showing a strong day-to-day variability. Our results were in harmony with the results obtained by Frederiksen et al. showing the risk of single spot urine analysis for phthalate metabolites and the corresponding correlation to certain health outcomes [23].

Conclusions

With new and ever more sensitive analytical methods becoming available, the amounts of quantitative analytical data from samples obtained in the context of epidemiological studies continues to increase. Whereas reproducibility and reliability of the methods themselves, if sufficiently validated, do not constitute a matter of concern, it is generally less clear how reliable the sample and the data gained from it are with respect to representing average or long-term measures of the respective study participant. In contrast to persistent organic pollutants with a high bioaccumulation potential, xenobiotic compounds with a large range of exposure levels and short metabolization and excretion rates are especially problematic in this regard. In such cases, instead of trying to increase cohort size, increasing the number of samples per study participant may produce a larger benefit as already stated by Fisher et al. since averaging individual concentration values will yield much more reliable estimates of a given person's general exposure and thereby increase the chance of discovering correlations with parameters likely to result from long-term effects such as medical outcomes [21].

Acknowledgements

The authors thank their colleagues Sandra Albrecht, Sven Baumann, Scarlett Gebauer, Carolin Graebsch, Joerg Hackermueller, Gabriele Heimpold, Markus Langhammer, Melanie Nowak, Susanne Pfeiffer, Ulrike Tzschoppe, Stefan Wicht, Brigitte Winkler, and Marleen Zurek for scientific discussion and technical assistance. This work was funded by the European Union and the Free State of Saxony via the European Regional Development Fund (EFRE) program No. 2007 DE 16 1 PO 004.

References

1. Koch HM, Lorber M, Christensen KL, Pälmke C, Koslitz S, et al. (2013) Identifying sources of phthalate exposure with human biomonitoring: results of a 48h fasting study with urine collection and personal activity patterns. Int J Hyg Environ Health 216: 672-681.

2. Langer S, Bekö G, Weschler CJ, Brive LM, Toftum J, et al. (2014) Phthalate metabolites in urine samples from Danish children and correlations with phthalates in dust samples from their homes and daycare centers. Int J Hyg Environ Health 217: 78-87.

3. Adamsson A, Salonen V, Paranko J, Toppari J (2009) Effects of maternal exposure to di-isononylphthalate (DINP) and 1,1-dichloro-2,2-bis(p-chlorophenyl)ethylene (p,p'-DDE) on steroidogenesis in the fetal rat testis and adrenal gland. Reprod Toxicol 28: 66-74.

4. Borch J, Axelstad M, Vinggaard AM, Dalgaard M (2006) Diisobutyl phthalate has comparable anti-androgenic effects to di-n-butyl phthalate in fetal rat testis. Toxicol Lett 163: 183-190.

5. Howdeshell KL, Furr J, Lambright CR, Rider CV, Wilson VS, et al. (2007) Cumulative effects of dibutyl phthalate and diethylhexyl phthalate on male rat reproductive tract development: Altered fetal steroid hormones and genes. Toxicological Sciences 99: 190-202.

6. Foster PM (2006) Disruption of reproductive development in male rat offspring following in utero exposure to phthalate esters. Int J Androl 29: 140-147.

7. Hauser R, Calafat AM (2005) Phthalates and human health. Occup Environ Med 62: 806-818.

8. Meeker JD, Sathyanarayana S, Swan SH (2009) Phthalates and other additives in plastics: human exposure and associated health outcomes. Philosophical Transactions of the Royal Society B-Biological Sciences 364: 2097-2113.

9. Thayer KA, Heindel JJ, Bucher JR, Gallo MA (2012) Role of environmental chemicals in diabetes and obesity: a National Toxicology Program workshop review. Environ Health Perspect 120: 779-789.

10. Hao C, Cheng X, Xia H, Ma X (2012) The endocrine disruptor mono-(2-ethylhexyl) phthalate promotes adipocyte differentiation and induces obesity in mice. Biosci Rep 32: 619-629.

11. Zhang Y, Meng X, Chen L, Li D, Zhao L, et al. (2014) Age and sex-specific relationships between phthalate exposures and obesity in Chinese children at puberty. PLoS One 9: e104852.

12. Feige JN, Gerber A, Casals-Casas C, Yang Q, Winkler C, et al. (2010) The Pollutant Diethylhexyl Phthalate Regulates Hepatic Energy Metabolism via Species-Specific PPAR alpha-Dependent Mechanisms. Environmental Health Perspectives 118: 234-241.

13. Feige JN, Gelman L, Rossi D, Zoete V, Métivier R, et al. (2007) The endocrine disruptor monoethyl-hexyl-phthalate is a selective peroxisome proliferator-activated receptor gamma modulator that promotes adipogenesis. Journal of Biological Chemistry 282: 19152-19166.

14. Kato K, Manori JS, John AR, Donald H, Nicole AM, et al. (2004) Mono(2-ethyl-5-hydroxyhexyl) phthalate and mono-(2-ethyl-5-oxohexyl) phthalate as biomarkers for human exposure assessment to di-(2-ethylhexyl) phthalate. Environmental Health Perspectives. 112: 327-330.

15. Koch HM, Preuss R, Angerer J (2006) Di(2-ethylhexyl)phthalate (DEHP): human metabolism and internal exposure-- an update and latest results. Int J Androl 29: 155-65.

16. Fromme H, Bolte G, Koch HM, Angerer J, Boehmer S, et al. (2007) Occurrence and daily variation of phthalate metabolites in the urine of an adult population. Int J Hyg Environ Health 210: 21-33.

17. Buckley JP, Palmieri RT, Matuszewski JM, Herring AH, Baird DD, et al. (2012) Consumer product exposures associated with urinary phthalate levels in pregnant women. J Expo Sci Environ Epidemiol 22: 468-475.

18. Wittassek M, Angerer J (2008) Phthalates: metabolism and exposure. Int J Androl 31: 131-138.

19. Hinz D, Simon JC, Maier-Simon C, Milkova L, Röder S, et al. (2010) Reduced

maternal regulatory T cell numbers and increased T helper type 2 cytokine production are associated with elevated levels of immunoglobulin E in cord blood. Clin Exp Allergy 40: 419-426.

20. Valvi D, Monfort N, Ventura R, Casas M, Casas L, et al. (2015) Variability and predictors of urinary phthalate metabolites in Spanish pregnant women. Int J Hyg Environ Health 218: 220-231.

21. Fisher M, Arbuckle TE, Mallick R, LeBlanc A, Hauser R, et al. (2015) Bisphenol A and phthalate metabolite urinary concentrations: Daily and across pregnancy variability. J Expo Sci Environ Epidemiol 25: 231-239.

22. Watkins DJ, Eliot M, Sathyanarayana S, Calafat AM, Yolton K, et al. (2014) Variability and predictors of urinary concentrations of phthalate metabolites during early childhood. Environ Sci Technol 48: 8881-8890.

23. Frederiksen H, Selma KK, Jørgensen N, Taboureau O, Jørgen HP, et al. (2013) Temporal variability in urinary phthalate metabolite excretion based on spot, morning, and 24-h urine samples: considerations for epidemiological studies. Environ Sci Technol 47: 958-967.

24. Meeker JD, Calafat AM, Hauser R (2012) Urinary phthalate metabolites and their biotransformation products: predictors and temporal variability among men and women. J Expo Sci Environ Epidemiol 22: 376-385.

25. Preau JL, Wong LY, Silva MJ, Needham LL, Calafat AM (2010) Variability over 1 week in the urinary concentrations of metabolites of diethyl phthalate and di(2-ethylhexyl) phthalate among eight adults: an observational study. Environ Health Perspect 118: 1748-1754.

26. Baird DD, Saldana TM, Nepomnaschy PA, Hoppin JA, Longnecker MP, et al. (2010) Within-person variability in urinary phthalate metabolite concentrations: measurements from specimens after long-term frozen storage. J Expo Sci Environ Epidemiol 20: 169-175.

27. Herberth G, Hinz D, Röder S, Schlink U, Sack U, et al. (2011) Maternal immune status in pregnancy is related to offspring's immune responses and atopy risk. Allergy 66: 1065-1074.

28. Hinz D, Bauer M, Röder S, Olek S, Huehn J, et al. (2012) Cord blood Tregs with stable FOXP3 expression are influenced by prenatal environment and associated with atopic dermatitis at the age of one year. Allergy 67: 380-389.

29. Becker K, Göen T, Seiwert M, Conrad A, Pick-Fuss H, et al. (2009) GerES IV: phthalate metabolites and bisphenol A in urine of German children. Int J Hyg Environ Health 212: 685-692.

30. Blount BC, Silva MJ, Caudill SP, Needham LL, Pirkle JL, et al. (2000) Levels of seven urinary phthalate metabolites in a human reference population. Environ Health Perspect 108: 979-982.

31. Fourth National Report on Human Exposure to Environmental Chemicals 2009.

32. Romero-Franco M, Hernández-Ramírez RU, Calafat AM, Cebrián ME, Needham LL, et al. (2011) Personal care product use and urinary levels of phthalate metabolites in Mexican women. Environ Int 37: 867-871.

33. Tranfo G, Papaleo B, Caporossi L, Capanna S, De Rosa M, et al. (2013) Urinary metabolite concentrations of phthalate metabolites in Central Italy healthy volunteers determined by a validated HPLC/MS/MS analytical method. Int J Hyg Environ Health 216: 481-485.

34. Lin S, Ku HY, Su PH, Chen JW, Huang PC, et al. (2011) Phthalate exposure in pregnant women and their children in central Taiwan. Chemosphere 82: 947-955.

35. Zeman FA, Boudet C, Tack K, Floch Barneaud A, Brochot C, et al. (2013) Exposure assessment of phthalates in French pregnant women: results of the ELFE pilot study. Int J Hyg Environ Health 216: 271-279.

36. CDC Fourth National Report on Human Exposure to Environmental Chemicals.

37. Koch HM, Rossbach B, Drexler H, Angerer J (2003) Internal exposure of the general population to DEHP and other phthalates - determination of secondary and primary phthalate monoester metabolites in urine. Environ Res 93: 177-185.

38. Silva M, Nicole AM, Carolyn CH, John AR, Kayoko K, et al. (2003) Improved quantitative detection of 11 urinary phthalate metabolites in humans using liquid chromatography–atmospheric pressure chemical ionization tandem mass spectrometry. Journal of Chromatography B 789: 393-404.

39. Wittassek M, Wiesmüller GA, Koch HM, Eckard R, Dobler L, et al. (2007) Internal phthalate exposure over the last two decades--a retrospective human biomonitoring study. Int J Hyg Environ Health 210: 319-333.

40. Wyss M, Kaddurah-Daouk R (2000) Creatine and creatinine metabolism. Physiol Rev 80: 1107-1213.

41. Silva MJ, Reidy JA, Preau JL, Samandar E, Needham LL, et al. (2006) Measurement of eight urinary metabolites of di(2-ethylhexyl) phthalate as biomarkers for human exposure assessment. Biomarkers 11: 1-13.

42. Heudorf U, Mersch-Sundermann V, Angerer J (2007) Phthalates: toxicology and exposure. Int J Hyg Environ Health 210: 623-634.

43. Koch HM, Christensen KL, Harth V, Lorber M, Brüning T (2012) Di-n-butyl phthalate (DnBP) and diisobutyl phthalate (DiBP) metabolism in a human volunteer after single oral doses. Arch Toxicol 86: 1829-1839.

44. Koch HM, Bolt HM, Angerer J (2004) Di(2-ethylhexyl)phthalate (DEHP) metabolites in human urine and serum after a single oral dose of deuterium-labelled DEHP. Arch Toxicol 78: 123-130.

45. Koch HM, Angerer J (2007) Di-iso-nonylphthalate (DINP) metabolites in human urine after a single oral dose of deuterium-labelled DINP. Int J Hyg Environ Health 210: 9-19.

46. Blount BC, Milgram KE, Silva MJ, Malek NA, Reidy JA, et al. (2000) Quantitative detection of eight phthalate metabolites in human urine using HPLC-APCI-MS/MS. Anal Chem 72: 4127-4134.

47. Frederiksen H, Jorgensen N, Andersson AM (2010) Correlations between phthalate metabolites in urine, serum, and seminal plasma from young Danish men determined by isotope dilution liquid chromatography tandem mass spectrometry. J Anal Toxicol 34: 400-410.

48. http: //www.umweltprobenbank.de/en/documents/investigations/results/analytes?analytes=10091&sampling_areas=&sampling_years=1988..2008&specimen_types=10004

49. http: //www.bfr.bund.de/cm/349/di_isobutyphthalate_in_food_contact_paper_and_board.pdf

50. Langer S, Charles JW, Andreas F, Gabriel B, Jørn T, et al. (2010) Phthalate and PAH concentrations in dust collected from Danish homes and daycare centers. Atmospheric Environment 44: 2294-2301.

51. Hauser R, Meeker JD, Park S, Silva MJ, Calafat AM (2004) Temporal variability of urinary phthalate metabolite levels in men of reproductive age. Environ Health Perspect 112: 1734-1740.

52. Cannarozzi G, Anisimova M, Liberles DA (2010) Finding the balance between the mathematical and biological optima in multiple sequence alignment. Trends in Evolutionary Biology 2: 39-48.

53. Jusko TA, Shaw PA, Snijder CA, Pierik FH, Koch HM, et al. (2014) Reproducibility of urinary bisphenol A concentrations measured during pregnancy in the Generation R Study. J Expo Sci Environ Epidemiol 24: 532-536.

Permissions

All chapters in this book were first published in JC&ST, by OMICS International; hereby published with permission under the Creative Commons Attribution License or equivalent. Every chapter published in this book has been scrutinized by our experts. Their significance has been extensively debated. The topics covered herein carry significant findings which will fuel the growth of the discipline. They may even be implemented as practical applications or may be referred to as a beginning point for another development.

The contributors of this book come from diverse backgrounds, making this book a truly international effort. This book will bring forth new frontiers with its revolutionizing research information and detailed analysis of the nascent developments around the world.

We would like to thank all the contributing authors for lending their expertise to make the book truly unique. They have played a crucial role in the development of this book. Without their invaluable contributions this book wouldn't have been possible. They have made vital efforts to compile up to date information on the varied aspects of this subject to make this book a valuable addition to the collection of many professionals and students.

This book was conceptualized with the vision of imparting up-to-date information and advanced data in this field. To ensure the same, a matchless editorial board was set up. Every individual on the board went through rigorous rounds of assessment to prove their worth. After which they invested a large part of their time researching and compiling the most relevant data for our readers.

The editorial board has been involved in producing this book since its inception. They have spent rigorous hours researching and exploring the diverse topics which have resulted in the successful publishing of this book. They have passed on their knowledge of decades through this book. To expedite this challenging task, the publisher supported the team at every step. A small team of assistant editors was also appointed to further simplify the editing procedure and attain best results for the readers.

Apart from the editorial board, the designing team has also invested a significant amount of their time in understanding the subject and creating the most relevant covers. They scrutinized every image to scout for the most suitable representation of the subject and create an appropriate cover for the book.

The publishing team has been an ardent support to the editorial, designing and production team. Their endless efforts to recruit the best for this project, has resulted in the accomplishment of this book. They are a veteran in the field of academics and their pool of knowledge is as vast as their experience in printing. Their expertise and guidance has proved useful at every step. Their uncompromising quality standards have made this book an exceptional effort. Their encouragement from time to time has been an inspiration for everyone.

The publisher and the editorial board hope that this book will prove to be a valuable piece of knowledge for researchers, students, practitioners and scholars across the globe.

List of Contributors

Lazzarini A, Floridi A, Pugliese L, Villani M, Cataldi S, Lazzarini R and Albi E
Laboratory of Nuclear Lipid BioPathology, CRABiON, Perugia, Italy

Michela Codini and Beccari T
Department of Pharmaceutical Science, University of Perugia, Italy

Ambesi-Impiombato FS and Curcio F
Department of Clinical and Biological Sciences, University of Udine, Italy

Nehad A Abdallah
Experiments and Advanced Pharmaceutical Research Unit (EAPRU), Faculty of Pharmacy, Ain Shams University, Egypt

Abdullah A Elshanawane and Lobna M Abdelaziz
Medicinal Chemistry Department, Zagazig University, Zagazig, Egypt

Mustafa S Mohram and Hany M Hafez
Quality control department, EIPICO, Egypt

Murthy TGK and Geethanjali J
Bapatla College of Pharmacy, Bapatla, Andhra Pradesh, India

Sohail Ahmed Soomro and Muhammad Yar Khuhawar
Institute of Advanced Research Studies in Chemical Sciences University of Sindh, Jamshoro, Pakistan

Shahid Hussain Soomro
Shaheed Mohatarma Benazeer Bhutto Medical University Larkana Sindh, Pakistan

Jeyathilakan N, Abdul Basith S and John L
Department of Veterinary Parasitology, Madras Veterinary College, Tamil Nadu Veterinary and Animal Sciences University, Chennai- India

Chandran NDJ
Department of Veterinary Microbiology, Madras Veterinary College, Tamil Nadu Veterinary and Animal Sciences University, Chennai- India

Dhinakarraj G
Department of Animal Biotechnology, Madras Veterinary College, Tamil Nadu Veterinary and Animal Sciences University, Chennai- India

Preet Kawal Kaur
IEC School of Pharmacy, IEC University, Kalujhanda, Solan, Baddi, Himachal Pradesh, India
University Institute of Pharmaceutical Sciences–UGC Centre for Advanced Studies, Panjab University, Chandigarh, India

Karan Vasisht and Maninder Karan
University Institute of Pharmaceutical Sciences–UGC Centre for Advanced Studies, Panjab University, Chandigarh, India

Jakimska A, Nagórski P, Kot Wasik A and Namieśnik J
Department of Analytical Chemistry, Faculty of Chemistry, Gdańsk University of Technology (GUT), G. Narutowicza Street 11/12, 80-233 Gdańsk, Poland

Śliwka Kaszyńska M
Department of Organic Chemistry, Faculty of Chemistry, Gdańsk University of Technology (GUT), G. Narutowicza Street 11/12, 80-233 Gdańsk, Poland

Gurinder Singh, Roopa S. Pai and Sanual Mustafa
Department of Pharmaceutics, Faculty of Pharmacy, Al-Ameen College of Pharmacy, Bangalore, Karnataka, India

Anar Rodriguez and Delphine Beukens
Laboratory of Biological and Medical Chemistry, Faculty of Pharmacy, Université Libre de Bruxelles (ULB), Boulevard du Triomphe, 1050 Brussels, Belgium

Frédéric Cotton
Laboratory of Biological and Medical Chemistry, Faculty of Pharmacy, Université Libre de Bruxelles (ULB), Boulevard du Triomphe, 1050 Brussels, Belgiu
Department of Clinical Chemistry, Erasme Hospital, Université Libre de Bruxelles, Route de Lennik 808 1070 Brussels, Belgium

Nicole Debouge and Béatrice Gulbis
Department of Clinical Chemistry, Erasme Hospital, Université Libre de Bruxelles, Route de Lennik 808 1070 Brussels, Belgium

Pedro AQ, Soares RF, Oppolzer D, Santos FM, Rocha LA, Gonçalves AM, Queiroz JA, Gallardo E and Passarinha LA
CICS-UBI-Centro de Investigacaoem Ciencias da Saude, Universidade da Beira Interior, 6201-506 Covilha, Portugal

Bonifacio MJ
Departamento de Investigacao e Desenvolvimento, BIAL 4745-457 Sao Mamededo Coronado, Portugal

Belissa E, Nino C, Bernard M, Henriet T, Surget E and Boccadifuoco G
Assistance Publique–Hôpitaux de Paris, Agence Générale des Produits et Equipements de Santé, Département de contrôle qualité et de développement analytique, France

Do B
Assistance Publique–Hôpitaux de Paris, Agence Générale des Produits et Equipements de Santé, Département de contrôle qualité et de développement analytique, France
Paris-SudUniversity, Faculty of Pharmacy, EA 401 Groupe Matériaux et Santé, France

Sadou-Yaye H
Assistance Publique – Hôpitaux de Paris, Groupe hospitalier Pitié-Salpétrière, Département Pharmacie, France

Yagoubi N
Paris-SudUniversity, Faculty of Pharmacy, EA 401 Groupe Matériaux et Santé, France

Saran S and Saxena RK
Technology Based Incubator, University of Delhi South Campus, Biotech Centre, Benito Juarez Road, New Delhi-110021, USA

Yadav S
Department of Microbiology, University of Delhi South Campus, Benito Juarez Road, New Delhi 110021, USA

Ezhilarasi K, Dhamodharan Umapathy and Vijay Viswanathan
Department of Biochemistry and Molecular Genetics, Prof. M. Viswanathan Diabetes Research Centre and M.V. Hospital for Diabetes (WHO Collaborating Centre for Research, Education and Training in Diabetes), No. 4, West Madha Church Road, Royapuram, Chennai 600013, Tamil Nadu, India

Sudha V, Geetha Ramachandran and Hemanth Kumar AK
Department of Biochemistry, National Institute for Research in Tuberculosis (Indian Council of Medical Research), No. 1, Mayor Sathiyamoorthy Road, Chetpet, Chennai 600031, Tamil Nadu, India

Rama Rajaram
Department of Biochemistry, Central Leather Research Institute, Adyar, Chennai, Tamil Nadu. India

Indira Padmalayam
Drug Discovery Division, Southern Research Institute, Birmingham, USA

Aikaterini I Piteni, Maria G Kouskoura and Catherine K Markopoulou
Laboratory of Pharmaceutical Analysis, Department of Pharmaceutical Technology, School of Pharmacy, Faculty of Health Sciences, Aristotle University of Thessaloniki, 54124 Thessaloniki, Greece

Seetharaman R and Lakshmi KS
Department of Pharmaceutical Analysis, College of Pharmacy, SRM University, Kattankulathur, Tamilnadu – 603203, India

Heyong Huang, Jiahong Zhou and Yuying Feng
Analysis and Testing Center, School of Geography Science, Nanjing Normal University, Nanjing 210023, China

Yan Zhou
Department of Chemistry and Environmental Science, Nanjing Normal University, Nanjing 210023, China

M Abdul Mottaleb
Center for Innovation and Entrepreneurship, 1402 North College Drive, Northwest Missouri State University, Maryville, Missouri 64468, USA
Department of Natural Sciences, 800 University Drive, Northwest Missouri State University, Maryville, Missouri 64468, USA

Michael K Bellamy, Musavvir Arafat Mottaleb and M Rafiq Islam
Department of Natural Sciences, 800 University Drive, Northwest Missouri State University, Maryville, Missouri 64468, USA

Wang Rui Guo, Su Xiao Ou, Wang Pei Long and Zhang Wei
Institute of Quality Standards and Testing Technology for Agricultural Products, Chinese Academy of Agricultural Science, Key Laboratory of Agrifood Safety and Quality, Ministry of Agriculture, Beijing, PR China

Xue Yan and Li Yu
Alltech Biological Products Co. Ltd., Beijing 100600, China

Bhatnagar P
Sir Padampat Singhania University, Udaipur, India
Macleods Pharmaceuticals Limited, Mumbai, India

Vyas D and Chakrabarti T
Sir Padampat Singhania University, Udaipur, India

Sinha SK
Macleods Pharmaceuticals Limited, Mumbai, India

Chhalotiya UK, Varsha LP, Dimal AS, Kashyap KB and Sunil LB
Department of Pharmaceutical Analysis, Indukaka Ipcowala College of Pharmacy, Gujarat, India

Shannon M Black, Sadia Muneem, Deborah Miller-Tuck and Prince A Kassim
Maryland Department of Health and Mental Hygiene, 201 West Preston Street, Baltimore, MD 21201 United States

Cavalheiro J, Tessier E, Donard OFX and Monperrus M
LCABIE-IPREM, CNRS UMR 5254 UPPA, Université de Pau et des Pays de l'Adour, Hélioparc Pau Pyrénées, 2, av. P. Angot, 64053 Pau cedex 9, France

Baltrons O
UT2A, Bâtiment de Vinci, Hélioparc Pau Pyrénées, 2, av. P. Angot, 64053 Pau cedex 9, France

Jain PS, Kale NK and Surana SJ
RC Patel Institute of Pharmaceutical Education and Research, Karwand Naka, Shirpur Dist, Dhule 425 405 (MS) India

Nehad A Abdallah
Experiments and Advanced Pharmaceutical Research Unit (EAPRU), Faculty of Pharmacy, Ain Shams University, Cairo, Egypt

Hesham Salem and Kholoud Ahmed
Pharmaceutical Analytical Chemistry Department, Faculty of Pharmacy, October University for Modern Sciences and Arts, Egypt

Safa M Riad and Mamdouh R Rezk
Analytical Chemistry Department, Faculty of Pharmacy-Cairo University, Kasr El-Aini Street, Egypt

Dharmendra D, Karan M, Bhoomi P and Rajshree CM
Quality Assurance Laboratory, Centre of Relevance and Excellence in Novel Drug Delivery system, Department of Pharmacy, Shree G.H.Patel Pharmacy building, Donor's Plaza, The Maharaja Sayajirao University of Baroda, Fatehgunj, Vadodara-390002, Gujarat, India

Pei-Shan Xie and Longgang Guo
Guangdong UNION Biochemical Development Ltd. Co. Guangzhou, China

Shuai Sun and Shunjun Xu
Guangdong ImVin Pharmaceutical Corporation, Guangzhou, China

Anil K Suresh
Department of Biotechnology, School of Life Sciences, Pondicherry University, Puducherry, India

Dale A Pelletier
Biosciences Division, Oak Ridge National Laboratory, Oak Ridge, Tennessee, USA

Mitchel J Doktycz
Biosciences Division, Oak Ridge National Laboratory, Oak Ridge, Tennessee, USA
Center for Nanophase Materials Science, Oak Ridge National Laboratory, Oak Ridge, Tennessee, USA

Ji-Won Moon and Tommy J Phelps
Environmental Sciences Division, Oak Ridge National Laboratory, Oak Ridge, Tennessee, USA

Eduard Rogatsky, Shaynah Browne, Min Cai, Harsha Jayatillake and Daniel Stein
Biomarker Analytical Resource Core, Einstein-Montefiore Institute for Clinical and Translational Research. Albert Einstein College of Medicine, USA

Alagar Raja M, Bhargav KS and Banji D
Department of Pharmaceutical Analysis & Quality Assurance, Nalanda College of Pharmacy, Nalgonda, India

Selva Kumar D
School of Pharmacy, Taylors University, Subangjaya, Malaysia

Ruan ED, Aalhus J and Juarez M
Lacombe Research Centre, Agriculture and Agri-Food Canada, 6000 C&E Trail, Lacombe, AB T4L 1W1, Canada

Karthikeyan R, Suresh Kumar K, Singaravadivel K and Alagusundaram K
Department of Food Microbiology, Indian Institute of Crop Processing Technology, Ministry of Food Processing Industries, Govt. of India, Thanjavur (TN), India

Ralph Feltens and Dirk K. Wissenbach
Department of Metabolomics, Helmholtz Centre for Environmental Research-UFZ, Leipzig, Germany

Martin von Bergen
Department of Metabolomics, Helmholtz Centre for Environmental Research-UFZ, Leipzig, Germany
Department of Proteomics, Helmholtz Centre for Environmental Research-UFZ, Leipzig, Germany
Department of Biotechnology, Chemistry and Environmental Engineering Aalborg University, Aalborg, Denmark

Stefan Roeder and Irina Lehmann
Department of Environmental Immunology, Helmholtz Centre for Environmental Research-UFZ, Leipzig, Germany

Wolfgang Otto
Department of Proteomics, Helmholtz Centre for Environmental Research-UFZ, Leipzig, Germany

Michael Borte
Children's Hospital, Municipal Hospital St. Georg Leipzig, Affiliated to the University of Leipzig, Germany

Index